Modern Spectroscopy

Modern Spectroscopy

Editor: Ryan Cooper

STATES
ACADEMIC PRESS
www.statesacademicpress.com

Published by States Academic Press
109 South 5th Street,
Brooklyn, NY 11249, USA
www.statesacademicpress.com

Modern Spectroscopy
Edited by Ryan Cooper

International Standard Book Number: 978-1-63989-365-2 (Hardback)

Cataloging-in-Publication Data

Modern spectroscopy / edited by Ryan Cooper.
p. cm.
Includes bibliographical references and index.
ISBN 978-1-63989-365-2
1. Spectrum analysis. 2. Spectroscope. I. Cooper, Ryan.
QD95 .M63 2022
543.085 8--dc23

Contents

Preface

The main aim of this book is to educate learners and enhance their research focus by presenting diverse topics covering this vast field. This is an advanced book which compiles significant studies by distinguished experts in the area of analysis. This book addresses successive solutions to the challenges arising in the area of application, along with it; the book provides scope for future developments.

Spectroscopy is the study of the absorption and emission of light and other radiation by matter as related to the dependence of these processes on the wavelength of the radiation. On the basis of the nature of interaction between the energy and the material, it can be classified into various types. A few of such types are absorption spectroscopy, emission spectroscopy, impedance spectroscopy and resonance spectroscopy. Spectroscopic analysis has contributed significantly to various areas of study within physics such as quantum mechanics, quantum electrodynamics, and special and general theories of relativity. The topics included in this book on modern spectroscopy are of utmost significance and bound to provide incredible insights to readers. Also included herein is a detailed explanation of the various concepts and applications of spectroscopy. Scientists and students actively engaged in this field will find this book full of crucial and unexplored concepts.

It was a great honour to edit this book, though there were challenges, as it involved a lot of communication and networking between me and the editorial team. However, the end result was this all-inclusive book covering diverse themes in the field.

Finally, it is important to acknowledge the efforts of the contributors for their excellent chapters, through which a wide variety of issues have been addressed. I would also like to thank my colleagues for their valuable feedback during the making of this book.

Editor

1

Raman Spectroscopic Characterization of Endodontic Biofilm Matrices

Tatiana Ramirez-Mora,[1] **Claudia Dávila-Pérez**,[2] **Fernando Torres-Méndez**,[2] **and Grettel Valle-Bourrouet**[3]

[1]*Restorative Department, Faculty of Dentistry, University of Costa Rica, San Pedro Montes de Oca 11801, Costa Rica*
[2]*Endodontics Postgraduate Program, Faculty of Dentistry, University of San Luis Potosí, San Luis Potosí 78290, Mexico*
[3]*Inorganic Chemistry Department, School of Chemistry, University of Costa Rica, San Pedro Montes de Oca 11801, Costa Rica*

Correspondence should be addressed to Tatiana Ramirez-Mora; tatiana.ramirez@ucr.ac.cr

Academic Editor: Austin Nevin

Endodontic persistent infections are often mediated by bacterial biofilms. This mode of bacterial growth is characterized by the presence of a matrix mainly composed of extracellular polymeric substances (EPSs) that protect the encased microorganisms. To establish better control and disinfection protocols, elucidation of the main components of biofilm matrices present in endodontic infections is required. The aim of the present study was to characterize the principal components of *E. faecalis*, *A. naeslundii*, and dual-species biofilm matrices by means of Raman spectroscopy and confocal scanning laser microscopy (CSLM) techniques. The total biomass of biofilms was quantified via crystal violet assays, and the monospecies biofilms showed higher biomass than the dual-species biofilms. Raman spectroscopy and confocal laser scanning microscopy were used to identify the biochemical composition and structure of the biofilm matrices. Spectra originating from the biofilms of two endodontic pathogens show the presence of carbohydrates, proteins, fatty acids, and nucleic acids in all samples; however, variation in the levels of expression of these biomolecules allows spectroscopic differentiation of the biofilms using principal component analysis. This study is the first attempt to identify the composition of monospecies and dual-species biofilms of endodontic origin. Our data provides an important approach to the understanding of molecular dynamics of endodontic infections.

1. Introduction

Microbiota diversity of persistent endodontic infections is very limited, and Gram-positive facultative bacteria are commonly isolated [1, 2]. *Enterococcus faecalis* and *Actinomyces* spp. represent two of the species most frequently isolated from specimens of this clinical condition [2, 3]. Both have the ability to form biofilms, which confers on them the ability to persist despite harsh environmental conditions such as root canal chemomechanical disinfection, intracanal medication, and even antibiotic therapy [4].

The occurrence of biofilms in infected root canals and elsewhere has been widely reported; morphological investigations of teeth with primary or persistent apical periodontitis have revealed the presence of biofilms not only in the main root canal but also within anatomical variations including lateral canals, isthmuses, and apical ramifications [5, 6]. Extraradicular biofilms attached to the external surface of the root in case of endodontic treatment failure have also been described [7, 8].

Bacterial biofilms are surface-associated communities that produce a variety of macromolecules. These so-called extracellular polymeric substances (EPSs), such as polysaccharides, proteins, fatty acids, and extracellular DNA (eDNA) [9], provide the structural and functional integrity of the biofilm matrix and are also involved in communication processes, genetic transfer, nutrition, and the interaction of the bacteria with their environment [10, 11]. The production and type of EPS is influenced by different factors such as bacterial species, environmental conditions, nutrient availability, adhesion surface, etc. [12]. These factors are the origin of high diversity and multiple types of biofilms, and this heterogeneity is the main cause of the pathogenesis of many chronic and life-threatening infections [13].

Since biofilm matrices vary greatly, many methods and techniques may be used to study their ultrastructure and chemical composition. Microscopy techniques such as scanning electron microscopy (SEM), atomic force microscopy (AFM), transmission electron microscopy (TEM), and confocal laser scanning microscopy (CLSM) are among the methods employed to determine biofilm matrix structure [14]. Recently, CSLM has become a very popular and useful tool for in situ assessment of biofilm structure and detection of matrix components using different staining protocols [15]. However, chemical information provided by CSLM can be limited because of missing interactions between stains and compounds, and it is also an expensive and highly time-consuming technique.

Raman microscopy, a molecular vibration-based spectroscopic technique, is a powerful analytical tool for the study of biological materials. It allows rapid, in situ, non-invasive acquisition of chemical and structural information through the generation of fingerprint spectra. Ivleva et al., Popp and coworkers, and other research groups have employed Raman spectroscopy for chemical analysis of biofilms, differentiation of planktonic and biofilm cells, discrimination of diverse species of bacteria in biofilms, and monitoring metabolic characteristics under different physiologic states [16–19].

In the field of endodontics, a previous study used Raman spectroscopy to analyse the chemical compositions of *E. faecalis* biofilms in different physiological states in which the matrix components differed according to the growth phase [20]. This study provided important information about monospecies biofilm composition; nevertheless, root canal infections are clearly mediated by monospecies, dual-species, or multispecies biofilms [8, 12, 21], and therefore it is necessary to investigate monospecies and dual species of different relevant root canal bacteria, especially those involved in persistent periapical pathosis. Data resulting from such studies could represent a valuable contribution to the establishment of a new understanding and the determination of strategies to prevent and eradicate biofilms from the root canal system. In this context, the present study was designed to combine Raman spectroscopy and confocal laser scanning microscopy techniques to obtain biochemical composition from monospecies and dual-species endodontic biofilms.

2. Materials and Methods

2.1. Bacterial Strains and Media. *E. faecalis* and *A. naeslundii* strains, from persistent endodontic infections, were used. Samples were incubated inside an anaerobic chamber (85% N_2, 10% H_2, 5% CO_2; Coy Laboratory Products, Grass Lake, MI, USA). Identification was performed using specific agar media and API biochemical tests. Confirmation of the identity and antibiotic resistance was performed with the VITEK 2 automated system (BioMeriéux®, Cambridge, MA, USA).

2.2. Biofilm Assays. Biofilm assays were conducted based on a previously reported method [22]. After confirmation of the strain purity by Gram staining and colony morphology, a single colony of each bacterial strain was inoculated in 10 mL of brain heart infusion (BHI) and cultured overnight at 37°C. For the monospecies suspension, the cell density was 10^9 cells/mL (OD = 1, as measured using a Pharmacia LKB Ultrospec III spectrophotometer). A mixed microbial suspension was prepared in a sterile falcon tube from equal volumes of *E. faecalis* and *A. naeslundii* suspensions (OD = 1) to grow the dual-species biofilms. For the biofilm formation, a 0.2 mL inoculum was transferred to a 24-well polystyrene tissue culture plate (Sarstedt Inc., Newton, NC, USA) containing 1.8 mL of BHI, and the biofilms were then grown under static and anaerobic conditions (85% N_2, 10% H_2, 5% CO_2; Coy Laboratory Products Inc., Grass Lake, MI, USA) for 14 days. The growth media was replaced every second day to remove dead cells and provide nutrition.

For biofilm EPS visualization of each sample via CSLM, glass coverslips (Thermo Fisher Scientific, Waltham, MA, USA) were placed in a 24-well culture plate and grown in triplicate.

2.3. Biofilm Quantification Assay. Formation of monospecies and dual-species biofilms was quantified using the crystal violet (CV) method [23] with slight modifications. *E. faecalis*, and *A. naeslundii* that had been grown overnight were diluted 1 : 100 in the BHI medium, and for each cell suspension, 200 μL was transferred to a sterile flat-bottomed 96-well polystyrene plate. For the dual-species biofilm, 100 μL of each cell suspension was added per well. After stationary anaerobic incubation at 37°C for 14 days, the broth was carefully aspirated, and then the wells were washed three times with 200 μL PBS and air-dried for 30 min at room temperature. The biofilms were then stained with 200 μL of 1% crystal violet, under constant agitation for 15 min at room temperature, and washed again with PBS (three times), and the bound crystal violet solubilized in 200 μL of 95% ethanol for 10 min. Thereafter, the optical density (OD) was measured at 570 nm (OD_{570}) using a microtiter plate reader (Bio-Tek Synergy HT). Biofilm mass was expressed as OD 570 nm values. The assay was repeated three times.

2.4. EPS Identification via CLSM. CLSM was used for the identification of EPSs within the biofilm matrix. After incubation for 14 days, the glass coverslips were washed twice with phosphate-buffered saline (PBS). Next, the biofilms were incubated in a staining mixture for 15 min in dark. Calcofluor and propidium iodide (Sigma-Aldrich Co. St. Louis, MO, USA) were used as fluorescent stains for polysaccharide and eDNA, respectively. Following this, the biofilms were washed in PBS to remove excess stain. Before the biofilms were completely dry, a mounting medium was applied. Confocal analysis was performed using an Olympus U-TB190 (100×) microscope objective under oil immersion. Confocal images of 1024 × 1024 pixels were acquired with the FV10-AV version 03.01 software (Olympus, Japan) and assembled using Adobe Photoshop CS3 (Adobe Systems).

2.5. Biofilm EPS Extraction. Biofilm EPSs were extracted as previously described [24] with slight modifications. Biofilms were recovered from the tissue flask by scraping (Falcon cell scraper, Corning Science Mexico, Reynosa, Tamps., Mexico), suspended in PBS (10 mL), transferred to polypropylene conical tubes (BD), and vortexed for 1 min. Next, the biofilm samples were sonicated in an ultrasonic bath (Branson 5800, Emerson, St. Louis, MO, USA) for 45 min, followed by vortexing for 2 min, and then centrifuged (3500 rpm; 15 min) (Sorvall Legend X1R, Thermo Fisher Scientific, Waltham, MA, USA). The supernatant fraction was recovered and filtered through 0.2 μm Acrodisc® syringe filters with Supor® Membrane (Pall Life Sciences, Deland, FL, USA).

Biofilm samples were transferred to quartz slides with a 10 μL inoculating loop. The smears were dried for 20 min with silica.

2.6. Spectroscopy Instrumentation and Measurement. All Raman spectra were recorded using a Raman microspectrometer (ProRaman-L, Enwave Optronics Inc., Irvine, CA, USA). A 50× microscope objective (Leica Microsystems Inc., Buffalo Grove, IL, USA) was used, and the samples were excited using 45–50 mW of a 785 nm diode laser. The Raman signal was collected in the spectral interval 600–1800 cm^{-1}; the integration time was 40 s, and spectral resolution was approximately 2 cm^{-1}. To record biofilm Raman spectra, on each smear of the quartz slides, a central focus point was selected, and around this centre, spectra were collected at ten points for each sample.

2.7. Spectral Preprocessing and Analysis. Spectra were processed via customized Raman processing software [25] and Origin 8.0 (OriginLab Corp., Northampton, MA, USA). Each Raman spectrum was first processed by using a median filter to remove cosmic rays and noise. Spectral intensities were normalized to the range of 0-1 arbitrary units, and an automated polynomial baseline correction method was applied. The Raman spectra were then analysed via multivariate techniques, namely, principal component analysis (PCA) and discriminant function analysis (DFA).

3. Results and Discussion

Biofilm quantification assay of the monoculture biofilms was compared to the dual-species biofilms. Both species *E. faecalis* and *A. naeslundii* had the ability to form biofilms under the experimental conditions. The criteria used to categorize biofilm formation capacity were based on the approach of previous researchers [26, 27]. According to them, both species can be categorized as "biofilm formers." Mean OD 570 nm values of monospecies biofilms were higher compared to the values of dual-species biofilms (Figure 1). It has been reported that, for bacteria that grow

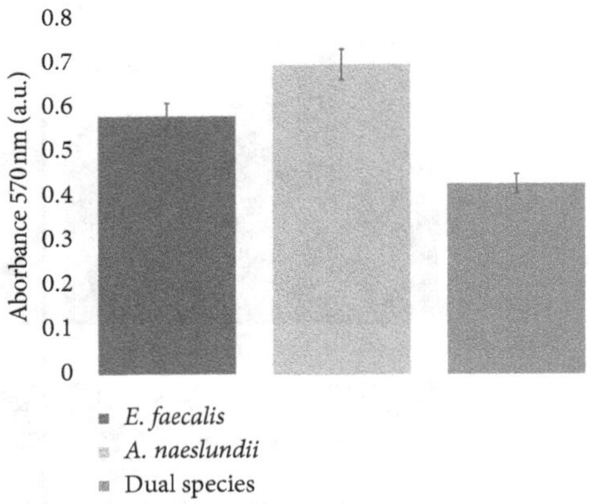

Figure 1: Crystal violet quantification of monospecies and dual-species biofilms of *E. faecalis* and *A. naeslundii* bacteria. Data represent mean ± SD of triplicate readings.

similarly well in the monospecies model, slight decreases can be expected for both components when grown in dual-species biofilms [28]. However, a recent study reported that when Gram-positive bacteria were cocultured with Gram negative, dual-species biofilms showed higher values compared to monospecies biofilms of Gram-negative bacteria [29]. It could be explained based on the fact that different bacterial cell-cell interrelation may exist when growing in cocultured biofilms, including competition, synergism, and nutrients availability [28, 30].

Our findings suggest the presence of the principal biomolecules that have been previously described as components of EPS biofilm matrices [11]. Furthermore, based on the intensity of the Raman bands, we were able to describe the EPS composition as dominated by a major protein fraction, with saturated fatty acids and polysaccharides occurring in a relatively smaller proportion, and a minor nucleic acid component. This composition reflects the characteristic of a mature biofilm as previously reported by other authors [20, 31, 32]. In another Raman spectroscopic study, Huang et al. found that the amide I region was more intense in the stationary phase than in the exponential phase [33]. Liu et al. reported that changes in nucleic acid and protein bands reflected the physiologic state of *E. faecalis* biofilms [20].

In the past, polysaccharides were considered to be almost the only structural component of the extracellular matrix, until further research elucidated another important structural biomolecule, eDNA [33]. Thus, in the present study, both components were characterized by means of CLSM. Calcofluor was used to perform polysaccharide staining, and propidium iodide for eDNA staining. CLSM images revealed the presence of polysaccharide clumps and eDNA fibres as components of the biofilm matrix. Mushroom-shaped structures and water channels, characteristics of mature biofilms, were also identified [34]. It has been proposed that eDNA can play distinct roles: first, as an adhesin in the early stages of biofilm formation, and second, as a structural component of the mature biofilm [35–37]. Figure 2 shows CSLM images of the developed biofilms at a mature stage, in

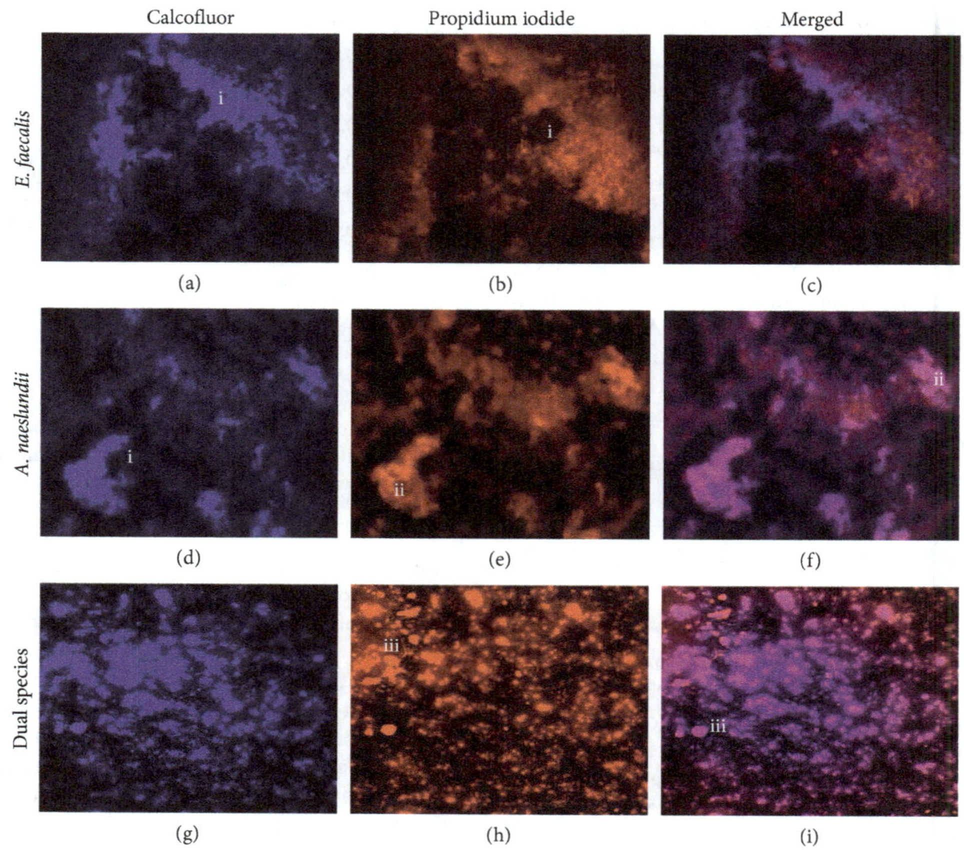

FIGURE 2: CSLM figures of biofilm matrices. Polysaccharides clumps (a, d, e) and eDNA aggregates (b, c, h). Structural features such as water channels (i), mushroom-like forms (ii), and eDNA fibres (iii) were identified.

which the presence of polysaccharides and eDNA as EPS matrix components was observed.

In spite of the similarity of the spectra of the different biofilms and the fact that both are Gram-positive bacteria, a PCA-DFA canonical plot of the Raman spectral profile data for the three biofilms suggests that *E. faecalis*, *A. naeslundii*, and the dual-species biofilm have extracellular matrices with different biochemical compositions; the cross-validation value was 95.1%. The main spectral differences are noticed in the spectral bands around 1270–1335 cm^{-1} and 1000–1125 cm^{-1}/800–980 cm^{-1}, which are assigned to protein and polysaccharides biomolecules, respectively (Figures 3(a) and 3(b)).

In a previous research, we found that biochemical composition among these biofilms species varies significantly from monospecies to dual species [38]. A major fraction of polysaccharides, proteins, and fatty acids was identified in the monospecies biofilms compared to those in the dual-species biofilms [38]. These results indicate that the two species influence each other at the biochemical level, and further data support the synergistic action between these two endodontic pathogens [39].

Figure 3(b) shows Raman spectra of the EPSs extracted from the three samples. Raman bands were assigned and interpreted based on previous research [15, 17, 40, 41]. Functional groups from the main biomolecules were identified (Figure 4). The bands in the region of 790–950 cm^{-1} can be attributed to side-group deformations (COH, CCH, and OCH) in biofilm-specific polysaccharides [15]. Raman spectra of the EPS fractions of the three biofilm samples showed variation in this region; the *E. faecalis* spectra were particularly different. Biofilm polysaccharides can vary greatly in their composition, which is mostly defined by factors such as genetic profile and environmental growth conditions. In Gram-positive cells, the presence and chemical structure of many rhamnose-containing cell-wall polysaccharides (RhaCWP) has been known for decades. The bands observed at 880–980, 1055, and 1075 cm^{-1} could be tentatively assigned to a combination of vibrational modes of rhamnose, galactose, and glucose [40]. Previous studies have reported the presence of these saccharides as part of *E. faecalis* as well as *Actinomyces* spp. [42–44].

The bands in the 1075–1125 cm^{-1} region can be assigned to the C-C and C–N stretching vibration contributions [15, 18, 19]. Amide I (the 1640–1680 cm^{-1} stretching vibration of C=O) and amide III (1250 cm^{-1}, associated with C–N stretching and N–H bending) bands were detected in the three biofilm spectra; however, the intensities of the bands varied significantly among them. The *A. naeslundii* spectra show an intense band around 1300 cm^{-1} (amide III), but this has also been assigned to the amino acid side-chain mode of tryptophan. It

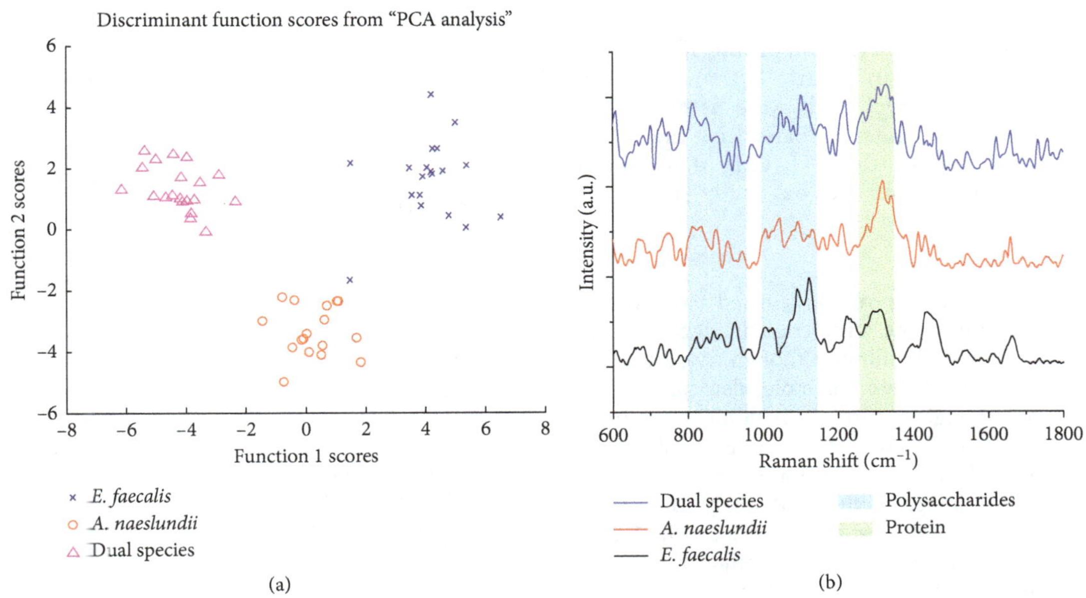

FIGURE 3: Raman spectroscopic analyses of *E. faecalis*, *A. naeslundii*, and dual biofilms (a). A PCA-DFA plot of all the Raman spectra from the biofilm developed from *E. faecalis* (asterisk), *A naeslundii* (circle), and dual species (triangle). The value of cross-validation was 95.1%. Principal spectral differences can be seen in three main ranges (b).

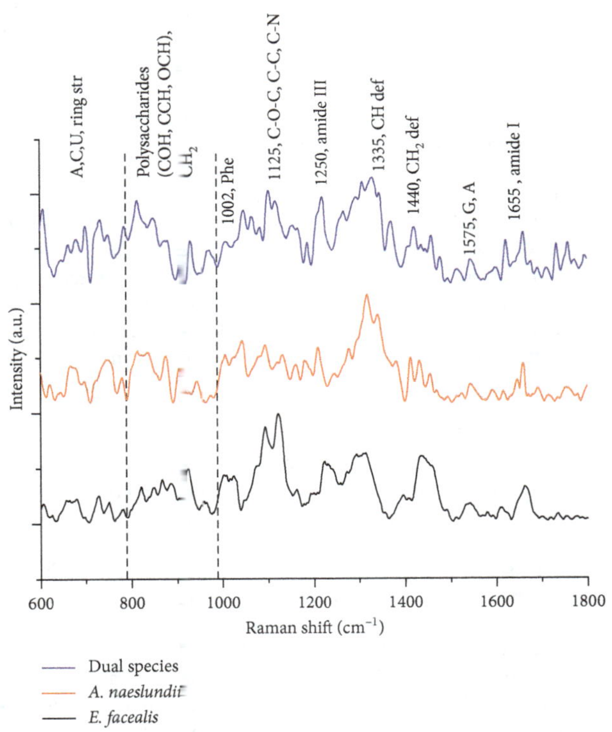

FIGURE 4: Raman spectra of three biofilm matrix samples. Functional groups from biomolecules were identified.

has been demonstrated that proteins contribute to initial adhesion and stability of the biofilm matrix [44]. Thus, there is a possibility that the bands related to proteins could represent lectins and amyloid-like proteins, a group of proteins that have been identified in many species implicated in biofilm formation [44–46]. These molecules can recognize carbohydrates in biofilm matrices and promote cell-to-cell communication. Raman spectra of lyophilized lectins show some bands at different positions, for instance, those around 1667 cm^{-1} and 1236 cm^{-1}, in the amide I and amide III bands, respectively [47]. Amyloids are involved in the building and stability of biofilm matrices. The term "amyloid" describes β-sheet-rich protein aggregates, which exhibit Raman bands mainly in the amide I (1665–1680 cm^{-1}) and amide III (1246–1270 cm^{-1}) regions [47, 48]. All of these bands were observed in the EPS samples analysed in the present study.

Since 2002, the functional role of eDNA in biofilm formation and structural integrity has been established [36], and the presence of bands attributable to the DNA backbone has been reported in previous research on biofilms [20]. Our findings reveal the presence of small bands around 728, 780, and 785 cm^{-1}, assigned to adenine, cytosine, and uracil, respectively [15]. Moreover, a band at 1575 cm^{-1} characteristic of guanine and adenine ring stretching [17] was identified in all our EPS samples.

The most distinctive features of the Raman spectra of lipids are related to the presence of the hydrocarbon chains, and for all lipids, these can be observed in the following regions: 1500–1400, 1300–1250, and 1200–1050 cm^{-1}, which is in the fingerprint region [49]. In the present study, bands around 1445 and 1500 cm^{-1} assigned to scissoring vibrations of the CH_2 group were identified. Furthermore, the Raman band at 1300 cm^{-1}, attributed to the twisting vibration of the CH_3 group, is particularly intense in the spectrum of palmitic acid [49]. The presence of fatty acids has been reported in bacterial biofilms, especially saturated fatty acids [15, 18, 26]. Moreover, in a fatty acid profile of the initial oral biofilm, palmitic (16 : 0), stearic (18 : 0), oleic (18 : 1n9c), and erucic (22 : 1n9c) acids were detected [50]. These fatty acids have intense bands in the 1400–1500 cm^{-1} region due to the CH_2/CH_3 deformation modes, similar to bands found in our study [15, 51].

4. Conclusions

As a preliminary study, our results showed variations in wave shape confirming the difference in the chemical composition of analysed biofilms, although more data are obviously required on the biofilm molecular composition when grown on biological surfaces including dentin or hydroxyapatite. It is of interest to examine metabolic activities of bacteria during the maturation of the biofilm process, especially in mixed populations where interspecies interactions show different chemical behavior [29, 38].

Raman spectroscopic and CSLM technique allowed the biochemical and structural characterization of monospecies and dual-species biofilm matrices. Our results provide valuable information about EPS matrix components of relevant biofilm-forming endodontic bacteria, elucidating the variability and complexity that prevail in microbial communities. These data represent the first approach, and further research will be necessary to confirm and clearly identify the EPS components that we tentatively identified in this study.

Conflicts of Interest

The authors declare that there are no conflicts of interest regarding the publication of this paper.

Acknowledgments

The study was approved by the Institutional Ethics Committee of the Faculty of Dentistry at San Luis Potosi University, San Luis Potosi, México (Approval Code CEI-FE-011-014).

References

[1] C. Zhang, J. Du, and Z. Peng, "Correlation between *Enterococcus faecalis* and persistent intraradicular infection compared with primary intraradicular infection: a systematic review," *Journal of Endodontics*, vol. 41, no. 8, pp. 1207–1213, 2015.

[2] L. C. de Paz, "Redefining the persistent infection in root canals: possible role of biofilm communities," *Journal of Endodontics*, vol. 33, no. 6, pp. 652–662, 2007.

[3] F. G. C. Signoretti, M. S. Endo, B. P. F. A. Gomes, F. Montagner, F. B. Tosello, and R. C. Jacinto, "Persistent extraradicular infection in root-filled asymptomatic human tooth: scanning electron microscopic analysis and microbial investigation after apical microsurgery," *Journal of Endodontics*, vol. 37, no. 12, pp. 1696–1700, 2011.

[4] L. Chavez de Paz, "Gram-positive organisms in endodontic infections," *Endodontic Topics*, vol. 9, no. 1, pp. 79–96, 2004.

[5] P. N. R. Nair, S. Henry, V. Cano, and J. Vera, "Microbial status of apical root canal system of human mandibular first molars with primary apical periodontitis after "one-visit" endodontic treatment," *Oral Surgery, Oral Medicine, Oral Pathology, Oral Radiology, and Endodontology*, vol. 99, no. 2, pp. 231–252, 2005.

[6] D. Ricucci and J. F. Siqueira, "Apical actinomycosis as a continuum of intraradicular and extraradicular infection: case report and critical review on its involvement with treatment failure," *Journal of Endodontics*, vol. 34, no. 9, pp. 1124–1129, 2008.

[7] L. Tronstad and P. T. Sunde, "The evolving new understanding of endodontic infections," *Endodontic Topics*, vol. 6, no. 1, pp. 57–77, 2003.

[8] G. Svensater and G. Bergenholtz, "Biofilms in endodontic infections," *Endodontic Topics*, vol. 9, no. 1, pp. 27–36, 2004.

[9] P. Stoodley, K. Sauer, D. G. Davies, and J. W. Costerton, "Biofilms as complex differentiated communities," *Annual Review of Microbiology*, vol. 56, no. 1, pp. 187–209, 2002.

[10] H.-C. Flemming, T. R. Neu, and D. J. Wozniak, "The EPS matrix: the "house of biofilm cells"," *Journal of Bacteriology*, vol. 189, no. 22, pp. 7945–7947, 2007.

[11] H.-C. Flemming and J. Wingender, "The biofilm matrix," *Nature Reviews Microbiology*, vol. 8, no. 9, pp. 623–633, 2010.

[12] S. George, A. Kishen, and P. Song, "The role of environmental changes on monospecies biofilm formation on root canal wall by *Enterococcus faecalis*," *Journal of Endodontics*, vol. 31, no. 12, pp. 867–872, 2005.

[13] R. M. Donlan, "Biofilms: microbial life on surfaces," *Emerging Infectious Diseases*, vol. 8, no. 9, pp. 881–890, 2002.

[14] E. Denkhaus, S. Meisen, U. Telgheder, and J. Wingender, "Chemical and physical methods for characterisation of biofilms," *Microchimica Acta*, vol. 158, no. 1-2, pp. 1–27, 2006.

[15] M. Wagner, N. P. Ivleva, C. Haisch, R. Niessner, and H. Horn, "Combined use of confocal laser scanning microscopy (CLSM) and Raman microscopy (RM): investigations on EPS–matrix," *Water Research*, vol. 43, no. 1, pp. 63–76, 2009.

[16] N. P. Ivleva, M. Wagner, H. Horn, R. Niessner, and C. Haisch, "Towards a nondestructive chemical characterization of biofilm matrix by Raman microscopy," *Analytical and Bioanalytical Chemistry*, vol. 393, no. 1, pp. 197–206, 2008.

[17] M. Harz, P. Rösch, and J. Popp, "Vibrational spectroscopy—a powerful tool for the rapid identification of microbial cells at the single-cell level," *Cytometry Part A*, vol. 75A, no. 2, pp. 104–113, 2009.

[18] D. Kusić, B. Kampe, A. Ramoji, U. Neugebauer, P. Rösch, and J. Popp, "Raman spectroscopic differentiation of planktonic bacteria and biofilms," *Analytical and Bioanalytical Chemistry*, vol. 407, no. 22, pp. 6803–6813, 2015.

[19] C. Sandt, T. Smith-Palmer, J. Pink, L. Brennan, and D. Pink, "Confocal Raman microspectroscopy as a tool for studying the chemical heterogeneities of biofilms in situ," *Journal of Applied Microbiology*, vol. 103, no. 5, pp. 1808–1820, 2007.

[20] H. Liu, Q. Xu, L. Huo, X. Wei, and J. Ling, "Chemical composition of *Enterococcus faecalis* in biofilm cells initiated from different physiologic states," *Folia Microbiologica*, vol. 59, no. 5, pp. 447–453, 2014.

[21] J. F. Siqueira, I. N. Rôças, and D. Ricucci, "Biofilms in endodontic infection," *Endodontic Topics*, vol. 22, no. 1, pp. 33–49, 2012.

[22] J. M. Guerreiro-Tanomaru, N. B. de Faria-Júnior, M. A. H. Duarte, R. Ordinola-Zapata, M. S. Z. Graeff, and M. Tanomaru-Filho, "Comparative analysis of *Enterococcus faecalis* biofilm formation on different substrates," *Journal of Endodontics*, vol. 39, no. 3, pp. 346–350, 2013.

[23] A. Kishen, M. Upadya, G. P. Tegos, and M. R. Hamblin, "Efflux pump inhibitor potentiates antimicrobial photodynamic inactivation of *Enterococcus faecalis* biofilm," *Photochemistry and Photobiology*, vol. 86, no. 6, pp. 1343–1349, 2010.

[24] M. Martins, P. Uppuluri, D. P. Thomas et al., "Presence of extracellular DNA in the *Candida albicans* biofilm matrix and its contribution to biofilms," *Mycopathologia*, vol. 169, no. 5, pp. 323–331, 2009.

[25] L. A. Reisner, A. Cao, and A. K. Pandya, "An integrated software system for processing, analyzing, and classifying Raman spectra," *Chemometrics and Intelligent Laboratory Systems*, vol. 105, no. 1, pp. 83–90, 2011.

[26] J. A. Mohamed, W. Huang, S. R. Nallapareddy, F. Teng, and B. E. Murray, "Influence of origin of isolates, especially endocarditis isolates, and various genes on biofilm formation by *Enterococcus faecalis*," *Infection and Immunity*, vol. 72, no. 6, pp. 3658–3663, 2004.

[27] J. A. T. Sandoe, J. Wysome, A. P. West, J. Heritage, and M. H. Wilcox, "Measurement of ampicillin, vancomycin, linezolid and gentamicin activity against enterococcal biofilms," *Journal of Antimicrobial Chemotherapy*, vol. 57, no. 4, pp. 767–770, 2006.

[28] Z. T. Wen, D. Yates, S.-J. Ahn, and R. A. Burne, "Biofilm formation and virulence expression by *Streptococcus mutans* are altered when grown in dual-species model," *BMC Microbiology*, vol. 10, no. 1, p. 111, 2010.

[29] J. Makovcova, V. Babak, P. Kulich, J. Masek, M. Slany, and L. Cincarova, "Dynamics of mono- and dual-species biofilm formation and interactions between *Staphylococcus aureus* and gram-negative bacteria," *Microbial Biotechnology*, vol. 10, no. 4, pp. 819–832, 2017.

[30] M. I. Klein, J. Xiao, B. Lu, C. M. Delahunty, J. R. Yates, and H. Koo, "*Streptococcus mutans* protein synthesis during mixed-species biofilm development by high-throughput quantitative proteomics," *PLoS One*, vol. 7, no. 9, Article ID e45795, 2012.

[31] F. Dubois-Brissonnet, E. Trotier, and R. Briandet, "The biofilm lifestyle involves an increase in bacterial membrane saturated fatty acids," *Frontiers in Microbiology*, vol. 7, p. 1673, 2016.

[32] Y. Chao and T. Zhang, "Surface-enhanced Raman scattering (SERS) revealing chemical variation during biofilm formation: from initial attachment to mature biofilm," *Analytical and Bioanalytical Chemistry*, vol. 404, no. 5, pp. 1465–1475, 2012.

[33] W. E. Huang, R. I. Griffiths, I. P. Thompson, M. J. Bailey, and A. S. Whiteley, "Raman microscopic analysis of single microbial cells," *Analytical Chemistry*, vol. 76, no. 15, pp. 4452–4458, 2004.

[34] A. M. T. Barnes, K. S. Ballering, R. S. Leibman, C. L. Wells, and G. M. Dunny, "Correlative ultrastructural analysis of extracellular DNA in early *Enterococcus faecalis* biofilms," *Microscopy and Microanalysis*, vol. 18, no. 2, pp. 8-9, 2012.

[35] A. M. Gonzalez, E. Corpus, A. Pozos-Guillen, D. Silva-Herzog, A. Aragon-Piña, and N. Cohenca, "Continuous drip flow system to develop biofilm of *E. faecalis* under anaerobic conditions," *Scientific World Journal*, vol. 2014, Article ID 706189, 5 pages, 2014.

[36] M. Sena-Vélez, C. Redondo, J. H. Graham, and J. Cubero, "Presence of extracellular DNA during biofilm formation by *Xanthomonas citri* subsp. *citri* strains with different host range," *PLoS One*, vol. 11, no. 6, Article ID e0156695, 2016.

[37] M. Okshevsky and R. L. Meyer, "The role of extracellular DNA in the establishment, maintenance and perpetuation of bacterial biofilms," *Critical Reviews in Microbiology*, vol. 41, no. 3, pp. 341–352, 2013.

[38] T. Ramirez-Mora, C. Retana-Lobo, and G. Valle-Bourrouet, "Biochemical characterization of extracellular polymeric substances from endodontic biofilms," *PLoS One*, vol. 13, no. 11, Article ID e0204081, 2018.

[39] L. E. Chávez de Paz, J. R. Davies, G. Bergenholtz, and G. Svensäter, "Strains of *Enterococcus faecalis* differ in their ability to coexist in biofilms with other root canal bacteria," *International Endodontic Journal*, vol. 48, no. 10, pp. 916–925, 2015.

[40] C. You-Peng, Z. Peng, G. Jin-Song, F. Fang, X. Gao, and C. Li, "Functional groups characteristics of EPS in biofilm growing on different carriers," *Chemosphere*, vol. 92, no. 6, pp. 633–638, 2013.

[41] J. De Gelder, K. De Gussem, P. Vandenabeele, P. De Vos, and L. Moens, "Methods for extracting biochemical information from bacterial Raman spectra: an explorative study on *Cupriavidus metallidurans*," *Analytica Chimica Acta*, vol. 585, no. 2, pp. 234–240, 2007.

[42] C. A. Guzmàn, C. Pruzzo, G. LiPira, and L. Calegari, "Role of adherence in pathogenesis of *Enterococcus faecalis* urinary tract infection and endocarditis," *Infection and Immunity*, vol. 57, no. 6, pp. 1834–1838, 1989.

[43] K. L. Palmer, P. Godfrey, A. Griggs et al., "Comparative genomics of enterococci: variation in *Enterococcus faecalis*, clade structure in *E. faecium*, and defining characteristics of *E. gallinarum* and *E. casseliflavus*," *mBio*, vol. 3, no. 1, article e00318-11, 2012.

[44] B. Rosan and B. F. Hammond, "Extracellular polysaccharides of *Actinomyces viscosus*," *Infection and Immunity*, vol. 10, no. 2, pp. 304–308, 1974.

[45] J. N. C. Fong and F. H. Yildiz, "Biofilm matrix proteins," *Microbiology Spectrum*, vol. 3, no. 2, pp. 1–23, 2015.

[46] A. Taglialegna, I. Lasa, and J. Valle, "Amyloid structures as biofilm matrix scaffolds," *Journal of Bacteriology*, vol. 198, no. 19, pp. 2579–2588, 2016.

[47] J. López-Ochoa, J. F. Montes-García, C. Vázquez et al., "*Gallibacterium* elongation factor-Tu possesses amyloid-like protein characteristics, participates in cell adhesion, and is present in biofilms," *Journal of Microbiology*, vol. 55, no. 9, pp. 745–752, 2017.

[48] A. Rygula, K. Majzner, K. M. Marzec, A. Kaczor, M. Pilarczyk, and M. Baranska, "Raman spectroscopy of proteins: a review," *Journal of Raman Spectroscopy*, vol. 44, no. 8, pp. 1061–1076, 2013.

[49] D. Kurouski, R. P. Van Duyne, and I. K. Lednev, "Exploring the structure and formation mechanism of amyloid fibrils by Raman spectroscopy: a review," *Analyst*, vol. 140, no. 15, pp. 4967–4980, 2015.

[50] K. Czamara, K. Majzner, M. Z. Pacia, K. Kochan, A. Kaczor, and M. Baranska, "Raman spectroscopy of lipids: a review," *Journal of Raman Spectroscopy*, vol. 46, no. 1, pp. 4–20, 2014.

[51] M. Reich, K. Kümmerer, A. Al-Ahmad, and C. Hannig, "Fatty acid profile of the initial oral biofilm (pellicle): an in-situ study," *Lipids*, vol. 48, no. 9, pp. 929–937, 2013.

2

Statistically Coherent Calibration of X-Ray Fluorescence Spectrometry for Major Elements in Rocks and Minerals

Surendra P. Verma,[1] Sanjeet K. Verma,[2] M. Abdelaly Rivera-Gómez,[3] Darío Torres-Sánchez,[4] Lorena Díaz-González,[5] Alejandra Amezcua-Valdez,[6] Beatriz Adriana Rivera-Escoto,[7] Mauricio Rosales-Rivera,[8] John S. Armstrong-Altrin,[3] Héctor López-Loera,[2] Fernando Velasco-Tapia,[9] and Kailasa Pandarinath[1]

[1]*Instituto de Energías Renovables, Universidad Nacional Autónoma de México, Temixco, Mor 62580, Mexico*
[2]*División de Geociencias, Instituto Potosino de Investigación en Ciencia y Tecnología, Camino a la Presa San José # 2055, Col. Lomas 4a Sec., San Luis Potosí, SLP 78216, Mexico*
[3]*Instituto de Ciencias del Mar y Limnología, Unidad de Procesos Oceánicos y Costeros, Universidad Nacional Autónoma de México, Circuito Exterior s/n, 04510 CDMX, Mexico*
[4]*Posgrado en Geociencias Aplicadas, Instituto Potosino de Investigación en Ciencia y Tecnología, Camino a la Presa San José # 2055, Col. Lomas 4a Sec., San Luis Potosí, SLP 78216, Mexico*
[5]*Centro de Investigación en Ciencias, Instituto de Investigación en Ciencias Básicas y Aplicadas, Universidad Autónoma del Estado de Morelos, Cuernavaca, Mor 62209, Mexico*
[6]*Posgrado en Ingeniería, Instituto de Energías Renovables, Universidad Nacional Autónoma de México, Temixco, Mor 62580, Mexico*
[7]*División de Materiales Avanzados, Instituto Potosino de Investigación en Ciencia y Tecnología, Camino a la Presa San José # 2055, Col. Lomas 4a Sec., San Luis Potosí, SLP 78216, Mexico*
[8]*Doctorado en Ciencias, Instituto de Investigación en Ciencias Básicas y Aplicadas, Universidad Autónoma del Estado de Morelos, Cuernavaca, Mor 62209, Mexico*
[9]*Universidad Autónoma de Nuevo León, Facultad de Ciencias de la Tierra, Ex-Hacienda de Guadalupe, Carretera Linares–Cerro Prieto km 8, Linares, N.L. 67700, Mexico*

Correspondence should be addressed to Surendra P. Verma; spv@ier.unam.mx

Academic Editor: Rafal Sitko

We applied both the ordinary linear regression (OLR) and the new uncertainty weighted linear regression (UWLR) models for the calibration and comparison of a XRF machine through 59 geochemical reference materials (GRMs) and a procedure blank sample. The mean concentration and uncertainty data for the GRMs used for the calibrations (Supplementary Materials) (available here) filewere achieved from an up-to-date compilation of chemical data and their processing from well-known discordancy and significance tests. The drift-corrected XRF intensity and its uncertainty were determined from mostly duplicate pressed powder pellets. The comparison of the OLR (linear correlation coefficient r~0.9523–0.9964 and 0.9771–0.9999, respectively, for before and after matrix correction) and UWLR models (r~0.9772–0.9976 and 0.9970–0.9999, respectively) clearly showed that the latter with generally higher values of r is preferable for routine calibrations of analytical procedures. Both calibrations were successfully applied to rock matrices, and the results were generally consistent with those obtained in other laboratories although the UWLR model showed mostly narrower confidence limits of the mean (slope and intercept) or lower uncertainties than the OLR. Similar sensitivity (~2.69–46.17 $kc{\cdot}s^{1}{\cdot}\%^{1}$ for the OLR and ~2.78–59.69 $kc{\cdot}s^{1}{\cdot}\%^{1}$ for the UWLR) also indicated that the UWLR could advantageously replace the OLR model. Another novel aspect is that the total uncertainty can be reported for individual chemical data. If the analytical instruments were routinely calibrated from the UWLR model, this action would make the science of geochemistry more quantitative than at present.

1. Introduction

All modern analytical instruments require some kind of calibration of the instrumental response (y-variable) as a function of the concentration (x-variable) [1–3]. This calibration is generally achieved through an ordinary least-squares linear regression (OLR) model. However, such a procedure is not strictly valid because all requirements for the statistical validity of the OLR model are not fulfilled. Usually, the assumptions "independent concentration variable x is error-free or less than one-tenth of the error in the dependent response variable y" and "error in y is homoscedastic' (i.e., equal errors for all y values) are not satisfied and, therefore, more sophisticated and statistically coherent regression procedures, such as weighted least-squares linear regression (WLR) models, should be used [4–18].

X-ray fluorescence (XRF) spectrometry is among the most popular analytical techniques for the determination of all major and some trace elements in rocks [4, 19–27]. Natural geochemical reference materials (GRMs) are commonly used for XRF calibrations and posterior characterization of those and other GRMs as well as of similar rock and mineral matrices [4, 19, 28–30]. As for most other analytical instruments, XRF spectrometers are also calibrated under the statistically incoherent OLR model.

To apply the WLR and compare it with the OLR, both central tendency (e.g., mean) and dispersion (e.g., confidence limits of the mean) estimates on both x-axis (concentration, generally expressed in the unit of % m/m, i.e., mass/mass unit expressed in percent) and y-axis (response, in this case XRF intensity, generally reported in the unit of $kc \cdot s^{-1}$, i.e., kilo counts per second) variables are required. More precise (and accurate) estimates of the central tendency will also be useful for both types of regressions. Therefore, precise concentrations of GRMs with the respective lowest possible "confidence limits of the mean" (referred hereafter as the "uncertainty" of the measured variable) [2, 17, 18] are required to apply the regression procedure. Sometimes, we had to use also the term "error" (instead of the uncertainty) because the use of the error is widespread in the literature.

We report the following five aspects: (a) evaluation of 59 GRMs to achieve the least possible uncertainties in the mean concentrations of all major elements (SiO_2 to P_2O_5); (b) the comparison of regression models (OLR and WLR) applied to net drift-corrected XRF intensities before the correction of matrix effects; (c) the second (or final) comparison of both models after achieving the matrix correction as well as for the estimation of sensitivities of the regression models; (d) application of the entire procedure to four GRMs treated as "unknown" samples and their comparison with the previous literature compilations; and (e) development of a computer program to achieve the abovementioned objectives. Thus, the regression equations (intercept and its uncertainty, slope and its uncertainty, and linear correlation coefficient values) for each constituent from SiO_2 to P_2O_5 and their application to similarly complex rock matrices are presented in this work.

2. Evaluation of Major Element Data for GRMs

A total of 59 GRMs (listed in alphabetical order in Table S1; this and four other tables are provided in Supplementary Materials), along with a procedure blank, were used in this study. The procedure blank was a pellet prepared in duplicate with only pure N,N′-Ethylene bis(stearamide) beads without any sample (Section 3). The individual data reported in earlier compilations [31–47] were first compiled in new databases.

The statistical parameters obtained in these early compilations could not be directly used for instrumental calibrations due to the following reasons: (i) the statistical methods used to achieve the statistical estimates were outdated (see [17, 18, 48, 49] for possible reasons), and the inferred statistical values were of low quality (high values of dispersion); (ii) there are still determinations reported during about 30 or more years (postcompilation years) that were not obviously available to those compilers; (iii) the precision of more recent determinations is likely to have improved due to the availability of online computers on most modern instruments; (iv) newer more reliable statistical techniques are now available for improving both precision and accuracy of the statistical inferences, e.g., the use of discordancy tests with the highest power and lowest swamping and masking effects [18, 48, 50–52]; and (v) importantly, new computer programs have been developed by our group [52–54], available at http://tlaloc.ier.unam.mx for download or online processing of data (after previous registration onto our server), which can be advantageously used for efficient processing of experimental databases.

The same kinds of objections are applicable even today for the originator's websites, such as https://gbank.gsj.jp/geostandards/welcome.html for Japanese GRMs or https://crustal.usgs.gov/geochemical_reference_standards for United States GRMs. The statistical information at these websites is based on early compilations (around 30 or more years ago). Furthermore, we were unable to use the recent work [55] because this paper reported significantly larger uncertainty values as compared to those achievable from our new validated statistical procedure [51–54]; besides, updated statistical information on the mean and its uncertainty was not available in [55] for many GRMs used in our work.

The initial databases were complemented by individual data from a large number of posterior publications (~480; Table S1), whose complete listing is available at our server http://tlaloc.ier.unam.mx under the heading of "Quality Control." These major element data were classified according to the analytical method groupings [56]. Data from each method group were considered as a univariate statistical sample. Appropriate discordancy and significance tests were applied from thoroughly automatized software UDASys2 [52] and UDASys3 (unpublished), which, in their "recommended procedure," apply the most powerful five (two new and three conventional) recursive tests with prior application of respective single-outlier tests having nil swamping and low masking effects [48, 57–60]. Although the application of discordancy tests is identical for both UDASys2 and UDASys3, the difference lies in that the latter applies the significance (ANOVA, F and t) tests in order to provide the final results automatically.

The resulting statistical information after the application of well-known discordancy tests at the strict 99% confidence level (mean and uncertainty values rounded according to the flexible rules [18]) is listed in Table S2. These GRM compositional data showed by far the lowest 99% uncertainty (Table S2), much lower than any existing compilation [31–47, 55]. We may also stress once again that this was achieved through an objective combination of discordancy tests having the highest performance and lowest swamping and masking effects [17, 18, 48, 53], i.e., from the methodology having the lowest type I and type II errors and the highest power.

Therefore, the population mean of these GRMs is now known within the narrowest possible 99% confidence limits of the mean to best represent the concentration (x) axis in the instrumental calibrations as suggested [2, 5, 7, 10, 17, 18, 53]. These data (in units of % m/m; Table S2) will also be useful for those who wish to achieve instrumental calibrations or simply use them for quality control of their results for rock and mineral matrices.

3. XRF Instrumentation and Intensity Measurements

A wavelength dispersive X-ray fluorescence (WDXRF) spectrometer Rigaku ZSX Primus II model (rhodium X-ray tube; 4 kW maximum power) was used for this work. We made the effort to best represent the response (y) axis (x-ray intensity in the units of kilo counts per second, $kc\cdot s^{-1}$) for the calibrations. For each GRM, duplicate (41 samples) or even triplicate (8 samples) pressed powder pellets were prepared. First, an appropriate amount of each GRM was dried overnight in an oven at about 105°C. For each pellet, accurately weighed 3.5 g of moisture-free GRM was thoroughly mixed with accurately weighed 3.0 g pure N,N′-ethylene bis(stearamide) beads, <840 μm as wax (Sigma-Aldrich), and stored in a desiccator. Pressed powder pellets were prepared at 20 $tons\cdot inch^{-2}$ pressure (about 310 MPa). However, for 10 GRMs, sufficient material was not available; therefore, only a single pellet could be prepared but the measured intensity uncertainty (u_{99}) at the 99% confidence level was increased by a factor of 2 to take into account the sample preparation variance. Similarly, accurately weighed 6.5 g of pure N,N′-ethylene bis(stearamide) beads, <840 μm as wax, was pressed to prepare a procedure blank sample. This was done in duplicate.

For the intensity measurements, the optimum instrumental conditions were first established through preliminary experiments prior to the routine measurements (Table S2). Each pellet was run at least 8 to 10 times in a random sequence, along with two drift monitors prepared from two volcanic rocks (basalt and rhyolite) from the San Luis Potosí Volcanic Field, San Luis Potosí (central Mexico).

The peak and background measuring conditions and time periods are also listed in Table S3. Appropriate mean drift corrections from two monitors were applied to all intensity measurements. Both monitors were run randomly 8 to 10 times each day. First, the expected monitor intensity was established as an average value of the first two days when the intensities were fairly stable and reproducible. Then, the average drift correction factors were calculated for each chemical element from the two monitors run in the XRF instrument periodically before and after a set of GRMs used for the calibration. These correction factors were then applied to the bracketed GRMs for the entire period of calibration, including the first two days and analysis of "unknown" samples.

Now, although the X-ray counts may obey a Poisson distribution, we are dealing with average values of count rates, which are likely to follow a normal distribution because of the central limit theorem. A normal distribution of measured intensities was also assured for each pellet from the application of discordancy tests as explained above for GRM concentrations. The intensity results for all pellets from a given GRM were then combined, the tests applied again to the combined data, and new mean and 99% uncertainty values were calculated for X-ray intensity of each GRM. This was done to take into account the variance of the sample preparation method, which was significantly higher than the instrumental variance of intensity measurements for individual pellets. The drift-corrected intensity values and their 99% uncertainties ($kc\cdot s^{-1}$) for all GRMs, along with the concentration data and their 99% uncertainties (% m/m), are listed in Table S2.

4. Regression Models

Two different regression models (OLR and UWLR) were used and compared in this work. The OLR model most frequently used for instrumental calibrations (x-axis concentration and y-axis response; GRM concentration and X-ray intensity, respectively, in XRF spectrometry) requires the following assumptions to be fulfilled [4, 7, 10, 12–18] (i) all errors are in the y-axis; (ii) x-axis is either error-free or has at most 10% error of the y-axis errors; (iii) errors in both axes are normally distributed; and (iv) errors in the y-axis are homoscedastic. Some or all of these assumptions are violated in most instrumental calibrations through the OLR model.

Thus, from the literature on the GRMs, it has been demonstrated that the concentration axis is not error-free (see non-zero uncertainties for all GRM concentrations in Table S2) [31–47, 51–53]. One can also clearly see that the errors in the intensity axis are not homoscedastic (see unequal, i.e., heteroscedastic uncertainties for any element in different GRMs in Table S2). For a heteroscedastic linear regression system, even if each error or noise term is still Gaussian, the OLR model is no longer the maximum likelihood estimate and consequently, it is no longer efficient [10]. The main advantage that the WLR has over the OLR is the ability to handle regression situations in which the data points are of varying quality as is the case with most instruments including the XRF spectrometers.

However, the major disadvantage of the WLR is that the approach is based on the assumption that the weights are known exactly. They can be estimated using several different equations or algorithms, but when the weights are produced from small numbers of replicated observations, the regression parameters can be unpredictably affected [10]. In the example of the XRF calibration that we are presenting, the numbers of observations were relatively large for both the x and y axes (concentration and X-ray intensity parameters). Besides, instead of the sample variance, we used the uncertainty values (that take into account the number of observations in the formula for uncertainty or confidence limits of the mean calculations) [2, 18] for estimating the weight factors. The problem of the sensitivity to outliers in the regression equations [10] was also appropriately handled by discordancy tests programmed in the UDASys and BiDASys software [53, 54, 61].

Therefore, although frequently used, the OLR model is not statistically correct or coherent. The statistically coherent WLR, especially the uncertainty-based WLR (UWLR [17]) model, should be used. The confidence level, such as 95% or 99% (significance level of 5% or 1%, respectively, or α of 0.05 and 0.01, respectively), can be explicitly expressed in the confidence limits of the mean or uncertainty used in the UWLR model as well as to estimate the weight factors [17]. We will deal with the 99% uncertainty to have the type I error small (about 1%). Unfortunately, software of most analytical instruments, including XRF spectrometers, allows only the OLR calibration. Therefore, any sophisticated regression model, such as the UWLR, will have to be applied outside the instrumental software. Thus, the probability concept (99% confidence level) can be explicitly used in the UWLR model for weight factors based on the inverse of the squared 99% uncertainty of the mean.

We now present a synthesis of the regression equations for instrumental calibrations [2, 10, 17, 18, 61].

4.1. Ordinary Least-Squares Linear Regression (OLR) Model.

Let us assume that we have a series of n reference materials or standard calibrators having individual mean concentrations x_i with respective uncertainties u_{x_i} where i varies from 1 to n. In order to calibrate an instrument, each of these n calibrators were run several times, obtaining individual mean responses y_i with respective uncertainties u_{y_i} where i varies from 1 to n. Thus, we have n bivariate concentration-response data pairs or calibrators $(x_i,\ y_i)$ with the respective uncertainties $(u_{x_i},\ u_{y_i})$.

We can apply the OLR model to these data for obtaining a calibration equation. The OLR fits a least-squares linear equation to the n pairs $(x_i,\ y_i)$ but does not take into account the respective uncertainties $(u_{x_i},\ u_{y_i})$.

The general regression equation for the OLR is as follows (the subscript O is for the OLR model):

$$y_O\left(\pm u_{y_O}\right) = b_O\left(\pm u_{b_O}\right) + \left[m_O\left(\pm u_{m_O}\right) \times x\right], \tag{1}$$

where m is the slope, u_m is the resulting uncertainty in the slope, b is the intercept, u_b is the resulting uncertainty in the intercept, x is the independent variable, y_O is the dependent variable from the OLR model, and u_{y_O} is the resulting uncertainty in y. The following equations allow the calculations of these parameters:

$$m_O = \frac{\sum_{i=1}^{n}\{(x_i-\overline{x})\times(y_i-\overline{y})\}}{\sum_{i=1}^{n}(x_i-\overline{x})^2}, \tag{2}$$

where $\overline{x}$ and $\overline{y}$ are, respectively, the mean values of the x and y variables:

$$u_{m_O} = \left\{\sqrt{\frac{\sum_{i=1}^{n}(y_i-\widehat{y}_i)^2}{(n-2)\sum_{i=1}^{n}(x_i-\overline{x})^2}}\right\} \times t_{(n-2)}^{\mathrm{CL}}, \tag{3}$$

where $\widehat{y}_i$ is the value of y_O for x_i in equation (1) and t is the Student's t test value for $(n-2)$ degrees of freedom, and the superscript CL is the confidence level, generally 95% or 99%:

$$b_O = \overline{y} - (m_O \times \overline{x}),$$

$$u_{b_O} = \left\{\sqrt{\frac{\sum_{i=1}^{n}(y_i-\widehat{y}_i)^2 \times \sum_{i=1}^{n}x_i^2}{n(n-2)\sum_{i=1}^{n}(x_i-\overline{x})^2}}\right\} \times t_{(n-2)}^{\mathrm{CL}}. \tag{4}$$

It is a general practice in most instrumental calibrations to ignore all uncertainties in equation (1) and use an OLR equation without any error (or uncertainty) as follows:

$$y_O = b_O + [m_O \times x]. \tag{5}$$

The resulting standard deviation values of repeat measurements of unknown samples are reported as the final errors. However, these are only partial errors because the errors in the calibration equation (1) are not taken into account. In this work, we will use equation (1) to report total errors (in fact, 99% uncertainties) for the OLR model.

4.2. Uncertainty Weighted Least-Squares Linear Regression (UWLR) Model.

For the UWLR model, the n pairs $(x_i,\ y_i)$ of calibrators as well as the respective uncertainties $(u_{x_i},\ u_{y_i})$ are taken into account in order to achieve the best least-squares linear fit.

The uncertainties u_{x_i} in the x-axis are first propagated to the y-axis, combined with the u_{y_i}, and the total uncertainty u_i values on the y-axis are used for the weighting factors [2, 10, 17, 18, 61]:

$$u_i = \sqrt{\left(m_O \times u_{x_i}\right)^2 + \left(u_{y_i}\right)^2}. \tag{6}$$

The weights are calculated from u_i as follows:

$$w_i = \frac{n \times (u_i)^{-2}}{\sum_{i=1}^{n}(u_i)^{-2}}, \tag{7}$$

where w_i values have the following property:

$$\sum_{i=1}^{n} w_i = n. \tag{8}$$

Thus, the UWLR fits a linear equation to the n pairs $(x_i,\ y_i)$ with the respective weighting factors w_i as follows (the subscript UW is for the UWLR model):

$$y_{UW}\left(\pm u_{y_{UW}}\right) = b_{UW}\left(\pm u_{b_{UW}}\right) + \left[m_{UW}\left(\pm u_{m_{UW}}\right) \times x\right]. \tag{9}$$

Note that this regression line will pass closer to the data with lesser uncertainty u_i. The intercept and slope variables and their uncertainties are calculated from the following equations:

$$m_{\mathrm{UW}} = \frac{\{\sum_{i=1}^{n}(w_i \times x_i \times y_i)\} - \{n \times \overline{x}_{\mathrm{UW}} \times \overline{y}_{\mathrm{UW}}\}}{\{\sum_{i=1}^{n}(w_i \times x_i^2)\} - \{n \times (\overline{x}_{\mathrm{UW}})^2\}}, \quad (10)$$

where $\overline{x}_{\mathrm{UW}}$ and $\overline{y}_{\mathrm{UW}}$ are, respectively, the weighted mean values of the x and y variables:

$$u_{m_{\mathrm{UW}}} = \left\{\sqrt{\frac{\sum_{i=1}^{n}(y_i - \widehat{y}_{i\mathrm{UW}})^2}{(n-2)\sum_{i=1}^{n}(x_i - \overline{x}_{\mathrm{UW}})^2}}\right\} \times t_{(n-2)}^{\mathrm{CL}}, \quad (11)$$

where $\widehat{y}_{i\mathrm{UW}}$ is the value of y_{UW} for x_i in equation (9):

$$b_{\mathrm{UW}} = \overline{y}_{\mathrm{UW}} - (m_{\mathrm{UW}} \times \overline{x}_{\mathrm{UW}}),$$

$$u_{b_{\mathrm{UW}}} = \left\{\sqrt{\frac{\sum_{i=1}^{n}(y_i - \widehat{y}_{i\mathrm{UW}})^2 \times \sum_{i=1}^{n}x_i^2}{n(n-2)\sum_{i=1}^{n}(x_i - \overline{x}_{\mathrm{UW}})^2}}\right\} \times t_{(n-2)}^{\mathrm{CL}}. \quad (12)$$

The best regression equation for a calibration curve should have the following characteristics (without distinguishing the subscripts O and UW): (i) intercept b small approaching to zero; (ii) slope m large; and (iii) both u_b and u_m small. Further, the quality of the regression, whether a calibration curve or any other bivariate relationship, is also expressed as the linear regression coefficient (r; r_{O} and r_{UW}, respectively, for the OLR and UWLR), which is ideally +1.00000 for a calibration curve [5, 18, 61].

5. Application of Regression Models for XRF Calibration

5.1. Original Drift-Corrected Net Intensities and GRM Concentrations: The First Set of Two Regression Equations for Each Element. The evaluations for both regression types on the drift-corrected net intensity-concentration (Int-Conc) relationships (Table S2) for all major elements from SiO_2 to P_2O_5 were performed (Table S4), for which the new online software BiDASys was used [61] at http://tlaloc.ier.unam.mx. BiDASys allows the application of the conventional OLR as well as the newly proposed UWLR model [17] and provides the output of all regression parameters in an Excel® file. Contrary to the common practice, we will refrain from showing the numerous x-y (variable x is drift-corrected net intensity "Int" and variable y is the GRM concentration "Conc") plots. This is because Table S4 statistically quantifies the visual interpretation of such diagrams. The quality parameters (standard errors se_b and se_m, uncertainty u_b and u_m, and linear correlation coefficient r and its squared value R^2 parameters) are reported in Table S4. Because we are using these several different quality parameters, the concern against the use of solely R^2 parameter [62] is not important for comparison purposes.

We will explain the implications of the statistical results for the first element SiO_2; the statistics for other elements (Table S4) can be similarly understood. The OLR regression equation from the first row of statistical information in Table S4 is as follows (after the element SiO_2, subscript O is for the OLR and p is for provisional concentration; note many decimal places are used for the regression variables in such equations, because these values are not final results, and we should not introduce rounding errors during the calculation stage):

$$C_{\mathrm{SiO_{2Op}}}\left(\pm u_{C_{\mathrm{SiO_{2Op}}}}\right) = 11.47071(\pm 4.90411) + \left[0.14325(\pm 0.01605) \times I_{\mathrm{SiO_2}}\left(\pm u_{I_{\mathrm{SiO_2}}}\right)\right]. \quad (13)$$

Similarly, the UWLR equation from the second row of statistical information in Table S4 is as follows:

$$C_{\mathrm{SiO_{2UWp}}}\left(\pm u_{C_{\mathrm{SiO_{2UWp}}}}\right) = -0.01316(\pm 2.83181) + \left[0.18539(\pm 0.00927) \times I_{\mathrm{SiO_2}}\left(\pm u_{I_{\mathrm{SiO_2}}}\right)\right]. \quad (14)$$

The implications of these regression equations can be understood from the comparison of the uncertainties of the intercept and slope, which are lower for the UWLR (equation (14)) than for the OLR (equation (13)). This means that the uncertainty of the calculated concentration will be lower for the UWLR than for the OLR. Correspondingly, the r value for the UWLR (0.99004, $n = 60$; $R^2 = 0.98017$) is much higher than that for the OLR (0.95229, $n = 60$; $R^2 = 0.90687$; Table S4). Similar trend in the r (and R^2) values was obtained for all other elements except MnO (Table S4).

5.2. Matrix-Effect-Corrected Intensities and GRM Concentrations: The Second Set of Two Regression Equations for Each Element. Matrix correction is certainly required because the abovementioned least-squares linear regression fits are far from "perfect" ($r \neq +1.00000$; in fact, $r < 1$; $n = 60$; $r =$ 0.95229–0.99638 for the OLR and $r =$ 0.97715–0.99760 for the UWLR; Table S4). There is a vast literature on the subject of matrix effects in XRF and their correction procedures [63–75]. In this study, the Lachance-Traill algorithm [73] was used for the matrix effect correction [53, 71]. This was done outside the XRF instrument software. In a review of the existing algorithms, Rousseau [63] showed that the Lachance-Traill algorithm could be considered as one of the most appropriate procedures for the matrix effect correction because other algorithms have limited application range or lack of accuracy. Thus, for each element from SiO_2 to P_2O_5, a system of overdetermined equations was solved and the resulting alpha coefficients were used to correct all intensities for matrix effects.

From the alpha coefficients, matrix-corrected intensities and improved concentration values for the GRMs and their uncertainties were calculated iteratively under the condition that the convergence parameter (absolute relative difference of the GRM calculated and input concentrations) for each compositional constituent (SiO_2 to P_2O_5) be minimized.

New regression equations for achieving the corrected concentrations were established from the relationship of the calculated GRM concentrations (ConcCalc) and the original GRM concentrations (Conc) given in Table S2, for which the online BiDASys software [61] was used at http://tlaloc.ier.unam.mx. These equations can be formulated from the regression coefficient values given in Table S4 (see ConcCalc-Conc rows corresponding to the OLR and UWLR). Again, we will highlight their significance for SiO_2 only.

The OLR regression equation from the third row of statistical information in Table S4 is as follows:

$$C_{SiO_{2O}}\left(\pm u_{C_{SiO_{2O}}}\right) = 1.87016(\pm 4.2048) + \left[0.96759(\pm 0.07369) \times C_{SiO_{2Oc}}\left(\pm u_{C_{SiO_{2Oc}}}\right)\right], \quad (15)$$

where the subscripts O and c stand for the OLR model and calculated concentration (ConcCalc), respectively.

Similarly, the UWLR equation from the fourth row of statistical information in Table S4 is as follows:

$$C_{SiO_{2UW}}\left(\pm u_{C_{SiO_{2UW}}}\right) = -0.02185(\pm 1.47356) + \left[1.00325(\pm 0.02701) \times C_{SiO_{2UWc}}\left(\pm u_{C_{SiO_{2UWc}}}\right)\right], \quad (16)$$

where the subscripts UW and c stand for the UWLR model and calculated concentration (ConcCalc), respectively.

Equations (15) and (16) show that the concentration values from the UWLR would be more reliable (lesser uncertainty values in both intercept and slope) than the OLR model. The r value is higher for the UWLR (0.99704, $n = 60$; $R^2 = 0.99408$; Table S4) than the OLR (0.97710, $n = 60$; $R^2 = 0.95472$).

After the matrix correction, in fact most regression equations are better because all r and R^2 values are higher for both OLR and UWLR than without the correction (Table S4; Figure 1 for r only). For the OLR, the matrix correction increased the r values ($n = 60$) from 0.95229–0.99638 ($R^2 = 0.90687 - 0.99277$) to 0.97710–0.99992 ($R^2 = 0.95472 - 0.99994$). Similarly, for the UWLR, this increase was from 0.97715–0.99760 ($R^2 = 0.95472 - 0.99521$) to 0.99704–0.99993 ($R^2 = 0.99408 - 0.99986$). Thus, after matrix correction, all r values increased for both OLR and UWLR. For the UWLR, the r values approached the ideal value of +1.00000 (Figure 1). One has to keep in mind that when the r values are close to the maximum possible value of 1 (the "ideal" fit), the improvement expressed by the actual (absolute) value of r will apparently be small. However, as long as the r value increases for the UWLR as compared to the OLR (Figure 1; Table S4), we can objectively infer that the UWLR is a better regression model than the OLR.

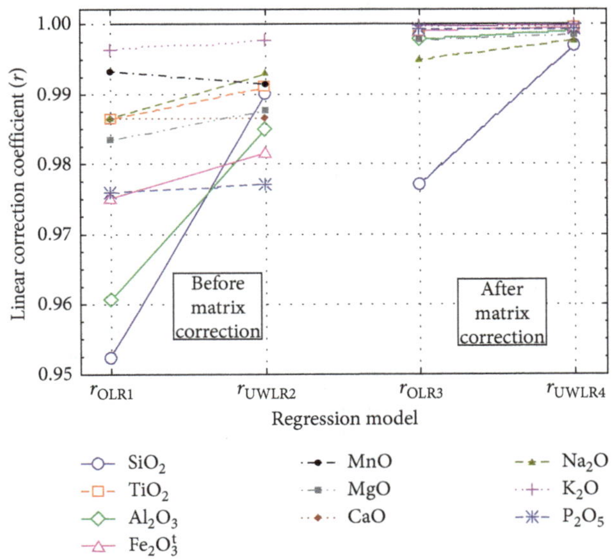

Figure 1: Linear correlation coefficient (r) values for the ordinary least-squares linear regression (OLR) and uncertainty-based weighted least-squares linear regression (UWLR) models for the XRF calibration of major elements (SiO_2 to P_2O_5) in rocks and minerals. OLR1: OLR model 1 for Int-Conc before matrix correction; UWLR2: UWLR model 2 for Int-Conc before matrix correction; OLR: OLR model 3 for ConcCalc-Conc after matrix correction; and UWLR4: UWLR model 4 for ConcCalc-Conc after matrix correction. Symbols are shown as inset. The horizontal line at the r value of 1 represents the "ideal" or "perfect" linear fit.

Before the matrix correction, the intercepts of the Int-Conc regression lines were closer to zero for the UWLR (range ~−0.013 to +0.011) than for the OLR (range ~−2.098 to +11.47) model (Table S4; Figure 2). The same is true for the intercept values (ConcCalc-Conc relationship) after the matrix correction (~−0.025 to +0.021 for the UWLR and ~−0.110 to +1.87 for the OLR).

Finally, the uncertainties on both intercept and slope parameters were mostly lower for the UWLR than the OLR (Table S4). We highlight these differences (lower uncertainties for the UWLR) from dimensionless (free of the measurement units) parameters δu_b and δu_m defined as follows:

$$\delta u_b = \frac{u_{b_O} - u_{b_{UW}}}{u_{b_{UW}}} \times 100,$$
$$\delta u_m = \frac{u_{m_O} - u_{m_{UW}}}{u_{m_{UW}}} \times 100. \quad (17)$$

Plots of these two parameters are presented in Figure 3. If $u_{b_O} > u_{b_{UW}}$, the δu_b will be positive, otherwise it will be negative. The same is true for δu_m. For the comparison of two models OLR and UWLR before the matrix correction, the uncertainty for the UWLR were lower than the OLR for 7 elements (positive δu_b and δu_m), whereas for after the matrix correction, it was so for 8 elements (out of 10; Figure 3). The exceptions were for 3 elements Mno, CaO, and P_2O_5 (negative δu_b and δu_m) for the uncertainties before matrix

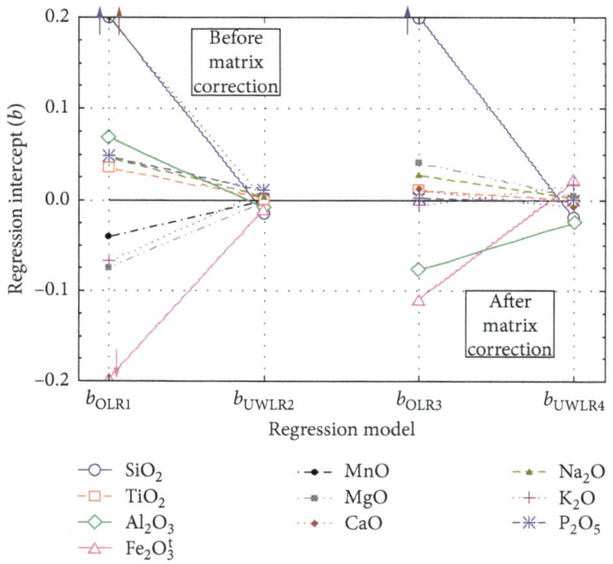

FIGURE 2: Intercept (b) values for the ordinary least-squares linear regression (OLR) and uncertainty-based weighted least-squares linear regression (UWLR) models for the XRF calibration of major elements (SiO_2 to P_2O_5) in rocks and minerals. Symbols are shown as inset. For abbreviations, see Figure 1. Note some intercept values plotted outside the graph; this is indicated by arrows next to the data point. The horizontal line at the intercept value of zero represents the "ideal" intercept.

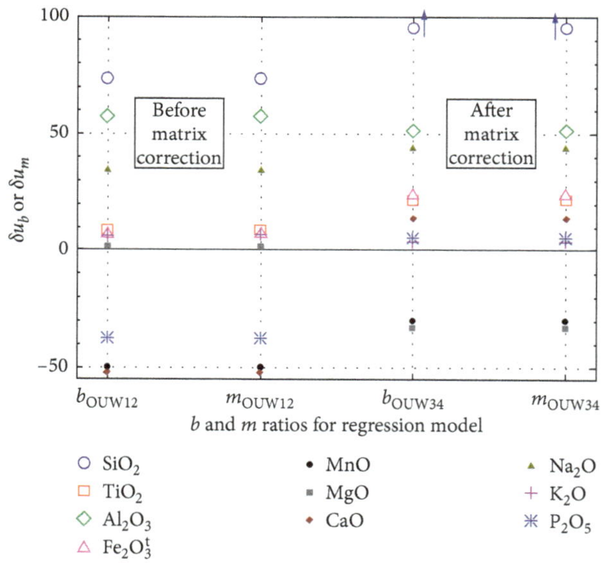

FIGURE 3: New parameters δu_b (for the intercept) and δu_m (for the slope) for the evaluation of intercept (b) and slope (m) of two regression models (OLR: ordinary least-squares linear regression and UWLR: uncertainty-based weighted least-squares linear regression) before (OUW12) and after (OUW34) the matrix correction. The horizontal solid line at the y value of zero represents the line with no difference in the uncertainties of the two models. The arrows indicate that these data plotted above the scale are used for the diagram.

correction and for 2 elements MnO and MgO for those after the matrix correction (Figure 3). Even for the exceptions of the elements MnO and MgO, the UWLR values should be usable (Table S4), i.e., it is not actually necessary to resort to the OLR model for these two exceptions (2 out of 10 cases). Thus, we can use the UWLR model for all purposes.

6. Sensitivities of Major Elements

We calculated the sensitivities as the slope of the Conc-IntCorr (GRM concentrations of Table S2 and matrix-corrected intensities of Table S5; see Supplementary Materials at http://tlaloc.ier.unam.mx) from the regression curve (line) for all 10 elements and for both models (Table 1). Because the r values are significantly high (all >0.961, $n = 60$; Table 1) and the residuals are randomly distributed (graphs not shown), the straight line is the most likely, statistically valid fit for the concentration-matrix-corrected intensity data [5, 17, 18]. Therefore, the slope of the regression line represents an average sensitivity value for a given element under the chosen working conditions (Table S3).

The intercept values were closer to zero (zero being the theoretically ideal intercept) for the UWLR regression (~−0.113 to +0.104; Table 1) as compared to the OLR (~−47.8 to +12.3; Table 1). The sensitivity values represented by the slopes of the regression lines (Table 1) were generally similar for both models (~2.69–46.17 kc·s^{-1} %$^{-1}$ for the OLR and ~2.78–59.69 kc·s^{-1}·%$^{-1}$ for the UWLR). The sensitivity actually depends on the measuring conditions (Table S3), which were the same for both models.

For the matrix-corrected intensity-concentration (IntCorr-Conc) regressions, the parameters are listed in Table 2. All intercepts for the UWLR model, without exception, were closer to zero as compared to the OLR model. This confirms the superiority of the UWLR model.

7. Application to Rock Matrices

The calibrations achieved in this work (Table S4) were applied to the analysis of four GRMs (attapulgite or Fuller's earth clay ATT1; bentonite clay CSB1; granite GH; and tonalite TLM1) taken as "unknown" samples. These GRMs, having similarly complex matrices as the calibration samples, were not included in the calibrations because their mean values were available only from early description or compilations (for ATT1 and CSB1 [76]; for GH [77]; and for TLM1 [78]). We were unsuccessful in complementing these "old" concentration values with newer ones for these GRMs. Therefore, these GRMs were used as unknown samples. They were analysed in exactly the same manner as the calibration samples.

All calculations for the unknown samples were done outside the instrumental software. The drift-corrected net intensities and the corresponding uncertainties were processed from the first set of two regression equations (Int-Conc OLR and UWLR models; Table S4) to obtain provisional concentration and uncertainty values. The

Table 1: Instrumental sensitivities (x-y: concentration-matrix-corrected intensity (Conc-IntCorr) regression model; mean and 99% uncertainty) for major elements.

Element	Regression		Number of data pairs (calibrators)	Regression equation parameters						Quality of regression equation	
	Variables x-y	Model		Intercept (b)	se_b	u_b	Slope (m)	se_m	u_m	r	R^2
SiO_2	Conc-IntCorr	OLR	60	−47.8441	13.1332	34.9772	6.3592	0.2401	0.6395	0.96105	0.92362
	Conc-IntCorr	UWLR	60	0.1035	5.3870	14.3471	5.3476	0.0985	0.2623	0.99157	0.98321
TiO_2	Conc-IntCorr	OLR	60	−0.2298	0.3000	0.7991	13.2255	0.2889	0.7694	0.98644	0.97307
	Conc-IntCorr	UWLR	60	−0.0188	0.2637	0.7024	12.3181	0.2539	0.6763	0.99113	0.98235
Al_2O_3	Conc-IntCorr	OLR	60	8.1398	4.1673	11.0987	6.1191	0.2269	0.6043	0.96237	0.92615
	Conc-IntCorr	UWLR	60	0.0594	3.0552	8.1369	7.0715	0.1663	0.4430	0.98557	0.97135
$Fe_2O_3{}^t$	Conc-IntCorr	OLR	60	12.2628	2.4739	6.5886	4.3943	0.1337	0.3561	0.97419	0.94904
	Conc-IntCorr	UWLR	60	0.0715	2.7723	7.3835	5.4403	0.1498	0.3991	0.98112	0.96259
MnO	Conc-IntCorr	OLR	60	2.0367	0.2963	0.7891	46.1686	0.6958	1.8531	0.99348	0.98700
	Conc-IntCorr	UWLR	60	0.0147	0.7535	2.0069	59.6875	1.7695	4.7127	0.99160	0.98328
MgO	Conc-IntCorr	OLR	60	1.1050	1.0231	2.7246	3.6928	0.0901	0.2400	0.98317	0.96662
	Conc-IntCorr	UWLR	60	0.0108	0.9641	2.5676	3.9863	0.0849	0.2262	0.98745	0.97505
CaO	Conc-IntCorr	OLR	60	−14.3055	3.5661	9.4974	15.4215	0.3371	0.8977	0.98643	0.97304
	Conc-IntCorr	UWLR	60	−0.0121	5.6252	14.9814	11.5666	0.5317	1.4160	0.98657	0.97332
Na_2O	Conc-IntCorr	OLR	60	0.0166	0.1553	0.4137	2.6888	0.0571	0.1521	0.98716	0.97449
	Conc-IntCorr	UWLR	60	−0.0111	0.1154	0.3073	2.7843	0.0424	0.1130	0.99344	0.98692
K_2O	Conc-IntCorr	OLR	60	1.2200	0.5896	1.5702	13.8296	0.1431	0.3811	0.99691	0.99383
	Conc-IntCorr	UWLR	60	−0.1130	0.5491	1.4623	14.4829	0.1333	0.3549	0.99793	0.99587
P_2O_5	Conc-IntCorr	OLR	60	−1.7376	0.5433	1.4469	39.1055	1.2932	3.4442	0.96972	0.94035
	Conc-IntCorr	UWLR	60	0.0626	0.7039	1.8747	29.4822	1.6756	4.4624	0.96926	0.93946

b, intercept; se, standard error; u, uncertainty at 99%; m, slope; r, linear correlation coefficient; R^2, squared linear correlation coefficient.

provisional concentrations were then used to obtain matrix corrections for each sample. The method was iteratively applied with the newer concentrations to obtain the final calculated concentration values (Table 3). These calculated concentration values were used to compute the final mean concentrations ($\bar{x}$) and 99% uncertainties of the mean (u_{99}) for each sample from the second sets of regression equations (ConcCalc-Conc, OLR and UWLR models; Table S4). The loss on ignition (LOI) was required to optimise the final results.

The results are listed in Table 3 and compared with the literature compilations [75–77]. On the other hand, because 99% uncertainties were not reported in the original compilations, they were computed for the comparison from the standard deviation, number of determinations, and appropriate two-sided t values at 99% confidence level [2, 18].

Firstly, although the mean concentration values determined by the OLR and UWLR models showed a general agreement, the 99% uncertainty values (u_{99}; Table 3) were generally lower for the UWLR models, which clearly

Table 2: Regression (x-y): matrix-corrected intensity-concentration (IntCorr-Conc) parameters.

Element	Regression		Number of data pairs (calibrators)	Regression equation parameters						Quality of regression equation	
	Variables x-y	Model		Intercept (b)	se_b	u_b	Slope (m)	se_m	u_m	r	R^2
SiO_2	IntCorr-Conc	OLR	60	10.8541	1.6762	4.4641	0.1452	0.0055	0.0146	0.96105	0.92362
	IntCorr-Conc	UWLR	60	−0.0148	0.9765	2.6006	0.1848	0.0032	0.0085	0.99157	0.98321
TiO_2	IntCorr-Conc	OLR	60	0.0344	0.0220	0.0587	0.0736	0.0016	0.0043	0.98644	0.97307
	IntCorr-Conc	UWLR	60	0.0020	0.0204	0.0542	0.0801	0.0015	0.0040	0.99114	0.98235
Al_2O_3	IntCorr-Conc	OLR	60	−0.1675	0.6763	1.8011	0.1514	0.0056	0.0150	0.96237	0.92615
	IntCorr-Conc	UWLR	60	−0.0070	0.4419	1.1769	0.1397	0.0037	0.0098	0.98557	0.97135
$Fe_2O_3^t$	IntCorr-Conc	OLR	60	−2.0576	0.5960	1.5873	0.2160	0.0066	0.0175	0.97419	0.94904
	IntCorr-Conc	UWLR	60	−0.0092	0.5458	1.4536	0.1772	0.0060	0.0160	0.98112	0.96259
MnO	IntCorr-Conc	OLR	60	−0.0412	0.0067	0.0178	0.0214	0.0003	0.0009	0.99348	0.98700
	IntCorr-Conc	UWLR	60	0.0001	0.0134	0.0356	0.0165	0.0007	0.0017	0.99160	0.98328
MgO	IntCorr-Conc	OLR	60	−0.0977	0.2748	0.7319	0.2618	0.0064	0.0170	0.98317	0.96662
	IntCorr-Conc	UWLR	60	−0.0014	0.2692	0.7168	0.2376	0.0063	0.0167	0.98745	0.97505
CaO	IntCorr-Conc	OLR	60	1.0561	0.2173	0.5789	0.0631	0.0014	0.0037	0.98643	0.97304
	IntCorr-Conc	UWLR	60	0.0064	0.4616	1.2293	0.0853	0.0029	0.0078	0.98657	0.97331
Na_2O	IntCorr-Conc	OLR	60	0.0447	0.0567	0.1511	0.3624	0.0077	0.0205	0.98716	0.97449
	IntCorr-Conc	UWLR	60	0.0054	0.0421	0.1122	0.3558	0.0057	0.0152	0.99345	0.98693
K_2O	IntCorr-Conc	OLR	60	−0.0720	0.0430	0.1146	0.0719	0.0007	0.0020	0.99691	0.99383
	IntCorr-Conc	UWLR	60	0.0082	0.0391	0.1042	0.0687	0.0007	0.0018	0.99793	0.99587
P_2O_5	IntCorr-Conc	OLR	60	0.0537	0.0128	0.0341	0.0241	0.0008	0.0021	0.96972	0.94035
	IntCorr-Conc	UWLR	60	0.0010	0.0202	0.0538	0.0320	0.0013	0.0033	0.96929	0.93952

b, intercept; se, standard error; u, uncertainty at 99%; m, slope; r, linear correlation coefficient; R^2, squared linear correlation coefficient.

indicates that this model should be used routinely, instead of the conventional OLR model. Secondly, there is also a general agreement among all mean values, especially for granite GH and tonalite TLM1. The two clay samples (ATT1 and CSB1) showed some differences with the preliminary values obtained by the originators of these GRMs [75]. These values for comparison were obtained in only one laboratory. The errors (uncertainties) reported in the literature were underestimated, because they did not include those resulting from the calibrations. Furthermore, the accuracy data of the originator's laboratory were not reported [75], such as the results for established GRMs and their comparison to other laboratories.

8. Computer Program XRFCalcUnknown

An online computer program JSpectrom_XRFCalcUnknown will be available at our server https://tlaloc.ier.unam.mx for use for unknown samples, which will guide other users to achieve the UWLR calibration outside of the instrumental software and its routine application to unknown samples. This program incorporates the iteration process to achieve reliable

Table 3: Application of the XRF calibrations for the determination of major elements (mean concentration in % m/m; mass/mass unit expressed in percentage, habitually called wt.% and 99% uncertainty u_{99} or simply u) in four rock samples (data reported as rounded value following flexible criteria proposed by [18]) and their comparison with the literature data.

Rock	Method (ref.)	SiO_2		TiO_2		Al_2O_3		$Fe_2O_3^t$		MnO		MgO		CaO		Na_2O		K_2O		P_2O_5	
		$\overline{x}$	u_{99}	$\overline{x}$	u_{99}	$\overline{x}$	u_{99}	$\overline{x}$	u_{99}	$\overline{x}$	u_{99}	$\overline{x}$	u_{99}	$\overline{x}$	u_{99}	$\overline{x}$	u_{99}	$\overline{x}$	u_{99}	$\overline{x}$	u_{99}
ATT1	OLR	61.7	6.1	0.4961	0.0173	10.00	0.49	4.246	0.292	0.02394	0.00188	5.373	0.377	2.139	0.086	0.083	0.094	0.7777	0.0378	0.8074	0.0134
	UWLR	62.30	2.23	0.4951	0.0143	9.666	0.321	4.402	0.235	0.02416	0.00269	6.17	0.52	2.144	0.076	0.058	0.065	0.7781	0.0366	0.7828	0.0125
	[76]	59.6	0.8	0.49	0.015	9.50	0.29	3.31	0.06	0.025	0.018	9.14	0.06	1.87	0.03	0.1	0.03	0.86	0.06	0.76	0.01
CSB1	OLR	52.7	5.5	0.1448	0.0157	24.60	0.71	3.493	0.289	0.01408	0.00188	1.523	0.270	1.117	0.085	2.126	0.119	0.4169	0.0374	0.0383	0.0060
	UWLR	52.76	2.04	0.1343	0.0130	24.40	0.48	3.564	0.233	0.01398	0.00268	1.604	0.398	1.105	0.075	2.123	0.083	0.4337	0.0361	0.0357	0.0057
	[76]	55.3	0.8	0.16	0.03	21.75	0.18	3.88	0.03	0.020	0.003	1.94	0.08	1.28	0.06	2.07	0.20	0.54	0.03	—	—
GH	OLR	73.3	6.5	0.0968	0.0157	13.98	0.53	1.336	0.285	0.05084	0.00190	0.074	0.262	0.843	0.084	3.618	0.151	5.302	0.059	0.0119	0.0059
	UWLR	73.62	2.38	0.0861	0.0129	13.714	0.349	1.466	0.229	0.05064	0.00271	0.038	0.390	0.828	0.074	3.625	0.105	5.216	0.057	0.0091	0.0057
	[77]	75.8	0.39	0.08	0.07	12.50	0.16	1.34	0.2	0.05	0.009	0.03	0.18	0.69	0.13	3.85	0.11	4.76	0.05	0.08	0.03
TLM1	OLR	57.7	5.7	0.8497	0.0200	17.38	0.58	7.637	0.306	0.11614	0.00199	3.972	0.319	6.807	0.100	2.675	0.130	1.7585	0.0403	0.1378	0.0063
	UWLR	57.67	2.11	0.8402	0.0165	17.046	0.386	7.682	0.247	0.11592	0.00281	4.281	0.447	6.821	0.088	2.667	0.090	1.742	0.0390	0.1336	0.0060
	[78]	58.85	0.35	0.820	0.013	17.48	0.25	7.67	0.15	0.105	0.013	3.3	0.08	6.67	0.08	2.98	0.07	1.67	0.08	0.15	0.013

concentrations as demonstrated in this work. Example input data files and a ReadMe document are provided to facilitate the application of JSpectrom_XRFCalcUnknown. One important aspect of the program is that for a sample to be identified as an "unknown" sample, the value of LOI (loss on ignition in percent) should be input in the first sheet of the measured intensity file.

A novel aspect of the present work is that total 99% uncertainty can be calculated for individual datum in a given sample (treated as unknown; Table 3). This innovation if put into practice can entirely change the geochemical literature, and in fact make geochemistry a more quantitative science. Further, if an appropriate GRM is analysed as unknown and the analytical data (both mean and total uncertainly) are reported along with the field samples, the data accuracy can be statistically judged from such reports.

9. Conclusions

The XRF spectrometer calibrated under both the OLR and UWLR models clearly showed that the UWLR provides more reliable results (lower uncertainty estimates) than the OLR model commonly practiced for most XRF instruments. The sensitivity and LOD values presented for both models also supported the use of the UWLR model. The UWLR model should therefore be used routinely in such calibrations. The use of a large number of well-characterized GRMs is also recommended for this purpose as illustrated in the present work. The application of our procedure was well documented for the analysis of similarly complex rock matrices. The reporting of total uncertainty values for individual datum is highly recommended for all future geochemical research. This work for the XRF shows that such a practice is easy to achieve in any other analytical calibration procedures. As the major conclusion, we can confirm that the statistically coherent WLR model was shown to perform better than the frequently used conventional statistically incoherent OLR model.

Conflicts of Interest

The authors declare that they have no conflicts of interest.

Acknowledgments

This work was supported through the Newton Advanced Fellowship Award (grant NA160116) of the Royal Society, U.K., to the second author (SKV) and from the sabbatical stay of SPV at IPICYT. We are grateful to the Nanoscience and Nanotechnology National Research Laboratory (LINAN), Carbon Nanostructures and Two-Dimensional Systems Laboratory at IPICYT, and Dr. Emilio Muñoz-Sandoval for providing access to the required facilities. M. Abdelaly Rivera-Gómez is grateful to CTIC and DGAPA for a postdoctoral fellowship at the ICML-UNAM. Darío Torres-Sánchez and Mauricio Rosales-Rivera thank CONACYT for the doctoral fellowship. The GRM compilation was initiated long ago in our group by the participation of R. González-Ramírez although the bulk of the work was carried out by the present authors. We are grateful to the IER-UNAM library personnel for efficiently providing some of the literature materials for compilation and to Alfredo Quiroz-Ruiz for the maintenance of the computing facility at IER-UNAM. Diego Villanueva-López helped us during the pressed powder pellet preparation and for checking the correctness of the information in GRM databases.

References

[1] H. R. Rollinson, *Using Geochemical Data: Evaluation, Presentation, Interpretation*, Longman Scientific Technical, Essex, UK, 1993.

[2] J. N. Miller and J. C. Miller, *Statistics and Chemometrics for Analytical Chemistry*, Pearson Prentice Hall, Essex, UK, 2010.

[3] R. G. Brereton, "Chemometrics in analytical chemistry. A review," *Analyst*, vol. 112, pp. 1635–1657 1987.

[4] M. Guevara, S. P. Verma, F. Velasco-Tapia, R. Lozano-Santa Cruz, and P. Girón, "Comparison of linear regression models for quantitative geochemical analysis: an example using x-ray fluorescence spectrometry," *Geostandards and Geoanalytical Research*, vol. 29, no. 3, pp. 271–284, 2005.

[5] P. R. Bevington, *Data Reduction and Error Analysis for the Physical Sciences*, Mc-Graw Hill Book Company, New York, NY, USA, 1969.

[6] D. York, "Least squares fitting of a straight line with correlated errors," *Earth and Planetary Science Letters*, vol. 5, pp. 320–324, 1969.

[7] A. H. Kalantar, "Weighted least squares evaluation of slope from data having errors in both axes," *Trends in Analytical Chemistry*, vol. 9, no. 5, pp. 149–151, 1990.

[8] K. L. Mahon, "The New "York" regression: application of an improved statistical method to geochemistry," *International Geology Review*, vol. 38, no. 4, pp. 293–303, 1996.

[9] M. E. Zorn, R. D. Gibbons, and W. C. Sonzogni, "Weighted least-squares approach to calculating limits of detection and quantification by modeling variability as a function of concentration," *Analytical Chemistry*, vol. 69, no. 15, pp. 3069–3075, 1997.

[10] N. R. Draper and H. Smith, *Applied Regression Analysis*, John Wiley & Sons, New York, NY, USA, 1998.

[11] A. Schick, "Improving weighted least-squares estimates in heteroscedastic linear regression when the variance is a function of the mean response," *Journal of Statistical Planning and Inference*, vol. 76, no. 1-2, pp. 127–144, 1999.

[12] A. Sayago, M. Boccio, and A. G. Asuero, "Fitting straight lines with replicated observations by linear regression: the least squares postulates," *Critical Review of Analytical Chemistry*, vol. 34, no. 1, pp. 39–50, 2004.

[13] A. Sayago and A. G. Asuero, "Fitting straight lines with replicated observations by linear regression: Part II. Testing for homogeneity of variances," *Critical Review of Analytical Chemistry*, vol. 34, no. 3-4, pp. 133–146, 2004.

[14] A. G. Asuero, A. Sayago, and A. G. González, "The correlation coefficient: an overview," *Critical Review of Analytical Chemistry*, vol. 36, no. 1, pp. 41–59, 2006.

[15] J. Tellinghuisen, "Weighted least-squares in calibration: what difference does it make?," *Analyst*, vol. 132, no. 6, pp. 536–543, 2007.

[16] S. P. Verma, L. Díaz-González, and R. González-Ramírez, "Relative efficiency of single-outlier discordancy tests for processing geochemical data on reference materials and application to instrumental calibration by a weighted least-squares linear regression model," *Geostandards and Geoanalytical Research*, vol. 33, no. 1, pp. 29–49, 2009.

[17] S. P. Verma, "Geochemometrics," *Revista Mexicana de Ciencias Geológicas*, vol. 29, no. 1, pp. 276–298, 2012.

[18] S. P. Verma, *Análisis Estadístico de Datos Composicionales*, Universidad Nacional Autónoma de México, CDMX, Mexico, 2016.

[19] P. J. Potts, *A Handbook of Silicate Rock Analysis*, Blackie, Glasgow, UK, 1987.

[20] P. J. Potts and P. C. Webb, "X-ray fluorescence spectrometry," *Journal of Geochemical Exploration*, vol. 44, no. 1-3, pp. 251–296, 1992.

[21] M. El Maghraoui, J.-L. Joron, J. Etoubleau, P. Cambon, and M. Treuil, "Determination of forty four major and trace elements in GPMA magmatic rock reference materials using x-ray fluorescence spectrometry (XRF) and instrumental neutron activation analysis (INAA)," *Geostandards Newsletter: Journal of Geostandards and Geoanalysis*, vol. 23, no. 1, pp. 59–68, 1999.

[22] K. Tani, H. Kawabata, Q. Chang, K. Sato, and Y. Tatsumi, "Quantitative analyses of silicate rock major and trace elements by X-ray fluorescence spectrometer: evaluation of analytical precision and sample preparation," *Frontiers in Research of Earth Evolution*, vol. 2, pp. 1–8, 2004.

[23] J. Enzweiler and M. A. Vendemiatto, "Analysis of sediments and soils by x-ray fluorescence spectrometry using matrix corrections based on fundamental parameters," *Geostandards and Geoanalytical Research*, vol. 28, no. 1, pp. 103–112, 2005.

[24] L. P. Bédard, "Neutron activation analysis, atomic absorption and x-ray fluorescence spectrometry review for 2004-2005," *Geostandards and Geoanalytical Research*, vol. 30, no. 3, pp. 183–186, 2006.

[25] K. Nakayama, Y. Shibata, and T. Nakamura, "Glass beads/x-ray fluorescence analysis of 42 components in felsic rocks," *X-Ray Spectrometry*, vol. 36, no. 2, pp. 130–140, 2007.

[26] W. Wu, T. Xu, Q. Hao, Q. Wang, S. Zhang, and C. Zhao, "Applications of x-ray fluorescence analysis of rare earths in China," *Journal of the Rare Earths*, vol. 28, pp. 30–36, 2010.

[27] H. Mashima, "XRF analyses of major and trace elements in silicate rocks calibrated with synthetic standard samples," *Natural Resources of Environment and Humans*, vol. 6, pp. 39–50, 2016.

[28] D. Robinson and M. C. Bennett, "XRF determination of 19 trace elements in international geochemical reference samples," *Geostandards Newsletter*, vol. 5, no. 2, pp. 175–181, 1981.

[29] S. P. Verma, T. Besch, M. Guevara, and B. Schulz-Dobrich, "Determination of twelve trace elements in twenty-seven and ten major elements in twenty-three geochemical reference samples by X-Ray fluorescence spectrometry," *Geostandards Newsletter*, vol. 16, no. 2, pp. 301–309, 1992.

[30] X. Wang, G. Li, Q. Zhang, and Y. Wang, "Determination of major/minor and trace elements in seamount phosphorite by XRF spectrometry," *Geostandards and Geoanalytical Research*, vol. 23, no. 1, pp. 81–88, 2004.

[31] K. Govindaraju, "1987 compilation report on Ailsa Craig Granite AC-E with the participation of 128 GIT-IWG laboratories," *Geostandards Newsletter*, vol. 11, no. 2, pp. 203–255, 1987.

[32] K. Govindaraju, "Report (1980) on three GIT-IWG rock reference samples: anorthosite from Greenland, AN-G; basalte d' Essey-la-Côte, BE-N; granite de Beauvoir, MA-N," *Geostandards Newsletter*, vol. 4, no. 1, pp. 49–138, 1980.

[33] E. S. Gladney and I. Roelandts, "1987 compilation of elemental concentration data for USGS BHVO-1, MAG-1, QLO-1, RGM-1, SCo-1, SDC-1, SGR-1, and STM-1," *Geostandards Newsletter*, vol. 12, no. 2, pp. 253–262, 1988.

[34] K. Govindaraju, "Report (1967-1981) on four ANRT rock reference samples: diorite DR-N, serpentine UB-N, bauxite BX-N, disthene DT-N," *Geostandards Newsletter*, vol. 6, no. 1, pp. 91–159, 1982.

[35] E. S. Gladney and I. Roelandts, "1987 compilation of elemental concentration data for USGS BIR-1, DNC-1 and W-2," *Geostandards Newsletter*, vol. 12, no. 1, pp. 63–118, 1988.

[36] S. Abbey, C. R. McLeod, and W. Liang-Guo, "FeR-1, FeR-2, FeR-3 and FeR-4 Four Canadian iron-formation samples prepared for use as reference materials," Geological Survey of Canada Report, Geological Survey of Canada, Ottawa, ON, Canada, 1983.

[37] K. Govindaraju, "Report (1973-1984) in two ANRT geochemical reference samples: granite GS-N and Potash Feldspar FK-N," *Geostandards Newsletter*, vol. 8, no. 2, pp. 173–206, 1984.

[38] E. S. Gladney, C. E. Burns, and I. Roelandts, "1982 compilation of elemental concentration data for the United States Geological Survey's geochemical exploration reference samples GXR-1 to GXR-6," *Geostandards Newsletter*, vol. 8, no. 2, pp. 119–154, 1984.

[39] E. S. Gladney and I. Roelandts, "1988 compilation of elemental concentration data for USGS geochemical exploration reference materials GXR-1 to GXR-6," *Geostandards Newsletter*, vol. 14, no. 1, pp. 21–118, 1990.

[40] K. Govindaraju, "Report (1984) on two GIT-IWG geochemical reference samples: albite from Italy, AL-I and Iron Formation sample from Greenland, IF-G," *Geostandards Newsletter*, vol. 8, no. 1, pp. 63–113, 1984.

[41] N. Imai, S. Terashima, S. Itoh, and A. Ando, "1994 compilation of analytical data for minor and trace elements in seventeen GSJ geochemical reference samples, "igneous rock series"," *Geostandards and Geoanalytical Research*, vol. 19, no. 2, pp. 135–213, 1995.

[42] S. Terashima, "Elemental concentrations in nine new GSJ rock reference samples," *Geostandards Newsletter*, vol. 14, no. 1, pp. 1–5, 1990.

[43] S. Terashima, S. Itoh, M. Ujiie, H. Kamioka, T. Tanaka, and H. Hattori, "Three new GSJ rock reference samples: rhyolite JR-3, gabbro JGb-2 and hornblendite JH-1," *Geostandards Newsletter*, vol. 17, no. 1, pp. 1–4, 1993.

[44] S. Abbey, "Reference materials: rock samples SY-2, SY-3, MRG-1," Energy, Mines and Resources Canada Report, Natural Resources Canada, Ottawa, ON, Canada, 1979.

[45] K. Govindaraju, "Report (1968-1978) on two mica reference samples: biotite Mica-Fe and phlogopite Mica-Mg," *Geostandards Newsletter*, vol. 3, no. 1, pp. 3–24, 1979.

[46] T. W. Steele and R. G. Hansen, "Major element data (1966-1978) for the six "Nimroc" reference samples," *Geostandards Newsletter*, vol. 3, no. 2, pp. 135–172, 1979.

[47] E. S. Gladney, E. A. Jones, E. J. Nickell, and I. Roelandts, "1988 compilation of elemental concentration data for USGS DTS-1, G-1, PCC-1, and W-1," *Geostandards Newsletter*, vol. 15, no. 2, pp. 199–396, 1991.

[48] V. Barnett and T. Lewis, *Outliers in Statistical Data*, John Wiley & Sons, Chichester, UK, 1994.

[49] K. Hayes, A. Kinsella, and N. Coffey, "A note on the use of outlier criteria in Ontario laboratory quality control schemes," *Clinical Biochemistry*, vol. 40, no. 3-4, pp. 147–152, 2007.

[50] S. P. Verma, L. Díaz-González, J. A. Pérez-Garza, and M. Rosales-Rivera, "Quality control in geochemistry from a comparison of four central tendency and five dispersion estimators and example of a geochemical reference material," *Arabian Journal of Geosciences*, vol. 9, p. 740, 2016.

[51] S. P. Verma, L. Díaz-González, J. A. Pérez-Garza, and M. Rosales-Rivera, "Erratum to: quality control in geochemistry from a comparison of four central tendency and five dispersion estimators and example of a geochemical reference material," *Arabian Journal of Geosciences*, vol. 10, p. 24, 2017.

[52] S. P. Verma, M. Rosales-Rivera, L. Díaz-González, and A. Quiroz-Ruiz, "Improved composition of Hawaiian basalt BHVO-1 from the application of two new and three conventional recursive discordancy tests," *Turkish Journal of Earth Science*, vol. 26, no. 5, pp. 331–353, 2017.

[53] S. P. Verma, R. Cruz-Huicochea, and L. Díaz-González, "Univariate data analysis system: deciphering mean compositions of island and continental arc magmas, and influence of underlying crust," *International Geology Review*, vol. 55, no. 15, pp. 1922–1940, 2013.

[54] M. Rosales Rivera, *Desarrollo de herramientas estadísticas computacionales con nuevos valores críticos generados por simulación computacional*, Universidad Autónoma del Estado de Morelos, Cuernavaca, Morelos, Mexico, 2018.

[55] K. P. Jochum, U. Weis, B. Schwager et al., "Reference values following ISO guidelines for frequently requested rock reference materials," *Geostandards and Geoanalytical Research*, vol. 40, no. 3, pp. 333–350, 2016.

[56] F. Velasco-Tapia, M. Guevara, and S. P. Verma, "Evaluation of concentration data in geochemical reference materials," *Chemie der Erde*, vol. 61, no. 1, pp. 69–91, 2001.

[57] B. Rosner, "On the detection of many outliers," *Technometrics*, vol. 17, no. 2, pp. 221–227, 1975.

[58] F. E. Grubbs and G. Beck, "Extension of sample sizes and percentage points for significance tests of outlying observations," *Technometrics*, vol. 14, no. 4, pp. 847–854, 1972.

[59] R. B. Jain and L. A. Pingel, "On the robustness of recursive outlier detection procedures to nonnormality," *Communications in Statistics - Theory and Methods*, vol. 10, no. 13, pp. 1323–1334, 1981.

[60] R. B. Jain, "Detecting outliers: power and some other considerations," *Communications in Statistics - Theory and Methods*, vol. 10, no. 22, pp. 2299–2314, 1981.

[61] M. Rosales-Rivera, L. Díaz-González, and S. P. Verma, "A new online computer program (BiDASys) for ordinary and uncertainty weighted least-squares linear regressions: case studies from food chemistry," *Revista Mexicana de Ingeniería Química*, vol. 17, no. 2, pp. 507–522, 2018.

[62] J. B. Willet and J. D. Singer, "Another cautionary note about R2: its use in weighted least-squares regression analysis," *American Statistician*, vol. 42, no. 3, pp. 236–238, 1988.

[63] R. M. Rousseau, "Corrections for matrix effects in X-ray fluorescence analysis—a tutorial," *Spectrochimica Acta Part B: Atomic Spectroscopy*, vol. 61, no. 7, pp. 759–777, 2006.

[64] T. Shiraiwa and N. Fujino, "Theoretical calculation of fluorescent x-ray intensities in fluorescent x-ray spectrochemical analysis," *Japanese Journal of Applied Physics*, vol. 5, no. 10, pp. 886–899, 1966.

[65] R. Rousseau and F. Claisse, "Theoretical alpha coefficients for the Claisse-Quintin relation for x-ray spectrochemical analysis," *X-Ray Spectrometry*, vol. 3, no. 1, pp. 31–36, 1974.

[66] R. M. Rousseau, "Fundamental algorithm between concentration and intensity in XRF analysis 2-Practical application," *X-Ray Spectrometry*, vol. 13, no. 3, pp. 121–125, 1984.

[67] R. M. Rousseau, J. P. Willis, and A. R. Duncan, "Practical XRF calibration procedures for major and trace elements," *X-Ray Spectrometry*, vol. 25, no. 4, pp. 179–189, 1996.

[68] A. J. Klimasara, "XRF analysis—theory experiment, and regression," in *Proceedings of X-Ray Conference (DXC) on Applications of X-Ray Analysis*, Denver, CO, USA, August 1997.

[69] B. Tan and W. Sun, "Correction method for the matrix effect in x-ray fluorescence spectrometric analysis," *X-Ray Spectrometry*, vol. 27, no. 2, pp. 95–104, 1998.

[70] B. I. Kitov, "Calculation features of the fundamental parameter method in XRF," *X-Ray Spectrometry*, vol. 29, no. 4, pp. 285–290, 2000.

[71] A. J. Klimasara, "Logical steps in the automated Lachamp-Triall XRF matrix correction method utilizing an electronic spreadsheet," in *Advances in XRF Analysis*, vol. 42, pp. 53–83, JCPDS-International Centre for Diffraction Data, Newtown Square, PA, USA, 2000.

[72] R. M. Rousseau, "Correction for long-term instrumental drift," *X-Ray Spectrom*, vol. 31, no. 6, pp. 401–407, 2002.

[73] G. R. Lachance and R. J. Traill, "A practical solution to the matrix problem in X-ray analysis," *Canadian Journal of Spectroscopy*, vol. 11, pp. 43–48, 1966.

[74] J. P. Willis and G. R. Lachance, "Comparison between some common influence coefficient algorithms," *X-Ray Spectrometry*, vol. 33, no. 3, pp. 181–188, 2004.

[75] J. P. Willis and G. R. Lachance, "A new approach to correcting theoretical emitted intensities for absorption and enhancement effects," *X-Ray Spectrometry*, vol. 33, no. 3, pp. 204–211, 2004.

[76] J. W. Hosterman and F. J. Flanagan, "USGS reference samples Attapulgite ATT-1 and Bentonite CSB-1," *Geostandards and Geoanalytical Research*, vol. 11, no. 1, pp. 1–9, 1987.

[77] K. Govindaraju, "1995 working values with confidence limits for twenty-six CRPG, ANRT and IWG-GIT geostandards," *Geostandards and Geoanalytical Research* vol. 19, pp. 1–32, 1995.

[78] F. J. Flanagan, "Rock reference samples, san marcos gabbro, GSM-1 and lakeview mountain tonalite, TLM-1," *Geostandards and Geoanalytical Research*, vol. 10, no. 2, pp. 111–119, 1986.

3

Use of Electrochemical Impedance Spectroscopy for the Evaluation of Performance of PEM Fuel Cells Based on Carbon Cloth Gas Diffusion Electrodes

Saverio Latorrata, Renato Pelosato, Paola Gallo Stampino, Cinzia Cristiani, and Giovanni Dotelli

Department of Chemistry, Materials and Chemical Engineering "Giulio Natta", Politecnico di Milano, Piazza Leonardo da Vinci 32, 20133 Milano, Italy

Correspondence should be addressed to Saverio Latorrata; saverio.latorrata@polimi.it

Academic Editor: Jau-Wern Chiou

Polymer electrolyte membrane fuel cells (PEMFCs) have attracted great attention in the last two decades as valuable alternative energy generators because of their high efficiencies and low or null pollutant emissions. In the present work, two gas diffusion electrodes (GDEs) for PEMFCs were prepared by using an ink containing carbon-supported platinum in the catalytic phase which was sprayed onto a carbon cloth substrate. Two aerograph nozzles, with different sizes, were used. The prepared GDEs were assembled into a fuel cell lab prototype with commercial electrolyte and bipolar plates and tested alternately as anode and cathode. Polarization measurements and electrochemical impedance spectroscopy (EIS) were performed on the running hydrogen-fed PEMFC from open circuit voltage to high current density. Experimental impedance spectra were fitted with an equivalent circuit model by using ZView software which allowed to get crucial parameters for the evaluation of fuel cell performance, such as ohmic resistance, charge transfer, and mass transfer resistance, whose trends have been studied as a function of the applied current density.

1. Introduction

Polymer electrolyte membrane fuel cells (PEMFCs) are interesting alternative energy generators because of their high output power density and conversion efficiency, which are associated with modularity, low working temperature (less than 100°C), and environmentally friendly emission. However, a commercial widespread of PEMFCs is still prevented by issues related to cost reduction, electrocatalyst activity, and material durability and stability [1–4].

The catalytic layer is the place where the redox reactions occur; thus, its structural and wetting properties are of paramount importance for the optimal operation of the fuel cell [5]. PEMFCs are usually platinum catalyzed, and the cost of the catalyst has therefore a great influence on the cost of these fuel cells. Thus, a massive use of FCs in the car industry entails finding a material alternative to a platinum catalyst. Many efforts have been done in this field [6, 7], but Pt-based catalysts still remain the state-of-the-art materials [3, 8, 9].

In the design of an active catalyst, Pt dispersion must be considered [10] in addition to Pt loading to maximize the active surface area. Furthermore, ionomer content and distribution, microstructure, and porosity are essential factors in providing a fast proton transfer rate from the anode to the cathode; of course, this has to be attained without increasing the mass transfer resistance to obtain high efficiency in the electrochemical conversion.

In a PEMFC, the catalyst layer can be applied directly to the electrolytic membrane (catalyst-coated membrane (CCM)) or, alternatively, to the gas diffusion medium (GDM) [11, 12]. In this latter case, the so-called gas diffusion electrode (GDE) is obtained, which is a multilayer structure with a quite complex composition. To prepare a GDE, the catalyst powder (i.e., Pt supported on C) is coated onto a gas diffusion medium (GDM). GDM usually consists of a sheet of macroporous carbon cloth or paper (usually called gas diffusion layer (GDL)) coated with a thin layer of carbon black and polytetrafluoroethylene, referred to as microporous layer (MPL). In the recent literature, the interest in the preparation of GDEs is increasing [13–18].

Catalytic layers are generally deposited onto the GDM by means of a variety of techniques, such as brush printing [19], ultrasonic spraying [13], and flexography printing [14] using an ink prepared by dispersing the Pt/C catalyst in a mixture of solvents. A lot of variables have to be considered in the preparation of the catalytic ink, such as nature and ratios of the components or mixing technique and mixing time. All these factors contribute to define the appropriate rheological behavior and the stability of the ink [20] for the specific deposition technique.

The performance of fuel cells and influence of different components on it are usually assessed by polarization curves which are useful to know the trend of cell potential and of the output power density as a function of the current density. Therefore, such curves are related to the macroscopic behavior of the whole device and give no precise information about the effect of inner components. In order to overcome this issue, electrochemical impedance spectroscopy (EIS) has to be carried out together with polarization curve measurements. EIS is a well-established technique for a complete analysis of electrochemical device behavior, but it has also been used in other fields of applied sciences in order to evaluate, for instance, the corrosion behavior of stainless steel [21] and different metallic alloys [22, 23] at various operating conditions and in different media and even for biomedical applications [24].

Especially for fuel cell applications, EIS is a powerful tool which allows a deep in situ kinetic analysis of catalytic phenomena as well as the separation of different processes contributing to overpotential [25] depending on frequency domain. Indeed, by performing EIS measurements, overall ohmic resistance (often referred to as high-frequency resistance (HFR)) of the whole fuel cell device, charge transfer resistance related to activation polarization occurring on catalytic surface, and mass transfer resistance due to diffusive limitations arising at high current density can be obtained. Moreover, depending on the evaluation of different relaxation time and reaction rates, anodic and cathodic contributions to the electrochemical process can be separated [26]. It can be sometimes expansive and complex but the information content of EIS is much higher than DC techniques or single frequency measurements, and it can test component properties and durability within an assembled device, such as a fuel cell [26].

The aim of the present work is to investigate, by EIS and polarization curves, the effect on fuel cell performance of two different GDEs prepared via spray coating technology. It has to be intended as a methodological work to understand the potentiality and the benefit of employing EIS together with polarization curves in assessing cell component quality and behavior.

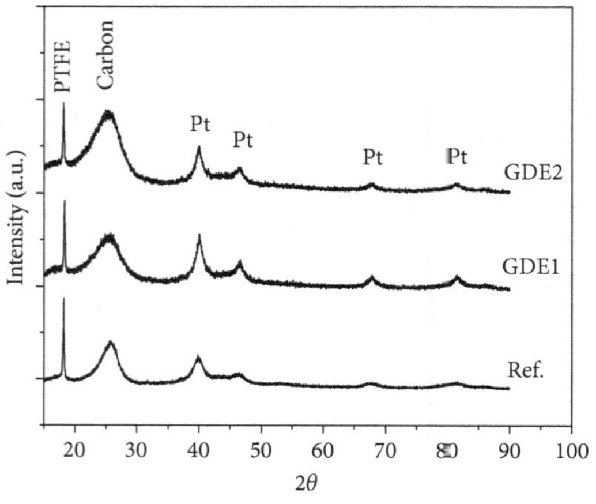

Figure 1: Comparison of the XRD spectra of GDE1, GDE2, and Ref.

The performances of the prepared GDEs, both at the anode and at the cathode side, were compared with those of a commercial GDE reference sample (E-TEK LT140) in a lab-scale single cell.

2. Materials and Methods

2.1. Sample Preparation. Commercial gas diffusion media (SCCG5–P10, SAATI) were used as substrates for the deposition of the catalytic layer. They consist of a carbon cloth-based GDL coated with a carbon microporous layer (MPL) containing 12 wt % of PTFE [27].

A catalytic ink to be sprayed onto the GDM was prepared according to reference [28]. A commercial (E-TEK XC-72) Pt/C catalytic powder was mixed with a 5 wt % Nafion® solution (Aldrich) and dispersed in a mixture of isopropanol and water upon stirring for 1 h and subsequently sonicated for 1.5 h. Two gas diffusion electrodes (referred to as GDE1 and GDE2) were obtained by spraying the ink onto the substrate; an aerograph with nitrogen as gas carrier was employed, and a Pt loading of 0.5 mg/cm^2 and 0.4 mg/cm^2 for GDE1 and GDE2, respectively, was obtained. Two nozzles with different apertures (diameter higher than 1 mm and lower than 1 mm for GDE1 and GDE2, resp.) were used and the properties of the two obtained GDEs were compared with reference electrode (Ref.), E-TEK LT140 with a catalytic loading (Pt/C) of 0.5 mg/cm^2.

After deposition, a thermal treatment in air at 70°C for 15 min was performed to strengthen and fix the catalytic layer to the support. For the cell testing, a commercial 50 μm thick Nafion 212 membrane (a perfluorocarbonsulfonic acid ionomer, supplied by Sigma-Aldrich) was used as electrolyte.

2.2. Structural and Morphological Analysis Phase compositions of the obtained GDEs were analyzed by XRD using a Bruker D8 Advance instrument. Spectra were recorded using the CuKα radiation ($\lambda = 1.54$ Å) in the 15–90 °2θ range, with a step of 0.02 °2θ and counting time of 12 s per step. Crystallite size of Pt catalysts were evaluated via the Scherrer equation using the integral breadth β of Pt *(111) (200) (220)*, and *(002)* reflections. The crystallite sizes calculated from

FIGURE 2: SEM micrographs of GDE1, GDE2, and Ref. at different magnifications.

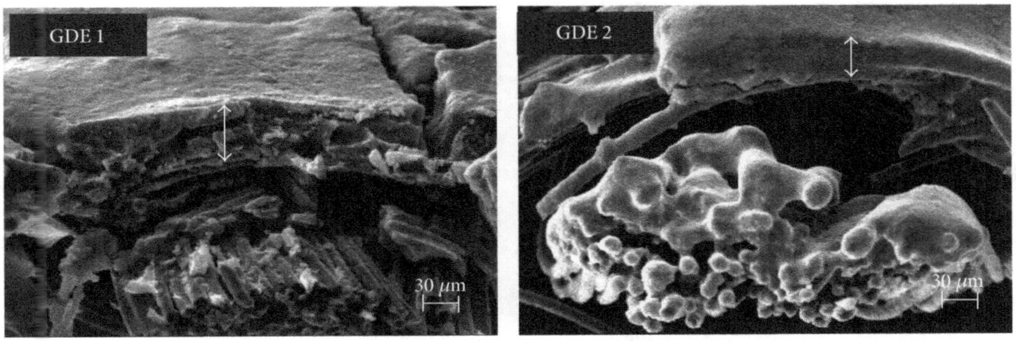

FIGURE 3: SEM images of the cross section of the coating layers of GDE1 and GDE2.

the single reflections have been averaged to obtain the reported values. For the sake of comparison, Ref. GDE was also characterized in the same way. Morphology observations of the GDEs were carried out with a Carl Zeiss EVO50VP scanning electron microscope (SEM) equipped with an energy dispersive spectrometer (EDS) for elemental analysis.

2.3. Electrochemical I-V Characterization. Electrochemical performances of GDE1 and GDE2 were tested in a single lab-scale cell (Fuel Cell Technologies) having a single serpentine plate at the anode and a triple parallel serpentine plate at the cathode side.

GDE1 was assembled alternately at the cathode and at the anode side producing such a MEA assembly: GDE1//Membrane//Ref.; the same setup was realized with GDE2 (GDE2//Membrane//Ref.). In the following, the two assemblies will be referred to as Assembly 1 and Assembly 2, respectively; for the sake of comparison, a third assembly with Ref. GDE at both electrodes was tested (Assembly Ref). When GDE1 works as an anode, the assembly will be labeled Assembly 1A; when GDE1 works as a cathode, it will be referred to as Assembly 1C; similarly for GDE2, the labels will be Assembly 2A and Assembly 2C. The active area was 25 cm^2 for all the assemblies. Pure hydrogen and air were fed at the anode and cathode, respectively. Two flow rate regimes were tested: (1) 0.2 Nl/min of hydrogen and 1.0 Nl/min of air, corresponding to a stoichiometric ratio $S = 1.2/2.4$ A/C at 1 A/cm^2 and (2) 0.5 Nl/min of hydrogen and 2.0 Nl/min of air, corresponding to a stoichiometric ratio $S = 2.9/4.8$ A/C at 1 A/cm^2 and were controlled and detected by a calibrated flow meter. The degree of humidity and the gas temperature were controlled by saturators and temperature controllers, respectively: the temperature of the cell was kept at 60°C and the inlet gases were fully humidified (RH A/C 100/100). An electronic load (RBL488-50-150-800) connected to the cell measured and controlled the voltage, the current, and the generated electric power. Polarization curves were recorded under potentiostatic mode in the voltage range from OCV to 0.1 V with steps of 0.05 V. At each step, the resulting current density was recorded (400 seconds per step, one point per second recorded). Current density values plotted in the steady-state polarization curves result from the averaging of the last 220 points recorded at each

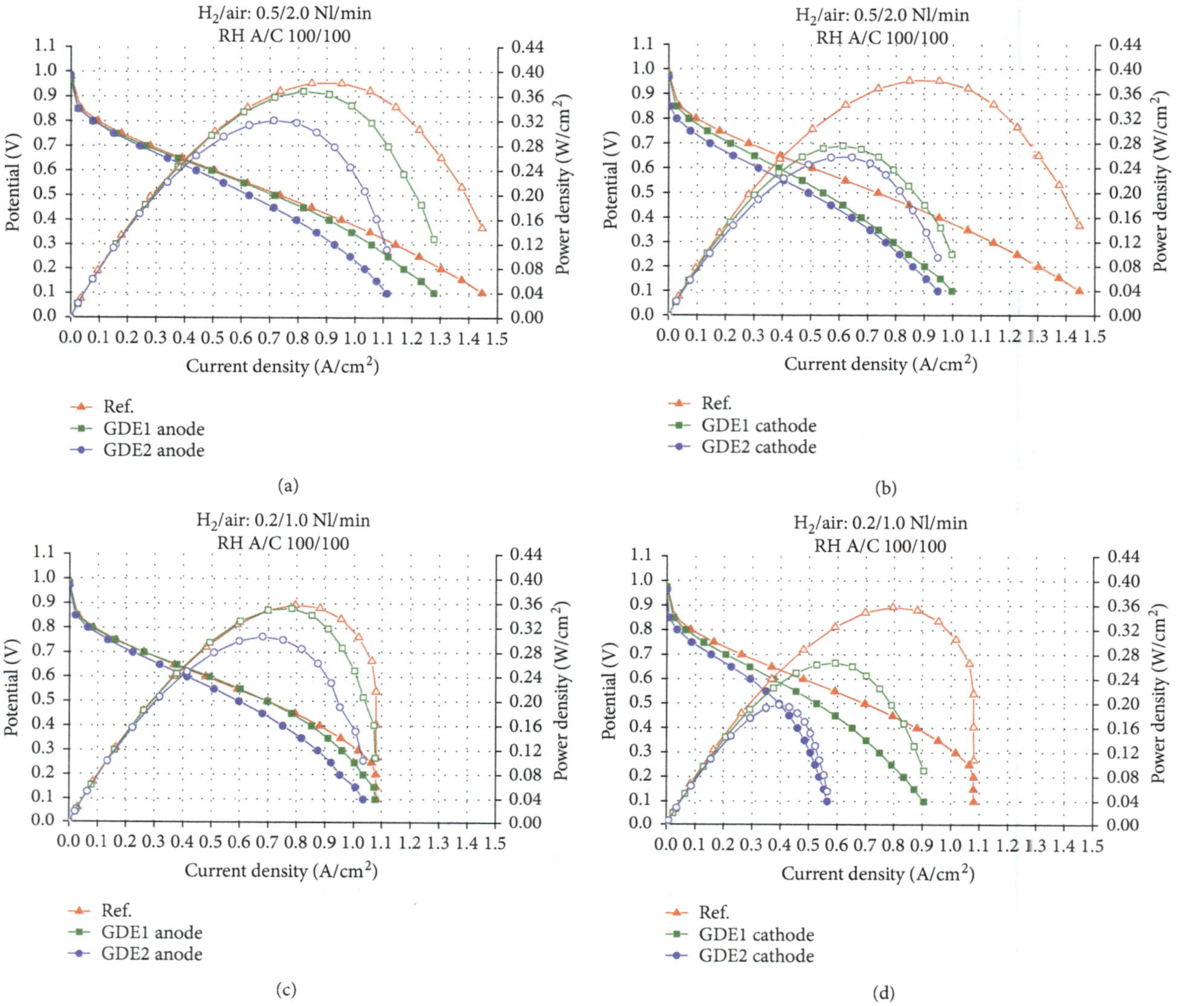

FIGURE 4: Polarization and power density curves at a high inlet gas flow rate (a, b)—0.5/2 Nl/min H_2/air—and at a low inlet gas flow rate (c, d)—0.2/1 Nl/min H_2/air.

step in order to minimize experimental artifacts due to transient phenomena [29].

2.4. Electrochemical Impedance Spectroscopy. Electrochemical impedance spectroscopy (EIS) of the running FC was performed using a frequency response analyzer (FRA) Solartron 1260, directly connected to the electronic load (RBL488-50-150-800). EIS was performed in galvanostatic mode [30] using an AC signal of amplitude 200 mA [31, 32]. Impedance spectra were collected by sweeping frequencies over the range 0.5 Hz–1 kHz, acquiring ten points per decade. The EIS spectra were recorded at OCV and from low to high current density (up to 1 A/cm^2). Five full spectra for each current density value were acquired, and the impedance spectrum finally used in the discussion was the result of an averaging procedure. The experimental spectra were modeled with equivalent circuits by employing the ZView® software (Scribner Associates). The equivalent circuit used [33, 34] is made of a resistance (R_s) in series with two parallel constant phase elements (CPE)/resistance circuits. R_s represents the ohmic losses, while the first CPE$_p$/R_p circuit models the activation polarization (i.e., charge transfer resistance) and the second one (CPE$_d$/R_d) is related to concentration losses due to mass transfer resistances. CPEs were used instead of pure capacitances to account for the capacitive losses that generally occur in porous electrodes [35].

3. Results

3.1. Microstructure and Morphology. In Figure 1, the XRD spectra of GDE1, GDE2, and Ref. are compared. The phase compositions of GDE1, GDE2, and Ref. samples are very similar and consist of a homogeneous mixture of amorphous carbon (JC-PDS 01-075-0444), crystalline PTFE, and crystalline Pt (JC-PDS 00-004-0802). Regardless of the nozzle aperture, the average Pt crystallite size is almost the same

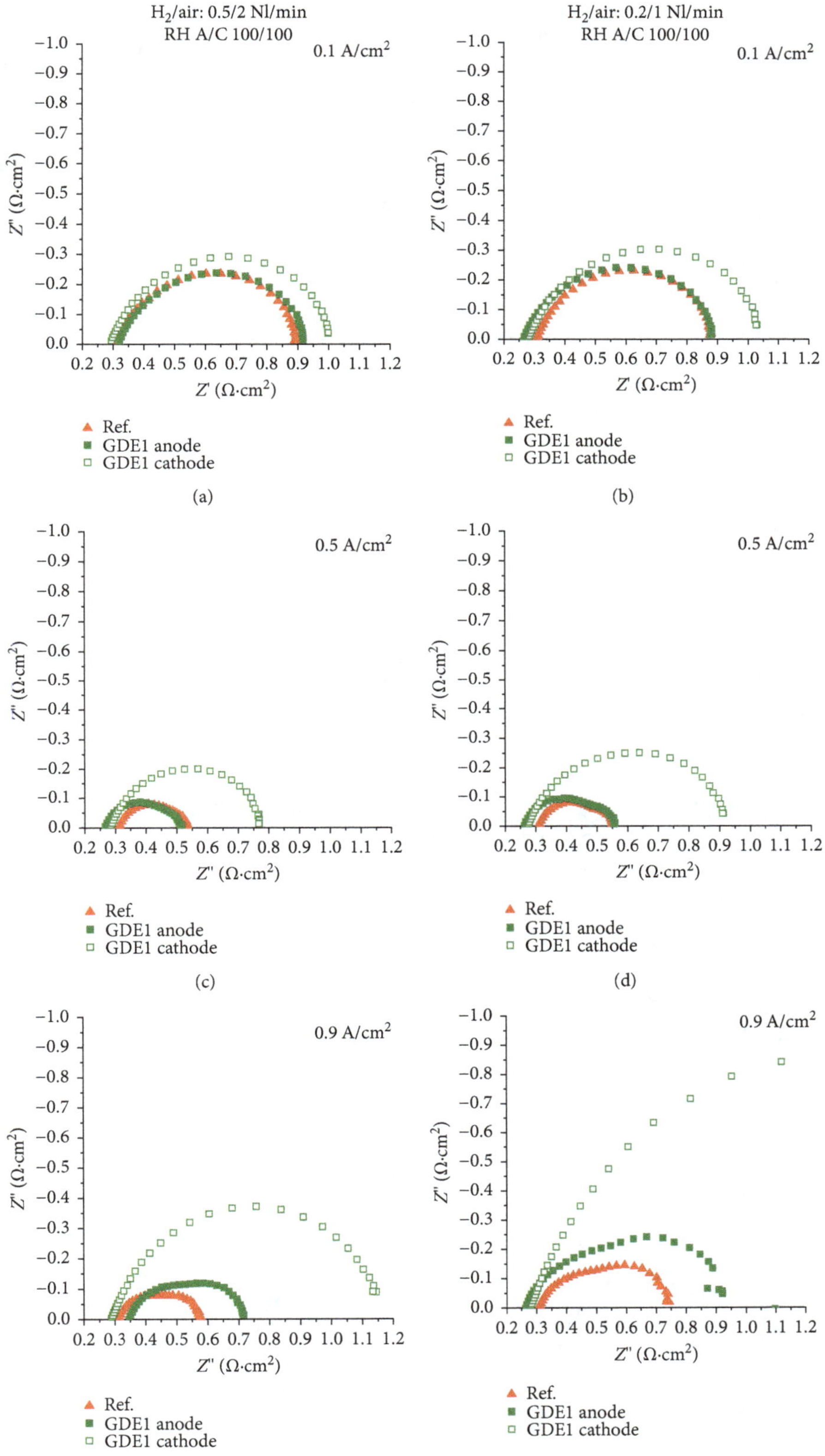

FIGURE 5: Example of impedance spectra acquired at different current densities (i.e., from 0.1 A/cm^2 to 0.9 A/cm^2) at high (a, c, e) and low (b, d, f) inlet gas flow rate for the reference sample, GDE1 anode, and GDE1 cathode.

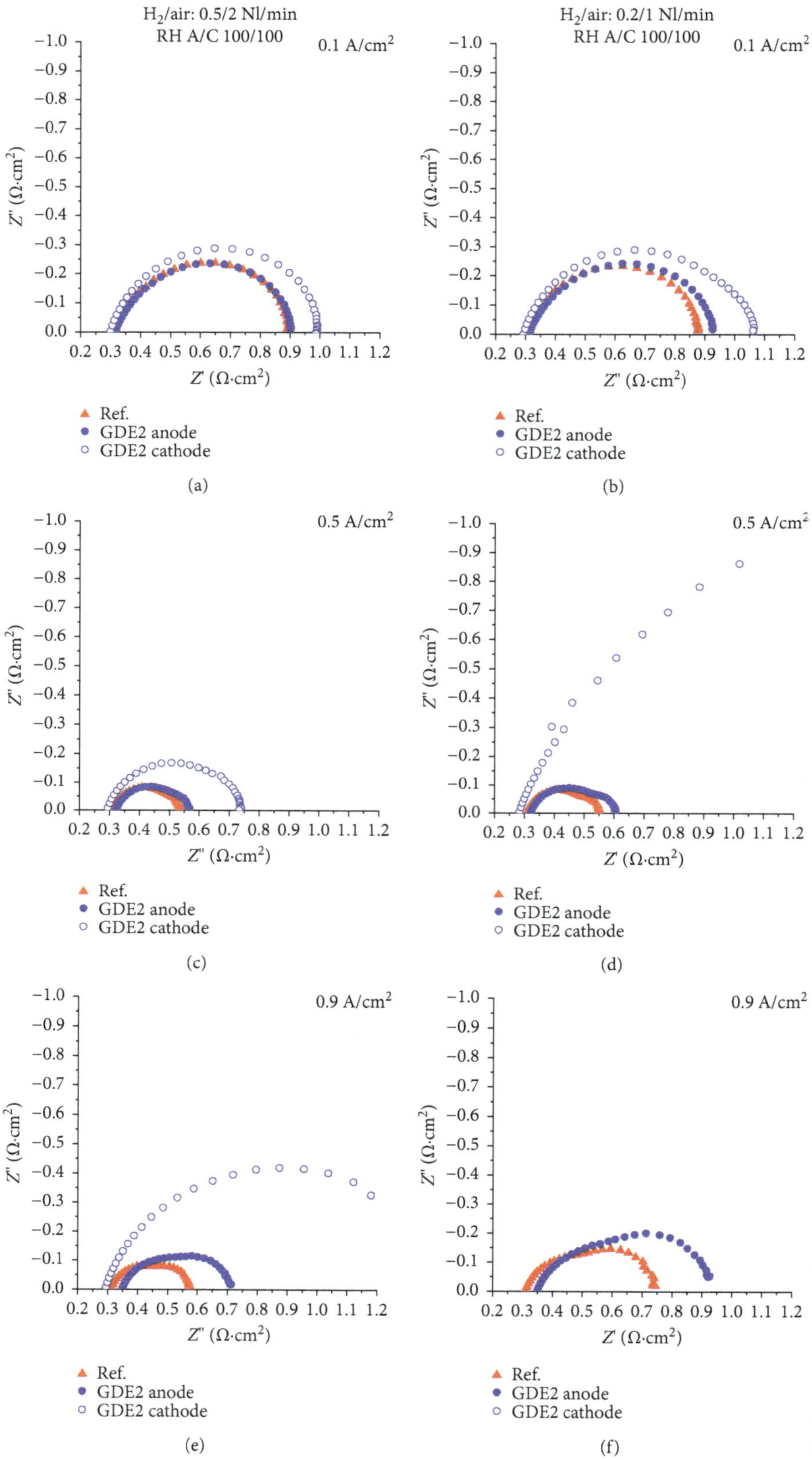

FIGURE 6: Example of impedance spectra acquired at different current densities (i.e., from 0.1 A/cm^2 to 0.9 A/cm^2) at high (a, c, e) and low (b, d, f) inlet gas flow rates for the reference sample, GDE2 anode, and GDE2 cathode.

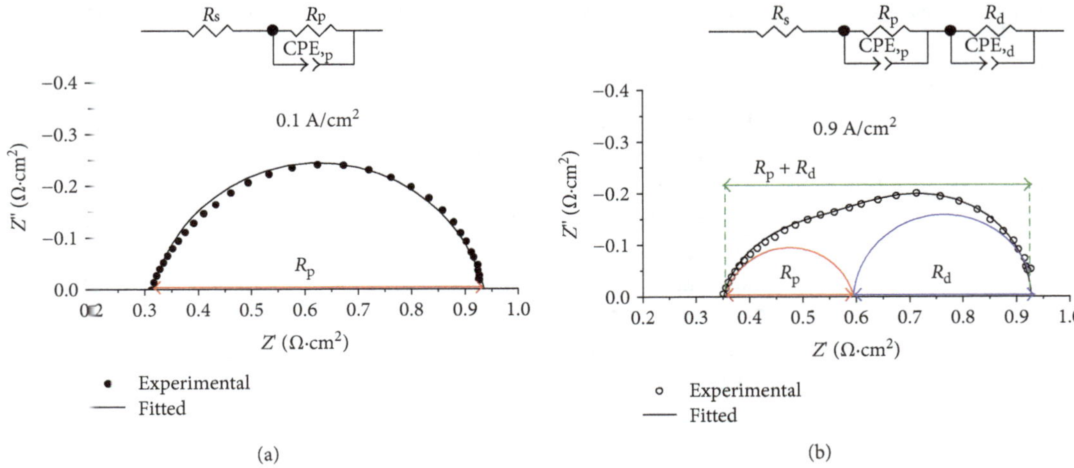

FIGURE 7: An example of modeling: equivalent circuits at low current density (a) and at high current density (b) for two spectra obtained for sample GDE2 anode at low inlet gas flow rate. R_p (a) and $R_p + R_d$ (b) are highlighted in order to clarify how the general output of the adopted procedure was obtained.

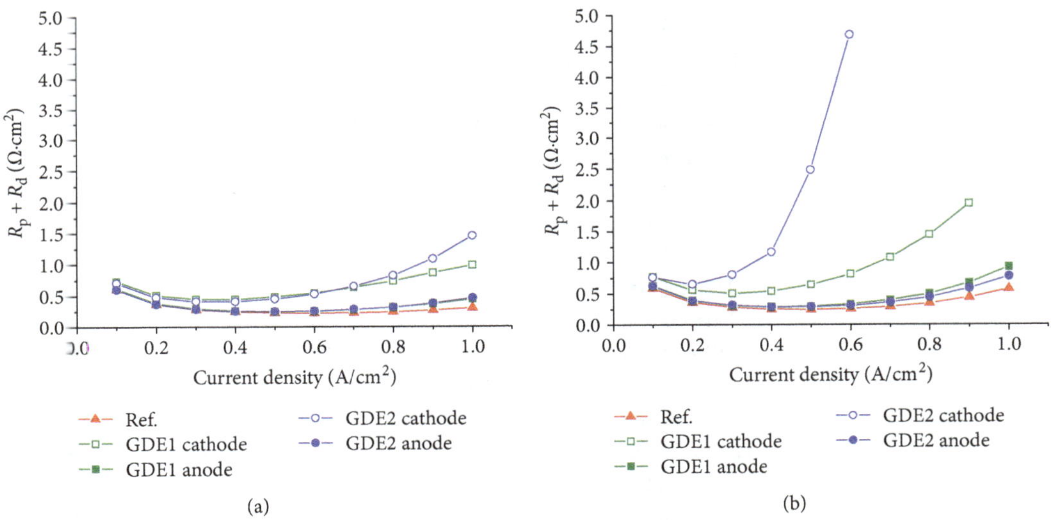

FIGURE 8: Sum of the charge transfer (R_p) and the mass transfer resistances (R_d) at a high inlet gas flow rate (a)—0.5/2 Nl/min H_2/air—and at a low inlet gas flow rate (b)—0.2/1 Nl/min H_2/air.

for the GDE1 and GDE2 samples: $d(\text{GDE1}) = 7\,\text{nm}$ and $d(\text{GDE2}) = 6\,\text{nm}$. Moreover, they are comparable with the crystallite size of Pt in Ref. sample, evaluated in the same way: $d(\text{Ref.}) = 5\,\text{nm}$. This value fairly agrees with the catalyst producer data (4 nm). As this size is reported to be appropriate for PEMFCs applications [36], both the ink preparation procedure and the adopted deposition technique allow for the attainment of GDEs with properties comparable to those of the commercial sample (Ref.).

A similar micromorphology was observed for the three GDEs: a fluffy-like powder is present in each sample (Figure 2) suggesting that the use of nozzles of different dimensions does not affect the compactness of the powders in the carbon layer. However, small differences can be observed on the covering degree of the GDL. Indeed, the Ref. surface is very smooth and no feature of the original warp and weft of the backing GDL is perceived. On the contrary, the feature of the GDL substrate is clearly evident in GDE1 and it is even more noticeable in GDE2. Nevertheless, the surfaces of GDE1 and GDE2 are more homogeneous than that of the Ref. sample that shows many cracks on the top layer. Accordingly, different thicknesses of the coating layers of the GDEs can be hypothesized: GDE1 and GDE2 coatings are probably thinner than the one of Ref.; indeed, thicker layers suffer larger mechanical stress during thermal treatment, producing a more cracked surface. Furthermore, GDE2 coating layer seems to be thinner than GDE1 even if both GDEs were obtained using the same ink formulation. Such a difference could be ascribed to different shear effects when nozzles of different apertures are used for the spray

TABLE 1: The fitting parameters for the impedance spectra at different current densities (i.e., from OCV to 0.9 A/cm^2) and high inlet gas flow rate (H_2/air: 0.5/2 Nl/min).

	Current density	Sample	OCV	0.1 (A/cm^2)	0.3 (A/cm^2)	0.5 (A/cm^2)	0.7 (A/cm^2)	0.9 (A/cm^2)
Ohmic resistance	R_s (Ω·cm^2)	Reference	0.317	0.315	0.313	0.311	0.311	0.311
		GDE1 cathode	0.291	0.291	0.286	0.285	0.285	0.286
		GDE1 anode	0.275	0.268	0.268	0.268	0.268	0.268
		GDE2 cathode	0.305	0.300	0.298	0.294	0.292	0.292
		GDE2 anode	0.321	0.321	0.321	0.322	0.326	0.339
Charge transfer circuit	R_p (Ω·cm^2)	Reference	6.350	0.596	0.276	0.199	0.176	0.174
		GDE1 cathode	5.598	0.727	0.236	0.208	0.241	0.109
		GDE1 anode	6.685	0.612	0.278	0.217	0.211	0.232
		GDE2 cathode	6.260	0.700	0.408	0.372	0.421	0.303
		GDE2 anode	6.369	0.596	0.280	0.209	0.222	0.276
	CPE$_p$-Q (F/cm^2)	Reference	0.047	0.035	0.038	0.041	0.047	0.048
		GDE1 cathode	0.059	0.044	0.054	0.065	0.097	0.129
		GDE1 anode	0.042	0.031	0.033	0.035	0.039	0.041
		GDE2 cathode	0.021	0.014	0.019	0.020	0.027	0.032
		GDE2 anode	0.043	0.032	0.034	0.034	0.047	0.067
	CPE$_p$-n	Reference	0.800	0.850	0.850	0.850	0.850	0.850
		GDE1 cathode	0.800	0.860	0.880	0.870	0.830	0.850
		GDE1 anode	0.800	0.850	0.850	0.850	0.850	0.850
		GDE2 cathode	0.800	0.880	0.860	0.880	0.850	0.880
		GDE2 anode	0.800	0.850	0.850	0.850	0.800	0.750
	C_p (F/cm^2)	Reference	0.035	0.018	0.017	0.0175	0.019	0.021
		GDE1 cathode	0.045	0.025	0.029	0.034	0.045	0.061
		GDE1 anode	0.031	0.015	0.014	0.015	0.016	0.018
		GDE2 cathode	0.013	0.007	0.008	0.010	0.013	0.017
		GDE2 anode	0.031	0.016	0.015	0.015	0.015	0.018
	f_p (Hz)	Reference	0.725	15.051	33.933	45.620	47.463	43.710
		GDE1 cathode	0.625	8.712	25.500	22.299	14.717	24.168
		GDE1 anode	0.768	17.110	40.159	48.738	46.540	38.558
		GDE2 cathode	1.959	30.661	45.629	41.178	29.298	30.677
		GDE2 anode	0.794	17.003	37.457	52.700	47.867	32.829
Mass transfer circuit	R_d (Ω·cm^2)	Reference	—	—	—	0.027	0.047	0.089
		GDE1 cathode	—	—	0.233	0.274	0.394	0.755
		GDE1 anode	—	—	0.012	0.028	0.064	0.137
		GDE2 cathode	—	—	—	0.075	0.228	0.782
		GDE2 anode	—	—	—	0.036	0.059	0.100
	CPE$_d$-Q (F/cm^2)	Reference	—	—	—	1.516	0.579	0.281
		GDE1 cathode	—	—	0.068	0.065	0.063	0.051
		GDE1 anode	—	—	1.959	0.985	0.390	0.184
		GDE2 cathode	—	—	—	0.341	0.095	0.036
		GDE2 anode	—	—	—	0.741	0.447	0.262
	CPE$_d$-n	Reference	—	—	—	0.900	1.000	1.000
		GDE1 cathode	—	—	1.000	1.000	1.000	0.950
		GDE1 anode	—	—	1.000	1.000	1.000	1.000
		GDE2 cathode	—	—	—	0.850	0.940	0.910
		GDE2 anode	—	—	—	1.000	1.000	1.000

TABLE 1: Continued.

Current density	Sample	OCV	0.1 (A/cm^2)	0.3 (A/cm^2)	0.5 (A/cm^2)	0.7 (A/cm^2)	0.9 (A/cm^2)
C_d (F/cm^2)	Reference	—	—	—	1.065	0.579	0.281
	GDE1 cathode	—	—	0.068	0.065	0.063	0.043
	GDE1 anode	—	—	1.959	0.985	0.390	0.184
	GDE2 cathode	—	—	—	0.179	0.074	0.025
	GDE2 anode	—	—	—	0.741	0.447	0.262
f_d (Hz)	Reference	—	—	—	5.435	5.851	6.331
	GDE1 cathode	—	—	10.082	8.924	6.361	4.867
	GDE1 anode	—	—	6.962	5.774	6.390	6.282
	GDE2 cathode	—	—	—	11.876	9.452	8.045
	GDE2 anode	—	—	—	6.014	6.069	6.096

procedure. A more narrow nozzle hole could result in higher shear effects that could produce a partial separation of the solid from the liquid in the ink during spraying. Such a modification will result in a lower solid load of the sprayed ink and therefore in a thinner coating layer in GDE2.

A direct measurement of the coating thickness has been attempted by SEM analysis of a sample cross section (Figure 3). A slightly higher thickness was found for GDE1 (about 40 μm) compared to GDE2 (<30 μm), in agreement with the observed coverage of the cloths.

3.2. I-V Characterization. In Figure 4, the polarization curves of the FC incorporating one of the prepared GDEs at the anode or cathode are compared with the reference assembly with Ref. GDEs at both electrodes.

The best electrochemical performances of GDE1 and GDE2 are achieved when working as anodes, while poor results are obtained if they are used as cathodes. Assembly 1 is always superior to Assembly 2 in any operating condition. Low flow rates drastically reduce the cell performances in any assembly here tested, but the effect is noticeably less marked in Assembly 1 (see Figures 4(a) versus 4(c) and Figures 4(b) versus 4(d)). Moreover, Assembly 1A's performance compares very well with that of Assembly Ref up to about 0.8 A/cm^2 (Figures 4(a) and 4(c)).

3.3. Electrochemical Impedance Spectroscopy. In Figure 5, some representative impedance spectra of Assembly 1A and 1C are compared with those of Assembly Ref, at increasing current density (CD), for the two flow rate regimes (high and low); analogously, impedance spectra of Assembly 2A and 2C, obtained at the same selected current densities, are shown in Figure 6.

The impedance spectra were modeled with $[_{\text{parallel}}R\text{-CPE}]_p\text{-}R_s$ (low CD) and $[_{\text{parallel}}R\text{-CPE}]_p\text{-}R_s\text{-}[_{\text{parallel}}R\text{-CPE}]_d$ (high CD) equivalent circuits to evaluate activation (R_p), diffusion (R_d), and ohmic (R_s) resistances [37]. In the low CD range, mass transfer limitations are negligible and for this reason there is no need of using two R/CPE parallel circuits to fit experimental spectra [34]. As an example, experimental and fitted impedance spectra of the running cell at low and high CD are shown in Figure 7, where the equivalent circuits used for the fitting are reported in the insets.

The worst performances of Assembly 1 (Figure 5) and Assembly 2 (Figure 6) primarily depend on the total internal resistance, namely, the sum of the two polarizations $R_p + R_d$; such sum is represented by the difference between right and left intercepts of the impedance spectrum with the real axis and reported in Figure 8 for tested samples.

In accordance with polarization curves, in any condition, $R_p + R_d$ increases upon increasing CD from 0.5 to 1.0 A/cm^2. Indeed, the total polarization resistance decreases somewhat from 0.1 to 0.3 A/cm^2 and this is usually ascribed to the lowering of the anode charge transfer resistance [37, 38]; as such, this phenomenon occurs in all the assemblies, reference included. The increase in total polarization resistance is clear when GDE1/GDE2 works as an anode, but it becomes dramatic when operating as a cathode, which explains the poor performances of Assembly 1 and 2 compared to Ref. in the high CD range.

On the contrary, the ohmic resistance R_s (Tables 1 and 2), that is, the high frequency intercept of the impedance spectrum, is comparable with the one of the reference assembly, or even lower; in any case, it is in the order of 0.3 $\Omega\cdot$cm^2 and does not vary appreciably upon increasing CD. The ohmic resistance R_s represents bulk material resistances and contact resistances [39], and among these, the highest is usually the membrane resistance. This, in turn, mainly depends on the internal humidity degree of the membrane itself; because the inlet gases were fully humidified, there is no reason for the ohmic resistance to vary with CD, and indeed it does not.

Using the fitting results reported in Table 1 (low flow rate) and Table 2 (high flow rate), the characteristic capacitance (C_p or C_d) and the relaxation frequency (f_p or f_d), which are useful parameters in identifying the polarization subprocesses, were calculated based on the following equations [40]:

$$C_i = \frac{(R_i Q_i)^{1/n_i}}{R_i},$$
$$f_i = \frac{(R_i Q_i)^{-1/n_i}}{2\pi}, \quad (1)$$
$$\frac{1}{Z} = Y = Q(jw)^n.$$

TABLE 2: The fitting parameters for the impedance spectra at different current densities (i.e., from OCV to 0.9 A·cm^{-2}) and low inlet gas flow rates (H_2/air: 0.2/1 Nl/min).

	Current density	Sample	OCV	0.1 (A/cm^2)	0.3 (A/cm^2)	0.5 (A/cm^2)	0.7 (A/cm^2)	0.9 (A/cm^2)
		Reference	0.312	0.312	0.312	0.313	0.313	0.313
		GDE1 cathode	0.285	0.285	0.285	0.276	0.276	0.276
Ohmic resistance	R_s (Ω·cm^2)	GDE1 anode	0.267	0.267	0.267	0.264	0.264	0.264
		GDE2 cathode	0.297	0.297	0.290	0.290	—	—
		GDE2 anode	0.316	0.316	0.311	0.319	0.333	0.351
		Reference	5.789	0.583	0.274	0.198	0.194	0.220
		GDE1 cathode	5.669	0.771	0.505	0.361	0.289	0.261
	R_p (Ω·cm^2)	GDE1 anode	6.271	0.625	0.305	0.233	0.251	0.299
		GDE2 cathode	6.748	0.759	0.623	0.584	—	—
		GDE2 anode	6.466	0.620	0.278	0.220	0.224	0.262
		Reference	0.042	0.038	0.042	0.046	0.050	0.055
		GDE1 cathode	0.063	0.049	0.053	0.064	0.086	0.088
	CPE$_p$-Q (F/cm^2)	GDE1 anode	0.045	0.033	0.035	0.037	0.040	0.045
		GDE2 cathode	0.030	0.0189	0.084	0.037	—	—
		GDE2 anode	0.046	0.032	0.032	0.035	0.039	0.043
		Reference	0.850	0.850	0.850	0.850	0.850	0.850
		GDE1 cathode	0.800	0.850	0.850	0.850	0.850	0.850
Charge transfer circuit	CPE$_p$-n	GDE1 anode	0.800	0.850	0.850	0.850	0.850	0.850
		GDE2 cathode	0.750	0.850	0.750	0.850	—	—
		GDE2 anode	0.800	0.850	0.850	0.850	0.850	0.850
		Reference	0.033	0.0196	0.0190	0.0199	0.0219	0.0253
		GDE1 cathode	0.048	0.028	0.028	0.033	0.045	0.045
	C_p (F/cm^2)	GDE1 anode	0.033	0.017	0.016	0.016	0.018	0.021
		GDE2 cathode	0.0176	0.009	0.031	0.019	—	—
		GDE2 anode	0.034	0.016	0.014	0.014	0.017	0.020
		Reference	0.832	13.901	30.618	40.490	37.447	28.578
		GDE1 cathode	0.580	7.439	11.370	13.391	12.188	13.514
	f_p (Hz)	GDE1 anode	0.772	15.236	33.191	42.838	35.531	25.123
		GDE2 cathode	1.341	23.459	8.133	14.625	—	—
		GDE2 anode	0.724	16.010	41.736	50.220	41.618	30.992
		Reference	—	—	—	0.042	0.091	0.209
		GDE1 cathode	—	—	—	0.280	0.783	1.682
	R_d (Ω·cm^2)	GDE1 anode	—	—	—	0.060	0.142	0.368
		GDE2 cathode	—	—	0.178	1.891	—	—
		GDE2 anode	—	—	0.037	0.067	0.131	0.314
		Reference	—	—	—	0.912	0.412	0.187
		GDE1 cathode	—	—	—	0.124	0.069	0.051
Mass transfer circuit	CPE$_d$-Q (F/cm^2)	GDE1 anode	—	—	—	0.598	0.253	0.113
		GDE2 cathode	—	—	0.161	0.033	—	—
		GDE2 anode	—	—	0.945	0.523	0.276	0.129
		Reference	—	—	—	1.000	1.000	1.000
		GDE1 cathode	—	—	—	1.000	1.000	1.000
	CPE$_d$-n	GDE1 anode	—	—	—	1.000	1.000	1.000
		GDE2 cathode	—	—	1.000	1.000	—	—
		GDE2 anode	—	—	1.000	1.000	1.000	1.000

Table 2: Continued.

Current density	Sample	OCV	0.1 (A/cm^2)	0.3 (A/cm^2)	0.5 (A/cm^2)	0.7 (A/cm^2)	0.9 (A/cm^2)
	Reference	—	—	—	0.912	0.412	0.187
	GDE1 cathode	—	—	—	0.124	0.069	0.051
C_d (F/cm^2)	GDE1 anode	—	—	—	0.598	0.253	0.113
	GDE2 cathode	—	—	0.161	0.033	—	—
	GDE2 anode	—	—	0.945	0.523	0.276	0.129
	Reference	—	—	—	4.184	4.239	4.068
	GDE1 cathode	—	—	—	4.585	2.939	1.844
f_d (Hz)	GDE1 anode	—	—	—	4.400	4.425	3.817
	GDE2 cathode	—	—	5.535	2.575	—	—
	GDE2 anode	—	—	4.528	4.537	4.379	3.934

In the high CD region where two subcircuits are needed to fit experimental data, the relaxation frequencies are quite well separated to point out that two different physical phenomena are occurring, that is, charge and mass transfers; on average, the two phenomena are more clearly distinguishable when GDE1 or GDE2 are used as anodes than as cathodes.

4. Discussion

The spray-coated GDE1 performs quite well in the low-medium CD range when it is used as an anode (Figures 4(a) and 4(c)), while it does not at the cathode side. The same trend is observed with GDE2, but in this case, the overall performances are somewhat lower.

Because the Pt load primarily affects the oxygen reduction reaction (ORR) occurring at the cathode, the somewhat lower amount of catalyst in GDE2 partly explains the poor performances of the assemblies tested with this GDE. However, the nominal Pt load and the Pt-particle size of the reference do not justify the worsening of the cell performances in the high CD region. Moreover, observing polarization curves at low flow rate (Figures 4(c) and 4(d)), it is clear that there are no flooding effects with the reference assembly and the curve drops to zero in correspondence of the limiting current ([illegible].1 A/cm^2) as calculated by the stoichiometric ratio. On the contrary, GDE1 and GDE2 polarization curves drop to zero much earlier and this may be explained by flooding of the electrodes. Indeed, flooding occurs mainly at the cathode side in a PEMFC, due to the excess water produced by the ORR; so, if a GDM is inadequate to manage water, this becomes particularly evident when it is used on the cathode side, as pointed out by high mass transfer resistances from EIS analysis. In conclusion, the combined effect of lower Pt load and scarce ability in water management seems to explain the poor performances shown by Assembly 2. Comparison of R_d values (Tables 1 and 2) between Assembly 1 or 2 and Assembly Ref corroborates this hypothesis; there is a marked difference in R_d values when GDE1 is used as an anode and a cathode, and even in the best case, they are larger than the corresponding ones obtained in the reference assembly. The same considerations apply to GDE2, but in this case, the differences are even larger. Reducing the flow rate does worsen mass transport and causes an increase in the R_d values, even in the case of the reference, more pronounced than the corresponding increase in R_p values.

5. Conclusions

Two GDEs were obtained by spray coating; the surface morphology and the catalytic Pt-particle average size are comparable with commercial samples, and they do not seem to depend on the nature of the substrate or on some preparation procedures (e.g., nozzle dimension). The two GDEs have slightly different Pt-load: 0.5 mg/cm^2 GDE1 and 0.4 mg/cm^2 GDE2. The worst electrical performances of the GDEs prepared in this work compared to those of a commercial sample can be mainly ascribed to the poor water management. This phenomenon is highlighted by the equivalent circuit analysis of impedance spectra collected during the fuel cell operation. As the catalytic load and the average Pt particle size do not substantially differ from the reference sample, the bad water management behavior of the GDEs should be ascribed to the porous structure of the GDM. Indeed, higher mass transfer contributions and total internal resistance were found for both GDE1 and GDE2 compared to the reference sample; this behavior was more evident when lab-made GDEs were employed in the cathodic side, where water management issues are more pronounced. Such findings were possible due to the employment of electrochemical impedance spectroscopy, which is the most powerful tool to be used in fuel cells field to highlight and easily distinguish all the different physical phenomena, both kinetic and diffusive, taking place within the running device.

Conflicts of Interest

The authors declare that there is no conflict of interest regarding the publication of this paper.

References

[1] O. Erdinc and M. Uzunoglu, “Recent trends in PEM fuel cell-powered hybrid systems: investigation of application areas,

design architectures and energy management approaches," *Renewable and Sustainable Energy Reviews*, vol. 14, no. 9, pp. 2874–2884, 2010.

[2] Y. Wang, K. S. Chen, J. Mishler, S. C. Cho, and X. C. Adroher, "A review of polymer electrolyte membrane fuel cells: technology, applications, and needs on fundamental research," *Applied Energy*, vol. 88, no. 4, pp. 981–1007, 2011.

[3] O. Groger, H. A. Gasteiger, and J. P. Suchsland, "Review–electromobility: batteries or fuel cells?," *Journal of the Electrochemical Society*, vol. 162, no. 14, pp. A2605–A2622, 2015.

[4] M. Zhou, H. L. Wang, and S. J. Guo, "Towards high-efficiency nanoelectrocatalysts for oxygen reduction through engineering advanced carbon nanomaterials," *Chemical Society Reviews*, vol. 45, no. 5, pp. 1273–1307, 2016.

[5] Y. M. Volfkovich, V. E. Sosenkin, and V. S. Bagotsky, "Structural and wetting properties of fuel cell components," *Journal of Power Sources*, vol. 195, no. 17, pp. 5429–5441, 2010.

[6] R. Othman, A. L. Dicks, and Z. Zhu, "Non precious metal catalysts for the PEM fuel cell cathode," *International Journal of Hydrogen Energy*, vol. 37, no. 1, pp. 357–372, 2012.

[7] Z. W. Chen, D. Higgins, A. P. Yu, L. Zhang, and J. J. Zhang, "A review on non-precious metal electrocatalysts for PEM fuel cells," *Energy & Environmental Science*, vol. 4, no. 9, pp. 3167–3192, 2011.

[8] G. F. Alvarez, M. Mamlouk, S. M. Senthil Kumar, and K. Scott, "Preparation and characterisation of carbon-supported palladium nanoparticles for oxygen reduction in low temperature PEM fuel cells," *Journal of Applied Electrochemistry*, vol. 41, no. 8, pp. 925–937, 2011.

[9] S. Sui, X. Y. Wang, X. T. Zhou, Y. H. Su, S. Riffatc, and C. J. Liu, "A comprehensive review of Pt electrocatalysts for the oxygen reduction reaction: nanostructure, activity, mechanism and carbon support in PEM fuel cells," *Journal of Materials Chemistry A*, vol. 5, no. 5, pp. 1808–1825, 2017.

[10] G. Diloyan, M. Sobel, K. Das, and P. Hutapea, "Effect of mechanical vibration on platinum particle agglomeration and growth in polymer electrolyte membrane fuel cell catalyst layers," *Journal of Power Sources*, vol. 214, pp. 59–67, 2012.

[11] L. Cindrella, A. M. Kannan, J. F. Lin et al., "Gas diffusion layer for proton exchange membrane fuel cells—a review," *Journal of Power Sources*, vol. 194, no. 1, pp. 146–160, 2009.

[12] S. Park, J.-W. Lee, and B. N. Popov, "A review of gas diffusion layer in PEM fuel cells: materials and designs," *International Journal of Hydrogen Energy*, vol. 37, no. 7, pp. 5850–5865, 2012.

[13] B. Millington, V. Whipple, and B. G. Pollet, "A novel method for preparing proton exchange membrane fuel cell electrodes by the ultrasonic-spray technique," *Journal of Power Sources*, vol. 196, no. 20, pp. 8500–8508, 2011.

[14] C. Bois, A. Blayo, D. Chaussy, R. Vincent, A. G. Mercier, and C. Nayoze, "Catalyst layers for PEMFC manufactured by flexography printing process: performances and structure," *Fuel Cells*, vol. 12, no. 2, pp. 199–211, 2012.

[15] C. Ruengkit and N. Tantavichet, "Influence of gas diffusion layer on Pt catalyst prepared by electrodeposition for proton exchange membrane fuel cells," *Thin Solid Films*, vol. 636, pp. 116–126, 2017.

[16] H. B. Li, P. P. Yang, Z. Y. Xie, M. Sun, and Q. Z. Huang, "Self-supported N-doped carbon nanonet integrated on carbon paper for gas diffusion electrode," *International Journal of Hydrogen Energy*, vol. 42, no. 12, pp. 8216–8223, 2017.

[17] I. Gatto, A. Stassi, V. Baglio et al., "Optimization of perfluorosulphonic ionomer amount in gas diffusion electrodes for PEMFC operation under automotive conditions," *Electrochimica Acta*, vol. 165, pp. 450–455, 2015.

[18] H. Su, T. C. Jao, S. Pasupathi, B. J. Bladergroen, V. Linkov, and B. G. Pollet, "A novel dual catalyst layer structured gas diffusion electrode for enhanced performance of high temperature proton exchange membrane fuel cell," *Journal of Power Sources*, vol. 246, pp. 63–67, 2014.

[19] İ. Fırtına, S. Güner, and A. Albostan, "Preparation and characterization of membrane electrode assembly (MEA) for PEMFC," *International Journal of Energy Research*, vol. 35, no. 2, pp. 146–152, 2011.

[20] S. Latorrata, P. G. Stampino, E. Amici, R. Pelosato, C. Cristiani, and G. Dotelli, "Effect of rheology controller agent addition to micro-porous layers on PEMFC performances," *Solid State Ionics*, vol. 216, pp. 73–77, 2012.

[21] E. F. Diaz, C. Cuevas-Arteaga, N. Flores-Garcia, S. M. Sintillo, and O. Sotelo-Mazon, "Corrosion performance of AISI-309 exposed to molten salts V_2O_5-Na_2SO_4 at 700°C applying EIS and R_p electrochemical techniques," *Journal of Spectroscopy*, vol. 2015, Article ID 826759, 12 pages, 2015.

[22] G. Bolat, D. Mareci, S. Iacoban, N. Cimpoesu, and C. Munteanu, "The estimation of corrosion behavior of NiTi and NiTiNb alloys using dynamic electrochemical impedance spectroscopy," *Journal of Spectroscopy*, vol. 2013, Article ID 714920, 7 pages, 2013.

[23] S. Godavarthi, J. Porcayo-Calderon, E. Vazquez-Velez, M. Casales-Diaz, D. M. Ortega-Toledo, and L. Martinez-Gomez, "Influence of the chemical composition in the electrochemical response of permanent magnets," *Journal of Spectroscopy*, vol. 2015, Article ID 356027, 16 pages, 2015.

[24] S. Sbrignadello, A. Tura, and P. Ravazzani, "Electroimpedance spectroscopy for the measurement of the dielectric properties of sodium chloride solutions at different glucose concentrations," *Journal of Spectroscopy*, vol. 2013, Article ID 571372, 6 pages, 2013.

[25] M. Ciureanu and R. Roberge, "Electrochemical impedance study of PEM fuel cells. Experimental diagnostics and modeling of air cathodes," *Journal of Physical Chemistry B*, vol. 105, no. 17, pp. 3531–3539, 2001.

[26] S. Latorrata, P. G. Stampino, C. Cristiani, and G. Dotelli, "Development of an optimal gas diffusion medium for polymer electrolyte membrane fuel cells and assessment of its degradation mechanisms," *International Journal of Hydrogen Energy*, vol. 40, no. 42, pp. 14596–14608, 2015.

[27] S. Latorrata, R. Balzarotti, P. G. Stampino, C. Cristiani, G. Dotelli, and M. Guilizzoni, "Design of properties and performances of innovative gas diffusion media for polymer electrolyte membrane fuel cells," *Progress in Organic Coatings*, vol. 78, pp. 517–525, 2015.

[28] S. Litster and G. McLean, "PEM fuel cell electrodes," *Journal of Power Sources*, vol. 130, no. 1-2, pp. 61–76, 2004.

[29] L. Omati, P. Gallo Stampino, G. Dotelli, D. Brivio, and P. Grassini, "Operative conditions effect on PEM-FC performance by in-situ and ex-situ analysis of gas diffusion media with different bulk textile structure," *International Journal of Hydrogen Energy*, vol. 36, no. 13, pp. 8053–8062, 2011.

[30] X.-Z. R. Yuan, C. Song, H. Wang, and J. Zhang, *Electrochemical Impedance Spectroscopy in PEM Fuel Cells: Fundamentals and Applications*, pp. 1–420, Springer, London, UK, 2010.

[31] N. Wagner and E. Gülzow, "Change of electrochemical impedance spectra (EIS) with time during CO-poisoning of the Pt-anode in a membrane fuel cell," *Journal of Power Sources*, vol. 127, no. 1-2, pp. 341–347, 2004.

[32] R. Ferrero, G. Dotelli, P. Gallo Stampino, and S. Latorrata, "Discussion of critical measurement issues of impedance spectroscopy on PEM fuel cells," in *IEEE International Workshop on Applied Measurements for Power Systems, AMPS*, pp. 97–102, Aachen, Germany, 2012.

[33] S. Asghari, A. Mokmeli, and M. Samavati, "Study of PEM fuel cell performance by electrochemical impedance spectroscopy," *International Journal of Hydrogen Energy*, vol. 35, no. 17, pp. 9283–9290, 2010.

[34] P. G. Stampino, S. Latorrata, D. Molina, S. Turri, M. Levi, and G. Dotelli, "Investigation of hydrophobic treatments with perfluoropolyether derivatives of gas diffusion layers by electrochemical impedance spectroscopy in PEM-FC," *Solid State Ionics*, vol. 216, pp. 100–104, 2012.

[35] R. Ramasamy, E. Kumbur, M. Mench, W. Liu, D. Moore, and M. Murthy, "Investigation of macro- and micro-porous layer interaction in polymer electrolyte fuel cells," *International Journal of Hydrogen Energy*, vol. 33, no. 13, pp. 3351–3367, 2008.

[36] H. Kim and B. N. Popov, "Development of novel method for preparation of PEMFC electrodes," *Electrochemical and Solid-State Letters*, vol. 7, no. 4, pp. A71–A74, 2004.

[37] N. Wagner, "Characterization of membrane electrode assemblies in polymer electrolyte fuel cells using a.c. impedance spectroscopy," *Journal of Applied Electrochemistry*, vol. 32, no. 8, pp. 859–863, 2002.

[38] P. Gallo Stampino, L. Omati, C. Cristiani, and G. Dotelli, "Characterisation of nanocarbon-based gas diffusion media by electrochemical impedance spectroscopy," *Fuel Cells*, vol. 10, no. 2, pp. 270–277, 2010.

[39] J. Wu, X. Yuan, H. Wang, M. Blanco, J. Martin, and J. Zhang, "Diagnostic tools in PEM fuel cell research: part I electrochemical techniques," *International Journal of Hydrogen Energy*, vol. 33, no. 6, pp. 1735–1746, 2008.

[40] F. Alcaide, E. Brillas, and P.-L. Cabot, "EIS analysis of hydroperoxide ion generation in an uncatalyzed oxygen-diffusion cathode," *Journal of Electroanalytical Chemistry*, vol. 547, no. 1, pp. 61–73, 2003.

4

Real-Time Release Testing of Herbal Extract Powder by Near-Infrared Spectroscopy Considering the Uncertainty around Specification Limits

Guolin Shi,[1] Bing Xu,[1,2] Xin Wang,[3] Zhong Xue,[4] Xinyuan Shi,[1,2] and Yanjiang Qiao[1,2]

[1]*Research Center of TCM Information Engineering, Beijing University of Chinese Medicine, Beijing 100029, China*
[2]*Beijing Key Laboratory for Production Process Control and Quality Evaluation of Traditional Chinese Medicine, Beijing 100029, China*
[3]*TianJin Children's Hospital, TianJin 300204, China*
[4]*School of Pharmacy, Hebei Medical University, Shijiazhuang 050017, China*

Correspondence should be addressed to Bing Xu; xubing@bucm.edu.cn and Yanjiang Qiao; yjqiao@263.net

Academic Editor: Alessandra Durazzo

The concept of real-time release testing (RTRT) has recently been adopted by the production of pharmaceuticals in order to provide high-level guarantee of product quality. Process analytical technology (PAT) is an attractive and efficient way for realizing RTRT. In this paper, near-infrared (NIR) determination of *cryptotanshinone* and *tanshinoneII*$_A$ content in *tanshinone* extract powders was taken as the research object. The aim of NIR analysis is to reliably declare the extract product as compliant with its specification limits or not. First, the NIR quantification method was developed and the parameters of the multivariate calibration model were optimized. The reliable concentration ranges covering the specification limits of two APIs were successfully verified by the accuracy profile (AP) methodology. Then, with the designed validation data from AP, the unreliability graph as the decision tool was built. Innovatively, the β-content, γ-confidence tolerance intervals (β-CTIs) around the specification limits were estimated. During routine use, the boundary of β-CTIs could help decide whether the NIR prediction results are acceptable. The proposed method quantified the analysis risk near the specification limits and confirmed that the unreliable region was useful to release the product quality in a real-time way. Such release strategy could be extended for other PAT applications to improve the reliability of results.

1. Introduction

Radix Salvia Miltiorrhizae is the dried root of *Salvia miltiorrhiza* Bge [1]. It is widely used in several therapy systems for the treatment of angina pectoris, coronary heart disease and myocardial infarction, atherosclerosis, chronic renal failure, and liver fibrosis [2]. *Tanshinone* extract, the important components in Radix Salvia Miltiorrhizae, is listed in the Chinese Pharmacopoeia (ChP, 2015 edition). The *tanshinone* extract powders were generally manufactured using a series of batch operations, including extracting, filtering, concentrate, washing, drying, and milling.

Traditionally, the quality of *tanshinone* extract powder was assured by laboratory testing after the manufacturing was completed. And two active pharmaceutical ingredients (APIs), i.e., the *cryptotanshinone* and the *tanshinoneII*$_A$, were assayed by the HPLC method. However, the HPLC analysis is time-consuming and requires labor-intensive protocols including sample collection, sample pre-treatment, sample analysis, and data processing procedures. Besides, this conventional approach was conducted on limited samples and had been at risk in providing qualified products to the public.

Since the promulgation of the process analytical technology (PAT) guidance in September 2004 [3], the American Food and Drug Administration (FDA) has encouraged the pharmaceutical manufacturers to adopt new technologies in pharmaceutical process, mainly for designing, analyzing, and controlling manufacturing through timely measurements of critical quality and performance attributes of raw and in-process materials and processes with goal of ensuring final product quality. Real-time release testing (RTRT) [4], which is advocated to substitute the end-point testing, is the ability to evaluate and ensure the quality of in-process and/or final product based on process data, which typically include a valid combination of measured material attributes and process controls. Advances in quality by design (QbD) have shown that the application of RTRT in any stage of the manufacturing process and any type of finished product may provide greater assurance of product quality than finished product testing alone [5].

The rapid spectroscopy techniques are important part of RTRT plans. Near-infrared (NIR) spectroscopy has proven to be effective for both qualitative [6–8] and quantitative analysis [9–12] in the pharmaceutical industry due to its high efficiency, nondestructive nature, and capacity to measure both physical and chemical properties with minimal or no sample preparation [13]. It is more and more considered as an attractive and promising analytical tool for PAT. Recently, the NIR spectroscopy has been introduced into the Chinese herbal medicine production processes, such as the extraction process of Epimedium brevicornum Maxim [14] and Qizhiweitong granules [15], the alcohol precipitation process of Reduning injection [16], the purification process of Aesculi semen extracts [17], the enrichment process of Danhong injection [18], the fluidizing drying process of Poria cocos formula granules [19], and the mixing process [20]. By summarizing the NIR analysis results of Chinese medicine materials from the reported literatures, it was found that the relative standard errors of prediction (RSEP) were in the range of 1.51% to 10.41% [13–19] and the prediction error of some herbal samples exceeded the general target acceptance criteria for bulk drug (2%) or dosage form (10%) [21]. Therefore, the operable region and unreliability region of these analytical methods should be judged to confirm their scope of application.

Conventionally, the performance of the NIR analytical method was evaluated by chemometric indicators [22–25], such as the root-mean-square error (RMSE) [26], the correlation coefficient (r), the ratio of performance to deviation (RPD) [27], which only gave the average level of information about errors and bias of NIR method and did not provide the uncertainty of each individual prediction over the range of measurement [28]. Therefore, more and more researchers adopted the accuracy profile (AP) approach to evaluate the risk or confidence of the NIR method [29–32]. The core of the AP methodology, in agreement with the ICH Q2A guidance, is to use the β-expectation tolerance interval (β-ETI) to integrate the trueness, the intermediate precision coefficient variation, as well as the β chance for future results [33]. Based on the validation results from the AP approach, Rozet et al. [34] further identified the unreliable region around the specification limit by intersection of the upper and lower β-ETIs with the specification limit. Such methods were successfully applied for HPLC-UV quantification of (R)-timolol impurity in (S)-timolol drug substance and for NIRS quantification of acetaminophen in the uniformity of dosage units (UDU) test [34, 35].

As reported by Saffaj et al., the β-content, γ-confidence tolerance interval (β-CTI) could provide a better estimate of measurement risk than β-expectation tolerance interval and gave the best guarantee concerning the decision of declaring a method as valid and reliable [36–38]. Our previous work also revealed that the overall uncertainty estimated by the β-CTI from the total error (bias and standard deviation) was similar to the overall uncertainty assessed from validation data according to the trueness, precision, and robustness experiments [39]. In this work, NIR was used as a rapid detective tool to assay the APIs content of *tanshinone* extract powder. The traditional figures of merit were used to optimize the multivariate calibration model. A full factorial design generating the validation data was used to calculate statistical intervals. The β-content, γ-confidence tolerance interval was for the first time used to develop the unreliable region around the specification limit of *tanshinone* extract powder, in order to increase the confidence when releasing the multicomponents natural product in a real-time way.

2. Experimental

2.1. Reagents and Materials. The *tanshinone* extract powders were purchased from Xi'an Changyue Phytochemistry Co., Ltd (Xi'an, China, lot: 140420), Xi'an Honson Biotechnology Co., Ltd. (Xi'an, China, Lot: 141029.) and Shanxi Undersun Bimedtech Co., Ltd (Shanxi, China, Lot: Udst130507). The *cryptotanshinone* reference standard (lot number: 110852–200806) and *tanshinoneII*$_A$ reference standard (lot number: 110776–200619) were purchased from the National Institutes for Food and Drug Control (Beijing, China). The acetonitrile and phosphoric acid of HPLC grade were purchased from the Thermo Fisher Scientific Inc. (Massachusetts, USA), and pure water was purchased from Wahaha Co., Ltd. (Hangzhou, China).

2.2. Acquisition of Spectroscopic Data. The sample was held in a circular sample cuvette with a solid cap, and the NIR spectra were collected in the integrating sphere diffuse reflectance mode with the Antaris Nicolet FT-NIR system (Thermo Fisher Scientific Inc., USA) at ambient temperature. Each spectrum was the average of 64 scans with 8 cm^{-1} resolution. The range of spectra was from 10000 to 4000 cm^{-1}. The background spectrum was taken daily in air.

2.3. Reference Method. The reference method used for *cryptotanshinone* and *tanshinoneII*$_A$ determination was HPLC assay recommended by the Chinese Pharmacopoeia (2015 Edition) for the extract of *Salvia miltiorrhiza* Bge. Firstly, samples were dissolved by methanol properly after NIR scanning. Then, the solution was filtered through a Millipore membrane filter with an average pore diameter of 0.45 μm. Finally, 10 μL of filtrate was injected into the HPLC system for analysis.

An Agilent 1100 series HPLC apparatus, equipped with a quaternary solvent delivery system, an auto sampler, a DAD detector, and HP workstation for data processing were used. The concentration of *cryptotanshinone* and *tanshinoneII*$_A$

were analyzed by reverse-phase chromatography on an Agilent XDB C18 column (4.6 × 250 mm, 5 μm) with gradient. The mobile phase A is acetonitrile, and the mobile phase B is phosphoric acid water (0.026%). The elution procedures are as follows: 0~25 min, 60%~90% A; 25~30 min, maintaining 90% A; 30~31 min, 90%~60% A; 31~40 min, 60%~60% A. The column temperature was 25°C, the flow rate was 1.2 mL·min^{-1}, and the detection wavelength at 270 nm was set.

2.4. Calibration and Validation Protocols. The experimental protocols were created for both calibration and validation sets in order to obtain a robust model. A total of 103 samples were collected in the calibration set. Four grams of *tanshinone* extract powder sample was weighed and then directly measured by NIR under the conditions specified in Section 2.2.

The external validation set was built with the same method as the calibration set. The validation protocol used the "8 × 5 × 3" full factorial experimental design. Eight different concentration levels of *cryptotanshinone*, i.e., 0.20%, 0.31% 0.50%, 1.18%, 2.05%, 2.26%, 5.29%, and 9.68%, were investigated. Eight different concentration levels of *tanshinoneII$_A$*, i.e., 0.10%, 0.15%, 0.36%, 0.54%, 2.04%, 6.03%, 18.76%, and 27.64%, were investigated. Each concentration level was performed in 5 replicates on 3 different days, resulting in 120 samples in the validation set for both two components. Moreover, all validation samples were from different batches of *tanshinone* extract powders to test the robustness of the NIR model.

2.5. NIR Method Development. To perform the quantitative determination of *cryptotanshinone* and *tanshinoneII$_A$* content in *tanshinone* extract powders, the partial least squares (PLS) regression was applied for the sake of linking the NIR spectra with the reference values analyzed by the HPLC method [39]. In order to improve the performance of the PLS model, a variety of spectroscopic data pretreatment methods were investigated to extract the useful information. For example, the first-order derivatives (1std) [40], the second-order derivatives (2ndd) [41] could be used to remove the baseline drift and decrease the overlapping. The multiplicative scatter correction (MSC) [42, 43] and the standard normal variate transformation (SNV) [44] could reduce the light scattering effects. The wavelet denosing of spectra (WDS) [45] and the Savitzky–Golay (SG) smoothing [46] can effectively eliminate the noise.

During the NIR method development process, correlation coefficients *r* for both the calibration and validation sets, the root-mean-square error of calibration (RMSEC), the root-mean-square error of cross-validation (RMSECV), the root–mean-square error of prediction (RMSEP), and RPD were used to evaluate and select the best NIR calibration model. The optimal latent variables (LVs) used to build PLS model were selected according to comprehensive consideration of RMSEC, RMSECV, RMSEP, and cumulative prediction error sum of square (PRESS) values.

2.6. The Unreliability Graph. The unreliability graph as a decision making tool is a 2D-graphical representation of tolerance intervals aiming at helping the analyst to decide whether an analytical result is reliable or not. For details about the theory, the authors are recommended to refer to the published literatures [36, 47].

2.6.1. Estimation of the β-Content, γ-Confidence Tolerance Interval. The "$I \times J \times K$" full factorial validation protocol was designed to obtain the validation dataset, where the effect of three aspects, i.e., conditions (I, $i = 1, 2, \ldots, m$), the number of repetitions (J, $j = 1, 2, \ldots, n$), and the level of concentrations (K, $k = 1, 2, \ldots, a$), were taken into account [48]. The β-content, γ-confidence tolerance interval can be expressed by the following formula:

$$P_{\overline{Y},\widehat{\sigma}_b^2,\widehat{\sigma}_e^2}\left(P_Y\left[\overline{Y} - k\widehat{\sigma}_M < Y < \overline{Y} + k\widehat{\sigma}_M \mid \overline{Y}, \widehat{\sigma}_M\right] \geq \beta\right) = \gamma. \quad (1)$$

In equation (1), $\overline{Y}$ is the average value of the results at each concentration level K; $\widehat{\sigma}_M$, $\widehat{\sigma}_b^2$, and $\widehat{\sigma}_e^2$, respectively means the intermediate precision, the interseries, and the intraseries variances; k denotes the coverage factor and is related to β and γ.

Mee's approach is used for estimating the β-content, γ-confidence tolerance interval as follows [36, 49]:

$$[L, U] = \left[\overline{Y} - k_C\widehat{\sigma}_M; \overline{Y} + k_C\widehat{\sigma}_M\right], \quad (2)$$

$$\text{with } k_C = \sqrt{\frac{\nu' \chi^2_{1;\beta}(\tau)}{\chi^2_{\nu';1-\gamma}}}, \quad (3)$$

$$\nu' = \frac{(R' + 1)^2}{\left((R' + (1/n))^2/(m-1)\right) + ((1 - (1/n))/mn)}, \quad (4)$$

$$\tau = \frac{1}{mnB'}, \quad (5)$$

$$B' = \frac{R' + 1}{nR' + 1}, \quad (6)$$

$$R' = \max\left[0, \ \frac{1}{n}\left(\frac{F}{F_\eta} - 1\right)\right], \quad (7)$$

where the lower (L) and upper (U) limits denote a specified proportion β of measured results that will fall within the interval $[L, U]$ at specified confidence level. k_C is an approximation to k. $\chi^2_{1;\beta}(\tau)$ denotes the β quantile of a noncentral chi-square distribution under the freedom degree of 1. τ means the noncentral parameter. $\chi^2_{\nu';1-\gamma}$ with degrees of freedom ν' denotes the $1-\gamma$ quantile of a noncentral chi-square distribution. And, F denotes the mean square ratio MS_B/MS_E. MS_B and MS_E, respectively, denote the mean square of the interseries and the intraseries variances. Under degree freedom $\nu_1 = m(n-1)$ and $\nu_1 = m - 1$, F_η is the 100η percentile of an F distribution. The recommended values of

η are 0.85, 0.905, and 0.975, which are corresponding to 0.90, 0.95, and 0.99 for γ, respectively [47].

The β-content, γ-confidence tolerance interval could also be written in a relative form:

$$[L(\%), U(\%)] = [\text{bias}(\%) - k_C \text{RSD}(\%), \text{bias}(\%) + k_C \text{RSD}(\%)],$$
$$\text{RSD}(\%) = \frac{\hat{\sigma}_{\mathcal{N}}}{\text{X}_\text{T}} \times 100, \tag{8}$$

where Xr is the theoretical value.

2.6.2. Establishment of the Accuracy Profile. In order to globally validate accuracy and robustness of the NIR quantitative method, the accuracy profile is developed as follows [50]:

(1) Set acceptance limits ± 20% for natural product in this paper
(2) Calculate the β-content, γ-confidence relative tolerance intervals [L (%), U (%)] for each concentration level based on equation (8) at a desired confidence level γ
(3) Construct a 2D-accuracy graph with the horizontal axis for the validation standards concentration and vertical axis for the relative tolerance interval limits [L (%), U (%)] and accuracy
(4) If [L (%), U (%)] at given concentration levels are within acceptance limits (±20%), it demonstrates that the developed method is accuracy and robustness; otherwise, the method cannot be accepted

2.6.3. Establishment of the Unreliability Graph. The unreliability graph was used as a decision making tool to increase the confidence of real-time release testing at the specification limit. The procedures for developing the unreliability graph are as follows [35, 39]:

(1) Set the specification limit (λ) according to the requirement.
(2) Calculate the [L, U] for each concentration level based on equation (2) at the desired confidence level γ.
(3) Develop a 2D graph with the horizontal axis for the validation standards concentrations and vertical axis for the observed concentrations.
(4) Locate the tolerance limits L and U for each validation concentration on the 2D graph. The tolerance limit L at each concentration level was connected into a broken line in turn. The same procedures were also performed on tolerance limits U.
(5) Make a straight line perpendicular to the horizontal axis at the specification limit (λ). The intersections of the lower and upper tolerance interval with the specification limit line are defined as L_{API} and U_{API}, respectively.
(6) Make two straight lines parallel to horizontal axis through the intersections. The area between the two parallel straight lines is the unreliability region around the specification limits.
(7) If the analytical results exceed the U_{API}, the target product can be immediately released; otherwise, it cannot be directly released.

2.7. Software. The SIMCA-P 11.5 (Umetrics, US) and Unscrambler 7.0 (CAMO, Norway) software were used to perform spectral pretreatments. The Matlab 7.0 (Mathwork, USA) with PLS Toolbox 2.1 (Eigenvector Research Inc., USA) was used to carry out the PLS regression. For calculation of the β-content, γ-confidence tolerance intervals, the Matlab codes were referred to [47].

3. Results and Discussion

3.1. NIR Method Development. In this study, raw spectra of 103 samples were obtained by NIR scanning of the extract powders, as shown in. Figure 1. It was difficult to observe the differences from the original spectra because the wave bands were seriously overlapped. Partial least square, one of the most commonly used chemometric methods, was applied to handle the overdetermined problem in the calibration process. And, the PLS1 algorithm was used predict the concentrations of each API in *tanshinone* extract. Before ascertaining the structure and finetuning the coefficients of PLS model, the Kennard–Stone (K-S) algorithm [51] was used to split the original data set into a calibration set (75 samples) and a test set (28 samples).

Then, various data preprocessing methods in Section 2.5 were used to extract useful information from the noisy spectral data. Tables 1 and 2 show the PLS modeling results in both calibration and prediction of *cryptotanshinone* and *tanshinoneII*$_A$ content, respectively. The PLS model based on the second-order derivative NIR absorption spectrum has the optimal results, where the RMSECV and RMSEP values were smallest and the RPD values were highest. This revealed that the robustness of the quantitative models with 2ndd pretreatment was satisfactory and the models had excellent predictive ability. Figure 2 shows the full spectra of all samples through 2ndd preprocessing method. It can be seen that this method can obviously eliminate the baseline drift, remove the background interference, and distinguish the absorption peaks. The characteristic absorption waveband was from 6400 cm^{-1} to 4000 cm^{-1}. Thereby, 2ndd was chosen as the preprocessing method.

The number of latent variables (LVs) was an important parameter and could directly affect the accuracy of the model. LVs listed in Tables 1 and 2 are optimized by the leave-one-out cross-validation. As can be seen from Figure 3, the RMSE and cumulative PRESS values gradually decreased and eventually did not change as the number of latent variables increased. Consequently, 10 LVs and 8 LVs were, respectively, used to establish the PLS models for *cryptotanshinone* and *tanshinoneII*$_A$, respectively. The RPD value in prediction of *tanshinoneII*$_A$ was 15.6, which was larger than that in prediction of *cryptotanshinone* (RPD = 5.3). The reason may be that the standard deviation of *tanshinoneII*$_A$ content in the test set was higher than that of *cryptotanshinone*.

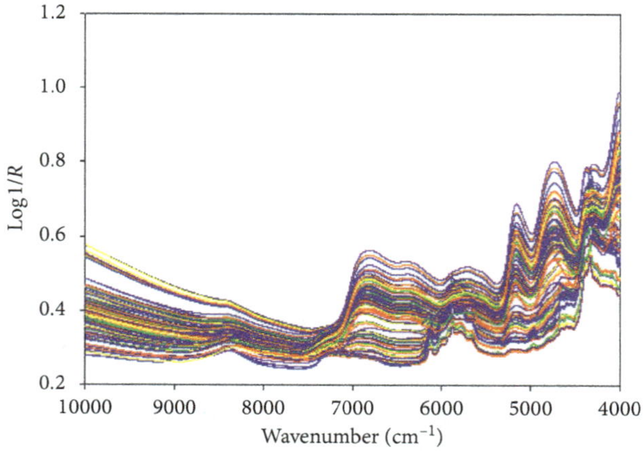

FIGURE 1: Raw NIR spectra of 103 samples.

TABLE 1: PLS model characteristics for *cryptotanshinone*.

Pretreatment	LVs	Calibration set				Validation set			
		r_{cal}	RMSEC	RMSECV	$BIAS_{cal}$	r_{val}	RMSEP	RPD	$BIAS_{val}$
Origin	11	0.9821	0.470	0.640	0.380	0.9712	0.5835	4.3	0.440
1std	10	0.990	0.344	0.557	0.261	0.9737	0.528	4.7	0.376
2ndd	10	0.9992	0.102	0.422	0.084	0.9818	0.466	5.3	0.341
S-G	11	0.9819	0.472	0.642	0.381	0.9709	0.587	4.2	0.441
MSC	12	0.9872	0.397	0.708	0.305	0.9761	0.536	4.6	0.399
SNV	12	0.9860	0.386	0.664	0.325	0.9713	0.583	4.3	0.420
WDS	11	0.9960	0.644	0.912	0.535	0.9516	0.755	3.3	0.623

TABLE 2: PLS model characteristics for *tanshinoneII$_A$*.

Pretreatment	LVs	Calibration set				Validation set			
		r_{cal}	RMSEC	RMSECV	$BIAS_{cal}$	r_{val}	RMSEP	RPD	$BIAS_{val}$
Origin	12	0.9840	1.322	2.012	0.999	0.9966	1.288	11.4	1.035
1std	10	0.9969	0.947	1.849	0.696	0.9970	1.141	12.8	0.807
2ndd	8	0.9991	0.512	1.775	0.376	0.9981	0.934	15.6	0.744
S-G	12	0.9938	1.338	2.040	1.012	0.9965	1.312	11.2	1.052
MSC	12	0.9970	0.938	2.023	0.739	0.9974	1.087	13.5	0.849
SNV	11	0.9973	0.891	1.674	0.622	0.9979	1.009	14.5	0.553
WDS	12	0.9896	1.736	2.478	1.292	0.9925	1.947	7.5	1.494

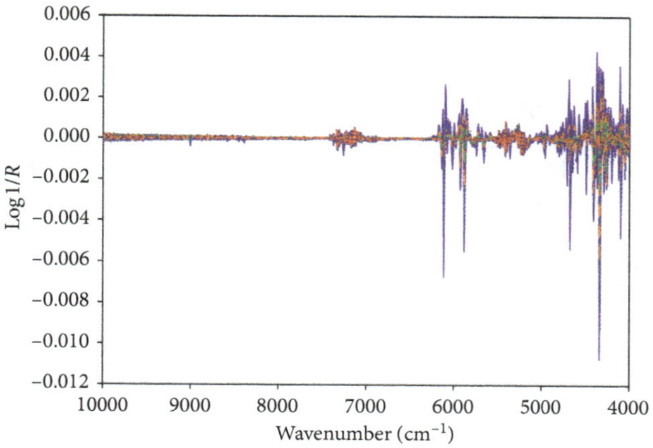

FIGURE 2: NIR spectra with the 2ndd preprocessing method.

3.2. NIR Method Validation. According to the ICH Points to consider (R2) document [52], the PLS quantitative model could be classified into high impact model, since the APIs content of *tanshinone* extract powder predicted by the multivariate calibration models were key indicators for quality control. Consequently, it is necessary to implement method validation to ensure the accuracy and robustness of the quantitative model. After 120 validation samples were prepared according to validation protocols illustrated in Section 2.4, the concentrations of *cryptotanshinone* and *tanshinoneII$_A$* were predicted by the developed NIR calibration model and are listed in appendix Tables S1 and S2, respectively.

The validation statistics for NIR quantitative method are shown in Tables 3 and 4 for *cryptotanshinone* and *tanshinoneII$_A$*, respectively. Taking the validation results of *cryptotanshinone* for example, the range of concentration

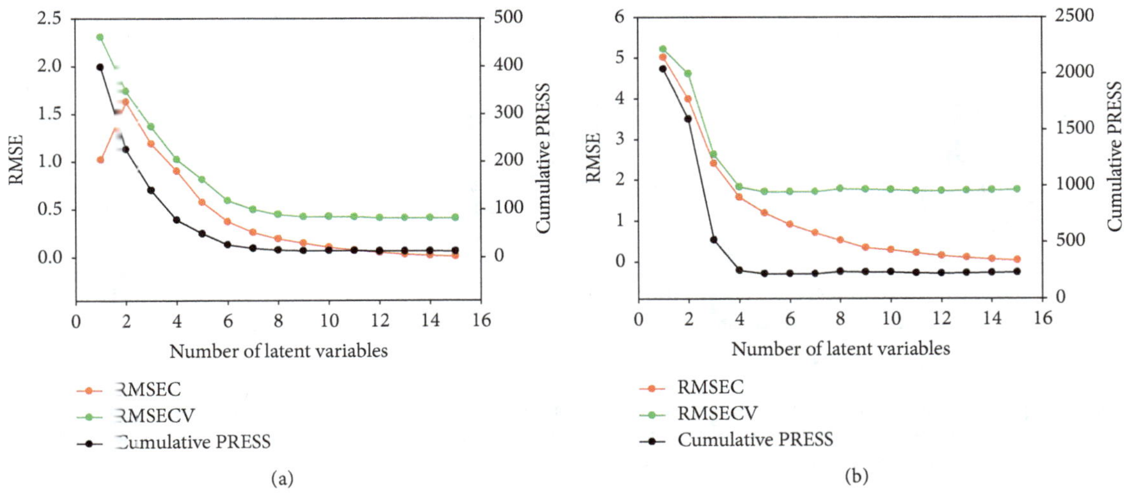

FIGURE 3: Calibration characteristics vs. number of latent factors. (a) The *cryptotanshinone* model; (b) the *tanshinoneII$_A$* model.

TABLE 3: The validation results for NIR determination of *cryptotanshinone*.

Theoretical concentration (%)	Mean calculated content (%)	Trueness		Precision	
		Relative bias (%)	Recovery (%)	Repeatability (%)	Intermediate precision (%)
0.20	0.40	100.00	200.00	73.58	92.85
0.31	0.46	46.88	146.88	46.34	52.73
0.50	0.69	38.93	138.93	29.16	35.68
1.18	0.86	−26.89	73.11	26.98	26.98
2.05	2.06	0.33	100.33	7.67	7.97
2.26	2.48	9.72	109.72	9.24	9.97
5.29	4.92	−6.99	93.01	3.48	4.28
9.68	9.51	−1.72	98.28	2.99	3.06

TABLE 4: The validation results of NIR determination of *tanshinoneII$_A$*.

Theoretical concentration (%)	Mean calculated content (%)	Trueness		Precision	
		Relative bias (%)	Recovery (%)	Repeatability (%)	Intermediate precision (%)
0.10	−0.20	−294.70	−194.70	390.80	390.80
0.15	−0.13	−184.44	−84.44	250.76	250.76
0.36	0.24	−32.96	67.04	87.85	93.70
0.54	0.85	58.02	158.02	47.40	47.40
2.04	1.60	−21.67	78.33	16.04	17.95
6.03	5.95	−1.34	98.66	6.37	6.37
18.76	18.34	−2.26	97.74	3.31	3.36
27.64	26.45	−4.30	95.70	1.63	1.91

studied in Table 3 can be divided into two parts. Part 1 and part 2 were the concentration range of [0.20–1.18]% and [2.05–9.68]%, respectively. In the first part, variances of trueness and precision were exceptionally severe, indicating that the precision and the accuracy of analytical method were anomalous. In part 2, it can be seen that the recovery was closed to 100% and the relative bias was small, indicating that the precision and accuracy of the quantitative model were acceptable within this range. With the same analysis procedures for *tanshinoneII$_A$*, it was easy to draw a conclusion that the contents of *tanshinoneII$_A$* within [0.10–2.04]% cannot be accurately detected since the precision and accuracy were anomalous. By contrast, the analytical method can accurately determine the *tanshinoneII$_A$* in the range [6.03–27.64]%. In conclusion, the NIR method can be used for the determination of *cryptotanshinone* and *tanshinoneII$_A$* both in the second parts of the concentration range.

According to Section 2.6.2, the accuracy profile (AP) was used to globally assess the NIR quantitative analysis method, as shown in Figure 4. The $\beta = 66.7\%$ and $\gamma = 90\%$ suggested by Saffaj's were applied to estimate the β-content, γ-confidence tolerance interval [36], and the results were listed in Tables 5 and 6. It was clearly seen from Figure 4(a) that the relative tolerance intervals for the first 4 concentrations visibly came out of the two acceptance limits, revealing that the contents within [0.20–1.18]% cannot be accurately

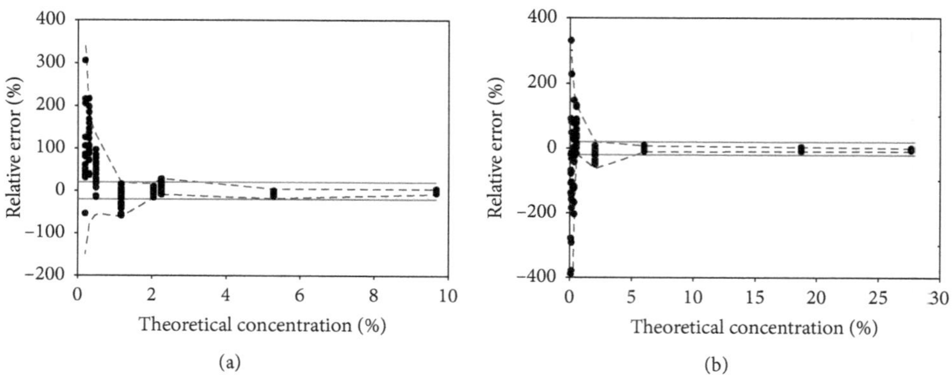

FIGURE 4: Accuracy profile for NIR quantitative methods. (a) Quantification of *cryptotanshinone* content; (b) quantification of *tanshinone*II_A content. The medium blue dashed lines are the 66.7% β-content, 90% γ-confidence tolerance intervals and the red lines are the acceptance limits (±20%); the black point at each concentration level represents the relative bias for each predictive value.

measured. Although at content of 2.26% the relative tolerance intervals slightly exceeded upper acceptance limit, the most contents measured by calibration model were acceptable and the recovery in this concentration was no more than 110%. It indicated that the PLS model could determine the contents of *cryptotanshinone* in this concentration level. However, the tolerance intervals of the last 4 concentration levels fell within the acceptance limits. Consequently, the method was considered to be valid in the concentration range within [2.05–9.68]%. By using the same analysis method, a conclusion can be clearly drawn that the NIR method was valid only for the last 3 concentration levels in detecting the *tanshinone*II_A, as shown in Figure 4(b). These findings are consistent with the validation results above.

TABLE 5: The β-content, γ-confidence tolerance intervals estimated for different concentration levels of *cryptotanshinone*.

Concentration level (%)	L	U	L (%)	U (%)
0.20	−0.097	0.895	−149.64	348.98
0.31	0.062	0.847	−79.99	173.75
0.50	0.23	1.15	−54.29	132.15
1.18	0.45	1.40	−61.29	18.81
2.05	1.72	2.39	−16.10	16.75
2.26	1.98	2.98	−12.33	31.68
5.29	4.33	5.51	−18.21	4.23
9.68	8.93	10.10	−7.80	4.35

3.3. Real-Time Release Testing for Tanshinone Extract Powders

3.3.1. The Specification Limit. According to the ChP (2015 Edition) [1], the minimum mass content of *cryptotanshinone* and *tanshinone*II_A in *tanshinone* extract powders are 2.1% and 9.8%, respectively. The specification limits were well located within the reliable concentration regions in Section 3.2, indicating that the developed NIR analysis method can be used to release the *tanshinone* extract powders in real time.

TABLE 6: The β-content, γ-confidence tolerance intervals estimated for different concentration levels of *tanshinone*II_A.

Concentration level (%)	L	U	L (%)	U (%)
0.10	−0.76	0.38	−864.48	275.14
0.15	−1.15	0.79	−864.79	423.90
0.36	−0.49	0.98	−236.06	170.13
0.54	0.45	1.26	−16.77	132.82
2.04	0.74	2.46	−63.86	20.52
6.03	5.39	6.50	−10.51	7.84
18.76	17.11	19.56	−8.80	4.27
27.64	25.14	27.77	−9.06	0.46

3.3.2. Unreliability Graph Development. By, respectively, connecting the tolerance limits *L* and *U* listed in Tables 5 and 6 at each concentration level, the unreliability graph could be drawn. Figures 5(a) and 6(a) show the unreliability graphs for *cryptotanshinone* and *tanshinone*II_A at full concentration ranges, respectively. The β-content, γ-confidence tolerance interval at the specification limit was then estimated. Taking *cryptotanshinone* for instance, a line passing through two points, i.e., the upper tolerance limits at 2.39% and 2.98%, was dropped, and the linear function was regressed as $y = 2.810 \times x - 3.370$ (x and y mean validation and observed concentration, respectively). The specification limit of *cryptotanshinone* was substituted into this function, and U_{API} of *cryptotanshinone* were calculated to be 2.53%. Then, another line passing through the lower tolerance limits at 1.72% and 1.98% was regressed as $y = 1.238 \times x - 0.818$. 2.1% was substituted into this function, and L_{API} of *cryptotanshinone* were calculated to be 1.78%. Similarly, as for *tanshinone*II_A, the tolerance interval around 9.8% was estimated to be from 8.86% to 10.37%. Figures 5(b) and 6(b) show the unreliable regions for *cryptotanshinone* and *tanshinone*II_A around the specification limits, respectively.

3.3.3. Real-Time Release Testing. The unreliable region around the specification limit can be seen as risk region or

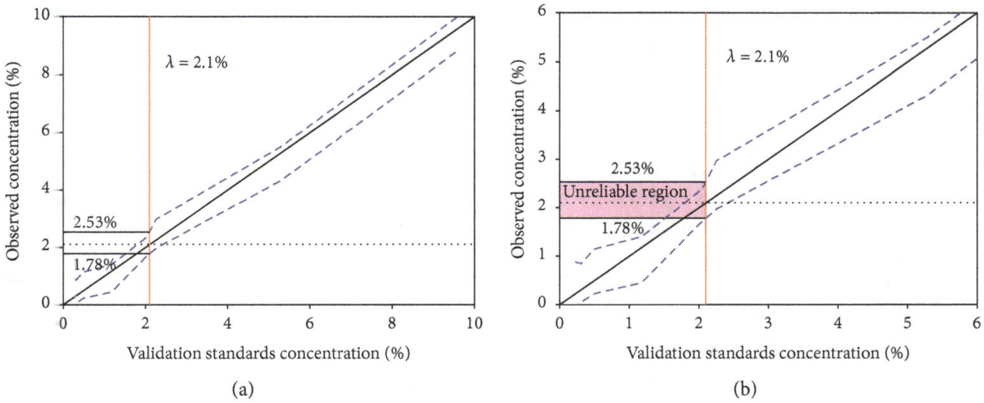

FIGURE 5: (a) The unreliability graph of *cryptotanshinone*. (b) The unreliable region estimated around the specification limit of *cryptotanshinone* in *tanshinone* extract. Blue dashed lines are the 66.7% β-content, 90% γ-confidence tolerance intervals; the diagonal continuous line is the identity line $y = x$; the red vertical straight line and the black horizontal dashed line are the API specification limits; the shaded region corresponds to the unreliable region.

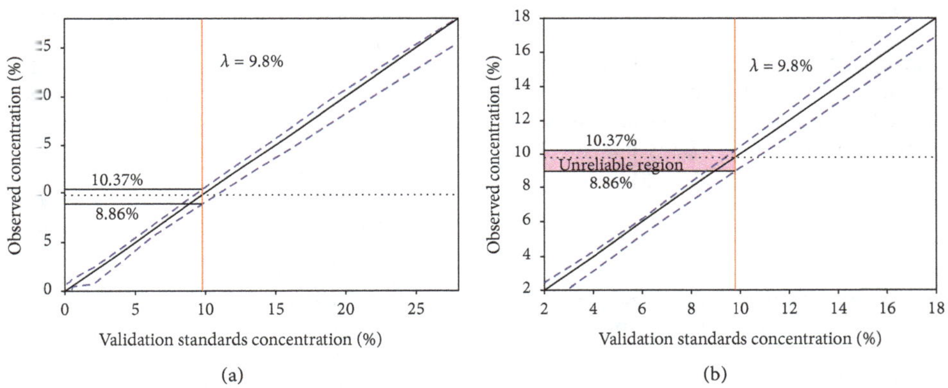

FIGURE 6: (a) The unreliability graph of *tanshinoneII*$_A$. (b) The unreliable region estimated around the specification limit of *tanshinoneII*$_A$ in *tanshinone* extract. Blue dashed lines are the 66.7% β-content, 90% γ-confidence tolerance intervals; the diagonal continuous line is the identity line $y = x$; the red vertical straight line and the black horizontal dashed line are the API specification limits; the shaded region corresponds to the unreliable region.

guard-banding, since there was a certain risk in the results of NIR quantitative analysis. During routine use, if the NIR analysis result is larger than the upper limit of the built unreliable region, it is assured that the content would satisfy the specification. And, for the analysis result within the unreliable region, it cannot accurately determine whether the content meets the specification or not. If the NIR analysis result is located under the lower limit of the unreliable region, it absolutely does not meet the target requirements.

For real-time release testing of *tanshinone* extract powders by NIR analytical method, the release standard was that the contents of *cryptotanshinone* and *tanshinoneII*$_A$ must be no less than 2.53% and 10.37%, respectively. Only in this way, the *tanshinone* extract powder can be directly released to the next pharmaceutical manufacture units or to markets. Otherwise, the *tanshinone* extract powders cannot be released. The NIR spectroscopy combined with the unreliability graph significantly increases the confidence about the compliance of the product in a real-time way.

4. Conclusions

In this paper, a new release strategy based on the unreliability graph methodology which incorporated the β-content, γ-confidence tolerance intervals, has successfully been achieved. Firstly, the *cryptotanshinone* and *tanshinoneII*$_A$ content in *tanshinone* extract powders were rapidly detected by NIR using the PLS quantitative model. And secondly, this quantitative model was validated by the accuracy profile. The NIR methods can accurately determine the *cryptotanshinone* in the range [2.05–9.68]% and the *tanshinoneII*$_A$ in the range [6.03–27.64]%. Finally, the release strategy with NIR quantitative model based on the unreliability graph was applied to real-time release test of *tanshinone* extract powders. The proposed approach offered a formal statistical framework to show when the analytical methods will provide daily results that can be used efficiently to make adequate decisions. Besides, the release strategy proposed can be applied to any quantitative analytical method and provide greater assurance for the quality of the final products, to achieve the purpose of real-time release.

Additional Points

(1) A release strategy was proposed to determine whether the analytical results were reliable in real-time release testing (RTRT). (2) The β-content, γ-confidence tolerance intervals were applied to establish the unreliability graph as the decision tool. (3) The new release strategy can be used for quality control of the complex system of Chinese medicine product.

Conflicts of Interest

There are no conflicts of interests regarding the publication of this manuscript.

Acknowledgments

The authors are thankful to the research funding supports from the National Science and Technology Major Projects (No. 2018ZX09201011-006, China) and Scientific Research Project of Beijing University of Chinese Medicine (No. 2019-JYB-JS-015).

References

[1] Chinese Pharmacopoeia Commission, *Chinese Pharmacopoeia 2015 Edition*, Beijing Chemical industry press, Beijing Shi, China, 2015.

[2] J. Luo, W. Song, G. Yang, H. Xu, and K. Chen, "Compound Danshen (Salvia miltiorrhiza) dripping pill for coronary heart disease: an overview of systematic reviews," *American journal of Chinese medicine*, vol. 43, pp. 25–43, 2015.

[3] Food and Drug Administration, *Guidance for Industry Guidance for Industry PAT–a Framework for Innovative Pharmaceutical Development, Manufacturing and Quality Assurance*, US Department of Health and Human Services, Food and Drug Administration, Center for Biologics Evaluation and Research, Rockville Google Scholar, Washington, DC, USA, 2004.

[4] European Commission, *EU Guidelines to Good Manufacturing Practice Medicinal Products for Human and Veterinary Use*, European Commission, Brussels, Belgium, 2009.

[5] H. F. Xu, X. Zhang, and Y. L. Ma, *Real time release testing extends to the overall manufacturing process*, FDA, Silver Spring, MD, USA, 2004.

[6] M. Blanco and I. Villarroya, "NIR spectroscopy: a rapid-response analytical tool," *TrAC Trends in Analytical Chemistry*, vol. 21, no. 4, pp. 240–250, 2002.

[7] H. Yang, J. Irudayaraj, and M. Paradkar, "Discriminant analysis of edible oils and fats by FTIR, FT-NIR and FT-Raman spectroscopy," *Food Chemistry*, vol. 93, no. 1, pp. 25–32, 2005.

[8] Y. Roggo, L. Duponchel, and J.-P. Huvenne, "Comparison of supervised pattern recognition methods with McNemar's statistical test," *Analytica Chimica Acta*, vol. 477, no. 2, pp. 187–200, 2003.

[9] M. Blanco and M. Alcalá, "Content uniformity and tablet hardness testing of intact pharmaceutical tablets by near infrared spectroscopy," *Analytica Chimica Acta*, vol. 557, no. 1-2, pp. 353–359, 2006.

[10] P. R. Wahl, G. Fruhmann, S. Sacher, G. Straka, S. Sowinski, and J. G. Khinast, "PAT for tableting: inline monitoring of API and excipients via NIR spectroscopy," *European Journal of Pharmaceutics and Biopharmaceutics*, vol. 87, no. 2, pp. 271–278, 2014.

[11] M. Blanco, M. Alcalá, J. M. González, and E. Torras, "A process analytical technology approach based on near infrared spectroscopy: tablet hardness, content uniformity, and dissolution test measurements of intact tablets," *Journal of Pharmaceutical Sciences*, vol. 95, no. 10, pp. 2137–2144, 2006.

[12] M. Otsuka and I. Yamane, "Prediction of tablet properties based on near infrared spectra of raw mixed powders by chemometrics: scale-up factor of blending and tableting processes," *Journal of Pharmaceutical Sciences*, vol. 98, no. 11, pp. 4296–4305, 2009.

[13] M. Jamrógiewicz, "Application of the near-infrared spectroscopy in the pharmaceutical technology," *Journal of Pharmaceutical and Biomedical Analysis*, vol. 66, pp. 1–10, 2012.

[14] X. U. Ding-Zhou, L. I. Jing, C. H. Liu, L. Yang, L. I. Yu, and M. Zhong, "Application of near-infrared spectroscopy technology in online detection of Epimedium brevicornum Maxim extraction process," *China Modern Medicine*, vol. 23, pp. 4–6, 2014.

[15] X. S. Meng, Y. Tan, Y. F. Zhang et al., *Method and Application for Rapid Detection of the Water Extraction Process of Qizhi weitong Granules by Near Infrared Spectroscopy*, CN106383096A, 2017.

[16] Y. X. Wang, H. J. Mi, C. L. Zhang et al., "Near infrared spectroscopy on-line and real-time monitoring of alcohol precipitation process of reduning injection," *China Journal of Chinese Materia Medica*, vol. 39, pp. 4608–4614, 2014.

[17] Y. L. Xue, A. X. Zuo, X. S. Liu et al., *An Automatic Judgment and Control Method in the Process of Extracting Liquid Receiving Eluent Aesculi*, CN103203123B, 2013.

[18] Y. Jin, Y. J. Wu, and X. S. Liu, *Online Detection Method for Double Effect Concentration Process of Danhong Injection*, CN103913433A, 2014.

[19] J. W. Zhang, Y. Y. Zhang, Y. Liu, M. Jia, and G. L. Zhao, *NIR On-Line Detection Method for Paeoniflorin Content in Extract of Paeonia Lactiflora Pall*, CN102058682A, 2011.

[20] H. Y. Chen, L. I. Qiong-Ya, J. L. Chen, L. J. Luan, and Y. X. Dai, "End point judgment of blending process of zhengtian pills by near infrared spectroscopy," *Chinese Journal of Experimental Traditional Medical Formulae*, vol. 22, pp. 13–16, 2016.

[21] V. P. Shah, K. K. Midha, S. Dighe et al., "Analytical methods validation: bioavailability, bioequivalence, and pharmacokinetic studies," *Journal of Pharmaceutical Sciences*, vol. 81, no. 3, pp. 309–312, 1992.

[22] M. Blanco and A. Peguero, "Analysis of pharmaceuticals by NIR spectroscopy without a reference method," *TrAC Trends in Analytical Chemistry*, vol. 29, no. 10, pp. 1127–1136, 2010.

[23] C. V. Liew, A. D. Karande, and P. W. S. Heng, "In-line quantification of drug and excipients in cohesive powder blends by near infrared spectroscopy," *International Journal of Pharmaceutics*, vol. 386, no. 1-2, pp. 138–148, 2010.

[24] H. Grohganz, D. Gildemyn, E. Skibsted, J. M. Flink, and J. Rantanen, "Towards a robust water content determination

of freeze-dried samples by near-infrared spectroscopy," *Analytica Chimica Acta*, vol. 676, no. 1-2, pp. 34–40, 2010.

[25] C. C. Corredor, D. Bu, and D. Both, "Comparison of near infrared and microwave resonance sensors for at-line moisture determination in powders and tablets," *Analytica Chimica Acta*, vol. 696, no. 1-2, pp. 84–93, 2011.

[26] L. Norgaard, A. Saudland, J. Wagner, J. P. Nielsen, L. Munck, and S. B. Engelsen, "Interval partial least-squares regression (iPLS): a comparative chemometric study with an example from near-infrared spectroscopy," *Applied Spectroscopy*, vol. 54, no. 3, pp. 413–419, 2000.

[27] R. Leardi and L. Nørgaard, "Sequential application of backward interval partial least squares and genetic algorithms for the selection of relevant spectral regions," *Journal of Chemometrics*, vol. 18, no. 11, pp. 486–497, 2005.

[28] A. C. Olivieri, N. M. Faber, J. Ferré, R. Boqué, J. H. Kalivas, and H. Mark, "Uncertainty estimation and figures of merit for multivariate calibration (IUPAC Technical Report)," *Pure and Applied Chemistry*, vol. 78, no. 3, pp. 633–661, 2006.

[29] P. Hubert, J.-J. Nguyen-Huu, B. Boulanger et al., "Harmonization of strategies for the validation of quantitative analytical procedures," *Journal of Pharmaceutical and Biomedical Analysis*, vol. 36, no. 3, pp. 579–586, 2004.

[30] Z. Wu, B. Xu, M. Du, C. Sui, X. Shi, and Y. Qiao, "Validation of a NIR quantification method for the determination of chlorogenic acid in *Lonicera japonica* solution in ethanol precipitation process," *Journal of Pharmaceutical and Biomedical Analysis* vol. 62, pp. 1–6, 2012.

[31] Y.-C. Feng and C.-Q. Hu, "Construction of universal quantitative models for determination of roxithromycin and erythromycin ethylsuccinate in tablets from different manufacturers using near infrared reflectance spectroscopy," *Journal of Pharmaceutical and Biomedical Analysis*, vol. 41, no. 2, pp. 373–384, 2006.

[32] J. Mantanus, E. Rozet, K. Van Butsele et al., "Near infrared and Raman spectroscopy as Process Analytical Technology tools for the manufacturing of silicone-based drug reservoirs," *Analytica Chimica Acta*, vol. 699, no. 1, pp. 96–106, 2011.

[33] P. Hubert, J.-J. Nguyen-Huu, B. Boulanger et al., "Harmonization of strategies for the validation of quantitative analytical procedures," *Journal of Pharmaceutical and Biomedical Analysis*, vol. 45, no. 1, pp. 82–96, 2007.

[34] E. Rozet, E. Ziemons, R. D. Marini, B. Boulanger, and P. Hubert, "Quality by design compliant analytical method validation," *Analytical Chemistry*, vol. 84, no. 1, pp. 106–112, 2011.

[35] E. Rozet, E. Ziemons, R. D. Marini, B. Boulanger, and P. Hubert, "Methodology for the validation of analytical methods involved in uniformity of dosage units tests," *Analytica Chimica Acta*, vol. 760, pp. 46–52, 2013.

[36] T. Saffaj and B. Ihssane, "Uncertainty profiles for the validation of analytical methods," *Talanta*, vol. 85, no. 3, pp. 1535–1542, 2011.

[37] T. Saffaj and B. Ihssane, "Response to comments on "Uncertainty profiles for the validation of analytical methods"," *Talanta*, vol. 94, pp. 361-362, 2012.

[38] T. Saffaj and B. Ihssane, "Remarks on "Reply to the responses to the comments on "uncertainty profiles for the validation of analytical methods" by Saffaj and Ihssane"," *Talanta*, vol. 106, pp. 155–157, 2013.

[39] Z. Xue, B. Xu, X. Shi et al., "Overall uncertainty measurement for near infrared analysis of cryptotanshinone in tanshinone extract," *Spectrochimica Acta Part A: Molecular and Biomolecular Spectroscopy*, vol. 170, pp. 39–47, 2017.

[40] V. J. Barclay, R. F. Bonner, and I. P. Hamilton, "Application of wavelet transforms to experimental spectra: smoothing, denoising, and data set compression," *Analytical Chemistry*, vol. 69, no. 1, pp. 78–90, 1997.

[41] M. C. Sarraguça and J. A. Lopes, "Quality control of pharmaceuticals with NIR: from lab to process line," *Vibrational Spectroscopy*, vol. 49, no. 2, pp. 204–210, 2009.

[42] T. Isaksson and T. Næs, "The effect of multiplicative scatter correction (MSC) and linearity improvement in NIR spectroscopy," *Applied Spectroscopy*, vol. 42, no. 7, pp. 1273–1284, 2016.

[43] I. Helland, T. Næs, and T. Isaksson, "Related versions of the multiplicative scatter correction method for preprocessing spectroscopic data," *Chemometrics and Intelligent Laboratory Systems*, vol. 29, no. 2, pp. 233–241, 1995.

[44] R. J. Barnes, M. S. Dhanoa, and S. J. Lister, "Standard normal variate transformation and de-trending of near-infrared diffuse reflectance spectra," *Applied Spectroscopy*, vol. 43, no. 5, pp. 772–777, 2016.

[45] D. Cozzolino, M. Kwiatkowski, M. Parker et al., "Prediction of phenolic compounds in red wine fermentations by visible and near infrared spectroscopy," *Analytica Chimica Acta*, vol. 513, no. 1, pp. 73–80, 2004.

[46] Y. de Micalizzi, N. Pappano, and N. Debattista, "First and second order derivative spectrophotometric determination of benzyl alcohol and diclofenac in pharmaceutical forms," *Talanta*, vol. 47, no. 3, pp. 525–530, 1998.

[47] T. Saffaj, B. Ihssane, F. Jhilal, H. Bouchafra, S. Laslami, and S. A. Sosse, "An overall uncertainty approach for the validation of analytical separation methods," *The Analyst*, vol. 138, no. 16, pp. 4677–4691, 2013.

[48] B. Govaerts, W. Dewé, M. Maumy, and B. Boulanger, "Pre-study analytical method validation: comparison of four alternative approaches based on quality-level estimation and tolerance intervals," *Quality and Reliability Engineering International*, vol. 24, no. 6, pp. 667–680, 2008.

[49] R. W. Mee, "β-Expectation and β-content tolerance limits for balanced one-way ANOVA random model," *Technometrics*, vol. 26, no. 3, pp. 251–254, 1984.

[50] Z. Xue, B. Xu, C. Yang et al., "Method validation for the analysis of licorice acid in the blending process by near infrared diffuse reflectance spectroscopy," *Analytical Methods*, vol. 7, no. 14, pp. 5830–5837, 2015.

[51] R. Galvao, M. Araujo, G. Jose, M. Pontes, E. Silva, and T. Saldanha, "A method for calibration and validation subset partitioning," *Talanta*, vol. 67, no. 4, pp. 736–740, 2005.

[52] ICH 2011, *ICH Points to Consider (R2), ICH-Endorsed Guide for ICH Q8/Q9/Q10 Implementation*, 2011, http://www.ich.org/fileadmin/Public_Web_Site/ICH_Products/Guidelines/Quality/Q8_9_10_QAs/PtC/Quality_IWG_PtCR2_6dec2011.pdf.

5

Using Compact ^{1}H NMR, NIR, and Raman Spectroscopy Combined with Multivariate Data Analysis to Monitor a Biocatalyzed Reaction in a Microreaction System

Robin Legner,[1,2] Alexander Wirtz,[2] and Martin Jaeger[2]

[1]*Department of Chemistry, University Duisburg-Essen, Universitaetsstraße 2, 45141 Essen, Germany*
[2]*Department of Chemistry and ILOC, Niederrhein University of Applied Sciences, Frankenring 20, 47798 Krefeld, Germany*

Correspondence should be addressed to Martin Jaeger; martin.jaeger@hs-niederrhein.de

Academic Editor: Rizwan Hasan Khan

Process analytical technology aims at process knowledge and process improvement, efficiency, and sustainability. A prerequisite is process monitoring. The combination of microreaction systems and spectroscopy proved suitable due to dimension and compound reduction and real-time monitoring capabilities. Compact ^{1}H NMR, NIR, and Raman spectroscopy were used to monitor the biocatalyzed hydrolysis and esterification of acetic anhydride to isoamyl acetate using immobilized *Candida antarctica* lipase B (CALB) in a microreaction system in real-time. To facilitate the identification of signals suitable for the extraction of concentration-time (*c-t*) graphs, 2D heterocorrelation spectra were generated through covariance transformations applied to 1D Raman, NIR, and NMR data. By means of this purely mathematical statistical procedure, the relevant signals of the process media were assigned to educts and products and thus made applicable for univariate data evaluation. The data obtained were interpreted in terms of a first-order kinetic model, and corresponding reaction rate constants were extracted. An alternative, elegant, and fit-for-automation approach for the kinetic analysis of the spectra was demonstrated in using multivariate curve resolution (MCR). The results of the univariate and multivariate approaches were comparable with regard to reaction rates and concentrations. While the manual integration of the ^{1}H NMR spectra followed by univariate analysis allowed to establish a concentration profile of the final product isoamyl acetate hence revealing more details, multivariate analysis was found more suitable for process automation.

1. Introduction

Since the joint initiative of regulatory authorities and the pharmaceutical industry, process analytical technology (PAT) has evolved from a mere production support discipline into a fast-growing research area [1–3]. It has established itself as an important element of industrial production processes [4]. Thus, process analytical instruments are located in or nearby large-scale processes and reactors. Process understanding results from their application, which in fact is process monitoring. The process knowledge obtained aims at ensuring a constant final product quality. It further enables process control and allows process optimization. The process may be improved towards efficiency, hence cost reduction, sustainability, and safety [1, 2]. Meeting the customer demand for reproducible product quality is an important prerequisite for the competitiveness of a company, which requires using facilities, resources, and energy in the most economical way [5].

Today's process analytical tools comprise a variety of methodologies and take advantage of their inherent features [2]. Spectroscopy, spectrometry, and chromatography represent highly specific methods [4]. Among them, Raman and near-infrared (NIR) play major roles [6–8]. More recently, nuclear magnetic resonance (NMR) spectroscopy has grown into focus [9, 10]. During process monitoring, the information sought is most often the quantity of a compound at a given time. Therefore, spectroscopic techniques providing even relatively lower resolution and sensitivity than those commonly required for structure elucidation purposes satisfy the requirements for process monitoring. Process monitoring can proceed in-line, on-line, at-line or—less desirable—off-line. For in-line monitoring, a probe is inserted into the reaction vessel. When the measurement is conducted on-line, a bypass is used. For at-line analysis, a sample is taken from the reaction and treated before analysis. Hence, in-, on-, and at-line monitoring types usually provide results in real-time [1, 2, 7, 11].

In order to compensate for the low resolution of, e.g., NIR or compact bench-top ^{1}H NMR spectrometers, suitable multivariate data analysis procedures have been devised and applied where univariate data processing did not prove sufficient [11–13]. Another strategy for enhancing spectral resolution has been developed by using covariance data processing [14–16]. This approach has been derived from statistical mathematics. Hence, different spectra that possess a common so-called perturbation dimension are correlated, where the spectral intensity represents a function of two independent variables such as wavenumber or frequency. This type of two-dimensional correlation spectroscopy has been introduced by Noda et al. for IR and also later for other vibrational and optical absorption methods [16–18]. For NMR spectroscopy, Brueschweiler et al. and Eads and Noda have reported on the use of covariance transformations to correlate frequency domains [19–21]. A 2D homocorrelation map is generated by transformation of a series of 1D spectra or a 2D spectrum from one spectroscopic technique. When two series of 1D spectra or two spectra of different spectroscopic techniques are correlated, a 2D heterocorrelation map results [22, 23]. Covariance methods are especially useful for the identification of spectral resonances suitable for process analysis.

Concentration-time (*c-t*) profiles from reaction monitoring can be obtained either by univariate methods or by multivariate methods such as the multivariate curve resolution alternating least squares (MCR-ALS) [24, 25]. Using this self-modelling curve resolution method, reaction kinetics can be determined directly from the recorded spectra series [26–28].

Isoamyl acetate belongs to flavoring substances, its odor being similar to both banana and pear. It is the main constituent of banana oil. Due to its odor, isoamyl acetate is used as artificial flavor [29, 30]. Its application as solvent and carrier, once known as aircraft dope, has vanished except for historical reproductions and scale models since airplanes are made of metal or carbon composites [31].

In this study, compact ^{1}H NMR, NIR, and Raman spectrometers were applied to monitor in real-time the biocatalyzed esterification to isoamyl acetate via the hydrolysis of acetic anhydride using immobilized *Candida antarctica* lipase B (CALB) in a microreaction system, cf. Scheme 1. The recorded spectra were transformed into 2D-heterocovariance correlation maps for the identification of specific resonances for univariate kinetic analysis. The results were compared to the kinetic parameters obtained from multivariate MCR-ALS analysis.

2. Theory

2.1. Multivariate Curve Resolution-Alternating Least Squares (MCR-ALS). Principal component nalysis (PCA) and partial least square regression (PLS-R) are among the most often used chemometric tools for process data analysis [32, 33]. Principal components and latent variables usually do not represent chemical and physical quantities. This is evenly true for the MCR-ALS algorithm. Physical-chemical knowledge on the process can be integrated into the chemometric analysis, as the number of components may be defined through the number of reactants and products [24, 25, 34]. From a data matrix **X** containing all spectral information as superpositions due to the reaction mixture, the information about the pure components, e.g., the spectra, can be extracted. In case of reaction monitoring, **X** represents spectral data. The data matrix can be decomposed into the small matrices **S** and **C** which consist of pure spectra and the associated concentration profiles and the residual matrix **E** as follows [25, 34–36]:

$$\mathbf{X} = \mathbf{C}\mathbf{S}^{\mathrm{T}} + \mathbf{E}. \tag{1}$$

The matrix **X** of dimensions $z \times s$ contains the recorded NMR spectra as rows with the respective intensity values. From reaction monitoring, spectral series represent a function of time. The transposed matrix $\mathbf{S}^{\mathrm{T}}$ of dimensions $a \times s$ contains the spectra of the pure components as rows a. In matrix **C** of dimensions $z \times a$, the contributions of the pure spectra in columns a correspond to the concentrations in the corresponding spectrum. The residual matrix **E** of dimensions $z \times s$ contains the spectral residues, which cannot be explained by the pure components. Residuals can represent random noise or systematic disturbances in the spectrum.

Matrices **C** and **S** can be computed by MCR-ALS. They are iteratively calculated from a first estimate for **C** or **S** in an alternating least-squares method as shown below [24, 35, 37]:

$$\hat{\mathbf{S}}^{\mathrm{T}} = \mathbf{C}^{+}\mathbf{X} = \left(\hat{\mathbf{C}}^{\mathrm{T}}\hat{\mathbf{C}}\right)^{-1}\hat{\mathbf{C}}^{\mathrm{T}}\mathbf{X}, \tag{2}$$

$$\hat{\mathbf{C}} = \mathbf{X}\mathbf{S}^{\mathrm{T}+} = \hat{\mathbf{S}}\mathbf{X}\left(\hat{\mathbf{S}}^{\mathrm{T}}\hat{\mathbf{S}}\right)^{-1}, \tag{3}$$

where $\wedge$ denotes the estimated value and + a pseudoinverse. The decomposition of **X** is achieved by iterative least-squares minimization of $\|\mathbf{E}\|$ under suitable constraints, i.e., a chemical kinetic model, nonnegativity in spectral, and concentration profiles as well as an initial guess of reagent and product concentrations. Since the uniqueness of the solution is determined by the secondary conditions, such as the chemical model, reaction order, and pure educt or product spectra, it is important to introduce as much physical-chemical knowledge of the system as possible into the constraints [34, 38].

2.2. Homo- and Heterocorrelation Spectroscopy. In contrast to NMR spectroscopy, vibrational spectroscopic techniques are not based on frequency pulse sequences that allow us to encode two frequency domains. While NMR spectroscopy is intimately associated with the Fourier transformation to generate 2D correlation spectra, a different approach was suggested by Eads and Noda and Frasinski et al. [21, 22, 39, 40] Covariance transformations were used to

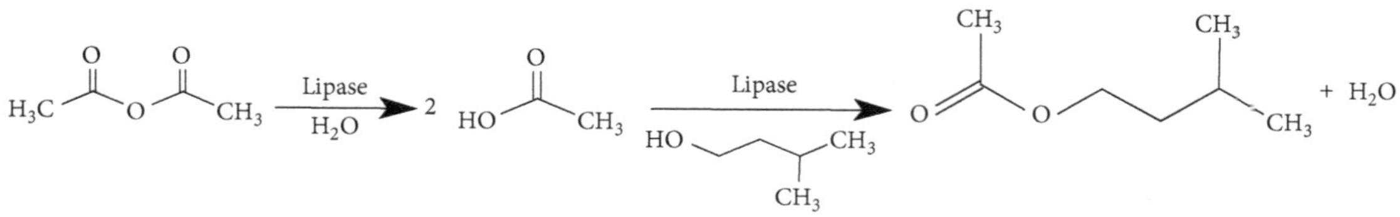

Scheme 1: Reaction scheme for the biocatalyzed hydrolysis of acetic anhydride followed by esterification with isoamyl alcohol to isoamyl acetate. Immobilized CALB catalyses both reaction steps as one-pot reaction in the microreaction system [29].

create a correlation map by recording one-dimensional spectra as responses to a quantitative, measurable perturbation applied to the sample in a systematic way [16, 18]. This perturbation may be temperature, pressure, concentration, stress, electrical field, and most importantly a chemical reaction [17, 41]. A correlation spectrum or map, $\mathbf{X}(\nu_1, \nu_2)$, is defined according to the following equation:

$$\mathbf{X}(\nu_1, \nu_2) = \langle \tilde{y}(\nu_1, t) \cdot \tilde{y}(\nu_2, t') \rangle, \quad (4)$$

where $\tilde{y}(\nu_1, t)$ is the spectrum affected by the perturbation, called the dynamic spectrum, and ν_1 and ν_2 are two different spectral variables at a given time or a fixed interval of the external variable t. The symbol $\langle\,\rangle$ denotes a suitable correlation function. The dynamic spectrum is computed by subtracting the average spectrum from each spectrum of a series of spectra. The quantitative comparison of spectral intensity variations can then be performed by means of statistical theory. Variations from a mean value are represented by covariances. Hence, the correlation map $\mathbf{X}(\nu_1, \nu_2)$ can be considered as the covariance matrix, when the series of dynamic spectra $\tilde{y}(\nu_1, t)$ is represented by a matrix. Assuming that $\mathbf{X}(\nu_1, \nu_2)$ is a complex number, it can be expressed according to the following equation consisting of the real part $\mathbf{\Phi}$ and the imaginary part $\mathbf{\Psi}$:

$$\mathbf{X}(\nu_1, \nu_2) = \mathbf{\Phi}(\nu_1, \nu_2) + i\mathbf{\Psi}(\nu_1, \nu_2). \quad (5)$$

For spectroscopic purposes, the relation between the covariance matrix and a spectrum, in particular 2D NMR spectra, was thoroughly proven [19, 20, 42]. The proof is based essentially on Parseval's theorem. The numerical calculation of the correlation maps has been described elsewhere [16, 20].

Due to the statistical nature of the approach, one series of spectra may be compared to another series of spectra originating from a different spectroscopic technique. This is referred to as heterocovariance spectroscopy [21, 41]. A heterocorrelation map will result according to the following equation:

$$\mathbf{X}(\mu_1, \nu_2) = \langle \tilde{x}(\mu_1, t) \cdot \tilde{y}(\nu_2, t') \rangle, \quad (6)$$

where μ_1 and ν_2 are the spectral variables from different spectroscopic techniques. Equation (6) can be rewritten in matrix notation, cf. Equation (7), where the square root is a consequence of Parseval's theorem [19, 20, 43]:

$$\mathbf{F} = \mathbf{C}^{1/2} = \left(\mathbf{S} \cdot \mathbf{G}^{\mathrm{T}}\right)^{1/2}, \quad (7)$$

where $\mathbf{F}$ is the 2D spectrum, $\mathbf{C}$ is the covariance matrix, and $\mathbf{S}$ and $\mathbf{G}$ are the two data matrices, such as a series of spectra with a common perturbation dimension. Cross-peaks in such correlation spectra have been reported as a help to disentangle crowded spectral regions [16, 43]. In this study, 2D heterocovariance correlation maps are computed from series of 1D ^{1}H NMR, Raman, and NIR spectra recorded during reaction monitoring.

3. Materials and Methods

3.1. Spectral Recording and Processing. NMR spectra were recorded on-line using a compact NMR picoSpin80 spectrometer with a proton Larmor frequency of 82 MHz (Thermo Fisher Scientific GmbH, Dreieich, Germany). It had a built-in flow cell and an electrical lock such that neat liquid process media were measured without further sample preparation or addition of deuterated solvents or reference standards. The sample cell possessed a total volume of 40 μL and an active volume of 40 nL. The operation temperature was 36°C. The spectrometer was controlled by Thermo Fisher PicoSpin software 0.9.3 running on the spectrometer control board and accessed via a web interface from a laptop computer. The free induction decay (FID) was recorded as 4092 data points that were zero filled to 8 k points prior to the Fourier transformation. The number of scans amounted to 16 and the pulse length was set to 60 μs corresponding to a 90° pulse. Spectra were recorded from 0.3 to 7 ppm. The bandwidth amounted to 4 kHz. All spectral processing was achieved through MestReNova 12.0.0 (Mestrelab Research S. L., Santiago de Compostela, Spain) on a laptop computer under Windows 7. The spectrum size after zero-filling and Fourier transformation (FT) resulted in 64 k. Pretreatment methods were used for spectra optimization. The spectra were binned with 0.004 ppm and normalized to the largest peak. Baseline correction was achieved using a Bernstein polynomial fit of 3rd order.

Raman spectra were recorded using the DXR™ SmartRaman spectrometer (Thermo Fisher Scientific GmbH, Dreieich, Germany) and processed using the Omnic software, version 9.2 (Thermo Fisher Scientific GmbH, Dreieich, Germany) running on a desktop PC under Windows 7. A Raman immersion probe from InPhotonics (InPhotonics, Inc., Norwood, MA, USA) was connected via a special module for fiber optics, allowing a distance between spectrometer and reaction vessel of up to 5 m. The probe was positioned at its focal point of 5 mm above the solution in the reaction vessel. Sixty-four spectra comprising a spectral range of 250–3500 cm^{-1} with a resolution of 5 cm^{-1} were acquired at each reaction time point. The laser was operated at 785 nm and 150 mW. For processing, the Omnic software

9.2 (Thermo Fisher Scientific GmbH, Dreieich, Germany) was used. During 1380 min, 50 spectra were recorded. All spectra were baseline corrected using an asymmetric 2nd order Huber function. The spectra were normalized to the highest band. For baseline correction and normalization, MatLab R 2017b (MathWorks, Inc., Natick, MA, USA) was used.

NIR measurements were carried out with the Antaris II FT-NIR Analyzer using the software Omnic, version 8.0 (Thermo Fisher Scientific GmbH, Dreieich, Germany). The spectral range was chosen from 4000–10000 cm^{-1} at a resolution of 4 cm^{-1} and the number of scans was set to 64. The spectrometer was equipped with a NIR transflection immersion probe Falcata (Hellma GmbH & Co. KG, Muellheim, Germany). All spectra were normalized after the baseline correction using an asymmetric Huber function 3rd order, again under MatLab R 2017b.

3.2. Microreaction Assembly. Stainless-steel microreaction components (Ehrfeld Mikroreaktionstechnik, Wendelsheim, Germany) were assembled according to Figure 1 [44, 45]. In an open storage vessel, the reactants, 20 mL of acetic anhydride (≥98%, Merck KGaA, Darmstadt, Germany), 40 mL of isoamyl alcohol (>98%, Merck KGaA, Darmstadt, Germany), and traces of distilled water for the benefit of the enzyme, were mixed and transferred via a peristaltic pump Smartline 100 (Knauer GmbH, Berlin, Germany) into the microreaction assembly at a rate of 5 mL/min. The excess of isoamyl alcohol ensured complete conversion of the anhydride. Immobilization of the enzyme *Candida antarctica* lipase B (Novozymes A/S, Bagsvaerd, Denmark) was achieved by mixing the enzyme with the epoxy acrylate resin Purolite® ECR8205F (Purolite GmbH, Ratingen, Germany) in 50 mM phosphate buffer and stirring for 20 h at ambient temperature. The immobilized enzyme was then transferred into the cartridge of the Ehrfeld reactor F 200. The filled volume amounted to 1.8 of the 2 mL cartridge volume. The reaction was initiated through heating the reactor to 35°C. The residence time of a solution fraction, approximately 2 mL in the reactor amounted to 2 min. The solution, then containing product, was recirculated into the storage vessel. The microreaction assembly was controlled through the software LabVision 2.10 (HiTec Zang GmbH, Herzogenrath, Germany).

The reaction mixture was diverted to the NMR spectrometer sample cell via a valve (Figure 1). Setting the valve, the flow was stopped after five seconds such that 16 spectra could be accumulated per sample. The following sample was introduced into the spectrometer via the bypass, transferring the previous one back to the vessel at the same time (cf. Figure 1). For Raman and NIR spectral recording, corresponding probes were attached a few millimeters above the surface of the reaction liquid in the storage vessel or was immersed into it (cf. Figure 2). The reaction was conducted over a period of 1380 min.

3.3. Covariance and Multivariate Data Analysis. The 2D correlation maps were generated using the program 2Dshige running on a personal computer under Microsoft Windows 7 [46]. Spectral data were exported as ASCII files from the spectral processing software listed above. To reduce the size of the data sets, 1D spectra were reduced by binning to 1 k data points for NMR, 732 data points for Raman, and 1 k data points for NIR. The data files were transformed into homo- or heterocorrelation maps and displayed as contour or color plots using appropriate threshold levels.

For MCR-ALS analysis, NMR, Raman, and NIR spectra were automatically imported into MatLab, version 2017b, wherein the MCR-ALS toolbox 2 was applied [35, 47]. Three latent variables or components were selected on the basis of visual inspection of the computational results. The non-negativity conditions for spectra and concentration profiles were set as constraints. Furthermore, a kinetic model $A \longrightarrow B$ was assumed. Other models tested were $A \longrightarrow B \longrightarrow C$ and $A + B \longrightarrow C + D$. It is noteworthy that the number of variables or components is not interchangeable with the number of real chemical components of the process system. An initial guess of the educt concentration ranged between 3.0 and 3.5 $mol{\cdot}L^{-1}$. As the starting value, the reaction rate constant was assumed to be $2.5{\cdot}10^{-3}\ min^{-1}$. The agreement between calculated and experimental concentration-time profile was taken as goodness of the obtained kinetic parameters.

4. Results and Discussion

4.1. Qualitative Reaction Monitoring. The enzyme catalysed acetic anhydride hydrolysis, and subsequent esterification with isoamyl alcohol was conducted in a microreaction platform. The enzyme was immobilized and located in a cartridge reactor as a heterogeneous catalyst. The process was monitored using on-line 1D ^1H NMR and in-line Raman and NIR spectroscopy. Spectra were recorded during the course of the reaction over 1380 min. One aim of process analysis, process understanding, is represented here by generating *c-t* plots and their mathematical description using chemical kinetic theory. It is straightforward to seek educt or product specific signals in spectra, which can be interpreted in terms of concentrations. Starting with the interpretation of the NMR spectra, the usefulness of covariance spectroscopy, especially heterocorrelation spectroscopy shall be demonstrated. Yet, analysts more familiar with vibrational spectroscopy might start from Raman or NIR spectra as well. Relevant educt and product signals could be identified in ^1H NMR spectra (Figure 2).

The hydrolysis of acetic anhydride to free acetic acid appeared in the NMR spectrum through the methyl resonances at 2.0 and 1.8 ppm where the latter signal stems from acetic anhydride and the former from acetic acid. The following esterification reaction of free acetic acid with isoamyl alcohol to isoamyl acetate could be recognized by the very weak methyl resonance at 3.8 ppm. The water resonance, which was observed at 4.1 ppm at the beginning of the reaction, shifted in the course of the reaction and was found at 6.5 ppm by the end of the reaction. The signal migrated due to the change in the pH induced by the liberation of acetic acid. Due to the high selectivity of NMR spectroscopy, individual and well-separated signals could be attributed. This

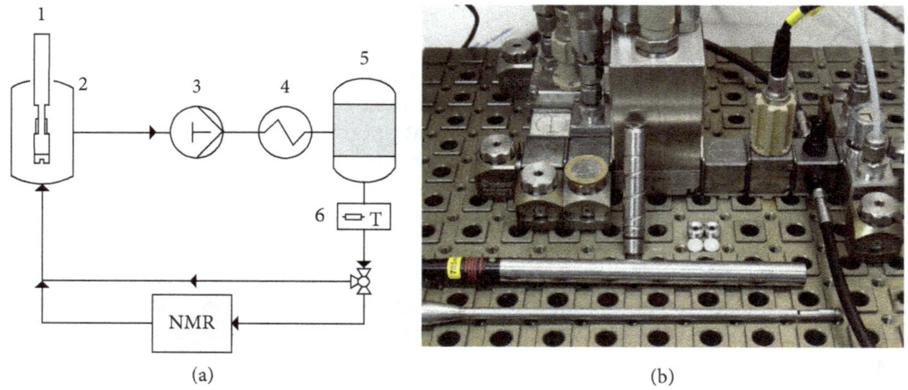

FIGURE 1: (a) Process flow-chart of the microreaction assembly: 1, Raman and NIR probe; 2, storage vessel containing the majority of the reaction mixture; 3, piston pump; 4, heat exchanger; 5, cartridge reactor; 6, temperature sensor; and then a compact ^{1}H NMR spectrometer is located in a bypass. (b) Picture of the microreaction platform (from left to right): inlet module, heat exchanger, cartridge reactor, T-sensor, optical flow cell, and outlet module. Raman probe and transflection NIR immersion probe (front).

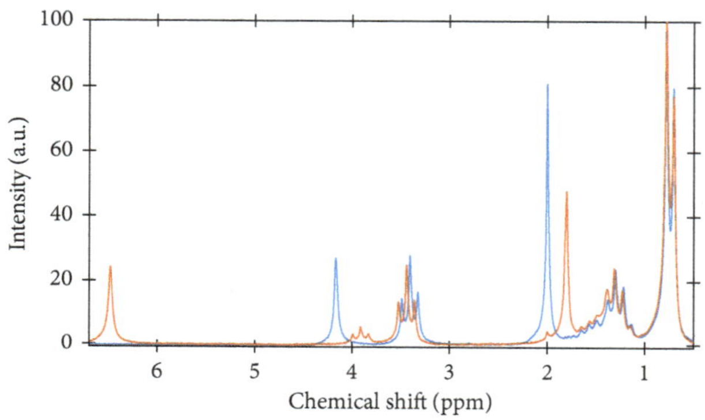

FIGURE 2: 1D ^{1}H NMR spectra of the biocatalyzed hydrolysis in the microreaction system. Spectrum at 0 min (blue) and at the end of process monitoring after 1380 min (brown).

is a prerequisite for the application of univariate data analysis. Hence, *c-t* curves were readily obtained for the reaction, (see Section 4.3).

While high resolution is intrinsic to high-field NMR spectrometers, the much smaller bench-top spectrometer operating at 82 MHz proton Larmor frequency provided sufficient resolution to obtain relevant educt and product bands. The following paragraph illustrates how the information gained from NMR analysis can be quickly transferred to Raman and NIR analyses.

4.2. 2D Correlation Spectroscopy to Support Reaction Monitoring. To easily discover signals or bands suitable for reaction monitoring, heterocorrelation maps were computed from NMR, Raman, and NIR spectra, as shown in Figure 3. Since spectral differences along with the reaction progress were most easily recognized and explained in ^{1}H NMR spectra, the gained information was transferred to the Raman and NIR spectra. While in other cases Raman or NIR spectrum prove more readily interpretable, the inverse approach works evenly well.

Heterocorrelation spectra do not contain diagonal peaks [16, 48]. Thus, all correlations observed related signal changes at a given NMR chemical shift to intensity changes at wavenumbers in Raman spectra. It is therefore by the sign of a correlation peak that educt and product signals can be attributed. Using selected successfully attributed NMR signals, i.e., 2.0 ppm for acetic anhydride, 3.8 ppm for isoamyl acetate, and 1.8 ppm for acetic acid, correlations with the bands at 800, 1700 and 1740, and 3000 cm^{-1} were observed (cf. Figure 3). In the region between 1740 and 1790 cm^{-1}, asymmetric and symmetric stretching vibrations of the two C=O groups of carboxylic acid anhydrides were detected. The product's methylene resonance at 3.8 ppm displayed only a very weak in-phase or positive correlation to the stretching C=O vibration from acetic acid at 1700 cm^{-1}. The positive sign is indicative for a product signal. The analogue applies for the correlations at 2.0 ppm, 2980 cm^{-1} and 1.8 ppm, 2980 cm^{-1}: The negative (blue) signal indicates an educt-product relationship, the positive (red) cross-peak a product-product one. In the range of 810–910 cm^{-1}, weak C-C stretching vibrations of acetates occurred. This signal was not directly apparent in the 1D Raman spectrum. Its counterpart however was visible in the NMR, which suggests that 2D correlation spectroscopy is also suitable for the identification of low-concentration byproducts [49].

The enzyme-catalysed hydrolysis of acetic anhydride could be followed through monitoring the C=O stretching vibration at 1740 cm^{-1} in Raman spectra and the methylene resonance in the range of 2.0 ppm in NMR spectra. The C=O stretching vibration displayed a negative correlation with the methylene resonance of acetic acid at 1.8 ppm, while the two educt and product signals run in opposite directions simultaneously. This is reflected by the positive correlations between 1740 cm^{-1} and 2.0 ppm, and between 1700 cm^{-1} and 1.8 ppm for the free acetic acid. Apart from a strong correlation in the range of the C-H vibrations at 3000 cm^{-1}, a moderate correlation in the range of 800 cm^{-1} from the substances involved in the hydrolysis could be recognized.

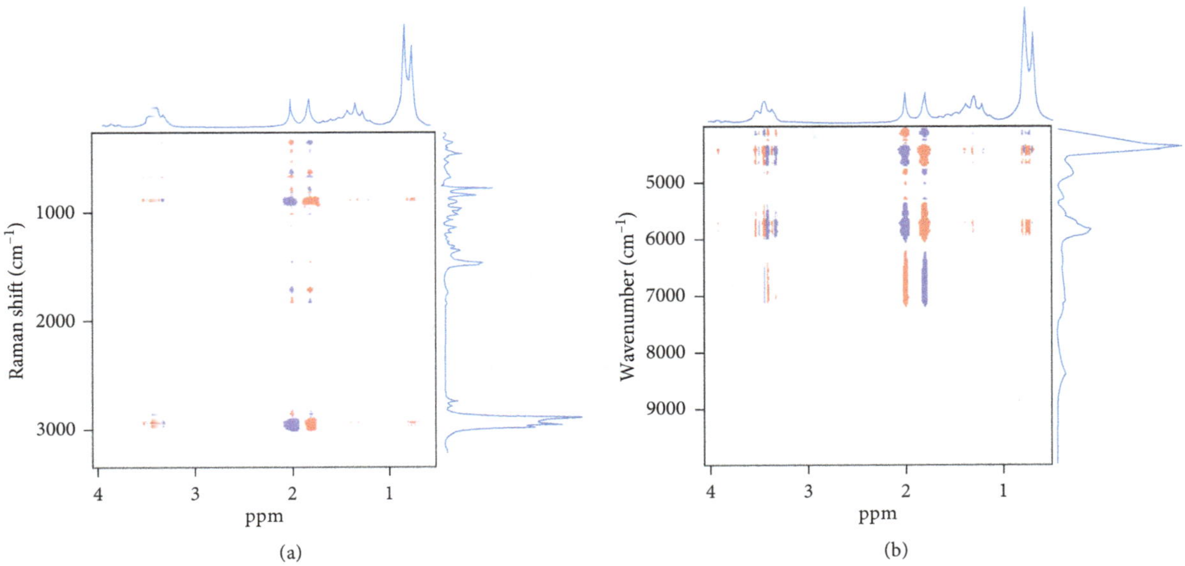

FIGURE 3: 2D ^{1}H NMR-Raman heterocorrelation spectra (a) of isoamyl acetate from immobilized lipase catalysis using in-line Raman spectroscopy (λ = 785 nm) and on-line 1D ^{1}H NMR. 2D ^{1}H NMR-NIR heterocorrelation map (b).

Firstly, a negative correlation of the acetic anhydride at 2.0 ppm resulted. Secondly, a positive correlation with the free acetic acid at 1.8 ppm was found [50].

In the ^{1}H NMR-NIR 2D heterocorrelation (cf. Figure 3(b)), signals reflecting the hydrolysis were observed, but a detailed assignment could not be achieved. Yet, from the 2D NMR-NIR correlation spectrum, the overtone and combination vibrations of C-H and O-H at 6250–7150 cm^{-1} and 4540–5000 cm^{-1} could be attributed by means of the respective positive and negative correlations to the methylene resonances at 2.0 and 1.8 ppm. It could also be recognized that a univariate evaluation would falsify the kinetic evaluation since observed bands represent superpositions of components [50].

In summary, the 2D heterocorrelation spectra helped to confirm the assignment of signals to an educt or a product by directly transferring the information from one spectroscopic technique to another. The 2D maps indicated the origin of vibrational bands from a single compound or as superposition. Heterocorrelation spectra were thus found to facilitate the identification and analysis of signals suitable for process monitoring. A selection of the 58 spectra recorded is shown in Figure 4 as stack plots for each spectroscopic method.

4.3. Quantitative Reaction Monitoring. Process analytics generally aims for the reception of concentration-time (c-t) data. Their extraction from spectra proceeded straightforward using the in-line and on-line monitoring at the microreaction assembly. Through spectral analysis using heterocovariance correlation plots, the signals corresponding to the educts and the product were identified as described above. Based on distinct signals, univariate analysis leading to c-t plots was applied. Nevertheless, series of spectra with the fully recorded spectral range were subjected to multivariate data analysis in the form of MCR-ALS.

Inspection of the NMR spectra (cf. Figure 4(a)) revealed that the water signal originating from the buffer system of the enzyme was well discernible. The spectra, in which the product signal was superimposed by the water shift, were discarded from kinetic evaluations. Since the area-under-the-curve of an NMR signal is proportional to the number of nuclear spins causing the signal, the actual relative concentration was determined by integrating the signals at 2.0, 1.8, and 3.8 ppm and by correcting them for the number of nuclei. To determine absolute concentrations, the internal standard method would need to be applied [51, 52]. Provided a sufficient linearity of the NMR amplifier, the absolute concentrations could be determined during the reaction since the initial concentrations of the reactants were known [51]. Due to nonlinearity of the amplifier, the relative composition of the mixture as obtained from direct integration of specific educt and product signals, hence univariate analysis, was plotted against reaction time (Figure 5).

While NMR spectra displayed a weak but interpretable signal of isoamyl acetate at 3.8 ppm, Raman and NIR spectra did not provide a corresponding band of sufficient intensity. Due to the choice of the process parameters, only a small amount of product was formed during the reaction period monitoring such that isoamyl ester product signals appeared weakly as seen by the NMR signal and its heterocorrelation at 3.8 ppm in the corresponding spectra. The first catalytic step, i.e., the hydrolysis of acetic anhydride to acetic acid, was monitored by the signals in the range of 2.0–1.8 ppm and the corresponding Raman band at 800 cm^{-1}. An acceleration of the reaction might be initiated by choosing a somewhat higher temperature. The esterification might occur eventually more prominent. The univariate evaluation of the signals from

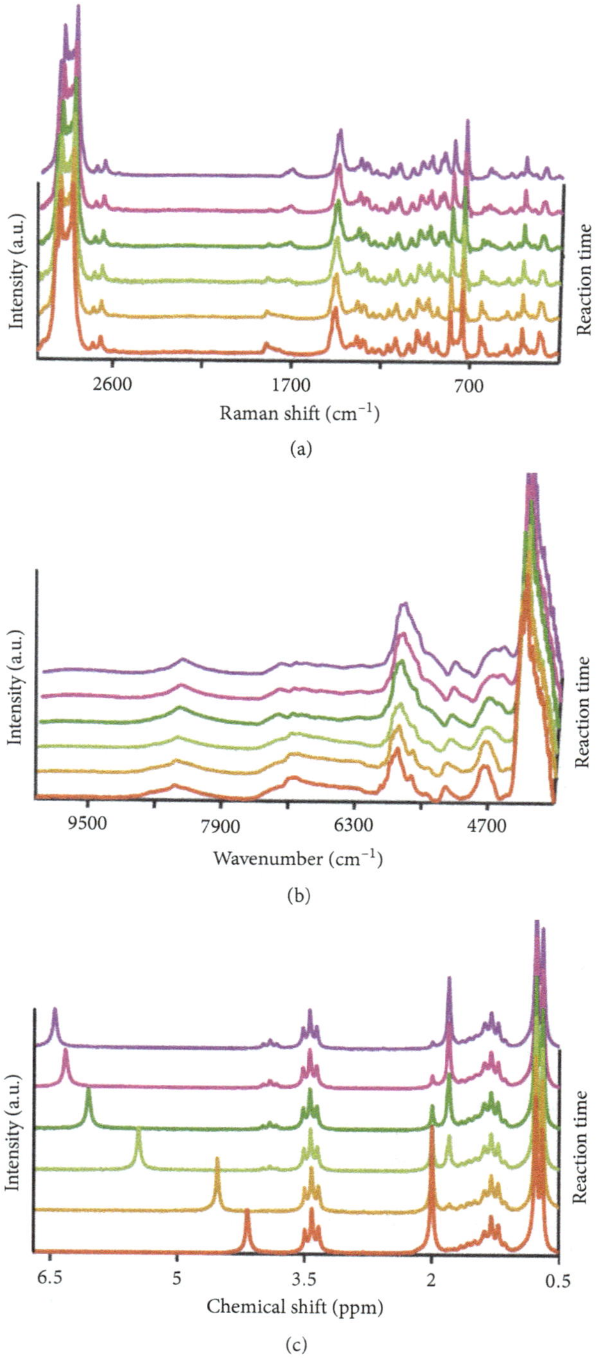

FIGURE 4: Raman (a), NIR (b), and ^{1}H NMR (c) spectra from on-line reaction monitoring of the enzyme catalysed reaction in the microreaction assembly. Spectra were recorded from the neat reaction mixture during 23 hours. NMR spectra, 16 accumulations, were acquired at 82 MHz and $T = 36°C$. NIR spectra were recorded using an immersion probe. Raman spectra, 64 accumulations, were acquired with laser excitation at 785 nm. Color coding refers to corresponding reaction times among the spectroscopic methods. The reaction was followed over a period of 1380 min.

acetic acid and acetic anhydride yielded comparable reaction rate constants from Raman and NMR spectroscopic monitoring, showing the values of 2.1 and $2.2 \cdot 10^{-3}$ min^{-1} for the transformation from acetic anhydride to acetic acid (cf. Table 1). Within Raman spectra, the bands at 1740 and 1700 cm^{-1} were used for educt and product monitoring, respectively. A trial to use univariate analysis on NIR data did not give satisfactory results. While envisaging automation for process control, MCR-ALS analysis is seemed to be an attractive method as identification of suitable bands would not

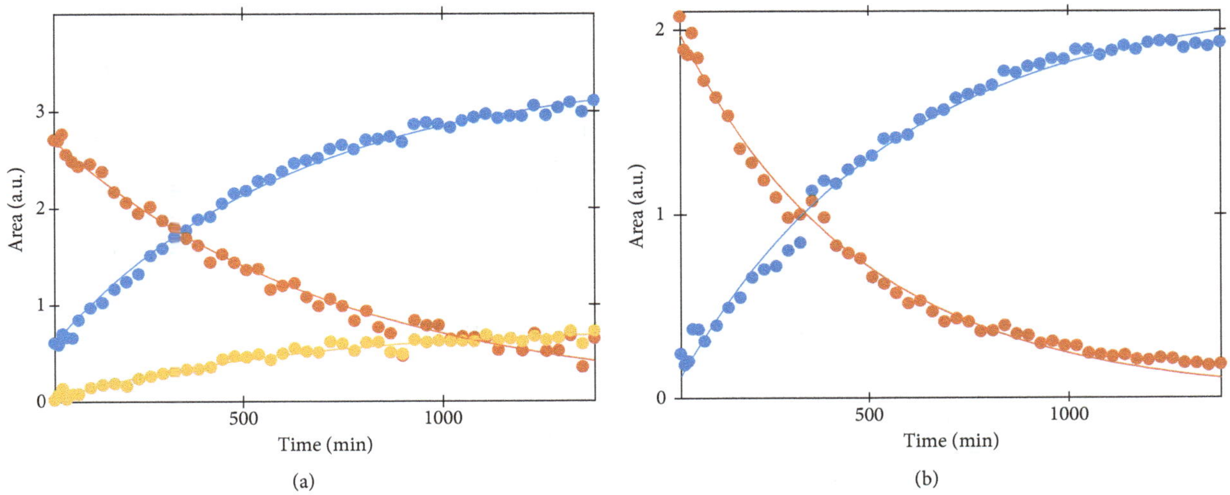

FIGURE 5: c-t plots as area-under-the-curve vs. time of the enzyme catalysed esterification from univariate data analysis and corresponding fits computed according to first-order reaction based on NMR (a) and Raman (b) spectra; acetic acid (blue), acetic anhydride (orange), and isoamyl acetate (yellow).

TABLE 1: Reaction order and rate constants according to ^{1}H NMR, Raman, and NIR spectroscopic methods for the enzyme-catalysed hydrolysis of acetic anhydride using univariate and multivariate spectral analyses.

	Univariate		Multivariate (MCR-ALS)	
	Reaction order	Rate constant ($min^{-1} \cdot 10^{3}$)	Rate constant ($min^{-1} \cdot 10^{-3}$)	Explained variance
^{1}H NMR	1	2.1	2.2	0.99
Raman	1	2.2	2.1	0.99
NIR	1	—	1.9	0.97

be required. The MCR-ALS analyses of NMR and Raman spectra are shown in Figure 6. Before computing the c-t curves, the first-order reaction mechanism, A ⟶ B, an initial guess of the rate constants, and the educt starting concentration were defined as boundary conditions [53].

The algorithm transforms spectral data into c-t data according to the predefined model, i.e., the first-order reaction. The correspondence of the extracted data to the model is visualized by the c-t curve. From Figure 6, it can be recognized that for Raman and NMR data, the agreement was very good. Only the NIR data did not prove of sufficient quality to yield good agreement to the model and thus minor scattering (cf. Figure 6(c)). The correspondence between c-t data and the first-order model was observed for all three spectroscopic techniques. Outliers, such as observed for the NIR analysis, may indicate disturbances, such as laboratory temperature change during sampling or measurement. Other reaction mechanisms were tested against the c-t data. Neither the second-order reaction A + B ⟶ C + D nor the follow-up reaction A ⟶ B ⟶ C displayed data fits with evenly good agreement as the first-order model. However, the low yield of the isoamyl acetate and therefore the resulting low signal-to-noise ratio influenced the performance of the algorithm so that no superior agreement with the models could be found.

Furthermore, the comparison of univariate to multivariate analyses showed a remarkable consistency with respect to the reaction order and the obtained rate constant. The results of the reaction monitoring of the enzymatic-catalysed hydrolysis reaction are summarized for univariate and multivariate analyses via MCR-ALS in Table 1.

All spectroscopic methods used revealed rate constants of a first-order chemical kinetic model. Because of enzyme catalysis, the reaction is likely to follow a pseudo-first-order mechanism, which would not be discernible from a first-order mechanism under the reaction conditions applied. Apparently, the catalysed hydrolysis proceeded faster or was preferred to the isoamyl ester formation. From univariate analysis of the NMR data (cf. Figure 5), a corresponding reaction constant of 1.3 $10^{-3}\,min^{-1}$ was obtained. No multivariate description of this subsequent esterification could be achieved since isoamyl acetate was observed only in the NMR spectrum after about 2 hours of reaction time, provided careful visual inspection. It is assumed that the MCR-ALS algorithm was not able to identify isoamyl signals and—in case of Raman—bands that possessed a sufficient signal-to-noise ratio. The signal intensity of the product proved too low for the determination of a reasonable c-t profile through the MCR-ALS algorithm, whereas a trained analyst was able to extract consistent data. Thus, the univariate data analysis offered in this case the advantage to quantify extremely small signals, which also requires a greater effort for the evaluation.

The enzyme is also known to suffer from acid pH which was a consequence of the formation of acetic acid. The pH of the process solution dropped below the working range of the biocatalyst of pH 5–9. It was hence assumed that the CALB lost activity and ceased to produce isoamyl acetate. As a consequence, the subsequent reaction would not significantly affect the determined rate constants.

It should be noted that the intensity of vibrational bands depends on oscillator strength and polarizability, which is in

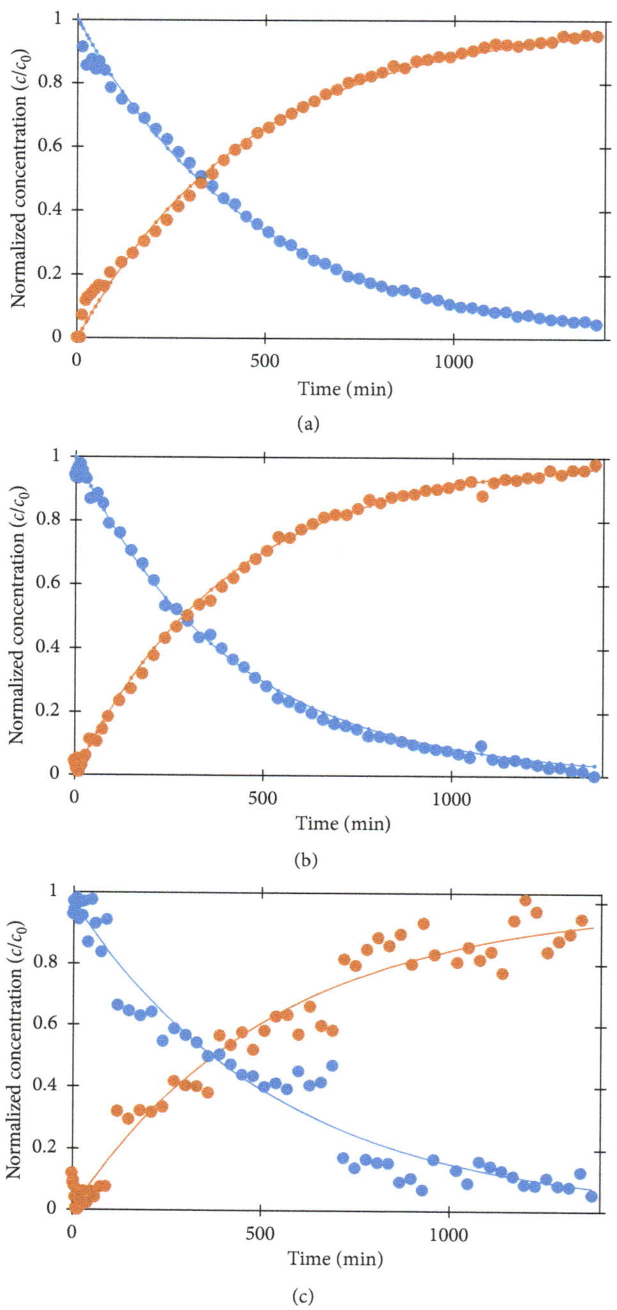

Figure 6: *c-t* profiles from MCR-ALS modelling for the enzyme-catalysed hydrolysis of acetic anhydride to acetic acid in the Ehrfeld microreaction system. Results on the basis of the ^{1}H NMR (a), Raman (b), and NIR (c) data. The computed data fits correspond to first-order reaction A ⟶ B (orange and blue lines), and the points are extracted data for product (red) and educt (blue).

general different for two different bands. Thus, the actual concentrations or concentration ratios of products and educts could not be derived from the Raman bands in contrast to the NMR resonances.

In summary, two approaches may be followed to obtain *c-t* data, which proved fit for process analytical technology: univariate analysis and MCR-ALS as a multivariate approach. Covariance spectroscopy, here in the form of heterocorrelation spectroscopy, was suitable to help identify specific signals of educts and products. Fundamental process understanding could be achieved through describing the *c-t* curves using kinetic models. Starting from the mathematical description of the data, it becomes possible to predict the course of the reaction and to quickly recognize deviations from the expected process. This knowledge may then be used to automate process control.

5. Conclusions

Analytical bench-top instruments, i.e., a compact ^{1}H NMR operating at 82 MHz, NIR and a Raman spectrometer, were successfully combined with a stainless-steel microreaction assembly. The CALB catalysed hydrolysis, and subsequent esterification of acetic anhydride to isoamyl acetate served as model for a heterogeneous biocatalysis reaction. NMR spectra provided distinct signals of all educts and the product, and NIR and Raman bands could not easily be assigned on visual inspection of the spectra. To facilitate the assignments, 2D heterocorrelation maps were computed from series of 1D NMR and 1D vibrational spectroscopic data using covariance transformations.

Quantitative analysis of the spectroscopic data was achieved using univariate and multivariate methods. The *c-t* data obtained from NMR, Raman, and NIR proved of sufficient quality to test a first-order or pseudofirst-order kinetic model yielding consistent results. The MCR-ALS algorithm was applied successfully for a biocatalyzed reaction. Yet, an improved signal-to-noise ratio, e.g., through increasing the number of acquisitions but thus prolongation of the measurement time, will enhance the recognition of low-abundant species. Furthermore, the MCR-ALS algorithm may be simultaneously applied to the data of the three analytical techniques. The knowledge on the process obtained by spectroscopic monitoring, heterocorrelation-based spectral interpretation, and univariate or multivariate kinetic analysis may allow future process control. For automation purposes, MCR-ALS was judged better suitable than univariate data treatment. Educt feed, reaction temperature, and flow rate should be optimized for best yields. These process parameters may then be adjusted automatically by detecting deviations from the corresponding reaction kinetics and counteracting through a feedback loop.

Data Availability

Data were recorded on NMR, Raman, and NIR spectrometers from Thermo Fisher Scientific. The raw data can be provided in the spectrometer file format. Data have been analyzed using MatLab. MatLab files can be obtained upon request from the corresponding author. Upon request, we can also include a link to download from the cloud Sciebo.

Conflicts of Interest

The authors declare that there are no conflicts of interest regarding the publication of this article.

Acknowledgments

The authors are grateful to Thermo Fisher Scientific for the vibrational and nuclear magnetic resonance spectrometers within a SEED project and A. Nickisch-Hartfiel and P. Naderwitz for technical and bio-technological expertise. R.L is very grateful for a grant from the German Academic Scholarship Foundation. The authors thank the Niederrhein University of Applied Sciences (HSNR) for financial support.

References

[1] United States Food and Drug Administration, *Guidance for Industry: PAT—A Framework for Innovative Pharmaceutical Development, Manufacturing, and Quality Assurance*, U.S. Department of Health and Human Services, Washington, DC, USA, 2004.

[2] European Medicines Agency, *Reflection Paper: Chemical, Pharmaceutical and Biological Information to be Included in Dossiers when Process Analytical Technology (PAT) is Employed*, European Medicines Agency, Canary Wharf, London, UK, 2006.

[3] K. A. Bakeev, *Process Analytical Technology*, John Weiley & Sons Ltd, Chichester, England, 2nd edition, 2010.

[4] W. Chew and P. Sharratt, "Trends in process analytical technology," *Analytical Methods*, vol. 2, no. 10, pp. 1412–1438, 2010.

[5] A. Chanda, A. M. Daly, D. A. Foley et al., "Industry perspectives on process analytical technology: tools and applications in API development," *Organic Process Research & Development*, vol. 19, no. 1, pp. 63–83, 2014.

[6] R. J. A. Do Nascimento, G. R. De Macedo, E. S. Dos Santos, and J. A. De Oliveira, "Real time and in situ near-infrared spectroscopy (NIRS) for quantitative monitoring of biomass, glucose, ethanol and glycerine concentrations in an alcoholic fermentation," *Brazilian Journal of Chemical Engineering*, vol. 34, no. 2, pp. 459–468, 2017.

[7] R. Schalk, F. Braun, R. Frank et al., "Non-contact Raman spectroscopy for in-line monitoring of glucose and ethanol during yeast fermentations," *Bioprocess and Biosystems Engineering*, vol. 40, no. 10, pp. 1519–1527, 2017.

[8] K. Dégardin, A. Guillemain, and Y. Roggo, "Comprehensive study of a handheld Raman spectrometer for the analysis of counterfeit medicines," *Journal of Spectroscopy*, vol. 2017, Article ID 3154035, 13 pages, 2017.

[9] B. Blümich and K. Singh, "Desktop NMR and its applications from materials science to organic chemistry," *Angewandte Chemie International Edition*, vol. 57, no. 24, pp. 2–17, 2017.

[10] M. F. Isaac-Lam, "Determination of alcohol content in alcoholic beverages using 45 MHz benchtop NMR spectrometer," *International Journal of Spectroscopy*, vol. 2016, Article ID 2526946, 8 pages, 2016.

[11] S. Kern, K. Meyer, S. Guhl et al., "Online low-field NMR spectroscopy for process control of an industrial lithiation reaction—automated data analysis," *Analytical and Bioanalytical Chemistry*, vol. 410, no. 14, pp. 3349–3360, 2018.

[12] F. Dalitz, M. Cudaj, M. Maiwald, and G. Guthausen, "Process and reaction monitoring by low-field NMR spectroscopy," *Progress in Nuclear Magnetic Resonance Spectroscopy*, vol. 60, pp. 52–70, 2012.

[13] N. Zientek, C. Laurain, K. Meyer, M. Kraume, G. Guthausen, and M. Maiwald, "Simultaneous 19F-1H medium resolution NMR spectroscopy for online reaction monitoring," *Journal of Magnetic Resonance*, vol. 249, pp. 53–62, 2014.

[14] D. A. Snyder, R. Bruschweiler, G.E. Morris, and J.W. Emsley, "Multidimensional correlation spectroscopy by covariance NMR," in *Multidimensional NMR Methods for the Solution State*, John Wiley & Sons Ltd, Chichester, UK, pp. 97–105, 2009.

[15] M. Jaeger and R. L. E. G. Aspers, "Covariance NMR and Small Molecule Applications," *Annual Reports on NMR Spectroscopy*, vol. 83, pp. 271–349, 2014.

[16] I. Noda, A. E. Dowrey, C. Marcott, G. M. Story, and Y. Ozaki, "Generalized two-dimensional correlation spectroscopy," *Applied Spectroscopy*, vol. 54, no. 7, pp. 236A–248A, 2000.

[17] I. Noda, "Generalized two-dimensional correlation method applicable to infrared, Raman, and other types of spectroscopy," *Applied Spectroscopy*, vol. 47, no. 9, pp. 1329–1336, 1993.

[18] I. Noda, "Two-dimensional infrared spectroscopy," *Journal of the American Chemical Society*, vol. 111, no. 21, pp. 8116–8118, 1989.

[19] N. Trbovic, S. Smirnov, F. Zhang, and R. Brüschweiler, "Covariance NMR spectroscopy by singular value decomposition," *Journal of Magnetic Resonance*, vol. 171, no. 2, pp. 277–283, 2004.

[20] R. Brüschweiler, "Theory of covariance nuclear magnetic resonance spectroscopy," *Journal of Chemical Physics*, vol. 121, no. 1, pp. 409–414, 2004.

[21] C. D. Eads and I. Noda, "Generalized correlation NMR spectroscopy," *Journal of the American Chemical Society*, vol. 124, no. 6, pp. 1111–1118, 2002.

[22] L. J. Frasinski, K. Codling, and P. A. Hatherly, "Covariance mapping: a correlation method applied to multiphoton multiple ionization," *Science*, vol. 246, no. 4933, pp. 1029–1031, 1989.

[23] I. Noda, "Modified two-dimensional correlation spectra for streamlined determination of sequential order of intensity variations," *Journal of Molecular Structure*, vol. 1124, pp. 197–206, 2016.

[24] L. R. Terra, M. N. Catrinck, and R. F. Teófilo, "MCR-ALS applied to the quantification of the 5-hydroxymethylfurfural using UV spectra: study of catalytic process employing experimental design," *Chemometrics and Intelligent Laboratory Systems*, vol. 167, pp. 132–138, 2017.

[25] W. Kessler and R. W. Kessler, "Multivariate curve resolution: a method of evaluating the kinetics of biotechnological reactions," *Analytical and Bioanalytical Chemistry*, vol. 384, no. 5, pp. 1087–1095, 2006.

[26] Y. Le Dréau, N. Dupuy, J. Artaud, D. Ollivier, and J. Kister, "Infrared study of aging of edible oils by oxidative spectroscopic index and MCR-ALS chemometric method," *Talanta*, vol. 77, no. 5, pp. 1748–1756, 2009.

[27] R. Tauler, "Multivariate curve resolution applied to second order data," *Chemometrics and Intelligent Laboratory Systems*, vol. 30, no. 1, pp. 133–146, 1995.

[28] R. J. Pell and X. Chen, "Multivariate curve resolution for understanding complex reactions," *Journal of Chemometrics*, vol. 28, no. 5, pp. 411–419, 2014.

[29] A. Güvenç, N. Kapucu, and Ü. Mehmetoğlu, "The production of isoamyl acetate using immobilized lipases in a solvent-free system," *Process Biochemistry*, vol. 38, no. 3, pp. 379–386, 2002.

[30] F. W. Welsh, W. D. Murray, R. E. Williams, and I. Katz, "Microbiological and enzymatic production of flavor and fragrance chemicals," *Critical Reviews in Biotechnology*, vol. 9, no. 2, pp. 105–169, 1989.

[31] C. H. Dickinson, J. Watson, and B. Wallace, "An impression method for examining epiphytic micro-organisms and its

application to phylloplane studies," *Transactions of the British Mycological Society*, vol. 63, no. 3, pp. 616–619, 1974.

[32] Z. P. Chen, J. Morris, A. Borissova et al., "On-line monitoring of batch cooling crystallization of organic compounds using ATR-FTIR spectroscopy coupled with an advanced calibration method," *Chemometrics and Intelligent Laboratory Systems*, vol. 96, no. 1, pp. 49–58, 2009.

[33] G. Mazarevica, J. Diewok, J. R. Baena, E. Rosenberg, and B. Lendl, "On-line fermentation monitoring by mid-infrared spectroscopy," *Applied Spectroscopy*, vol. 58, no. 7, pp. 804–810, 2004.

[34] W. Kessler and R. W. Kessler, "Multivariate curve resolution—integration von Wissen in chemometrische modelle," *Chemie Ingenieur Technik*, vol. 82, no. 4, pp. 441–451, 2010.

[35] J. Jaumot, N. Escaja, R. Gargallo, C. González, E. Pedroso, and R. Tauler, "Multivariate curve resolution: a powerful tool for the analysis of conformational transitions in nucleic acids," *Nucleic Acids Research*, vol. 30, no. 17, p. 92e, 2002.

[36] R. Tauler, "Multivariate curve Resolution : theory and applications," *Trends in Analytical Chemistry*, vol. 23, pp. 70–79, 2004.

[37] W. Kessler, *Multivariate Datenanalyse für die Pharma-, Bio- und Prozessanalytik*, Wiley-VCH, Weinheim, Germany, 2006.

[38] A. L. Pomerantsev and O. Y. Rodionova, "Process analytical technology: a critical view of the chemometricians," *Journal of Chemometrics*, vol. 26, no. 6, pp. 299–310, 2012.

[39] L. J. Frasinski, A. J. Giles, P. A. Hatherly, J. H. Posthumus, M. R. Thompson, and K. Codling, "Covariance mapping and triple coincidence techniques applied to multielectron dissociative ionization," *Journal of Electron Spectroscopy and Related Phenomena*, vol. 79, pp. 367–371, 1996.

[40] R. L. E. G. Aspers, P. E. T. J. Geutjes, M. Honing, and M. Jaeger, "Using indirect covariance processing for structure elucidation of small molecules in cases of spectral crowding," *Magnetic Resonance in Chemistry*, vol. 49, no. 7, pp. 425–436, 2011.

[41] M. Jaeger and R. Legner, "Homo- and hetero-covariance NMR spectroscopy and applications to process analytical technology," in *Spectroscopic Analyses—Developments and Applications*, E. Sharmin and F. Zafar, Eds., pp. 55–80, 2017.

[42] B. W. Hu, P. Zhou, I. Noda, and G. Z. Zhao, "An NMR approach applicable to biomolecular structure characterization," *Analytical Chemistry*, vol. 77, no. 23, pp. 7534–7538, 2005.

[43] B. Hu, J. P. Amoureux, and J. Trebosc, "Indirect covariance NMR spectroscopy of through-bond homo-nuclear correlations for quadrupolar nuclei in solids under high-resolution," *Solid State Nuclear Magnetic Resonance*, vol. 31, no. 4, pp. 163–168, 2007.

[44] W. Ehrfeld, V. Hessel, and H. Loewe, "Microreactors—new technology for modern chemistry," *Organic Process Research & Development*, vol. 5, no. 1, p. 89, 2001.

[45] R. Legner, S. Häfner, M. Voigt et al., "Micro process analytical technology—monitoring chemical processes in micro-reactors using miniature spectrometers," *GIT*, vol. 3-4, pp. 38–40, 2016.

[46] S. Morita, *2Dshige (c)*, Kwansei-Gakuin University, Nishinomiya, Japan, 2004-2005.

[47] J. Jaumot, A. de Juan, and R. Tauler, "MCR-ALS GUI 2.0: new features and applications," *Chemometrics and Intelligent Laboratory Systems*, vol. 140, pp. 1–12, 2015.

[48] Y. Mee Jung and I. Noda, "New approaches to generalized two—dimensional correlation spectroscopy and its applications," *Applied Spectroscopy Reviews*, vol. 41, no. 5, pp. 515–547, 2006.

[49] I. Noda, "Progress in two-dimensional (2D) correlation spectroscopy," *Journal of Molecular Structure*, vol. 799, no. 1–3, pp. 2–15, 2006.

[50] G. Socrates, *Infrared and Raman Characteristic Group Frequencies*, John Wiley & Sons Ltd., Chichester, UK, 3rd edition, 2004.

[51] G. Wider and L. Dreier, "Measuring protein concentrations by NMR spectroscopy," *Journal of the American Chemical Society*, vol. 128, no. 8, pp. 2571–2576, 2006.

[52] F. Malz and H. Jancke, "Validation of quantitative NMR," *Journal of Pharmaceutical and Biomedical Analysis*, vol. 38, no. 5, pp. 813–823, 2005.

[53] N. Viereck, L. Nørgaard, R. Bro, and S. B. Engelsen, "Chemometric analysis of NMR data," *Modern Magnetic Resonance*, pp. 1833–1843, 2008.

6

Free-Electron Screening Mechanism of the Shallow Impurity Breakdown in n-GaAs: Evidences from the Photoelectric Zeeman and Cyclotron Resonance Spectroscopies

O. Z. Alekperov[1] and E. P. Nakhmedov[1,2]

[1]*Institute of Physics, Azerbaijan National Academy of Sciences, Huseyn Javid Avenue 33, AZ-1143 Baku, Azerbaijan*
[2]*Faculty of Physics, Moscow State University, Baku Branch, Str. Universitetskaya 1, AZ-1144 Baku, Azerbaijan*

Correspondence should be addressed to E. P. Nakhmedov; enver.nakhmedov@msu.az

Academic Editor: Jau-Wern Chiou

A novel breakdown (BD) mechanism of shallow impurity (SI) under the electric field at low temperatures is suggested for n-GaAs samples with the donor concentrations $10^{14}\ \mathrm{cm}^{-3} \le N_D \le 10^{16}\ \mathrm{cm}^{-3}$ and the compensation degree $0.3 \le K = N_A/N_D \le 0.8$ with acceptors of concentration N_A in the external magnetic fields up to $H = 6.5T$, oriented parallel or perpendicular to the external electric field. Diagnosis of the BD mechanism was performed by SI Zeeman (mainly from the ground state 1s to $2p_{+1}$ and other excitation states) and cyclotron resonance photoelectric spectroscopy (PES) methods in the wide interval of the electric field including the BD region too. The obtained results reveal that the BD electric field ε_{BD} does not correlate with K and the carrier's mobility μ of the samples, which contradict to the well-known impact ionization mechanism (IIM). A serious discrepancy with IIM is that ε_{BD} does not almost depend on the magnetic field up to $H = 6.5T$ when $\varepsilon \| \mathbf{H}$ though the SI ionization energy increases two times. The cyclotron resonance (CR) measurements show that the line width does not depend on the electric field for $\varepsilon < \varepsilon_{BD}$, indicating the lack of free-carrier (FC) heating in contradiction with IIM. A considerable decrease of the free carriers' capture cross section (CCS) area by ionized SI centers with a subsequent increase in the FC concentration n is observed by means of PES investigation of the $1s \longrightarrow 2p_{+1}$ and CR lines in the electric fields $\varepsilon \le \varepsilon_{BD}$ and at different magnetic fields, applied along ($\mathbf{H} \| \varepsilon$) the electric field or perpendicular ($\mathbf{H} \perp \varepsilon$) to the electric field. The slope of the $1s \longrightarrow 2p_{+1}$ line intensity on the electric field for $\varepsilon \| \mathbf{H}$ does not depend on the magnetic field, which is valid for ε_{BD} too. Various effects determined in the PES measurements at $\varepsilon = \varepsilon_{BD}$, such as a drastic narrowing of the $1s \longrightarrow 2p_{+1}$ and CR lines, a shift of the CR line to higher magnetic fields, and disappearing of the lines to higher excited SI states, were clarified to be a result of screening of the SI Coulomb potential by free carriers. The FC screening at the BD reduces the potential fluctuation and its influence to the PES line shape of $1s \longrightarrow 2p_{+1}$ and other excited states. It is shown that an increase in the FC concentration reduces the CCS, which can be assumed as the main factor along with the increase in the ionization coefficient for the SI breakdown in the electric field. The screening length r_s of the SI Coulomb potential decreases with the increasing FC concentration, reducing the CCS; the latter seems to vanish completely at $r_s = a_B^*$ (a_B^* is the effective Bohr radius), when high screening results in vanishing of all the bound states of the Coulomb potential. Note that this limit is similar to the Mott transition. Many experimental facts and our calculation of the CCS support the suggested mechanism for the SI breakdown. The well-known IIM is valid for samples with SI concentrations $N \ll 10^{13}\ \mathrm{cm}^{-3}$ and takes place at very high electric fields.

1. Introduction

The gallium arsenide is one of the most utilized semiconductors in the modern electronics technology. n-GaAs epitaxial layers are the widely used heterojunctions for the investigation of the comprehensive class of 2D modern electronic and spintronic phenomena and for fabrication of different devices on their basis such as Gann diodes, photodetectors operating in the wide range of frequency, and high-frequency field transistors. [1–6]. In order to improve the electrophysical characteristics of these devices, fabricated by using high-purity semiconductors, it is necessary to know the chemical nature and the relative concentration of residual impurities. Many of these devices operate at low temperatures and sufficiently high electric fields when shallow impurities (SI) breakdown (BD) takes place (see, for review, e.g., [7, 8]). Therefore, more careful investigation of the SIBD mechanism in n-GaAs is essential.

The low-temperature submillimeter wave photoelectric spectroscopy (PES) of SI is the most sensitive method for identification of the SI contents in semiconductors [9]. Significant information on SI can be obtained from the photoconductivity spectral line width and the line broadening mechanism. The line width of the SI PES was shown [10, 11] to be determined by the concentrations of the major and compensated impurities as well as their distribution. The difference of ground state energies $E(1s)$ for different SI atoms, which is called a chemical shift (CS) or central cell correction, is a result of deviation (ΔV) of SI Coulomb potential $V(r) = e/(\epsilon_0 r) + \Delta V$ at small distances from pure Coulomb form. The value of the CS for SI in n-GaAs can be estimated to be $\Delta E \sim (a_0/a_B^*)^2 \epsilon_0 Ry^* \approx 0.1$ meV, where a_0 and a_B^* are the unit cell size and the effective Bohr radius, respectively, and Ry^* is the effective Rydberg. This means that the energies of all SI in n-GaAs lay in the interval of $Ry^* \pm \Delta E = (5.7 \pm 0.1)$meV; therefore, CS correction to higher excited states of SI ($E(2p_{+1})$ and $E(3p_{+1})$, ...) is much smaller than that for 1s state and can be neglected. For the purest semiconductors with donor concentrations of $N_D < 10^{14}\,cm^{-3}$, the photoexcitation lines of impurities, the intensity of which is proportional to the concentration of the corresponding impurity, are broadened due to the different values of the CS, and the broadening value for n-GaAs samples is the same order as the energy distance between impurities. Each impurity atom is considered in this case to be isolated, and the line width broadening is determined only by the charged impurity mechanism. As we will show in the following section, the quadratic Stark effect and the quadrupole-gradient shifting of neutral impurities' levels are responsible for the inhomogeneous broadening of the charged impurities. Note that such a small value of CS, which is the same order as the impurity photoexcitation line width, is inherent to the most of A_3B_5 semiconductors (InSb, GaAs, and InP). However, the difference of SI ionization energies for different impurities in Ge and Si is in the order of Ry^*. That is why the CS in these materials does not affect the line width.

For n-GaAs samples with the SI concentrations higher than $10^{14}\,cm^{-3}$, the picture is significantly different [12]. First of all, our experiments have determined that the line width for these samples does not correlate with the SI concentration, so that for a sample with smaller concentration, the line width may be larger than that of more doped samples (Table 1). Our investigations show that the line width of these samples is determined not only with the SI concentration but with the potential fluctuation due to the inhomogeneous SI donors and acceptors distribution too. Such an inhomogeneous SI distribution creates a potential fluctuation, which gives an additional contribution to the line width. At lower temperatures, close to zero, the electrons are captured by impurities, providing a correlated distribution of electrons, which is realized by minimizing the total Coulomb energy of the system of the charged donors, acceptors, and electrons [13]. At higher temperatures, when $T \gg T_c = e^2/\epsilon_0 r_m k_B$ with $r_m = (4\pi/3N_I)^{1/3}$ being the mean distance between charged impurities, the electrons are activated, and their distribution becomes random. The $1s \longrightarrow 2p_{+1}$ transition line width in this case depends on the charged impurity concentration $N_I = 2N_A$ and does not depend on the compensation $K = N_A/N_D$. Instead, the line width in the real experiments, which are realized at lower temperatures, the electrons distribution is correlated, since the acceptors are charged, taking an electron from the nearest donor [11], and the former strongly depends on K. A transformation from the correlated to the random distribution can be reached not only by increasing temperature but also by applying an external electric field.

The impurities in a strongly inhomogeneous sample gather into cluster with a higher impurity concentration. Hall measurement yields only the mean value of the impurity concentration. Since these clusters are optically active, they are responsible for the radiation absorption. The impurity orbitals of the neighboring donors in the cluster are overlapped. Therefore, the $1s \longrightarrow 2p_{+1}$ transition is allowed not only to the excited $2p_{+1}$ state of the owner atom but also to the neighboring impurity $2p_{+1}$ states. Note that just such kind of transitions cause potential fluctuations broadening of $1s \longrightarrow 2p_{+1}$ transition line width. The line width for these transitions depends on the radiation intensity also [14, 15]. The transition between different impurity states may provide an additional inhomogeneous broadening mechanism for inhomogeneous impurity distribution at higher concentrations. Therefore, identification of the lines becomes difficult due to inhomogeneous line broadening, which increases significantly at the SI concentrations higher than $10^{14}\,cm^{-3}$.

It is clear that the width of the impurity photoelectric excitation line has to be reduced in order to increase the resolution of the PES, which needs an investigation of the inhomogeneous broadening mechanisms of PES of SI. An additional interband illumination of samples causes a narrowing of SI PES line width [16] due to decreasing of charged impurities concentration. Nevertheless, our experiments showed that narrowing of the $1s \longrightarrow 2p_{+1}$ line in PES by interband illumination is effective only in samples with high compensations $K = N_A/N_D > 0.5$. We found that the SI PES line widths are considerably narrowed at the SI breakdown electric field.

The aim of this work is to investigate a mechanism of the electric field breakdown and of the broadening of the photoelectric excitation spectra lines of shallow impurities and the cyclotron resonance (CR) in n-GaAs samples with the donor concentrations $N_D \geq 10^{14}\,cm^{-3}$ and intermediate compensations. Effects of the electric breakdown to the line shapes of $1s \longrightarrow 2p_{+1}$ transition allow one to develop a method of determination of the impurity contents and their relative concentration in the highly doped samples.

TABLE 1: Characteristics of different n-GaAs samples, grown by liquid-phase epitaxy (LPE) and vapor-phase epitaxy (VPE); the experimentally measured ΔE and theoretically predicted δ values of the line widths of $1s \longrightarrow 2p_{+1}$ donor lines at different cyclotron frequencies.

Samples	N_D ($\times 10^{14}$ cm^{-3})	$K = N_A/N_D$	μ(78 K) ($\times 10^4$ (cm^2/V·s))	ΔE (meV) $H \approx 36$ kOe	ΔE (meV) $H \approx 61$ kOe	δ (meV)	$\delta/\Delta E$
1 VPE	4.33	0.52	10	0.064	0.07	0.214	3.06
2 VPE	7.35	0.65	7.3	0.07	0.08	0.270	3.38
3 LPE	8.67	0.53	7.3	0.15	0.16	0.271	1.69
4 VPE	1.46	0.38	5.7	0.085	0.09	0.297	3.03
5 LPE	4.22	0.36	8.6	0.136	0.14	0.194	1.39
6 VPE	4.31	0.52	10	0.062	0.065	0.214	3.29
7 VPE	High purity	—	—	0.012	0.008		

According to the existence theories [7, 8, 17–20], the SI breakdown occurs by means of the impact ionization of the neutral impurities by hot electrons, heated in the electrical field to energies more than $2E_i$. The idiom of "impact ionization" was introduced in physics firstly by Townsend [21] in order to explain the effects of an electrical discharging in gas. Furthermore, Ioffe argued [22] that the impact ionization may be a reason of the electric breakdown in solid insulators. The electron multiplication effect in Ge and Si p-n junctions [23] as well as in n-Ge and p-Ge [24, 25] was attributed to the impact ionization of impurities by free charge carriers. We present in this work a new mechanism of the SI PES line broadening and of the electric field breakdown mechanism of SI in a n-GaAs sample, consisting of the charged impurities screening with free electrons, which is an alternative one to the impact ionization. This SIBD mechanism takes place at higher concentrations of the SI when $N_D > 10^{13}$ cm^{-3}. A brief description of this mechanism was mentioned firstly in [26]. The nonhomogeneity degree of the impurity distribution in samples with similar concentrations of the major and the compensated impurities was shown to be determined by the width of $1s \longrightarrow 2p_{+1}$ SI PES in the linear part of the current-voltage characteristics (CVC) and by the electric field dependence of the width in the prebreakdown electric field.

The main scattering mechanism of electrons in the samples under our experimental condition is scattering on the ionized impurities. However, the investigation of the CR line width at the prebreakdown electric fields shows no correlation with ionized impurity concentration as it must be $\omega_c\tau \sim H\mu$. A reason of the CR line width broadening and different values of the cyclotron mass under the same experimental conditions is established to be a potential fluctuation due to nonhomogeneous distribution of the impurities. Drastic increase of the free-electron concentration at BD results in two kinds of CR line shift; the first shift is connected with influence of the plasma oscillation as a result of enhancement of the free-electron concentration, and it increases with decreasing magnetic field. The second shift, which takes place at higher magnetic fields, increases with the magnetic field due to decrease of influence of the fluctuation potential on CR line shape as a result of free-carrier screening. Note that $\Delta H < 0$ is at the plasma shift, while $\Delta H > 0$ is at the screening shift of the CR line. It is worthy to note that a comparison of the CR line shapes between breakdown and prebreakdown electric fields allows us to characterize a degree of the nonhomogeneous distribution of the impurities. The dependence of the CR line width on the electrical field around the breakdown $\varepsilon \approx \varepsilon_{BD}$ is experimentally observed to be nonhomogeneous: the CR line is considerably narrowed at the beginning part of the breakdown, whereas the line width increases and exceeds several times its own prebreakdown value. The analysis of the experimental data establishes that the breakdown takes place by means of the current filament formation in the sample. At the beginning stage of the breakdown, the released free electrons in the current filament screen the charged impurities and reduce their influence on the cyclotron motion of the electrons. With further increase in the electron concentration with the current, the electron-electron correlations dominate over the electron-impurity interactions, which leads to the broadening of the CR line width. The reason for the increase in the CR intensity with the electric field is established to be an increase in the free-electron concentration in the first Landau level and in the life time of the photoexcited electrons in the first Landau level due to reduction of the capture coefficient at the impurity centers.

The novelties can be summarized as follows:

(1) The PES of the Zeeman lines of n-CaAs samples with SI donor concentrations smaller than that corresponding to Mott transition 10^{14} cm^{-3} < $N_D < 10^{16}$ cm^{-3} and with the intermediate compensations $0.3 \le K = N_A/N_D = 0.3 \le 0.8$ at the breakdown electric field undergoes to drastic narrowing of the PES lines $1s \longrightarrow 2p_{+1}$ and $1s \longrightarrow 3p_{+1}$.

(2) All PES lines, corresponding to transitions from the ground state of the donors to the quasidiscrete excited states higher than $3p_{+1}$, disappear at the prebreakdown region of CVC.

(3) Drastic narrowing of the CR line width in PES occurs at the breakdown region of CVC of samples.

(4) All these experimental facts can be explained by screening of the charged impurity potential by free electrons released in the breakdown process. Calculations of the $1s \longrightarrow 2p_{+1}$ transition line shape for the two main line broadening mechanisms, namely, the quadratic Stark and the quadrupole-gradient effects, by replacing the surrounding neutral impurity charged impurities Coulomb potential with the screened Coulomb potential, $(e/\epsilon_0 r)\exp(-r/r_s)$, explain the narrowing of the FES line. The calculations yield that the $1s \longrightarrow 2p_{+1}$ line is narrowed several times for the value of the screening length

$r_s = (4\pi e^2 n^{-1} \epsilon_0^{-1})^{1/2}$ corresponding to free-electron concentration n smaller than the neutral impurity concentration.

(5) Two different kinds of shifts of the CR line at the breakdown takes place, and both of them are connected with abrupt increase of free-carrier concentration n at breakdown. The first shift ($\Delta H_{CR} < 0$), which increases with decrease of CR magnetic field value H_{CR}, is due to free-carrier plasma influence on CR line shape. The second shift ($\Delta H_{CR} > 0$), which increases with H_{CR}, is due to the screening of charged impurities potential fluctuations with free carriers.

(6) Two kinds of the CR are observed at small electric fields corresponding to the linear region of CVC in PES. The first is the free carriers CR line, at slightly smaller magnetic field with very low intensity in comparison with that of the broader line of the pseudo-free-carrier CR of electrons localized on fluctuation of potential CR (FPCR). With increase of the electric field, activation of carriers from the potential fluctuation to the zeroth Landau level causes strong increase in the free-carrier CR in comparison to the FPCR broad line. Such kind of CR electric field dependence is typical for samples with inhomogeneous distribution of impurities. So this fact can be used in characterization of impurity distribution in samples of n-GaAs with nearly equal N_D and K.

(7) The intensity $\Delta\sigma/\sigma$ of photoelectric excitation spectra $1s \longrightarrow 2p_{+1}$ and CR lines was shown to increase with the prebreakdown electric field due to the decrease in the capture cross section $\alpha(\varepsilon)$ and increase in the coefficient of thermal ionization from excited states $\beta(\varepsilon)$. So the concentration of free electrons was shown to increase exponentially at prebreakdown electric fields.

Our experiments have been done at different values of the magnetic field, $H = 37$ kOe and $H = 61$ kOe, and for two mutual orientations of the electric and magnetic fields, $\varepsilon \| \mathbf{H}$ and $\varepsilon \perp \mathbf{H}$. The experimental results confirm that a reason an increase in the $1s \longrightarrow 2p_{+1}$ line intensity in the electric field is a decrease in the capture cross section of the excited electrons by the impurity centers. Furthermore, the capture cross section for given values of the electric and magnetic fields depends on their mutual orientations: it takes a minimal value for $\varepsilon \| \mathbf{H}$ and a maximal value for $\varepsilon \perp \mathbf{H}$.

2. Experiment and Measurement Methods

The experiments were provided in six samples of n-GaAs epitaxial layers with the residual donor concentrations (10^{14} cm^{-3} $< N_D <$ 10^{16} cm^{-3}) and compensation degree $0.3 \le K = 0.3 \le 0.8$, grown on the high-resistive p-GaAs substrate by using the liquid-phase epitaxy (LPE) and vapor-phase epitaxy (VPE) methods. The samples of thickness, varying in the interval from 15 μm to 150 μm, have the surface area of several mm^2. Hall measurements of the samples [27] are provided to determine N_D, K, and the mobility μ. The submillimeter laser magnetic spectrometer was used to investigate the SI PES and CR line shapes. Submillimeter gas-discharge lasers operating on the flowing regime of $H_2O + H_2$, $D_2O + D_2$, and HCN$(CH_4 + N_2)$ vapors were used as a radiation source with the radiation wavelengths of 118.6 μm, 78.4 μm, 84.3 μm, and 337 μm. Photoconductivity spectra of CR and SI PES were registered at fixed quantum energies of lasers as a function of the magnetic field with scanning of magnetic field intensity up to $H = 65$ kOe of superconducting solenoid, setting in liquid helium cryostat at $T = 4.2$ K. The cross modulation method was used in registration of photoconductivity with modulation of radiation intensity at frequency 750 Hz. Alternating signal of photoconductivity was measured from load resistor R_L series connected with sample with resistivity of R_S by using the lock-on method. Voltage drops from R_L and R_S were used in CVC registration, respectively. Line shapes of CR and SI PES were studied at different regions of samples' CVC. At electric fields smaller than the breakdown one $\varepsilon < \varepsilon_{BD}$, the measurements were done in the constant voltage regime, when $R_L \ll R_S$ for different values of the electric field. In this case, the photosignal is proportional to $\Delta\sigma$ under the radiation. At prebreakdown and at "candle"-like region $\varepsilon \approx \varepsilon_{BD}$, the measurements are performed in a constant current regime, when R_L is much greater than sample resistivity R_S, $R_L \gg R_S$. In constant voltage regime, the PC signal is proportional to the change of conductivity $\Delta U \sim \Delta\sigma$ under the radiation, and as a result, PC line shape corresponds to real line shape at $\varepsilon < \varepsilon_{BD}$. However, it is not suitable at $\varepsilon \approx \varepsilon_{BD}$ due to the instability of electric parameters of sample. Constant current condition gives the possibility to detect line shapes at the candle-like region of CVC at different free-carrier concentrations. In this case, PC of sample $\Delta U \sim \Delta\rho = 1/\sigma_{dark} - 1/\sigma_{light} = (e\Delta n\mu)/(\sigma_{dark}^2 + \sigma_{dark} e\Delta n\mu)$, where σ_{dark} is a dark conductivity. At small dark conductivity $\sigma_{dark} \ll e\Delta n\mu$ (which corresponds to the small electric field), the PC signal would have a saturation with increase of Δn, as a result PC spectra do not correspond to true line shape. In the opposite case, which corresponds to the breakdown region ($\sigma_{dark} > e\Delta n\mu$), photosignal is proportional to the number of resonance transition Δn, and it corresponds to the true line shape. The analysis of PC signals ΔU at the breakdown region shows that the saturation of ΔU on the radiation intensity takes place at sufficiently high Δn comparable with impurity concentration. This means that the line widths of CR and $1s \longrightarrow 2p_{+1}$ transitions obtained at the breakdown in the current regime are only broader than the true line shapes. This is important for us because our conclusions partially are based on strong decreasing of these line shapes at SI breakdown. It is obvious that N-like regions of CVC has to be observed at constant voltage, whereas S-like regions of CVC is observed only at constant current regimes. Nevertheless, CVC of samples obtained at a constant current regime should not show an S-like region, as illustrated in CVC at different magnetic fields in Figure 1. Note that in the case of CVC recording in constant voltage regime at $\varepsilon \approx \varepsilon_{BD}$, the voltage drop switches from the sample

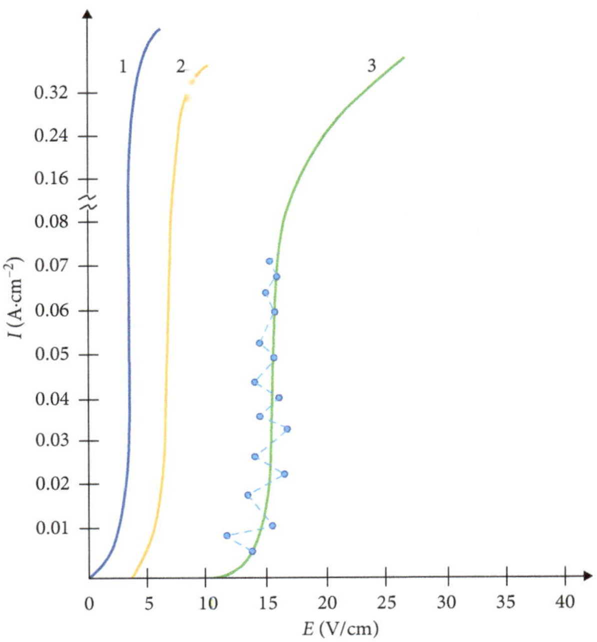

FIGURE 1: CVC of n-GaAs samples measured at different magnetic fields in the strictly constant current regime at $R_L > 100R_S$ (solid curves 1 to 3) and in the pseudocurrent regime when $R_L \approx 10R_S$ (dotted curves in 3).

to load resistor at the beginning of the breakdown, and this causes a decrease of ε on the sample; and as a result of this, an external voltage drop returns back to the sample after time τ_{cap}, when the excited free carriers are captured back to impurity sites again. This corresponds to the first cycle of the oscillation, which is known as low-frequency current oscillations on sample at impurity breakdown and other switching effects. The amplitude of these oscillations decreases with the increasing load resistance, and it disappears at the strong constant current regime. The points on the CVC line, obtained at th strong constant current condition, correspond to load resistance value nearly 10 times greater than the resistivity of sample at the beginning of the breakdown. As shown in Figure 1, fluctuations of voltage in the sample decrease at a strong current regime.

The parameters of the samples and their line width characteristics at different magnetic fields are given in Table 1. The line width ΔE was measured at half height of the line in magnetic field units ΔH, and then it was transformed into energetic units $\Delta E = (\partial E_{1s\longrightarrow 2p_{+1}}/\partial H)\Delta H$. The rate of $1s \longrightarrow 2p_{+1}$ transition energy increase in magnetic field $\partial E_{1s\longrightarrow 2p_{+1}}/\partial H \approx 0.18$ meV/kOe was determined at two different values of the magnetic field $H \approx 36.5$ kOe and $H \approx 60.5$ kOe. The preliminary CVC measurements have been done at the magnetic fields $\mathbf{H}$ corresponding to the top of the photoexcited spectra as well as in the electric fields ε crossed and parallel to the magnetic field. CVC of a typical sample, measured in the crossed magnetic and electric fields, for different magnetic fields is presented in Figure 1 at either fixed voltage (solid lines) or fixed current (dotted lines) regimes. A jump in the CVC at the electric field $\varepsilon = \varepsilon_{BD}$ indicates a breakdown of the neutral impurities. It is worthy that e_{BD} increases linearly with the magnetic field at $\varepsilon \perp \mathbf{H}$, whereas it does not depend practically on H at $\varepsilon \| \mathbf{H}$, as shown in Figure 2.

3. Influence of the Electric Field Breakdown on the SI Photoelectric Spectroscopy

According to the conventional explanation [9–11, 13], the main mechanism of the SI photoexcitation line broadening in the magnetic field is the quadratic Stark effect and the electric field gradient of the quadrupole interactions of charged impurities. Our experiments show that there exists an additional line broadening mechanism in the samples with high impurity concentrations, which is caused by the potential fluctuations. As shown in Table 1, there is no correlation of the PES line width neither with SI concentration nor with the degree of the compensation for $N_D \approx 10^{14}$ to 10^{15} cm^{-3} and $K \approx 0.3$ to 0.8. This evidence can be understood, providing that an inhomogeneous SI distribution plays an essential role in the line broadening. At high impurity concentrations, the impurity states are broadened forming an impurity band. In this case, a transition takes place not only from the 1s ground state to the $2p_{+1}$ excited state of a given neutral donor but also to the $2p_{+1}$ excited state of nearest-neighboring charged donors due to partially overlapping of the excited states' wave functions at high SI concentrations. The distribution function of the charged SI, randomly distributed in the sample, was proposed to be in the Gaussian form [28]:

$$P(e\varphi) = \frac{1}{\delta\sqrt{2\pi}}\exp\left(-\frac{e^2\varphi^2}{2\delta^2}\right), \quad (1)$$

where δ is the half width of Gaussian distribution, $\delta = 0.29e^2N_D^{1/3}K^{1/4}\epsilon_0^{-1}$ characterizes the theoretically predicted value of the line width E_0, and δ increases with the impurity concentration and the compensation degree. This line broadening mechanism dominates at $N_D > 10^{14}$ cm^{-3} in n-GaAs, [12]. The values of δ for the samples under investigation with $\epsilon_0 = 12.5$ are given in Table 1. Although the magnitudes of δ is of the same order as ΔE, they are 1.5 and 3 times greater than the experimental values of the $1s \longrightarrow 2p_{+1}$ line widths for VPE and LPE samples, respectively.

A discrepancy between the theoretical and experimental values of the PES line width seems to be explained by screening of the charged impurity potential by electron gas. The screening of δ can be qualitatively described as

$$\delta^* = \delta\,\exp\left(-\frac{r_m}{r_s}\right), \quad (2)$$

where $r_m = (4\pi/3N_I)^{1/3}$ is the mean distance between the charged impurities $N_I \approx 2N_A$ and $r_s = (\epsilon_0 k_B T/4\pi e^2 n)^{1/2}$ is the Debye screening radius. For the samples in Table 1, $N_A \approx (2-4)\times 10^{14}$ cm^{-3} and one can estimate $r_m \approx 20a_B$. At SI BD, around ~10% of all neutral donors are ionized, which yields $r_s \approx (10-20)a_B$ at $T \approx 4.2$ K. These values

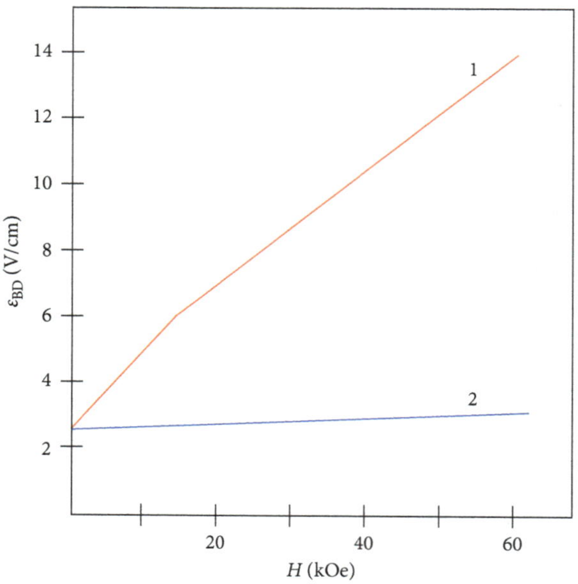

FIGURE 2: Dependence of the breakdown electric field ε_{BD} on the magnetic field H for different n-GaAs samples when the electric ε_{BD} and magnetic **H** fields are (1) perpendicular $\varepsilon_{BD} \perp \mathbf{H}$ and (2) parallel $\varepsilon_{BD} \| \mathbf{H}$ each other. A weak dependence of ε_{BD} on H may be caused by an error in the setting of a rotation angle of the sample.

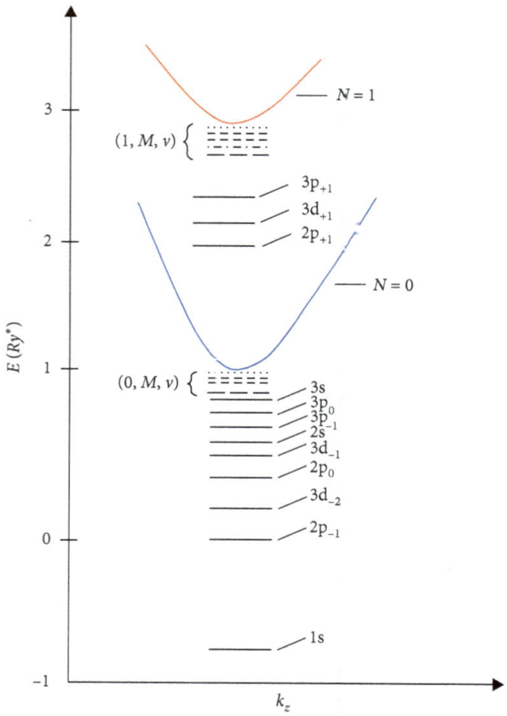

FIGURE 3: The shallow donor levels in n-GaAs in the presence of the magnetic field corresponding to $\gamma = 1$. The $2p_{+1}$ excited impurity state lies higher than the $N = 0$ zeroth Landau level under $N = 1$. The zero energy is labeled from the bottom of the conduction band at $H = 0$.

reduce the value of δ to $\delta^* \approx (0.14-0.20)\delta$, which is in consistence with the experimentally measured value.

The requirement that δ and the experimentally measured half line width ΔE take the same values imposes a condition that the intraimpurity transitions have to be neglected at all, and interimpurity transitions not only between the nearest-neighboring donors but between any two impurities have to be realized with equal probability. δ takes larger values yielding wider line width where a local donor concentration is higher in the inhomogeneous impurity distribution even for a sample with smaller N_D. Therefore, the value $\Delta E/\delta$ as well as $(\Delta E - \delta)$ can serve as a measure of SI inhomogeneous distribution degree in samples with nearly equal values of N_D and K. One concludes that if a sample with higher SI concentration has smaller line width then it has more homogeneous distribution of the donors. The value of $\delta/\Delta E$ is a suitable parameter for comparison of a sample quality, so that the higher the $\delta/\Delta E$ the better the quality of a sample in comparison with other ones with less homogeneous SI distribution. This parameter is shown in Table 1 for $H \approx 61$ kOe only since neither δ nor ΔE depends on the magnetic field.

The energy distance or gap between the 1s ground state and $2p_{+1}$ excited state of electron in the impurity increases with the magnetic field, and the resonance $1s \longrightarrow 2p_{+1}$ transition occurs when the gap becomes equal to the radiation energy. At $H > 20$ kOe, which corresponds to the intermediate magnetic fields $0 < \gamma = \hbar\omega_c/2Ry^* < 1$ with $\omega_c = eH/m^*c$ being the cyclotron frequency ($\gamma = 1$ in n-GaAs corresponds to $H \approx 65$ kOe), the $2p_{+1}$ level of shallow impurities in n-GaAs lies higher than the zeroth Landau level ensuring a transition for the photoexcited electrons to the conduction band (Figure 3). Therefore, there is not a necessity to argue other mechanism, like field-induced tunneling or impact ionization from the excited $2p_{+1}$ state to explain a generation of the PES signal.

It is worthy to note that the line shape of high purity samples in the PES is asymmetric (Figure 3 in [12]). The line width at a lower energy or higher magnetic field side of the $1s \longrightarrow 2p_{+1}$ transition line is considerably wider in accordance with the quadratic Stark effect broadening mechanism. Instead, all other samples have symmetric line shape. The line width for the $1s \longrightarrow 2p_{+1}$ transition of the purest sample is 1.5 times narrowed when the magnetic field increases from $H \approx 36$ kOe to ~ 51 kOe, which confirms also a validity of the quadratic Stark broadening mechanism [10].

PES of the SI donors at the breakdown electric field is measured at the constant current regime, which ensures stability of the electrical parameters of samples during PES registration. The impurity PES line at sufficiently higher current density, corresponding to the deep breakdown regime at $\varepsilon \approx \varepsilon_{BD}$, has the following features:

(i) The transition lines $1s \longrightarrow 2p_{+1}$ and $1s \longrightarrow 3p_{+1}$ are narrowed from 5 up to 10 times with the increasing current density for all samples under investigation (Figure 4).

(ii) The $1s \longrightarrow 2p_{+1}$ line acquires an asymmetric form while a narrowing takes place, and the half width becomes wider at higher magnetic fields. This

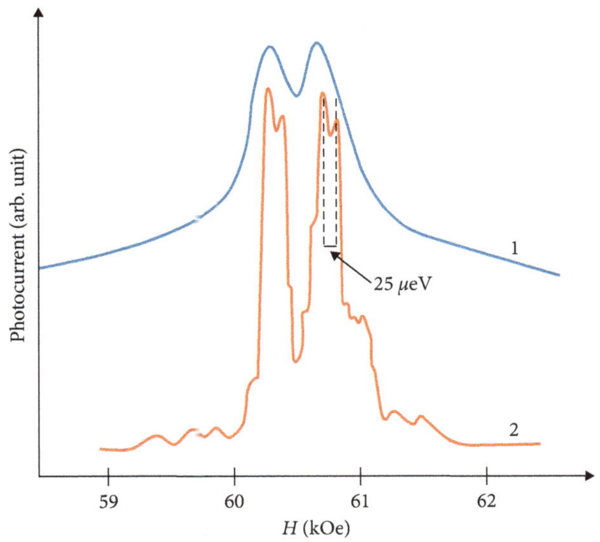

Figure 4: Shape of the 1s $\longrightarrow$ $2p_{+1}$ line of n-GaAs sample 6 at the radiation energy of $E = 14.7$ meV and $T = 4.2$ K, $K = 0.5$ for (1) $\varepsilon < \varepsilon_{BD}$ in the constant field regime and (2) $\varepsilon \approx \varepsilon_{BD}$ in the constant current regime at $J =$ [illegible]7 μA. The PES lines are strongly narrowed at the breakdown, the fine structure of each line and new lines with smaller intensities appear at the tails of the main line.

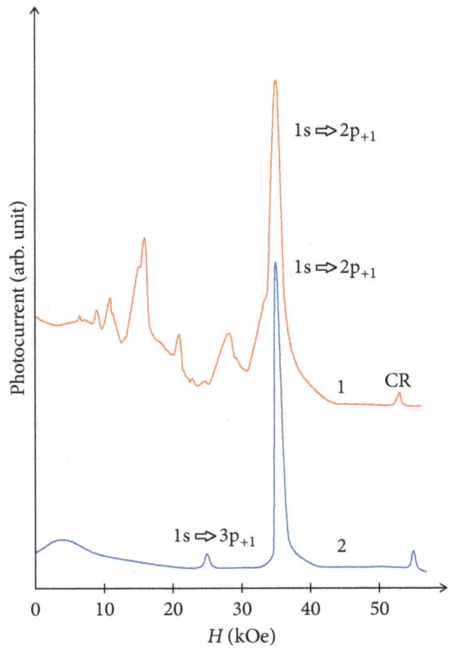

Figure 5: Photoelectric spectra of n-GaAs samples at the radiation energy $E = 10.45$ meV and $T = 4.2$ K for (1) $\varepsilon < \varepsilon_{BD}$ and (2) $\varepsilon \approx \varepsilon_{BD}$.

behavior is similar to that observed for a pure sample.

(iii) The lines in the photoelectric spectra, corresponding to the transitions from the 1s ground state to the quasi-discrete Boyle-Howard's $\{N, M, \nu\}$ states [29], determined by the main quantum number N corresponding to the Landau levels, the magnetic quantum number M, and the number ν of the bounded discrete Coulomb levels (the Coulomb quantum number, Figure 3), disappear under $N \geq 1$ with the increasing breakdown current (Figure 5); i.e., apart from the 1s $\longrightarrow$ $2p_{+1}$ transition, the line 1s $\longrightarrow$ $3p_{+}$ disappears too.

(iv) Intensity of the spectra at the tails strongly decreases.

(v) The new lines in the PES emerge due to narrowing of the 1s $\longrightarrow$ $2p_{+1}$ line, which correspond to residue SI donors with smaller concentrations (Figure 4).

(vi) Narrowing of the 1s $\longrightarrow$ $2p_{+1}$ line results in an appearance a fine structure of each donor line in consequence of interactions of an electron spin with the external magnetic field (Figure 4). The revealing of the spin splitting in the PES line makes the line structure very similar to those observed in the purest samples.

The above specified features of the PES at the breakdown electric field can be explained by assuming that the free electrons, released at the breakdown, screen the Coulomb potential of charged impurities. Note that a screening influences to excited states of neutral donors much more, resulting in their disappearance.

The estimations show that the line broadening due to the excitation finite life time is negligibly small, $\sim 10\,\mu$eV [30], in comparison to the experimentally observed data even in an ultrapure n-GaAs samples. Therefore, the homogeneous broadening mechanism can be neglected for the samples under investigation.

The nonhomogeneous PES broadening is determined as a shift between the transition energy of the impurity states in crystal and that of an isolated hydrogen-like impurity atom. In this case, the PES line shape is determined by statistic distribution of the optical transitions from the ground state to the excited ones for all impurity atoms. The nonhomogeneous broadening of the PES line can be caused by the following factors [10, 11, 31]: (i) the interactions between the neutral impurity atoms, (ii) quadrupole-gradient interactions of the charged and neutral impurity atoms, and (iii) the quadratic Stark effect of the electric field was created by the randomly distributed charged impurities on neutral impurities. Note that the linear Stark effect is absent in the magnetic field [10].

Interactions between the neutral donors have a dipole-dipole interaction origin, which broaden the line width symmetrically in comparison with an unperturbed transition line. The dipole-dipole interactions mediated half width of the line in the absence of the magnetic field and are given [9] by $\Delta_{d-d} \approx 6N_D^0 a_B^{*3} \epsilon_i$, which can be estimated to vary $2 \leq \Delta_{d-d} \leq 20\,\mu$eV for the neutral donor concentration $10^{14}\,\text{cm}^{-3} \leq N_D^0 \leq 10^{15}\text{cm}^{-3}$. This value in the intermediate magnetic field, $0.5 \leq \gamma \leq 1.0$, is much more smaller due to squeezing of the electronic wave functions in the excited states.

We show that the distribution functions of quadratic electric field $P_1(\varepsilon^2)$ and of the electric field gradient $P_2(\partial\varepsilon_z/\partial z)$, which determine the quadratic Stark effect and the gradient broadening of PES line, respectively, are squeezed due to screening of the Coulomb potential of surrounding charged impurities by free electrons. Calculations of the distribution functions of ε^2 and $\partial\varepsilon_z/\partial z$ on the neutral donors are performed according to the method developed by Shklovskii and Efros [32]. In order to realize the ground state of a doped compensated semiconductor, the coordinates of N donors and KN acceptors are generated in a cube of size $L = (N/N_D)^{1/3}$ by means of the random numbers generator. Coulomb energy of the system is calculated by random distribution of $(1-K)N$ electrons over the donors. The minimal energy of the system, corresponding to the ground state, is reached by permuting the electrons over the filled and unfilled donor atoms. For a given realization of the ground state, $\varepsilon^2(m)$ and $\partial\varepsilon_z(m)/\partial z$ at the mth neutral donor are calculated by summing up the electric field and its gradient created by the N charged donors and KN acceptors [11, 33]:

$$\varepsilon^2(m) = \left[-\sum_{i=1}^{N}(1-n_i)\frac{\mathbf{R}_{im}}{R_{im}^3} + \sum_{j=1}^{KN}\frac{\mathbf{R}_{jm}}{R_{jm}^3}\right]^2, \tag{3}$$

$$\frac{\partial\varepsilon_z(m)}{\partial z} = \sum_{i=1}^{N}\frac{(1-n_i)}{R_{im}^3}\left(1-\frac{3Z_{im}^2}{R_{im}^2}\right) - \sum_{j=1}^{KN}\frac{1}{R_{jm}^3}\left(1-\frac{3Z_{jm}^2}{R_{jm}^2}\right), \tag{4}$$

where $n_i = 1$ if a donor is neutral and $n_i = 0$ if it is ionized; $\mathbf{R}_{im} = \mathbf{R}_i - \mathbf{R}_m$ is a distance between the mth donor and all other impurities. The dimensionless radius vectors $\mathbf{R}_i$, the square of the electric field ε^2, and its derivative $\partial\varepsilon_z/\partial z$ are expressed in units $N_D^{-1/3}$, $(eN_D^{2/3}/\epsilon_0)^2$, and eN_D/ϵ_0, respectively. The directions of the Cartesian coordinate system are chosen to be along the cube axes. ε^2 and $\partial\varepsilon_z/\partial z$ are calculated according to equations (3) and (4) for each realization and averaged over these realizations. In our calculations, $N = 800$ donor atoms in the cube are chosen and the parameters are averaged over 30 realizations of donor and acceptor coordinates. The electrons in the impurities are distributed by correlated manner, corresponding to zero temperature, when the acceptors are charged by taking an electron from the nearest-neighboring donors. The distribution functions of ε^2 and $\partial\varepsilon_z/\partial z$ for the sample with $N_D = 4\times10^{14}\ \text{cm}^{-3}$ and $K = 0.5$ are presented, in Figures 6 and 7, respectively.

At the breakdown electric field, free electrons with the concentration n, determined by the current density $j = en\mu_c\varepsilon_{BD}$, are generated in the conduction band. Beside n free electrons, there appear the same amount of positively charged donors, and the charge concentration becomes $N_I = 2KN_D + n$. The free-electron concentration n even at strong breakdown $\varepsilon = \varepsilon_{BD}$ amounts to a few percent of the total neutral donors $n \ll N_I \approx 2KN_D$, which was confirmed by the experimental studies of the plasma shift of the CR line in n-GaAs [34].

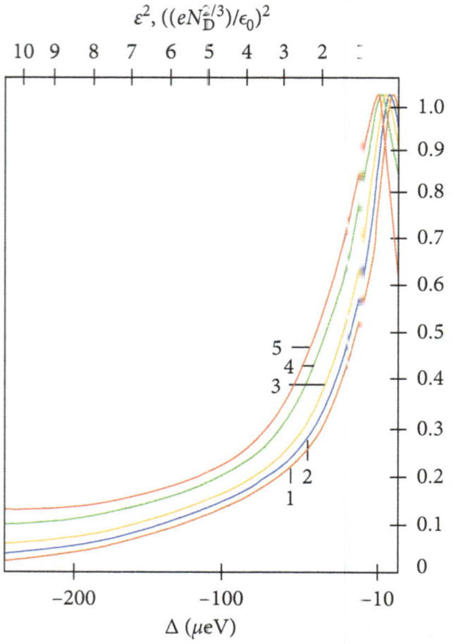

FIGURE 6: Shape of the $1s \longrightarrow 2p_{+1}$ line, determined by the quadratic Stark effect at $H = 0$, $N_D a_B^{*3} = 4\times10^{-4}$, $K = 0.5$, $T = 0$, for different impurity potentials: (1) $r_s = 10a_B^*$; (2) $r_s = 20a_B^*$; (3) $r_s = 30a_B^*$; (4) $r_s = 40a_B^*$; and (5) Coulomb potential. ε^2 is given in unit of $(eN_D^{2/3}/\varepsilon_0)^2$.

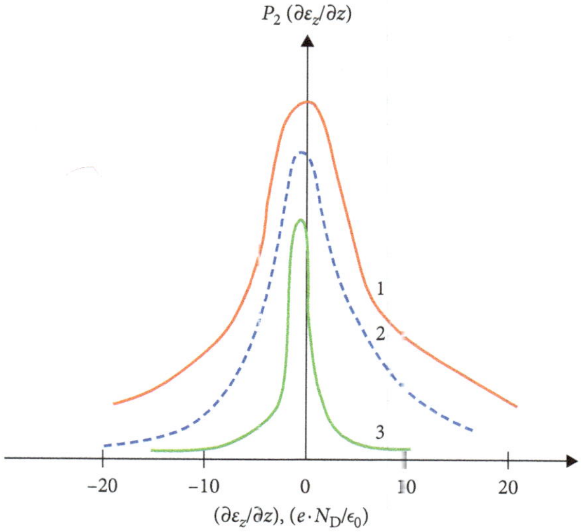

FIGURE 7: Distribution of $P_2(\partial\varepsilon_z/\partial z)$ on neutral donor at $N_D a_B^{*3} = 4\times10^{-4}$ and $K = 0.5$ for different impurity potentials (1) Coulomb potential at $T = \infty$ (noncorrelated case), (2) Coulomb potential at $T = 0$, and (3) screened potential at $T = 0$ with $r_s = 10a_B^*$. The gradient $\partial\varepsilon_z/\partial z$ is given in unit of eN_D/ϵ_0.

It is necessary to take into account screening of the Coulomb potential of charged impurities by the free electrons at the breakdown. At enough high concentration n of the free electrons, when the Debye radius $r_D = (\epsilon_0 k_B T/4\pi e^2 n)^{1/2}$ is smaller than the mean distance R_i between the neutral donors and the charged impurities, $r_D \le R_I$, the distribution probabilities ε^2 and $\partial\varepsilon_z/\partial z$ can be calculated by replacing in equations (3) and (4) the Coulomb potential $e/\epsilon_0 r$ by the screened potential $(e/\epsilon_0 r)\exp(-(r/r_D))$. This replacement yields

$$\varepsilon^2(m) = \left[-\sum_{i=1}^{N}(1-n_i)\frac{\mathbf{R}_{im}}{R_{im}^2}\left(\frac{1}{R_{im}}+\frac{1}{r_D}\right)e^{-(R_{im}/r_D)} + \sum_{j=1}^{KN}\frac{\mathbf{R}_{jm}}{R_{jm}^2}\left(\frac{1}{R_{im}}+\frac{1}{r_D}\right)e^{-(R_{im}/r_D)}\right]^2, \quad (5)$$

$$\frac{\partial\varepsilon_z(m)}{\partial z} = \sum_{i=1}^{N}\frac{(1-n_i)}{R_{im}^3}\left(2-\frac{5Z_{im}^2}{R_{im}^2}-\frac{Z_{im}^2}{r_D R_{im}}-\frac{Z_{im}^2}{r_D^2}\right)e^{-(R_{im}/r_D)} - \sum_{j=1}^{KN}\frac{1}{R_{jm}^3}\left(2-\frac{5Z_{jm}^2}{R_{jm}^2}-\frac{Z_{jm}^2}{r_D R_{jm}}-\frac{Z_{jm}^2}{r_D^2}\right)e^{-(R_{jm}/r_D)}. \quad (6)$$

The distribution functions calculated for n-GaAs at $N_D = 4\times10^{14}\,\text{cm}^{-3}$ and $K = 0.5$ are depicted in Figures 6 and 7 for different values of the screening radius.

The line width caused by the interaction of the quadrupole moment of a neutral impurity with the electric field gradient can be calculated according to the following expression [10]:

$$\Delta = 2.53(\Delta Q)Ry^*(N_I a_B^* 3), \quad (7)$$

where ΔQ is a difference of the quadrupole moments of the ground state and an excited state. Note that a dependence of quadrupole moments of several states on the magnetic field for an isolated hydrogen-like atom is presented in [10]. Estimation of Δ for a sample with $N_D \approx 4\times10^{14}\,\text{cm}^{-3}$ and $K = 0.5$ in the magnetic field when $\gamma \sim 0.5$ yields 1 μeV and 0.7 μeV, respectively, for noncorrelated and correlated electron distributions.

In Figure 8, we show the dependence of the half width ΔE, caused by the quadratic Stark effect, on the screening radius, calculated for the 1s $\longrightarrow$ $2p_{+1}$ line according to the following formula:

$$\Delta E = 75.75(2Ry^*)(N_D a_B^* 3)^{4/3}\varepsilon^2. \quad (8)$$

All these calculations show that the distribution probabilities ε^2 and $\partial\varepsilon_z/\partial z$ are squeezed and their long tails disappear gradually with decrease in the screening radius or increase in the free-electron concentration. In order to narrow the 1s $\longrightarrow$ $2p_{+1}$ line, e.g., 4 times, the Debye radius has to be $r_D \sim 10a_B^*$ (Figure 8) which corresponds to $n \sim 2\times10^{13}\,\text{cm}^{-3}$ for the free-electron concentration or 10% of the total neutral donor concentration. Possible ionization of such a high concentration of neutral donors at $\varepsilon = \varepsilon_{BD}$ seems to be doubtful. Nevertheless, we have to note that (i) although the 1s $\longrightarrow$ $2p_{+1}$ line shape was calculated for zero magnetic field, the experiments were done at strong magnetic fields when $\gamma = \hbar\omega_c/2Ry^* \sim 1$, and (ii) the most line narrowing occurs in the "candle"-like region of the sample CVC. It is known that a current filament with a high carrier concentration, much more than the average bulk concentration, is formed in this region. We think that the main

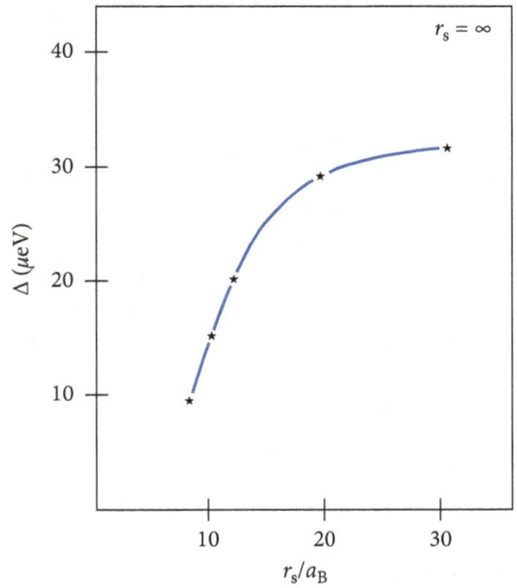

FIGURE 8: The numerical dependence of the 1s $\longrightarrow$ $2p_{+1}$ line width at half-height, caused by the quadratic Stark effect, on the screening radius at $H = 0$, $N_D a_B^{*3} = 4\times10^{-4}$, $K = 0.5$, and $T = 0$.

contribution to the current comes from the filaments, which determines the PES line shape at the breakdown. At higher concentration of the free electrons, not only the charged ions surrounding neutral impurities are screened, but the screening affects on the bound state spectrum of neutral atoms too. The screened Coulomb potential $(e/\epsilon_0 r)\exp(-r/r_s)$, in difference from the "bare" Coulomb one $e/(\epsilon_0 r)$, has the finite number of the discrete states. Investigation of the screened Coulomb potential spectrum [35] in the absence of a magnetic field shows that all discrete states higher than $N \geq 4$ disappear for the screening radius $r_s = 10a_B^*$, and the states $N \geq 3$ disappear for $r_s = 9a_B^*$. The quasi-discrete states of the Coulomb potential in a magnetic field are higher excited states with large radii; so they are more sensitive to the screening of free electrons. Therefore, a disappearance of the line corresponding to the transitions from the 1s ground state to the quasi-discrete $\{N, M, \nu\}$ excited states at the breakdown electric field (Figure 5) must be explained as a result of the screening.

The narrowing of the 1s $\longrightarrow$ $2p_{+1}$ PES line at the breakdown electric field reveals a fine structure of the line due to the spin splitting of particular donor's states especially for donor with higher concentrations. Similar splitting of the shallow impurities 1s $\longrightarrow$ $2p_{+1}$ lines in GaAs was observed in [36, 37]. The splitting in the samples with the parameters similar to those in our samples was reported [37] at 119 μm wavelength due to the narrowing of the 1s $\longrightarrow$ $2p_{+1}$ line by means of an additional illumination of the sample with a fundamental absorption edge. The authors of [36] argue that the spin splitting of 1s and $2p_{+1}$ states differ each other in the magnetic field with a small difference in the 1s $\longrightarrow$ $2p_{+1}$ transition energies, e.g., different g-factor for 1s and $2p_{+1}$ states, whereas a reason of the splitting is assumed in [37] to be an exchange interaction between neutral donors. We have observed this splitting in one sample at different values of the magnetic field, $H = 36.5$ kOe and $H = 61$ kOe, by narrowing the $1s \longrightarrow 2p_{+1}$ line at the breakdown electric field. The measurements at different magnetic fields allow us to determine how the splitting depends on the magnetic field. The value of the splitting at $H = 61$ kOe and at the absorption energy of $\epsilon_{qu} = 14.7$ meV was measured to be $\sim 25\,\mu$eV. The splitting value can be estimated to be proportional to the difference of the Landé g-factors corresponding to the 1s ground state g_{1s} and the excited $2p_{+1}$ state $g_{2p_{+1}}$, provided that the splitting is caused by the different g-factors, $\Delta g = (|g_{1s}| - |g_{2p_{+1}}|) = \Delta E/\mu_B H \approx 0.07$. This value of Δg is very close to the difference of the g-factors of 0th and 1th Landau levels, determined from the CR splitting ($\Delta g \approx 0.074$) at the magnetic field $H = 61$ kOe [30]. It is necessary to note that the similar splitting of the shallow donors line $1s \longrightarrow 2p_{+1}$ measured by the photoexcited spectroscopy method [38, 39] has not been observed in enough pure GaAs samples up to the magnetic fields ~ 100 kOe. The reason of this fact may be that the g-factors, corresponding to the 1s and $2p_{+1}$ states, take very close values while both states are placed under $N = 0$ Landau level. Nevertheless, this fact does not deny to explain the line splitting by means of the exchange interactions. Indeed, the exchange interactions should vanish at low donor concentrations due to exponential weakening of the interactions. On the contrary, the exchange interactions are sensitive to the magnetic field. We think that a clarification of a correct mechanism of the fine structure of the PES line at the breakdown needs further experimental investigations as well as theoretical studies of a magnetic field dependence of the exchange interactions.

3.1. Dependence of $1s \longrightarrow 2p_{+1}$ Line Intensity on Prebreakdown Electric Field. As mentioned above, the PES signal at the prebreakdown electric field must be detected in the constant voltage regime. PES signal is determined by $\Delta U = AU\Delta\sigma$, where A is a circuit factor, and it is not changed when the line intensity varies with voltage U in all possible acceptable intervals. Therefore, a dependence of the intensity on the electric field can be learned by analyzing a dependence of $\Delta\sigma(U) \sim \Delta U/U$ on applied voltage. A mechanism of the SI PES line intensity enhancement with an electric field for a $1s \longrightarrow 2p_{+1}$ transition close to breakdown would shed a light on the breakdown mechanism too.

An increase in the PES line intensity with the electric field has been observed and analyzed in various works [40–42], where this effect was interpreted either as increase of the hopping conductivity from the excited states [40] or with impact ionization [41] by free carriers from the excited states $2p_{+1}$ and $2p_{-1}$. In [42], an increase in PES was explained by electric field ionization (tunneling) of electrons from optically excited states of the SI atom to the conduction band. In [40–42], SI final states are far from the conduction band minimum, so $E_c - E_{SI}(2p_1)$ is much higher than $k_B T$. We consider the case when optically excited states $2p_{+1}$ and $3p_{+1}$ of SI are higher than $N = 0$ conduction subband minimum. In n-GaAs, such a situation is realized at higher magnetic fields satisfying the condition $\gamma = \hbar\omega_c/2Ry^* > 0.3$, when the $2p_{+1}$ level crosses the $N = 0$ Landau level. In this case, the localized $2p_{+1}$ state is degenerated with $E(N = 0, k_Z) = \hbar\omega_c/2 + (\hbar^2 k_z^2/2m^*)$, where $\hbar^2 k_z^2/2m^*$ can be estimated as difference between $E(1s \longrightarrow 2p_{+1})$ transition energy of FIR laser and the value of $\hbar\omega_c/2$ for the $N = 0$ conduction subband (Figure 3). These differences for n-GaAs at FIR lasers wavelengths $\lambda \approx 119\,\mu$m and $84\,\mu$m are about 1 meV and 4 meV, respectively. This means that electric field stimulation mechanisms must be neglected in transferring electrons from the localized $2p_{+1}$ state into the conduction band, which would take place through more fast process of recombination by emitting acoustic and optic phonons. The $1s \longrightarrow 2p_{+1}$ transition intensity was investigated for both $\varepsilon \| \mathbf{H}$ and $\varepsilon \perp \mathbf{H}$ mutual directions of electric and magnetic fields at $H \approx 36.5$ kOe and $H \approx 61$ kOe using fixed quantum energies of FIR lasers $\hbar\omega \approx 10.45$ meV and $\hbar\omega \approx 14.71$ meV, respectively. The intensity dependence $\Delta\sigma(U)/\Delta\sigma(0)$ on the applied voltage at $U < U_{BD}$ for $1s \longrightarrow 2p_{+1}$ transitions is shown in Figure 9. The value of $\Delta\sigma(0)$ is obtained by extrapolation of $\Delta\sigma(U)$ from small values of U to zero. In order to clarify the reason of increase in $\Delta\sigma(U)$ with U, one considers PES signal $\Delta\sigma(U) = e\mu_c \Delta n(U)$, where $\Delta n(U)$ is the concentration of nonequilibrium electrons in the conduction band at $1s \longrightarrow 2p_{+1}$ photoexcitation of SI and μ_c is $N = 0$ conduction subband mobility of electrons. It is important to note that heating of electrons takes place at electric fields mach higher than the breakdown field value. The criterion for heating of the electrons is that the CR line width should be asymmetrically broadened because of a transition from $(N = 0, k_z) \longrightarrow (N = 1, k_z)$, with $k_z \neq 0$, which takes place, as will be shown in the next section, at fields much stronger than the breakdown field.

As shown in [43], heating of electrons in n-GaAs samples at almost the same SI concentrations and under the conditions similar to our experimental ones takes place at electric fields much higher than those in our experiments. Concentration of nonequilibrium electrons $\Delta n(U)$ at $1s \longrightarrow 2p_{+1}$ photoexcitation from SI to the conduction band can be expressed as a multiplication of the rate $G_{1s \longrightarrow N=0}$ of such excitation and the time $\tau(H,U)$ necessary for an photoexcited electron to spend in the conduction band, before it would be captured back:

$$\Delta n(U) = G_{1s \longrightarrow N=0}\,\tau(H,U). \tag{9}$$

Generation rate of electrons from the ground state 1s into the conduction band $N = 0$ through the $2p_{+1}$ state can be written as a product of the neutral donor concentration N_D^0, radiation intensity P, and the probabilities $W_{1s \longrightarrow 2p_{+1}}$ and $W_{2p_{+1} \longrightarrow N=0}$ for the $1s \longrightarrow 2p_{+1}$ transition and the isoenergetic transition from the $2p_{+1}$ state to the $N = 0$ Landau zone, respectively:

$$G_{1s \longrightarrow N=0} = P\,N_D^0 W_{1s \longrightarrow 2p_{+1}} W_{2p_{+1} \longrightarrow N=0}. \tag{10}$$

It is obvious that all abovementioned quantities do not depend on the external electric field. The only parameter which does depend under our experimental conditions on the external electric field according to [44, 45] is $\tau(H,U) = \tau_{k_z \longrightarrow 0} + \tau_{cap}(H,U)$. This relaxation time consists

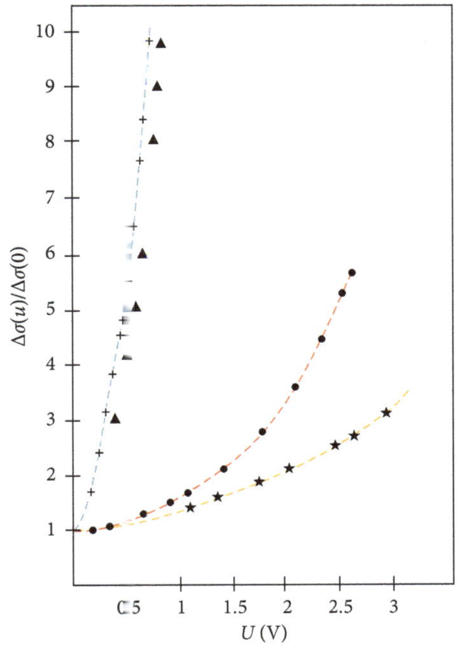

FIGURE 9: Dependence of $\Delta\sigma(U)/\Delta\sigma(0)$ on the electric field U at the top of $1s \longrightarrow 2p_{+1}$ line at $T = 4.2$ K is shown for (1) $H = 36.5$ kOe by +-points; for (2) $H = 61.8$ kOe by ▲-points when $\varepsilon \| \mathbf{H}$ and for (3) 36.5 kOe by •-points; for (4) $H = 61.8$ kOe by ∗-points when $\varepsilon \perp \mathbf{H}$ in the same sample.

of two parts. The first part $\tau_{k_z \longrightarrow 0}$ is the relaxation time of an electron, excited from the $2p_{+1}$ localized state to the Landau subband of $N = 0$ with the $k_z \neq 0$ state of the conduction subband and fell subsequently to the bottom with $k_z = 0$ of the conduction subband by many-fold irradiating acoustic phonons with characteristic energy increased in the magnetic field up to $\epsilon_{ph} = (\hbar\omega_c m^* s^2)^{1/2}$. There is no need for $\tau_{k_z \longrightarrow 0}$ to depend on the external electric field because of absence of heating. The second part is the cascade capture time from the $N = 0$ subband to the SI Coulomb potential excited state settled from bottom up to $k_B T$ energy distance [44]. Then, the relative increase of PES signal reads the following:

$$\frac{\Delta\sigma(H,U)}{\Delta\sigma(H,U \longrightarrow 0)} = \frac{\tau_{k_z \longrightarrow 0} + \tau_{cap}(H,U)}{\tau_{k_z \longrightarrow 0} + \tau_{cap}(H,0)}. \quad (11)$$

Let us consider the relation (11) in two limiting cases, $\tau_{k_z \longrightarrow 0} \gg \tau_c(H,0)$ and $\tau_{k_z \longrightarrow 0} \ll \tau_c(H,0)$. In the first case, $\Delta\sigma(\varepsilon)/\Delta\sigma(0) \approx 1$, and there would be no electric field dependence of the $1s \longrightarrow 2p_{+1}$ line intensity. In the second case,

$$\frac{\Delta\sigma(H,U)}{\Delta\sigma(H,0)} \approx \frac{\tau_{cap}(H,U)}{\tau_{cap}(H,0)}, \quad (12)$$

which means that the reason of $1s \longrightarrow 2p_{+1}$ PES line intensity enhancement is an increase of the capture time in the electric field. The experimental results of $1s \longrightarrow 2p_{+1}$ line intensity dependence on the electric field are shown in Figure 9. As it is seen from Figure 9, the power-like increase in the intensity with the electric field is considerably slowed down with the increasing magnetic field for a case when the electric field and magnetic field directions are orthogonal each other, whereas the intensity increases with the electric field in the case of parallel electric and magnetic fields much sharper than that of the case of orthogonal electric and magnetic fields, and furthermore, $\delta\sigma(U)/\Delta\sigma(0)$ increase with U does not depend in this case on the magnetic field strength. In the first glance, it seems that all results can be easily explained with carriers' heating mechanism, which is more effective in the case of $\varepsilon \mathbf{H} k_z$ than that in the case of $\varepsilon \perp \mathbf{H}$. However, in order to understand the SI electric field breakdown mechanism, it is important to know that the common reason of the $1s \longrightarrow 2p_{+1}$ line intensity increase as well as superlinear increase of the current and increase of cyclotron resonance intensity at prebreakdown electric fields in CVC is the capture time increase (or the capture cross section decrease) with the electric field. One can conclude that, although the carriers' heating is extremely important to explain the impact ionization, this mechanism is negligible for a moderately doped n-GaAs as our samples even if it takes place at all. There are two reasons, which increase the capture time in the electric field. The first one is a destruction of all discrete states of the SI Coulomb center up to the value $\epsilon_\varepsilon \approx 2(e^3\varepsilon/\epsilon_0)^{1/2}$ [46] and, as a consequence, a reduction of the capture efficiency of the excited carriers by traps.

In difference from the $\varepsilon = 0$ case, electron cannot be captured when, by wandering over excited states of impurity, it appears in distance $k_B T$ below the conduction band. At $\varepsilon \neq 0$, it would have a probability to tunnel back by field ionization. So, the tunnel ionization in the electric field will prevent diffusive descent of excited carriers to the 1s ground state. The influence of this mechanism on capture time is more effective at highly excited states of carriers [31]. It is clear that impurity levels at which the carriers are practically captured would be shifted down in the energetical scale with the increasing electric field. We must explain why does the relative intensity decrease with increasing H in the case of $\varepsilon \perp \mathbf{H}$, while it becomes unchanged for $\varepsilon \| \mathbf{H}$. Note that this is in consistent with another experimental fact that the breakdown ε-field much weakly depends on H for $\varepsilon \| \mathbf{H}$, as it is shown in Figure 9.

The wave functions of the SI localized states are compressed in a plane normal to the magnetic field within the cyclotron radius $l_c = (c\hbar/eH)^{1/2}$; nevertheless, they are almost unchanged along the direction of $\mathbf{H}$. Therefore, it is natural to expect that depth of the SI energy states' spreading, as well as tunnel ionization in the case of $\varepsilon \| \mathbf{H}$ would be much more than those in the case of $\varepsilon \perp \mathbf{H}$. So, the relative intensity increase $\Delta\sigma(U)/\Delta\sigma(0)$ in the electric field for the $1s \longrightarrow 2p_{+1}$ transition is because of the decrease of capture cross section of photoexcited carriers.

4. Cyclotron Resonance Measurements of n-GaAs in Dependence on the Electric Field

4.1. Influence of Shallow Impurities on the Electronic Effective Mass in n-GaAs. The CR measurements were performed in n-GaAs in a large interval of the electric field including the postbreakdown electric field, which allows us to understand deeper many features of the CR line shape. The CR is the well-known method to determine the carrier's effective mass $m_c^* = eH_{cr}/c\omega$ in semiconductor at the H_{cr} resonance magnetic field for a given ω radiation frequency. Usually a measured effective mass differs from this definition, e.g., in n-GaAs, due to particularly nonparabolic zone structure and plasma oscillation of the free electrons. The cyclotron frequency ω_c is shifted from its resonance value ω due to the plasma oscillation [47] with frequency $\omega_p = (4\pi ne^2 m_0^* - 1\epsilon_0^{-1})^{1/2}$ as follows:

$$\omega_c = \omega\left[1-\left(\frac{\omega_p}{\omega}\right)^2\right]. \tag{13}$$

Then, the plasma shift of the CR line in the magnetic field at a fixed value of ω is as follows:

$$\frac{\Delta H}{H_{cr}} = \frac{\Delta\omega_c}{\omega} = -\frac{\omega_p^2}{\omega^2}, \tag{14}$$

which means that the plasma oscillation shifts the CR line to lower values of the magnetic field, which is indicated by the minus sign in the front of the last term. A correction to the cyclotron mass ($\sim\Delta H$), caused by the plasma shift, is significant at small magnetic fields and decreases as $\sim H^{-1}$ with the increasing magnetic field. Estimation of the free-electron concentration from the expression (14), which determines the plasma shift of the CR line, provides

$$n = \frac{m_0^* \omega^2 \epsilon_0 \Delta H}{4\pi e^2 H_{cr}}. \tag{15}$$

Then, the correction to the cyclotron mass can be estimated according to the above considerations to be not more than 0.03% at $H \approx 21$ kOe and for $n \approx 10^{12}$ cm^{-3}. A negligible correction to the effective mass from the plasma oscillation confirms the experimental fact that, at the prebreakdown electric field, no shift in the CR lines is observed while the free-electron concentration increases.

Nevertheless, at the electric fields, higher than the breakdown value $\varepsilon > \varepsilon_{BD}$, and $H \approx 21$ kOe, a correction to the effective mass due to the plasma shift becomes essential [34]. A considerable shift in the CR line to the small magnetic field (Figure 10) and a reduction of the cyclotron mass (Figure 11) are observed under these conditions. Estimation of the electron concentrations according to equation (15) yields $n \approx 4 \times 10^{13}$ cm^{-3} for the measured value of shift $\Delta H \approx 0.1/0.15$ kOe at the electric field $\varepsilon \approx 2\varepsilon_{BD}$. This value constitutes 10% of the total number of the neutral donors, under which condition the plasma corrections become observable. Note that the estimated value of the electron concentration corresponds to the current filament; therefore, it may be overestimated because the CR signal is received only from the filament. A considerable asymmetric line shape of the CR to smaller values of the magnetic fields in Figure 10 at postbreakdown fields is a result of the free-electron distribution inside the filament.

Correction to the cyclotron mass m_c^* due to a nonparabolic conduction band results in increase of the effective

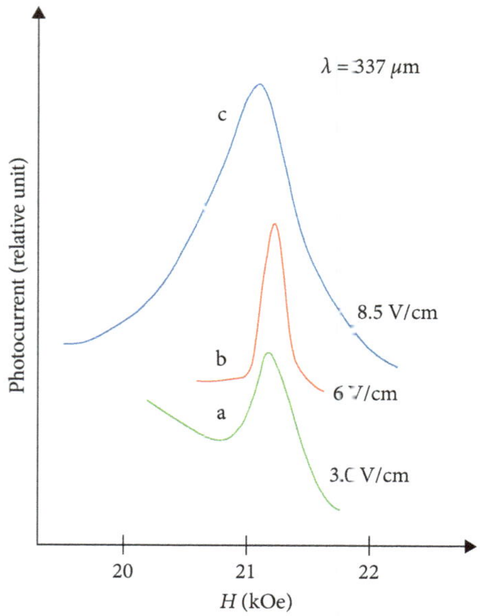

FIGURE 10: Shape of the CR line for the sample I at $T = 4.2$ K with $\lambda \approx 337\,\mu$m and $\varepsilon_{BD} \approx 5\,V\cdot$cm^{-1} at different electric fields: (a) $\varepsilon < \varepsilon_{BD}$, (b) $\varepsilon \approx \varepsilon_{BD}$, and (c) $\varepsilon > \varepsilon_{BD}$. As it is seen, the plasma shift is essential for the electric field higher than the breakdown value (the case (c)), where the line shifts to the left side.

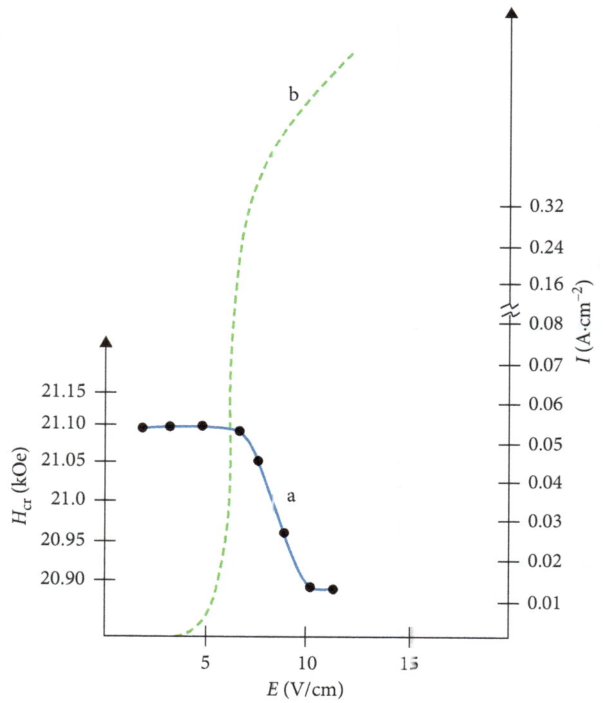

FIGURE 11: (a) Plasma shift of the resonance magnetic field H_{cr} in dependence on the electric field at $T = 4.2$ K with $\lambda \approx 337\,\mu$m. The curve (b) shows the CVC of the sample in the steady current regime ($R_S \ll R_L$).

TABLE 2: Effective masses of n-GaAs samples under investigation for the radiation wavelengths $\lambda = 119\,\mu m$ and $\lambda = 337\,\mu m$.

Samples	$N_D\,(\times 10^{14}\,cm^{-3})$	$N_I = 2N_A\,(\times 10^{14}\,cm^{-3})$	m^*_{CR}/m_0 $\lambda = 119\,\mu m$	m^*_{CR}/m_0 $\lambda = 337\,\mu m$
1 VPE	4.33	4.5	0.06770 ± 0.00002	0.06625 ± 0.00002
2 VPE	7.35	9.6	0.06764 ± 0.00005	0.06657 ± 0.00005
3 LPE	8.67	9.0	0.06777 ± 0.00005	0.6662 ± 0.00005
4 VPE	1.46	1.2	0.06760 ± 0.00005	0.06635 ± 0.00005
5 LPE	4.22	3.0	0.06700 ± 0.00005	0.06500 ± 0.00005
6 VPE	4.31	4.5	0.06750 ± 0.00002	0.06620 ± 0.00002

mass with the increasing CR magnetic field [45] according to the following equation:

$$m^*_c(H) = m^*_c(0)\left(1 + \frac{4\hbar\omega_c}{E_g}\right)^{1/2}. \quad (16)$$

As it is seen from Table 2, the CR mass m^*_c (61 kOe) is greater than m^*_c (21 kOe) (corresponding to the radiation wavelengths $\lambda = 119\,\mu m$ and $337\,\mu m$) more than ~2%. Nevertheless, according to equation (16), $m^*_c(61\,kOe) = m^*_c(21\,kOe) \cdot 1.00875$. So, assuming that the experimentally observed difference of the effective masses is due to the nonparabolicity, the difference would be 0.874%, which is much smaller than the experimental observation. So, neither the plasma shift nor the band nonparabolicity can not be a reason of the CR masses as in the same as well in the different magnetic fields.

The effective mass in the CR experiments in n-GaAs was measured by other authors too, where the measured effective masses varies in a wide interval from $m^* = 0.0665m_0$ [48, 49] up to $m^* = 0.07m_0$ [50] under the identical conditions of the CR experiments. This difference amounts 5% of the effective electron mass.

It is worthy to note that the effective cyclotron mass of electrons differ each other more that 0.5% at the pre-breakdown electric field even in the samples with close physical properties, which is much more than the experimental error, and it increases with the impurity concentrations.

It is necessary to stress out an existence a contribution to the effective mass due to the polaronic effects. Indeed, an electron mass is renormalized by means of the polaron constant α as [51, 52]

$$m_{pol} \approx m^*\left(1 + \frac{\alpha}{6}\right), \quad (17)$$

where $\alpha = (e^2/\hbar)((1/\epsilon_\infty) - (1/\epsilon_0))(m^*/2\hbar\omega_{LO})^{1/2}$ with $\hbar\omega_{LO}$ being a LO phonon energy and $\hbar\omega_{LO} \approx 36\,meV$. The effective dielectric constant ϵ^*, existing in the polaron theory, is defined as

$$\frac{1}{\epsilon^*} = \frac{1}{\epsilon_\infty} - \frac{1}{\epsilon_0}, \quad (18)$$

where ϵ_∞ and ϵ_0 are high-frequency and static-dielectric constants, respectively. According to the percolation theory of Efros and Shklovskii [53], the static dielectric constant in a system with randomly distributed metallic and dielectric regions diverges on approaching a critical percolation threshold value of the metallic region's volume fraction, where an insulator-metal phase transition takes place. In the insulator phase, ϵ_0 takes small values, comparable with ϵ_∞, yielding relatively high value for the effective dielectric constant ϵ^*. In this case, the polaron constant α takes small values, which does not change the effective mass. The static-dielectric constant ϵ_0 diverges [53] in the breakdown regime, which can be considered as an insulator-metal phase transition too. As a result, the polaron constant α increases due to decrease in ϵ^*, yielding a considerable enhancement of the effective mass. Estimation of α for GaAs far from the breakdown regime yields $\alpha \approx 0.06$ by taking $\epsilon_0 \approx 12.5$ and $\epsilon_\infty \approx 10.9$. The effective mass of an electron takes the value $m_{pol} = m^*(1 + (0.006/6)) \approx 0.06767$, which is consistent with our experimental results (Table 2). The static dielectric constant increases up to value of $\epsilon_0 \approx 10000$ inside the "candle"-like region of CVC at the breakdown, reducing the effective dielectric constant $\epsilon^* \approx 10.9$. Therefore, the polaron mass increases as $m_{pol} = m^*(1 + 0.32247/6) \approx 0.0706$. Such an enhancement of the effective mass around the breakdown regime may explain our experiment (Figure 12); nevertheless, the polaronic mechanism fails to explain a narrowing of the line width according to Figure 12. On the contrary, the CR line shifts to the left side at $\lambda \approx 337\,\mu m$, showing that the effective mass decreases in this case Figure 10. This fact cannot be explained too within the polaronic model. It is worthy to note that the polaronic mechanism may explain many experiments in semiconductors with ionic conductivity and in the presence of a metal-insulator phase transition, where the dielectric constant diverges at the structural and metal-dielectric phase transitions, respectively.

Observation of different values of the effective mass can be explained by influence of the nonhomogeneous spatial distribution of the impurities on the parameters of electron in the conduction band bottom. The CR measurements in the electric field, leading to shallow impurities breakdown, confirm this assumption.

Shape of the CR line at the radiation wavelength $\lambda \approx 119\,\mu m$ for different values of the electric field, including the breakdown field, is depicted in Figure 13. It is clearly seen from this picture that the CR lines shift to higher values of the magnetic field ($\Delta H > 0$) at $\varepsilon \approx \varepsilon_{BD}$. Although similar shift observed in, e.g., n – InSb [54], CdHgTe [55], and n-GaAs [43] crystals, has been explained as a result of electrons heating in the electric field when a resonance absorption in the nonparabolic band occurs at higher magnetic field, our investigations of the CR lines and CVC at different magnetic fields such as 21 kOe ($\lambda \approx 337\,\mu m$), 43 kOe ($\lambda \approx 172\,\mu m$), and 61 kOe ($\lambda \approx 119\,\mu m$) contradict to the electrons heating mechanism for the CR line shift [56].

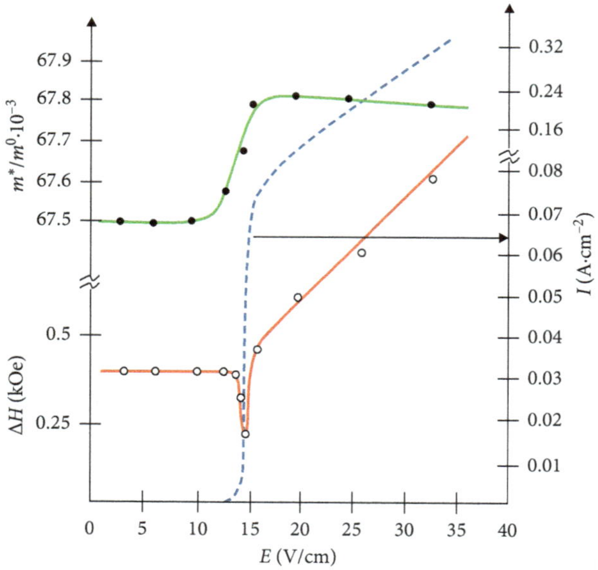

FIGURE 12: Dependence of the cyclotron mass and the CR line width of the sample I on the electric field at $T = 4.2$ K and $\lambda \approx 119\,\mu$m. The solid dots and the open dots indicate the cyclotron mass and the line width, correspondingly. The dashed curve signifies the CVC of the sample I which is measured in the constant current regime.

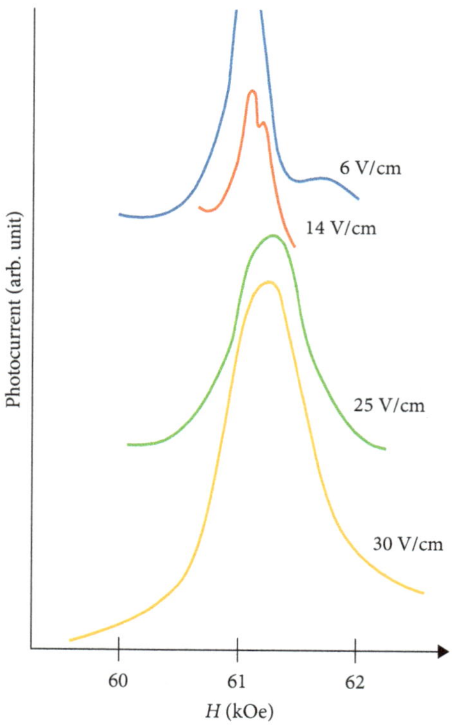

FIGURE 13: Shape of the cyclotron resonance lines of the sample I at $T = 4.2$ K, $\lambda \approx 119\,\mu$m, $\varepsilon_{BD} \approx 14\,$V·cm^{-1}, and at different electric fields. The measurements were done in the voltage regime far from the breakdown, and in the current regime at the breakdown.

The dependence of the effective mass m^* on the electric field ε is depicted in Figure 12. The CVC for this sample is shown by a dashed curve at the magnetic field $H \approx 61$ kOe, corresponding to the cyclotron resonance. A shift in the cyclotron resonance line and the corresponding increase in the effective mass are seen from the picture to take place around the breakdown electric field, and it disappears at $\varepsilon > \varepsilon_{BD}$. Hence, this fact indicates that a shift in the CR line width is caused not by the electric field but by a strong increase in the free-electron concentration due to the breakdown of neutral impurities at $\varepsilon = \varepsilon_{BD}$.

A shift of the cyclotron line from the resonance value H_{cr} takes different values for different samples, and it amounts up to ~0.6%, e.g., our sample 5 at $\lambda = 119\,\mu$m. The shift fails at relatively small magnetic fields ($H = 21$ kOe), and its value increases with the magnetic field from $H = 43$ kOe to $H = 61$ kOe [34].

The free electrons of a semiconductor, placed in the magnetic field, lie at zeroth Landau level at low temperatures, $k_B T \ll \hbar\omega_c$, and the peak in the CR (solid curve in Figure 14) corresponds to a transition from a state with the maximal density of states (DOSs) $\rho_0(E)$ in this level to the state in the first Landau level. Fluctuations of the charged impurities' density result in a randomly distributed potential of impurities, which smears the electronic states in the crystal. Therefore, the maximum of the DOS in the magnetic field shifts; the values of the shift ϵ_0^0 and ϵ_0^1 from the DOS maxima $\hbar\omega_c/2$ and $3\hbar\omega_c/2$ for a crystal depend on the impurity distribution and on the degree of nonhomogeneity of this distribution. This fact allows us to determine a degree of the nonhomogeneity in the impurity distribution [57]. The shift decreases with increasing the number N of the Landau level due to the increase in the cyclotron radius $r_{cr}^N = [c\hbar(2N+1)/He]^{1/2}$, i.e., the value of the DOS peak shift for zero Landau level is higher than that for the first one. Thus, the CR line peak shifts to the higher values of the energy; the value of the CR line intensity shifts for semiconductors with similar concentrations of donor, and acceptors may differ each other and depend on impurity distribution in a sample. The higher the nonhomogeneity of the distribution, the bigger is the line shift (Figure 15 and Table 2). This method allows us to determine a degree of nonhomogeneity in a given sample. Note that a difference between the CR masses for different samples considerably decreases at the breakdown, which supports our model.

4.2. PES Intensity Dependence of the CR Line on Prebreakdown Electric Field and Influence of Potential Fluctuations on the CR Line Shape. PES for SI 1s $\longrightarrow$ $2p_{+1}$ transition must be nonzero at arbitrary small electric field because the final state $2p_{+1}$ in our experiments is higher than $N = 0$ Landau level. As a result, an electron photoexcited to $2p_{+1}$ appears in the $N = 0$ conduction band after its first relaxation step. However, for CR to be detected, the existence of sufficient free-carrier concentration is obligatory condition. We must consider the dependence of the CR intensity on ε-field up to

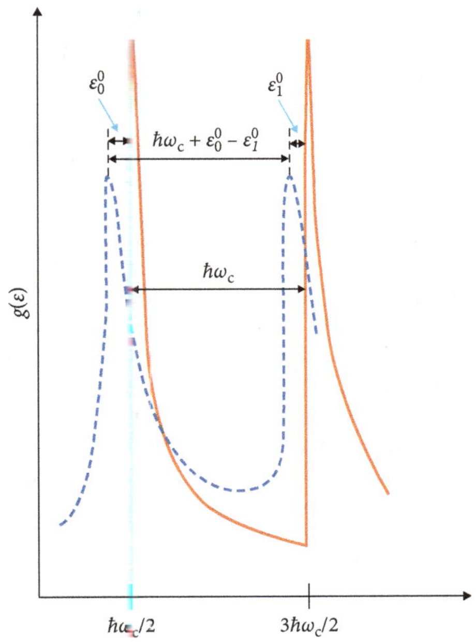

FIGURE 14: The density of electronic states of the zeroth and the first Landau levels in an ideal semiconductor (solid line) and in the presence of impurities (dashed line). ε_0^0 and ε_1^0 denote the shifts of the zeroth and first Landau levels due to the SI fluctuation potential.

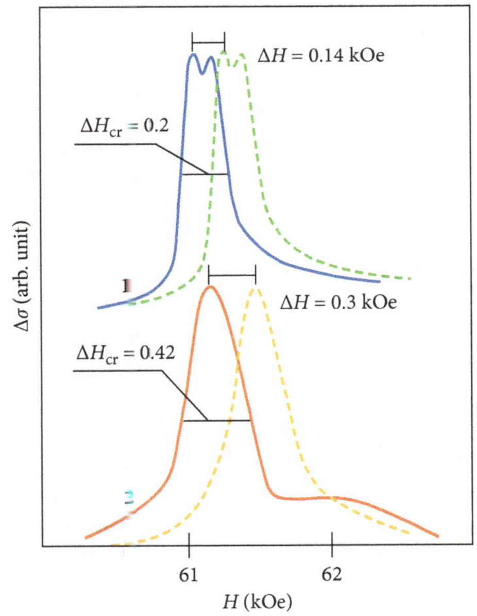

FIGURE 15: The prebreakdown line-width (solid lines) and the line-shift of the CR spectroscopy at breakdown (dashed lines) for sample 1 (curve 1) and sample 6 (curve 2) at $\lambda \approx 119\,\mu$m.

its prebreakdown value $\varepsilon < \varepsilon_{BD}$. In the linear region of CVC, where the electric field-induced enhancement of the free electrons is negligibly small, the free-electron concentration $n(T)$ in magnetic field H can be expressed ($n(T) \ll N_D$) [50] as

$$n(T) = N_c \frac{(N_D - N_A)}{2N_A} \frac{\hbar\omega_c}{k_B T} \exp\left(-\frac{E_i}{k_B T}\right), \quad (19)$$

where $N_c = 2(m^* k_B T/2\pi\hbar^2)^{3/2}$ is the electronic density of sates in the bottom of the conduction band, $E_i = Ry^* + (\hbar\omega_c/2) - [E_{1s}(H) - E_{1s}(0)] \approx Ry^* + \hbar\omega_c/2$ is the ionization energy of SI in magnetic field which can be found from Figure 3. The prefactor $\hbar\omega_c/k_B T$, which is added empirically to the expression (19), corrects the free-electron concentration in the presence of magnetic field. The amount of thermally excited free electrons at $T = 4.2$ K would be only few hundred electrons by taking into account the volume of epitaxial n-layer of n-GaAs in magnetic fields corresponding to $\gamma \approx 1$. The ionization energy of SI donors at $\gamma \approx 1$ is $E_i \approx 10.5$ meV which is two times greater than Ry^*. This is a reason why the CR signal corresponding to the linear part of CVC is almost comparable with the noise and it is very hard to detect it not only in PES but in the transmission experiments too, without additional band gap illumination. Thus, one concludes that the concentration of thermally excited free electrons in n-GaAs samples at $T = 4.2$ K under the magnetic field at least at $H = 61$ kOe is insufficient to detect the CR. On the other hand, the value of the electric field, necessary to observe the CR, increases with magnetic field approximately linearly. All these facts ensure that a necessary concentration of free electrons for observation of CR is provided by excitation of free electron from neutral impurity atoms by means of electric field. Our investigation of dependence of PES intensity of the CR spectra on the external electric field at different magnetic fields also leads to this conclusion.

Figures 16 and 17 depict a dependence of the CR intensities on the electric field for fixed radiation wavelengths $\lambda \approx 337\,\mu$m and $\lambda \approx 119\,\mu$m at magnetic fields $H \approx 21$ kOe and $H \approx 61$ kOe, correspondingly, for two mutual orientations of ε and $\mathbf{H}$ fields when $\varepsilon \| \mathbf{H}$ and $\varepsilon \perp \mathbf{H}$. It can be seen from Figures 16 and 17 that $\Delta\sigma$ increases strongly with the applied voltage U in the process of cyclotron absorption of radiation, as it takes place also in the PES of SI. Nevertheless, in difference from SI $1s \longrightarrow 2p_{+1}$ PES experiments, the line intensity dependence on ε in the CR is negligibly weak for both $\varepsilon \| \mathbf{H}$ and $\varepsilon \perp \mathbf{H}$ orientations, and it does not change with increasing H especially for $\varepsilon \perp \mathbf{H}$. There is a critical value ε_{cr} of the electric field, behind of which the CR line intensity increases with H.

In order to understand all these effects, one should analyze the change in the static conductivity $\Delta\sigma$ of samples under cyclotron absorption of the radiation. PES of the CR at prebreakdown electric fields was investigated in constant voltage regime, so PES signal $\Delta U = AU\Delta\sigma$, where $\Delta\sigma = n_0(\mu_1 - \mu_0)J(\hbar\omega_c)$, n_0 is the free-carrier concentration, $\Delta\mu = \mu_1 - \mu_0$ is the difference of the mobilities in the first and zeroth Landau levels (LL), correspondingly, and $J(\hbar\omega_c)$ is the radiation intensity at the CR magnetic field. The CR line intensity is determined as a product of $\Delta U/U$ and CR line width, the latter of which does not depend on electric field up to the breakdown field. In order to detect the CR in PES experiment, it is necessary to determine the mobilities

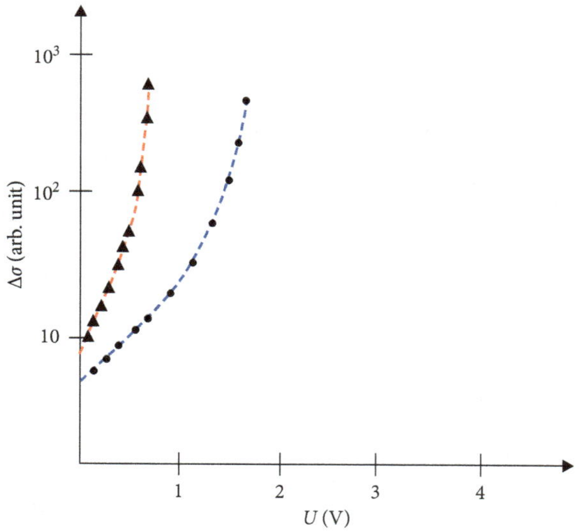

FIGURE 16: Dependence of the CR PES lines of the sample 6 on the applied voltage U at $T = 4.2$ K, $\lambda \approx 337\,\mu$m, $H = 21$ kOe, where the curves marked by ▲ and • correspond to the case of $\varepsilon \| \mathbf{H}$ and $\varepsilon \perp \mathbf{H}$, correspondingly.

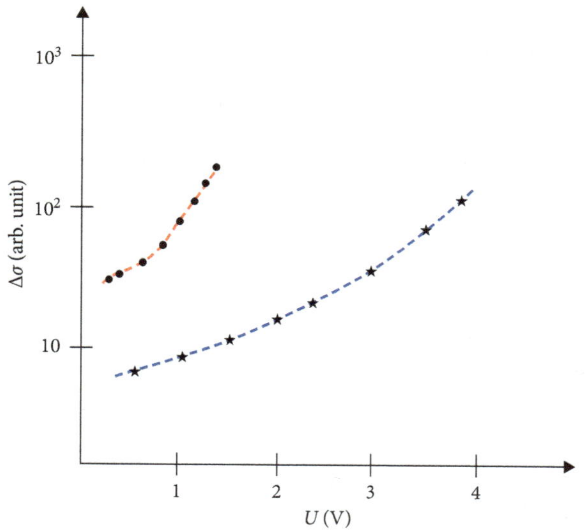

FIGURE 17: Dependence of the CR PES line intensity on the applied voltage U of the sample 6 at $T = 4.2$ K, $\lambda \approx 119\,\mu$m, $H = 61$ kOe, where the curves marked by • and $*$ correspond to the case of $\varepsilon \| \mathbf{H}$ and $\varepsilon \perp \mathbf{H}$, correspondingly.

difference at two adjacent LL. At low temperatures, the dominant mechanism of free-electron scattering is a scattering on ionized impurities ($\mu \propto \epsilon^{3/2}$). So, the mobilities' difference at CR in the absence of the external electric field ($k_z = 0$) or in the absence of free carriers' heating at prebreakdown field is given by

$$\Delta\mu \propto \left[\left(\frac{3\hbar\omega_c}{2}\right)^{3/2} - \left(\frac{\hbar\omega_c}{2}\right)^{3/2}\right]. \quad (20)$$

This means that CR-induced conductivity change is due to not only the free-carrier concentration redistribution between LLs, but also the mobility difference between two adjacent Landau bands. The mobility difference in electric fields leading to heating of electrons ($k_z \neq 0$) reads as

$$\Delta\mu \propto \left[\left(\frac{3\hbar\omega_c}{2} + \frac{\hbar^2 k_z^2}{2m^*}\right)^{3/2} - \left(\frac{\hbar\omega_c}{2} + \left(\frac{\hbar^2 k_z^2}{2m^*}\right)\right)^{3/2}\right], \quad (21)$$

where the selection rules for the CR ($\Delta N = 1$, $\Delta k_z = 0$) are taken into account [58]. Supposing that the heating takes place at prebreakdown electric fields for $H = 61$ kOe, when the heating causes an increase in the energy $\hbar^2 k_z^2/2m^* = 30\text{ K} \approx 2.5$ meV. Then, $\Delta\mu$ is calculated from the relations (20) and (21), which would differ from each other only about 10%. Therefore, an increase in the line intensity is provided not by an increase in the mobility but by an increase in the free-electron concentration. The above-described heating of the free carriers has to increase the CR line-width asymmetrically to higher magnetic field side from the CR line maximum, which is not the case at prebreakdown electric fields, and takes place only at $\varepsilon > (3/4)\varepsilon_{\text{bd}}$. This is confirmed by two facts. The first, the experimental observations, that at the breakdown electric field $\varepsilon = \varepsilon_{\text{bd}}$, a strong narrowing takes place not only in $1\text{s} \longrightarrow 2\text{p}_{+1}$ line but also in CR line shape, refusing the heating of free carries in external electric field as a breakdown mechanism. Indeed, this fact contradicts to IIM, where heating of free carriers in prebreakdown electric field is the necessary condition. Second, increase of CR intensity at prebreakdown electric field is not due to increase of mobility difference $\Delta\mu = \mu_1 - \mu_0$ of free electrons at CR transition but the increase of carrier concentration no at zero $N = 0$ LL.

Provided that the ionization energy decreases in external electric field according to Pool–Frankel effect, the free-electron concentration increase Δn would differ from formula (19) by a factor $\beta(\varepsilon)$ [46], which takes into account the cascade ionization of donor from 1s state to the excited states of SI:

$$\beta(\varepsilon) = \frac{\beta(0)}{1 + 0.98\,(E_\varepsilon/k_B T)} \exp\left(\frac{E_\varepsilon}{k_B T}\right), \quad (22)$$

where $E_\varepsilon = 2(e^3\varepsilon/\epsilon_0)^{1/2}$ is an ionization energy reduction of the SI Coulomb potential in the electric field ε. Its value at breakdown electric field $E_\varepsilon \approx 0.1$ meV for n-GaAs is much smaller than SI ionization energy E_i in CR magnetic field. Two times increase of the electric field ε from 1 to 2 V·cm^{-1} results in increase of CR intensity, determined by the electron concentration (19) multiplied to (22), at least 2 times while experimental increase according to Figures 16 and 17 is much more at prebreakdown fields. So, an enormous increase of free-electron concentration cannot be explained according to the ground state ionization. At low temperatures $k_B T \ll E_i$, an essential part of the carriers is localized on fluctuations of the SI Coulomb potentials ΔE_{fl}, which may be identified to the half-width of the Gaussian distribution δ (equations (1) and (2)) with the characteristic value $\Delta E_{\text{fl}} \approx 0.29 e^2 N_D^{1/3} K^{1/4} \epsilon_0^{-1}$ [59–61] For the samples used in our experiments, with the parameters N_D and K given in Table 1, the fluctuation potential varies in the interval $\Delta E_{fl} \approx 0.1/0.2$ meV, which is smaller than Ry^*, and it increases with compensation K. Prebreakdown electric field would easily increase the free-carrier concentrations, by releasing the electrons hooked on fluctuations. The fluctuation potential distorts basically the DOS on the $N = 0$ Landau level because it has inherited properties of the conduction band in the absence of the magnetic field. Influence of the SI potential fluctuations on the conduction band for higher LLs ($N = 1, 2, \ldots$) is negligibly small due to their larger CR radii. At low temperature $k_B T \ll Ry^*$, the fluctuation potential vanishes $\Delta E_{fl} = 0$ in the absence of compensating acceptors $K = 0$. Fluctuations of potential arise as a result of charged impurities influence on conduction band electrons: the negatively charged impurities increase and the positively charged ones decrease the energy of free carriers, concerning to conduction band gap at $k = 0$. In a magnetic field H directed along z-axis, CR radius for N-th LL is $r_{cr}^N = [c\hbar\ 2N + 1)/eH]^{1/2}$. Estimation of the CR radius in magnetic field corresponding to $\gamma \approx 1$ yields for zeroth and first orbits, respectively, $r_{cr}^0 \approx a_B^*$ and $r_{cr}^1 \approx 1.41\, a_B^*$. Note that r_{cr} growth with N would increase the probability of finding of charged donor-acceptor pairs, which decreases the influence of fluctuations of potential. The influence of the potential fluctuations on CR line shape will be reduced with decrease of the magnetic field by the same reason. Figure 18 displays the CR line shapes for the same sample at $H = 21$ kOe and $H = 61$ kOe with increasing electric field from its small values, corresponding to linear part of CVC, up to the prebreakdown electric field in Voight geometry ($\mathbf{k} \perp \mathbf{H}$) of measurements of the CR, when applying of $\varepsilon \| \mathbf{H}$ is possible.

The CR reveals a resonance absorption around the CR line from the fluctuation potential, which has not been observed in other experiments in n-GaAs samples. As it is seen from Figure 18, apart from the CR line at $H \approx 61$ kOe, a wide absorption band is observed at higher fields, which contains several lines with low resolutions. Such kind a resonance absorption band was observed for the long wavelength illumination ~337 μm (Figure 18(a)) of the sample 6, where less resolution results in an asymmetric broadening of the resonance CR line at higher magnetic fields. An electric field transfers the electrons from the absorption band of the fluctuation potential to the conduction band. Therefore, the intensity of the CR line increases in the electric field rather rapidly than that of the absorption band. Note that the effect is more precisely observed in the electric fields parallel to the magnetic field. The same tendency is observed also under the illumination of samples in the region of fundamental absorption in the presence of a constant electric field (Figure 19).

Finally, we would like to discuss the low-temperature breakdown mechanism of SI by an electric field. The free-carrier concentration n increases in the electric field ε not only due to an increase in the ionization coefficient $\beta(\varepsilon)$ but also a decrease in the capture coefficient $\alpha(\varepsilon)$. The value of n can be estimated from the condition of balance between the capture and the thermal ionization as [26]

$$n(\varepsilon)\alpha N_D^+ = \beta N_D^0, \tag{23}$$

where $N_D^+ = N_A + n$ and N_D^0 are the concentrations of charged and neutral donors. β increases with electric field and diverges at $\varepsilon = \varepsilon_{BD}$, instead α decreases with increasing ε and vanishes at $\varepsilon = \varepsilon_{BD}$, resulting in sharp enhancement of the electron concentration

$$n(\varepsilon) = \frac{N_D^0(\varepsilon)}{N_D^+(\varepsilon)} \frac{\beta(\varepsilon)}{\alpha(\varepsilon)}. \tag{24}$$

The concentration n reaches a critical value by approaching the breakdown electric field, when the charged donors become strongly screened as $(e/\epsilon_0 r)e^{-(r/r_s)}$. Further increase in n decreases the capture cross section of the charged donor drastically because of screening, resulting in an avalanche of increase of the concentration. So, the current shows an avalanche-like increase with the electric field as $j(\varepsilon) = en(\varepsilon)\mu(\varepsilon)\varepsilon$. We should to note that the above-described breakdown mechanism is valid for the intermediate concentration $10^{14}\ \mathrm{cm}^{-3} \le N_D \le 10^{16}\ \mathrm{cm}^{-3}$ of SI in a crystal. For higher concentrations of the impurities, the screening radius decreases approaching to the Bohr radius, $r_s \approx a_B^*$, when all impurities become ionized. The condition $r_s \approx a_B^*$ corresponds also to Mott transition when $a_B^* N_D^{1/3} \approx 0.25$ is satisfied for the donor concentration $N_D \approx 2 \times 10^{16}\ \mathrm{cm}^{-3}$. The CVC characteristics of the semiconductor in this case turns from a "candle"-like shape to a superlinear dependence on the electric field. This means that the breakdown should not be observed for samples with the donor concentrations higher than $N_D = N_D^{cr} \approx 2 \times 10^{16}\ \mathrm{cm}^{-3}$.

Appearance of a fluctuation band can be explained providing a nonhomogeneous distribution of the impurities in a semiconductor. Fluctuations in the impurities' distribution provide a realization of a specific potential $V(\mathbf{r})$, so-called a fluctuation potential. This random potential is assumed to distribute according to Gauss law, [59, 60].

$$W(V) \propto \exp\left(-\frac{\int V^2(\mathbf{r})d\mathbf{r}}{2\gamma}\right), \tag{25}$$

where γ is the mean-square fluctuation potential, which characterizes a nonhomogeneous distribution of impurities. Theoretical studies of a free-electron motion in the field of a random potential like 25 and under a magnetic field show a possibility of bound-electronic states below the conduction band bottom. A resonance absorption of a radiation was shown [60] to be possible in the dipole approximation by electrons bonded in the fluctuation potential bottom. Furthermore, the energy of the resonance absorption differs from that of the cyclotron absorption $\hbar\omega_c$, and this difference increases with the magnetic field.

Majority of the conduction band's electrons are localized at low temperature and in the presence of a magnetic field in the dips of the fluctuation potential. The number of electrons in the conduction band $N = 0$ increases with the electric

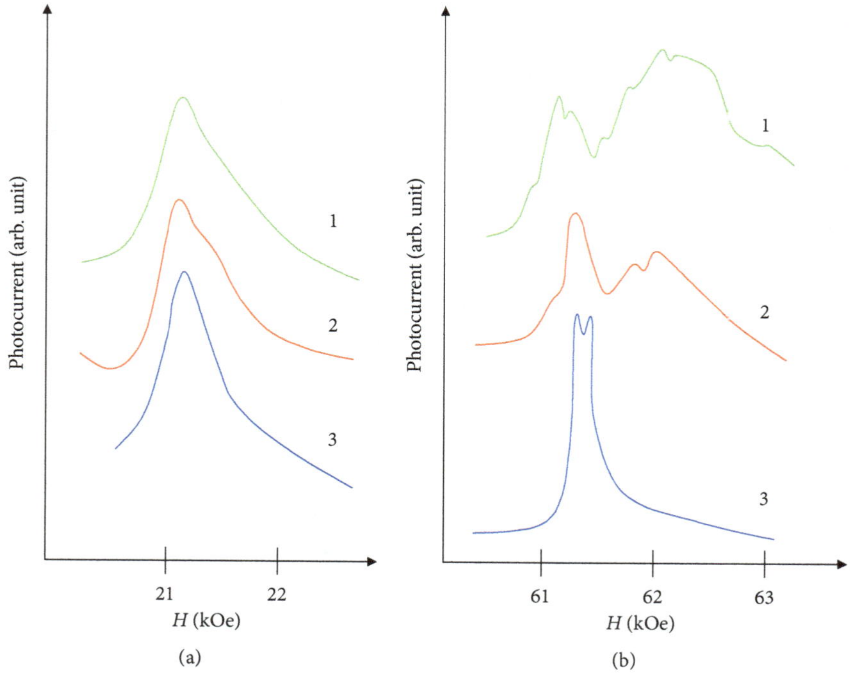

FIGURE 18: Low temperature ($T = 4.2$ K) structures of the CR lines, measured for the sample 6 at different electric fields $\varepsilon_1 < \varepsilon_2 < \varepsilon_3 < \varepsilon_{BD} = 2$ V/cm corresponding to curves 1, 2, and 3, and aligned parallel to the magnetic field $\varepsilon \| \mathbf{H}$, for (a) $\lambda \approx 337\,\mu$m and (b) $\lambda \approx 119\,\mu$m. The broad line in 1*b* is due to the transition indicated by arrow $\epsilon_0^0 \longrightarrow \epsilon_1^0$ in Figure 14, the intensity of which decreases with increasing the pre-breakdown electric field.

field due to excitation of electrons from the neutral impurities, which results in enhancement of the CR intensity. On the other hand, the capture probability of the band electrons in the dips of the fluctuation potential decreases with increasing electric field, since the latter suppresses the fluctuation dips and releases the electrons from these dips.

The similar picture is realized in the process of the illumination, which increases again the band electrons due to band-to-band excitation of electrons by illumination.

One can conclude that a nonhomogeneous distribution of the impurities in the sample and its fluctuation potential result in the alternative resonance lines near the CR line.

4.3. Effects of the Electric Field on the CR Line Width. Dependence of the CR lines on the magnetic and electric fields, temperature, and the concentrations of neutral and ionized impurities yields an essential signature on the broadening mechanism of the lines. If the CR line broadening is determined by scattering of the current carriers in a semiconductor, then the momentum relaxation time τ_{CR} is expressed through the resonance magnetic field H_p, the half-width $\Delta H_{1/2}$ and the radiation frequency ω as [62]

$$\tau_{CR} = \frac{2H_p}{\Delta H_{1/2}\omega}. \quad (26)$$

Although τ_{CR} is determined by all scattering mechanisms, the main scattering mechanisms in n-GaAs at $T = 4.2$ K are scattering on acoustic phonons as well as on neutral and ionized impurities. Since the number of the ionized

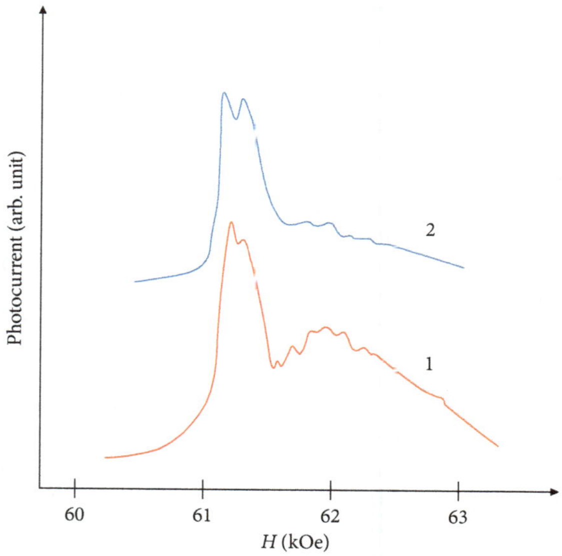

FIGURE 19: Structure of the CR line of the sample 6 at $\lambda \approx 119\,\mu$m and $T = 4.2$ K. (1) $\varepsilon = 1$ V/cm without an illumination of the sample, and (2) $\varepsilon = 1$ V/cm under illumination in the region of fundamental absorption.

impurities is much more than that of the neutral impurities, one can neglect scattering on the neutral impurities. Two experimental facts demonstrate that the scattering on the ionized impurities dominates over scattering on the acoustic phonons. First, the electron mobility $\Delta\mu = \mu_1 - \mu_0 > 0$ is changed in the cyclotron absorption, which is confirmed by equal polarities of the CR photo-signal and of the photo-excited impurities. The energy dependencies of the mobility for ion $\mu_{ion} \sim E^{3/2}$ [52] and acoustic phonon $\mu_{ph} \sim E^{-(1/2)}$ [63] scattering mechanisms differ from each other. Since the electron energy increases amount of $\hbar\omega_c$ in the CR, the scattering mechanism coincides with the ion-scattering one mechanism when $\Delta\mu > 0$, and it contradicts to the phonon scattering mechanism when $\Delta\mu < 0$. Second, according to the perturbation theory, the CR line should be narrowed with increasing magnetic field either $\sim H^{-(1/4)}$ [64] or $\sim H^{-1}$ [65, 66] for the dominant ion-scattering mechanism, and it should broaden as $\sim H$ [63, 67] for the dominant phonon-scattering mechanism. The CR line measurements in n-GaAs [30, 49] with the impurity concentration one order smaller than that in our samples were reported to show the line narrowing in the magnetic fields from 21 kOe to 61 kOe, confirming that a scattering on the ionized impurities is a dominant mechanism at $T = 4.2$ K even at low impurity concentrations. Note that the charged impurities either scatter the electrons or their potential shifts the cyclotron frequency of the electrons. The relaxation time for the nonadiabatic electron scattering on the charged impurities, when the cyclotron radius l_c is smaller that the radius a of an impurity potential, $l_c < a$, was calculated [65] to be

$$\tau^{-1} = \frac{2\pi e^4 N_I}{\hbar\omega_c \epsilon_0^2}\left(2m^* k_B T\right)^{-(1/2)}. \quad (27)$$

On the other hand, a broadening of the CR line [64], due to the cyclotron frequency shift $\Delta\omega_c = (V_{xx} + V_{yy})/2m^*\omega_c$ with V_{xx} and V_{yy} being the second derivative of the impurity potential at the cyclotron orbit center, is given as

$$\tau^{-1} = 0.915\left(\frac{e^2}{\epsilon_0}\right) m^{*-(1/4)} \omega_c^{1/4} \hbar^{-(3/4)} N_I^{1/2}. \quad (28)$$

The relaxation time, determining from the CR line width according to equation (26), seems to be one order higher than that determining from equations (27) and (28). One peculiarity of the CR measurements in n-GaAs is that there is not a monotonical dependence of the line width on the ionized impurities concentration. The CR lines in Figure 20 given for two samples with similar physical parameters show an increase in the line width with decreasing the impurity concentrations. As it is seen from these curves, the CR line of sample 5 is wider than that of sample 6, though the concentration of the ionized impurities in sample 5 is less than that in sample 6.

The considerable fact is an existence of a correlation between the width and the value of shift of the CR lines at the prebreakdown electric fields. A broadening of Landau levels due to the impurity potential fluctuations results in an additional broadening of the CR lines and shifting of their peak positions. Therefore, the CR line widths of two samples with the same concentrations of the ionized impurities can differ from each other, so that the higher the nonhomogeneity of the impurity distribution, the wider the CR line.

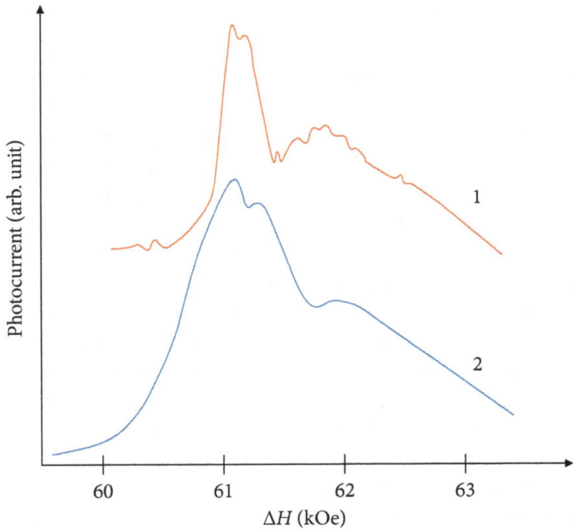

FIGURE 20: CR line shapes of two samples, 5 (curve 2) and 6 (curve 1), with similar values of the neutral and ionized impurity concentrations. The electric field takes prebreakdown values; $\lambda \approx 119\,\mu$m and $T = 4.2$ K. (1) $N_D = 4.3 \times 10^{14}$ cm^{-3} and $N_I = 4.5 \times 10^{14}$ cm^{-3}; (2) $N_D = 4.2 \times 10^{14}$ cm^{-3} and $N_I = 3.0 \times 10^{14}$ cm^{-3}.

The CR line width is defined in the energetic unit as $\Delta E = (\partial E/\partial H)\Delta H$, where the change rate of the cyclotron energy with a magnetic field $\partial E/\partial H$ can be estimated to be for the first two Landau levels 0.17 meV/kOe. This value is comparable with the fluctuation potential of charged impurities, the correct expression of which is given for a small compensation as [28]

$$E_0 = 0.29 e^2 N_A^{1/3} K^{1/2} \epsilon_0^{-1}. \quad (29)$$

This expression takes a value of $0.1\,\text{meV} \le E_0 \le 0.2\,\text{meV}$ for the samples under investigation.

Broadening of the Landau levels due to the potential fluctuations results in a broadening of the CR line in addition to the broadening due to the scattering of electrons on the ionized impurities. Therefore, the wider the CR line for the samples with equal concentration of the charged impurities, the higher the nonhomogeneous distribution of impurities.

Influence of the distribution nonhomogeneity on the width and the peak position of the CR line is essential only for higher magnetic fields when the cyclotron radius of the zeroth Landau level l_c^0 is smaller than the potential size a_B^* of an impurity, $l_c^0 < a_B^*$. At smaller H, a cyclotron orbit with larger radius contains several charged impurities with opposite signs, the effects of which cancel each other in the averaging.

The free-electron concentration increases several orders at the breakdown electric field, corresponding to a sharp enhancement of the current at $\varepsilon = \varepsilon_{BD}$ in the CVC, which changes strongly the CR line shapes. The CR line narrows approximately two times in the close vicinity of the prebreakdown electric field (Figures 12 and 13), which is observed at the radiation wave length $\lambda \approx 119\,\mu$m as well as at

$\lambda \approx 337\,\mu$m with increasing electric field. Recalling that the CR line broadening is caused by the ionized impurity potential, narrowing the line width at the breakdown field can be understood as a reduction of the potential fluctuations due to the screening of the ionized atoms by free electrons released at the breakdown. The photoelectric spectrum of the CR lines, presented in Figure 21, is measured in the steady current regime with its different values and at $\lambda \approx 337\,\mu$m, which shows instead a broadening approximately three times with increasing the value of the current. A similar measurement at $\lambda \approx 119\,\mu$m ($H \approx 61$ kOe) at the breakdown field $\varepsilon \approx \varepsilon_{BD} = 15\,\text{V}\cdot\text{cm}^{-1}$ yields the similar results. Since the broadening occurs at a steady current, it rules out the impact ionization mechanism by means of the electron heating. The theories of the CR line broadening, suggesting either $\Delta H_{1/2} \sim <\omega\tau> \sim N_I^{-(1/2)}$ [64] or $\Delta H_{1/2} \sim <\omega\tau> \sim N_I^{-1}$ [65, 66] on the base of the impurity ionization mechanism, cannot explain the measured broadening, the value of which needs several times increase in the concentration. The observed broadening of the CR line seems to be caused by electron-electron correlations, which act similar to the scattering on the ionized impurities. Indeed, an increase in the free-electron concentration with the current J of cross-section area S can be estimated according to

$$n \sim \frac{J}{S}\frac{1}{e\mu\varepsilon_{BD}}. \quad (30)$$

The free electrons of a small concentration at the beginning step of the breakdown screen the charged impurities potential, weakening their effect on the CR line broadening. Further increase in the free-electron concentration changes the scattering mechanism from the electron-impurity to the electron-electron mechanism. The free-electron concentration in the current filament reaches the ionized impurity concentration which is several orders higher than that of the average bulk concentration of the impurities.

One experimental evidence is that the CR line width in the broadening regime depends also on the radiation intensity. The radiation intensity for the CR line, shown by the curve 5 in Figure 21, is 7 times higher than that of the curve 4, which results in strong reduction in the line width. In this case, the free-electron concentration considerably decreases due to an increase in the current filament cross section resulting in a narrowing of the line width. Note that an asymmetry in the line width broadening is observed at higher breakdown fields. The asymmetry at $\varepsilon \approx 2\varepsilon_{BD}$ reaches 20% around the line maximum (Figure 13), which seems to occur due to an electron heating by high electric field. The electron heating enhances the electron concentration with $k_z \neq 0$. Thus, the resonance condition $\hbar\nu = E(1,k_z) - E(0,k_z)$ for a nonparabolic zone is satisfied at higher magnetic fields, yielding an asymmetric broadening.

Recently, the luminescence spectra of shallow impurities in the n- and p-type Si, Ge, and GaAs semiconductors, where the ionization energies vary from five meV to several dozen meV, have been intensively studied in order to develop

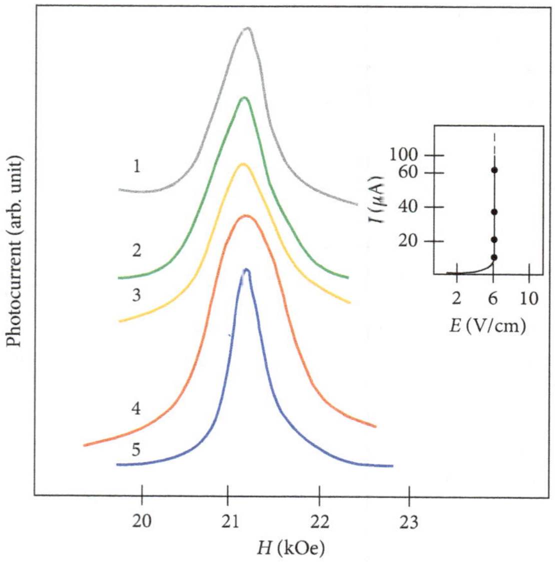

FIGURE 21: CR line shape measured in the constant current regime at $\varepsilon \approx \varepsilon_{BD}$ and $\lambda \approx 337\,\mu$m. The corresponding values of the current are shown in the inset of the CVC curve. The value of the current corresponding to the curve 5 is the same as that in the curve 4, but the radiation intensity is 7 times lower.

tunable THz-radiation sources [68–74]. The luminescence in these semiconductors is stimulated under electrical breakdown of shallow impurities in the constant current regime at low temperatures when $k_B T \ll R_y$. Nonequilibrium electrons and holes, created in the breakdown of the shallow impurities, are captured firstly in excited states (2p, 3p, etc.,) of impurities and then are radioactively recombined to the ground state 1s. The energy of the radiation quanta in this case is distributed so that the luminescence maximum corresponds to 2p $\longrightarrow$ 1s transition; it decreases sharply to zero at small energies and reaches to the conduction and valence zones edges at higher energies that correspond to donors and acceptors. In this case, the integral intensity of the luminescence decreases due to short-length tail of the intensity curve with increasing the current flowing through the sample [73]. A deep breakdown is realized with increasing the free-electron concentration n, which leads to disappearance of highly excited states of impurities due to screening, as it occurs in our case (Figure 1 in [73] similar to Figure 5 in this paper), with free electrons. Indeed, a narrowing and long-wavelength shifting in 2p $\longrightarrow$ 1s transition in the luminescence spectra under the breakdown has been observed in several experiments [73]. However according to [69, 72], the luminescence spectra disappear at higher-impurity concentrations, corresponding to the Mott transition at $N(a_B^*)^3 > 0.25$, which is determined also with disappearance of the impurity bound states up to 1s state. The listed facts observed from the luminescence spectra under the impurity breakdown agree well with the suggested mechanism of the shallow impurities breakdown in semiconductors.

5. Conclusions

In this paper, we suggest a new mechanism of SI breakdown in n-GaAs samples particularly for an intermediate donor concentration $10^{12}\,\mathrm{cm}^{-3} < N_D < 10^{16}\,\mathrm{cm}^{-3}$ and for the compensation degree $0.3 < K = N_A/N_D < 0.8$ when the electric field is varied in wide interval from the linear part of CVC up to 3-4 times higher than the SI BD field. All the obtained results contradict the impact ionization mechanism of the breakdown. The main contradiction with the IIM is the absence of carrier heating at pre-BD and BD electric fields. In difference from the *N*-like CVC dependence, which is observed in constant voltage regime [75], *S*-like CVC dependence should be observed in the heating mechanism due to strong increase of carriers' mobility $\mu(E) \propto E^{3/2}$ in the scattering process of electrons from charged impurities. CVC characteristics of our samples have shown no *S*-like dependence in the constant current regime measurements. The other fact, which contradicts to the IIM, is that the BD electric field practically does not depend on the magnetic field in the case of parallel electric and magnetic fields $\varepsilon \| \mathbf{H}$ although the impurity ionization energy increases two times at the magnetic field $\gamma = \hbar\omega_c/2Ry^* \approx 1$.

We showed that the PES line-width of $1s \longrightarrow 2p_{+1}$ transition is determined not only by the SI concentration and the compensation degree but also by the potential fluctuation, which is provided by nonhomogeneous distribution of the charged impurities. The potential fluctuation disappears at the breakdown electric field $\varepsilon = \varepsilon_{BD}$, when the released electrons screen the charged impurities, which may be caused also with quadratic-Stark effect and gradient-broadening mechanism of PES. Disappearance of Zeeman spectra, corresponding to the transitions from the ground state to highly excited Boyle-Howard's states up to $1s \longrightarrow 3p_{+1}$ at the BD, also confirms the screening mechanism of the ionized impurities by free carriers at RD. The PES line-width is strongly narrowed at $\varepsilon = \varepsilon_{BD}$.

For small concentration of impurities $N_D < 10^{12}\,\mathrm{cm}^{-3}$, the above-described mechanism is not valid, since the free-electron concentration is insufficient for screening of the charged impurities. Nevertheless, the conventional impact ionization mechanism may work and the breakdown occurs at much higher (several order higher than that of for our samples) electric fields.

At higher concentrations of the SI, when $N_D > 10^{16}\,\mathrm{cm}^{-3}$, the mean distance between the donors becomes comparable with the Bohr radius. In this case, all donors become ionized, and therefore a Mott's regime is realized. Therefore, further ionization of the impurities is not expected. The CVC in this case should not display a "candle-" like jump, and the current will increase superlinearly with the electric field.

A breakdown of the SI in n-GaAs under electric field was investigated in the work in two different configurations of the magnetic field, when the latter is parallel and perpendicular ($\mathbf{H} \| \varepsilon$ and $\mathbf{H} \perp \varepsilon$) to the electric field. For $\mathbf{H} \perp \varepsilon$, the breakdown field ε_{BD} strongly depends on the magnetic field at $H = 61\,\mathrm{kOe}$, corresponding to $\gamma = \hbar\omega_c/2Ry^* \approx 1$, as a result of two-time increase of the ionization energy of SI. Instead, CVC investigation for the case of $\mathbf{H} \| \varepsilon$ showed practically no dependence of ε_{BD} on the magnetic field. Indeed, magnetic field does not change the wave function shape along the electric field when it is parallel to the electric field. A weak dependence on the higher magnetic field is a result of localization of the electronic wave in the perpendicular to the electric field plane, which reduces a transition of an electron to the excited states of the neighboring impurities.

Investigations of PES of $1s \longrightarrow 2p_{+1}$ transition as well as CR line shape at prebreakdown electric fields have shown no correlation between the physical characteristics and the compensation *K*.

Our CR investigation of n-GaAs samples yields an effective electron mass value, which varies in a wide interval. Moreover, our measurements reveal that an effective electron mass in samples even with small SI concentration is higher than that in dirtier samples. At the breakdown, the effective electron masses in all samples become comparable with each other, which means that the potential fluctuation plays an essential role. Therefore, in order to provide more correct data for an electron effective mass, it is necessary to measure the latter at the breakdown region.

Conflicts of Interest

The authors declare that they have no conflicts of interest.

Acknowledgments

This work was supported by the Science Development Foundation under the President of the Republic Azerbaijan (grant no. EIF-KETPL-2-2015-1(25)-56/01/1). The authors acknowledge A. Najafov for the technical supports.

References

[1] H.-L. Chen, C. Himwas, A. Scaccabarozzi et al., "Determination of n-type doping level in single GaAs nanowires by cathodoluminescence," *Nano Letters*, vol. 17, no. 11, pp. 6667–6675, 2017.

[2] J. Ma, P. Shi, X. Qian, Y.-X. Shang, and Y. Ji, "Spin noise spectroscopy in N-GaAs: spin relaxation of localized electrons," *Chinese Physics Letters*, vol. 34, no. 7, article 077202, 2017.

[3] N. Vainorius, S. Kubitza, S. Lehmann, L. Samuelson, K. A. Dick, and M.-E. Pistol, "Temperature dependent electronic band structure of wurtzite GaAs nanowires," *Nanoscale*, vol. 10, no. 3, pp. 1481–1486, 2018.

[4] P. S. Sokolov, M. Y. Petrov, K. V. Kavokin et al., "Subsecond nuclear spin dynamics in n-GaAs," *Physical Review B*, vol. 99, no. 7, article 075307, 2019.

[5] A. J. Ritenour, S. Levinrad, C. Bradley, R. C. Cramer, and S. W. Boettcher, "Electrochemical nanostructuring of n-GaAs photoelectrodes," *ACS Nano*, vol. 7, no. 8, pp. 6840–6849, 2013.

[6] M. Schubert, T. Hofmann, and C. M. Herzinger, "Generalized far-infrared magneto-optic ellipsometry for semiconductor layer structures: determination of free-carrier effective-mass, mobility, and concentration parameters in *n*-type GaAs," *Journal of the Optical Society of America A*, vol. 20, no. 2, p. 347, 2003.

[7] K. Seeger, *Semiconductor Physics: An Introduction*, Springer-Verlag, Berlin, Germany, 9th edition, 2004.

[8] E. I. Zavaritskaya, "Elektricheskie i opticheskie svoystva poluprovodnikov," in *Proceedings of the Physics Academy of Sciences USSR*, P. N. Lebedeva, Ed., p. 41, Nauka, Moscow, Russia, 1966.

[9] S. M. Kogan and T. M. Lifshits, "Photoelectric spectroscopy—a new method of analysis of impurities in semiconductors," *Physica Status Solidi (a)*, vol. 39, no. 1, pp. 11–39, 1977.

[10] D. M. Larsen, "Inhomogeneous line broadening in donor magneto-optical spectra," *Physical Review B*, vol. 8, no. 2, pp. 535–552, 1973.

[11] S. M. Kogan and V. L. Nguyen, *Soviet Physics—Solid State*, vol. 15, p. 44, 1981.

[12] O. Z. Alekperov, "The influence of fluctuations on shallow impurity line broadening and quality diagnose of n-GaAs," *Physica Status Solidi (b)*, vol. 214, no. 1, pp. 69–79, 1999.

[13] B. I. Shklovskii and A. L. Efros, *Soviet Physics—Semiconductors*, vol. 14, p. 540, 1980.

[14] J. Blinovsky and J. Mycielski, "Theory of absorption of electromagnetic radiation by hopping in n-type silicon and germanium," *Physical Review*, vol. 136, no. 1, pp. 266–271, 1964.

[15] J. Blinovsky and J. Mycielski, "Theory of absorption of electromagnetic radiation by hopping in n-type silicon and germanium II," *Physical Review*, vol. 140, no. 3, pp. 1024–1030, 1965.

[16] V. Y. Ivanov, *Physics of $A^{III}B^{IV}$ Compounds*, p. 104, 1979.

[17] J. Yamashita, "Low-temperature electrical breakdown in germanium," *Journal of the Physical Society of Japan*, vol. 16, no. 4, pp. 720–732, 1961.

[18] J. Spangler, A. Brandi, and W. Prettl, "Sequence of different types of nonlinear current oscillation in n-GaAs," *Applied Physics A Solids and Surfaces*, vol. 48, no. 2, pp. 143–147, 1989.

[19] A. Dargys and J. Kundrotas, "Piezoelectric-interaction-limited capture process in A_3B_5 compounds," *Physica Status Solidi (b)*, vol. 200, no. 2, pp. 509–518, 1997.

[20] P. Rodin, U. Ebert, A. Minarsky, and I. Grechov, "Theory of superfast fronts of impact ionization in semiconductor structures," *Journal of Applied Physics*, vol. 102, no. 3, article 034508, 2007.

[21] J. S. Townsend, "The conductivity produced in gases by the motion of negatively-charged ions," *Nature*, vol. 62, no. 1606, pp. 340-341, 1900.

[22] A. F. Ioffe, "Ллектрическая и мечаническая Црочность и молекулярные силы," *Uspekhi Fizicheskih Nauk*, vol. 8, no. 2, pp. 141–154, 1928.

[23] K. G. Mc Kay and K. B. Mc Afee, "Electron multiplication in silicon and germanium," *Physical Review*, vol. 91, no. 5, pp. 1079–1084, 1953.

[24] S. H. Koenig and G. R. Gunther-Mohr, "The low temperature electrical conductivity of n-type germanium," *Journal of Physics and Chemistry of Solids*, vol. 2, no. 4, pp. 263–283, 1957.

[25] N. Sclar and E. Burstein, "Impact ionization of impurities in germanium," *Journal of Physics and Chemistry of Solids*, vol. 2, no. 1, pp. 1–23, 1957.

[26] O. Z. Alekperov, "Capture of carriers by screened charged centres and low-temperature shallow-impurity electric field breakdown in semiconductors," *Journal of Physics: Condensed Matter*, vol. 10, no. 38, pp. 8517–8521, 1998.

[27] O. V. Emelyanenko, T. S. Logunova, D. N. Nasledov, A. A. Telegin, and Z. I. Tchugueva, *Soviet Physics—Semiconductors*, vol. 10, p. 1280, 1976.

[28] A. L. Efros, B. I. Shklovskii, and I. Y Yanchev, "Impurity conductivity in low compensated semiconductors," *Physica Status Solidi (b)*, vol. 50, no. 1, pp. 45–52, 1972.

[29] W. S. Boyle and R. E. Howard, "Transition to the high field limit in the Zeeman spectra of germanium donors," *Journal of Physics and Chemistry of Solids*, vol. 19, no. 3-4, pp. 181–188, 1961.

[30] H. R. Fetterman, D. M. Larsen, G. E. Stillman, P. E. Tannenwald, and J. Waldman, "Field-dependent central-cell corrections in GaAs by laser spectroscopy," *Physical Review Letters*, vol. 26, no. 16, pp. 975–978, 1971.

[31] D. M. Larsen, "Inhomogeneous broadening of the Lyman-series absorption of simple hydrogenic donors," *Physical Review B*, vol. 13, no. 4, pp. 1681–1691 1976.

[32] B. I. Shklovskii and A. L. Efros, *Soviet Physics—Semiconductors*, vol. 13, no. 3, 1979.

[33] A. L. Efros, N. V. Lien, and B. I. Shklovskii, "Impurity band structure in lightly doped semiconductors," *Journal of Physics C: Solid State Physics*, vol. 12, no. 10, pp. 1869–1881, 1979.

[34] O. Z. Alekperov, V. G. Golubev, V. I. Ivanov-Omskii, and A. S. Mekhtiev, *Physica Status Solidi (b)*, vol. 120, no. 2, pp. 179–181, 1983.

[35] K. M. Raussel and R. F. O'Connel, *Physical Review A*, vol. 9, p. 52, 1974.

[36] R. A. Stradling, L. Eaves, R. A. Hoult, N. Muira, P. E. Simmonds, and C. C. Bradley in *Proceedings of the International Conference on GaAs and Related Compounds*, p. 65, Boulder, CO, USA, 1972.

[37] V. Y. Ivanov and T. M. Lifshitz, *Soviet Physics—Semiconductors*, vol. 17, p. 2050, 1983.

[38] D. M. Korn and D. M. Larsen, "Stark effect on donor transitions in high-purity GaAs," *Solid State Communications*, vol. 13, no. 7, pp. 807–810, 1973.

[39] T. S. Low, G. E. Stillman, A. Y. Cho, H. Morkoc, and A. R. Calawa, "Spectroscopy of donors in high purity GaAs grown by molecular beam epitaxy," *Applied Physics Letters*, vol. 40, no. 7, pp. 611–613, 1982.

[40] A. C. Carter, G. P. Carver, R. J. Nicholas, J. C. Portal, and R. A. Stradling, "Evidence for a contribution to the extrinsic photoconductive signal by hopping through excited states of the donors in silicon and CdTe," *Solid State Communications*, vol. 24, no. 1, pp. 55–60, 1977.

[41] E. A. Kurakova and V. I. Sidorov, *Soviet Physics—Semiconductors*, vol. 9, p. 1286, 1975.

[42] S. M. Kogan and A. F. Polupanov, *Journal of Experimental and Theoretical Physics*, vol. 65, p. 169, 1987.

[43] G. Lindeman, W. Seidenbusch, R. Lassing, J. Edlinger, and E. Cornik, *International Conference Wien*, p. 473, 1983.

[44] V. N. Abakumov, V. I. Perel, and I. N. Yassievich, *Journal of Experimental and Theoretical Physics*, vol. 45, pp. 354–360, 1978.

[45] B. Lax, *Journal of Magnetism and Magnetic Materials*, vol. 11, no. 6, pp. 1–15, 1979.

[46] V. N. Abakumov, L. N. Kreshchuk, and I. N. Yassievich, *Journal of Experimental and Theoretical Physics*, vol. 47, pp. 535–539, 1978.

[47] E. D. Palik, G. S. Picus, S. Teitler, and R. F. Wallis, "Infrared cyclotron resonance in InSb," *Physical Review*, vol. 122, no. 2, pp. 475–481, 1961.
[48] J. R. Waldman, H. R. Fetterman, R. E. Tannenwald, and C. C. Wolfe, *Bulletin of the American Physical Society*, vol. 16, p. 1415, 1971.
[49] R. A. Stradling, *Research in Quantum Physics (Semiconductor Physics)*, Annual Reports of Clarendon Laboratory, Oxford, UK, 1972.
[50] H. J. A. Bluyssen, J. C. Maan, T. B. Tan, and P. Wyder, "Study of cyclotron-resonance-induced conductivity in n-GaAs," *Physical Review B*, vol. 22, no. 2, pp. 749–760, 1980.
[51] Y. A. Firsov, *Polarons*, Nauka, Moscow, Russia, 1975.
[52] A. I. Anselm, *Introduction to the Semiconductors Theory*, Nauka, Moscow, Russia, 1978.
[53] A. L. Efros and B. I. Shklovskii, "Critical behaviour of conductivity and dielectric constant near the metal-non-metal transition threshold," *Physica Status Solidi (b)*, vol. 76, no. 2, pp. 475–485, 1976.
[54] O. Matsuda and E. Otsuka, "Cyclotron resonance study of conduction electrons in n-type indium antimonide under a strong magnetic field-II," *Journal of Physics and Chemistry of Solids*, vol. 40, no. 11, pp. 819–829, 1979.
[55] V. G. Golubev, V. I. Ivanov-omskii, and T. I. Kropotov, *Soviet Physics—Journal of Experimental and Theoretical Physics*, vol. 4, p. 553, 1978.
[56] O. Z. Alekperov, V. G. Golubev, and V. I. Ivanov-Omskii, *Soviet Physics—Semiconductors*, vol. 17, p. 155, 1983.
[57] O. Z. Alekperov, V. G. Golubev, V. I. Ivanov-Omskii, and A. S. Mekhtiev "Method of the semiconductor materials analyzing by means of cyclotron resonance," Patent No 3845907/31-25, 1986.
[58] J. M. Ziman, *Principles of the Theory of Solids*, Cambridge University Press, Cambridge, UK, 1972.
[59] B. I. Shklovskii and A. L. Efros, *Electronic Properties of Doped Semiconductors*, Vol. 45, Springer Series in Solid-State Sciences, Berlin, Germany, 1984.
[60] L. B. Ioffe and A. I. Larkin, *Journal of Experimental and Theoretical Physics*, vol. 54, p. 556, 1981.
[61] L. A. Blumenfeld, *Problems of Semiconductor Physics*, V. L. Bonch-Bruevich, Ed., p. 515, Moscow, Russia, 1957.
[62] M. Fukai, H. Kawamura, K. Sekido, and I. Imai, "Line-broadening of cyclotron resonance due to lattice and neutral impurity scattering in silicon and germanium," *Journal of the Physical Society of Japan*, vol. 19, no. 1, pp. 30–39, 1964.
[63] V. K. Arora and H. N. Spector, "Quantum-limit cyclotron resonance linewidth in semiconductors," *Physica Status Solidi (b)*, vol. 94, no. 2, pp. 701–709, 1979.
[64] H. Kawamura, H. Saji, M. Fukai, K. Sekido, and I. Imai, "Cyclotron resonance line broadening due to carrier-carrier interaction in germanium," *Journal of the Physical Society of Japan*, vol. 19, no. 3, pp. 288–296, 1964.
[65] A. Kawabata, "Theory of cyclotron resonance line width," *Journal of the Physical Society of Japan*, vol. 23, no. 5, pp. 999–1006, 1967.
[66] Y. A. Gurevich, *Journal of Experimental and Theoretical Physics*, vol. 39, p. 322, 1974.
[67] V. K. Arora, M. A. Al-mass'ari, and M. Prasad, "Superoperator theory of cyclotron resonance linewidth in semiconductors," *Physica B+C*, vol. 106, no. 3, pp. 311–319, 1981.
[68] Y. P. Gousev, I. V. Altukhov, K. A. Korolev et al., "Widely tunable continuous-wave THz laser," *Applied Physics Letters*, vol. 75, no. 6, pp. 757–759, 1999.
[69] S. G. Pavlov, R. K. Zhukavin, E. E. Orlova et al., "Stimulated emission from donor transitions in silicon," *Physical Review Letters*, vol. 84, no. 22, pp. 5220–5223, 2000.
[70] N. N. Zinovev, A. V. Andrianov, V. Y. Nekrasov et al., "Terahertz injection electroluminescence in multiperiod quantum-cascade AlGaAs/GaAs structures," *ournal of Experimental and Theoretical Physics Letters*, vol. 74, no. 2, pp. 100–102, 2001.
[71] N. N. Zinovev, A. V. Andrianov, V. Y. Nekrasov et al., *Soviet Physics—Solid State*, vol. 36, pp. 226–229, 2002.
[72] T. N. Adam, R. T. Troeger, S. K. Ray, P.-C. Lv, and J. Kolodzey, "Terahertz electroluminescence from boron-doped silicon devices," *Applied Physics Letters*, vol. 83, no. 9, pp. 1713–1715, 2003.
[73] A. V. Andrianov, A. O. Zakhar'in, I. N. Yassievich, and N. N. Zinov'ev, "Terahertz electroluminescence under conditions of shallow acceptor breakdown in germanium," *Journal of Experimental and Theoretical Physics Letters*, vol. 79, no. 8, pp. 365–367, 2004.
[74] W. Quade, G. Hüpper, E. Schöll, and T. Kuhn, "Monte Carlo simulation of the nonequilibrium phase transition in p-type Ge at impurity breakdown," *Physical Review B*, vol. 49, no. 19, pp. 13408–13419, 1994.
[75] S. M. Ryvkin, *Photoelectric Effects in Semiconductors*, Vol. 15, Springer Series in Solid State Sciences, Berlin, Germany, 1964.

7

Feasibility Study of the use of Handheld NIR Spectrometer for Simultaneous Authentication and Quantification of Quality Parameters in Intact Pineapple Fruits

Charles L. Y. Amuah,[1] **Ernest Teye**,[2] **Francis Padi Lamptey**,[2] **Kwasi Nyandey**,[1] **Jerry Opoku-Ansah**,[1] **and Peter Osei-Wusu Adueming**[1]

[1]*University of Cape Coast, College of Agriculture and Natural Sciences, School of Physical Sciences, Department of Physics, Laser and Fibre Optics Centre, Cape Coast, Ghana*
[2]*University of Cape Coast, College of Agriculture and Natural Sciences, School of Agriculture, Department of Agricultural Engineering, Cape Coast, Ghana*

Correspondence should be addressed to Ernest Teye; teyernest@gmail.com

Academic Editor: Wee Chew

The potential of predicting maturity using total soluble solids (TSS) and identifying organic from inorganic pineapple fruits based on near-infrared (NIR) spectra fingerprints would be beneficial to farmers and consumers alike. In this study, a portable NIR spectrometer and chemometric techniques were combined to simultaneously identify organically produced pineapple fruits from conventionally produced ones (thus organic and inorganic) and also predict total soluble solids. A total of 90 intact pineapple fruits were scanned with the NIR spectrometer while a digital refractometer was used to measure TSS from extracted pineapple juice. After attempting several preprocessing techniques, multivariate calibration models were built using principal component analysis (PCA), *K*-nearest neighbor (KNN), and linear discriminant analysis (LDA) to identify the classes (organic and conventional pineapple fruits) while partial least squares regression (PLSR) method was used to determine TSS of the fruits. Among the identification techniques, the MSC-PCA-LDA model accurately identified organic from conventionally produced fruits at 100% identification rate. For quantification of TSS, the MSC-PLSR model gave Rp = 0.851 and RMSEC = 0.950 °Brix, and Rc = 0.854 and RMSEP = 0.842 °Brix at 5 principal components in the calibration set and prediction set, respectively. The results generally indicated that portable NIR spectrometer coupled with the appropriate chemometric tools could be employed for rapid nondestructive examination of pineapple quality and also to detect pineapple fraud due to mislabeling of conventionally produced fruits as organic ones. This would be helpful to farmers, consumers, and quality control officers.

1. Introduction

Pineapple (*Ananas comosus* (L.) Merr) is the most economically significant crop in the family Bromeliaceae with exceptional juiciness, vibrant tropical flavour, and immense health benefits. Pineapple fruit is a good source of vitamin C, fiber, and other minerals. It also contains sugar, bromalin (protein-digesting enzyme), citric acid, malic acid, vitamins A and B, and excellent amount of fiber [1]. Quality evaluation and assurance of pineapple fruits before export, during processing, and in the fresh market is a required activity to ensure quality and safety. These are normally based on internal quality traits such as total soluble solids (TSS), firmness, and acidity. However, TSS (°Brix) has been established as the most vital internal quality indicator. For instance, TSS is among the most important internal quality attribute in determining fruit maturity and harvesting time as well as assessing and grading postharvest quality fruits [2].

At present, the conventional technique for the determination of the internal quality parameters of fruits such as pineapple involves destructive means. This method is usually cumbersome and wasteful. It requires specialised equipment, elaborate procedures, and trained personnel, which results in high analysis costs and does not allow for the entire fruits produced to be analyzed [3]. Oftentimes, a representative sample is used to predict the entire fruits and this normally leads to misjudgement. Therefore, rapid nondestructive prediction of TSS in pineapple would be of great value in determining the best harvesting time with vital consequence on its eating quality. This would be timely in meeting the ever-increasing consumer demand for consistent high-quality fruits.

Furthermore, the high price enjoyed by selling organically produced pineapples as against conventional ones has led to mislabeling also known as food fraud to gain undue financial advantage. Studies have revealed that consumer demand for organic food is growing and organic agriculture is more profitable due to higher price farmers receive for their produce [4, 5]. Organic production in general is a system which excludes the use of synthetic fertilizer, pesticides, growth regulators, and other agrochemicals [6]. However, the techniques used for detecting the differences (organically produced fruits such as pineapples from conventional ones) are relatively expensive, destructive, and time consuming and often require elaborate sample preparation. Hence, simultaneous detection of pineapple fruit quality parameters (authenticity and quantification) using portable NIR spectrometer and chemometric technique would be very beneficial. This could offer rapid examination of pineapple fruit quality.

Near-infrared spectroscopy provides an alternative for the determination of pineapple quality parameters. This method provides rapid and nondestructive detection of food quality and safety. It has found its use for qualitative and quantitative analysis in the food industry. NIR spectroscopy is simple, rapid, and nondestructive and requires minimal or no sample preparation. According to other authors, NIR spectroscopy could be one of the most commonly used techniques due to its speediness, noncontact, and low operating cost [7–9]. For fruit quality measurements, NIR spectroscopy has been used to detect nitrate levels in intact pineapple [10], soluble solids content and acidity in kiwifruit [11], and pear internal quality indices [12]. Furthermore, miniaturization of NIR spectroscopy has resulted in the development of commercial handheld or portable spectroscopic systems that offer additional speed, simplicity, and sensitivity. It also presents an ideal tool for agrifood quality evaluation for *in situ* measurements due to its portability [13, 14]. Other researchers have used handheld or portable NIR spectrometers to determine fruit quality parameters such as TSS, TA, and sugar content in fruits [7, 15, 16]. Furthermore, Cayuela and Weiland used two portable spectrometers to predict several quality parameters in intact oranges [17] while Sánchez et al. studied by improving the performance of portable NIR instrument for intact nectarines [18]. However, and to the best of our knowledge, no researcher has investigated the feasibility of using portable NIR spectrometer coupled with chemometric techniques to simultaneously discriminate organically produced pineapples and conventional ones and also predict total soluble solids (TSS) nondestructively.

Therefore, the main objective of this study was to evaluate the potential of using portable NIR spectrometer for rapid and nondestructive identification and quantification of pineapple quality parameters such as TSS of intact pineapple fruits. The specific objectives were to determine the best multivariate technique for identification and to predict TSS of intact pineapple fruits.

2. Materials and Methods

2.1. Pineapple Fruit Samples. In this study, 90 pieces of sugarloaf pineapple fruits at different maturity stages were obtained directly from pineapple farmers in the Central Region of Ghana and transported to the University of Cape Coast, School of Agriculture Teaching and Research Laboratory. These fruits comprise 30 pieces of organically produced pineapple fruits and 60 pieces of conventionally produced pineapple fruits. The fruits were then stored at 26°C (±1°C) for two days before measurements were taken.

2.2. Sample Spectra Acquisition. The spectrum of each pineapple was collected in the reflectance mode using a handheld spectrometer (SCIO™) with spectra range between 740 nm and 1070 nm in a 1 nm resolution for spectra data recording. For each fruit, the lower part was scanned three times after rotating it at 120°. The scanning was done at an ambient temperature of 26 ± 1°C with a humidity of 60%. Figure 1 shows the setup of the scanning processing using SCIO NIR spectrometer.

2.3. Reference Measurements (TSS/°Brix). Total soluble solids (TSS) contents were determined using a digital refractometer (model: PAL-1, °Brix range of 0–35%; Atago, Tokyo, Japan) according to the methods described by others [15, 16, 19]. For each pineapple fruit, the base was selected and juiced. About 1.0 ml juice was then taken for TSS measurement with a digital refractometer. Triplicate measurements were performed and the results expressed as °Brix.

2.4. Data Partition. The raw dataset (from the 90 samples) after preprocessing with suitable techniques was divided into two subsets, calibration set (data from 68 samples) for developing the model and prediction set (data from 22 samples) for evaluating the predictive ability of the constructed models. To avoid bias, 75% of data from both organic and inorganic samples were selected as the calibration set while the remaining data were selected as the prediction set. As shown in Table 1, the members in each set were selected in order to come to a 3/1 division of calibration set/prediction set.

2.5. Software Device. Spectra data recordings stored in a cloud-based dataset with their corresponding reference value for time of scanning were downloaded using a research license of SCIO lab and imported to MATLAB version 9.5.0 (Mathworks Inc., USA) with Windows 10 Basic for data

FIGURE 1: Setup of scanning processing.

processing for all preprocessing treatments and multivariate algorithms.

2.6. Spectra Preprocessing Techniques. The average raw spectra of the pineapple samples are shown in Figure 2(a) while the other pretreatment spectra are also shown in Figures 2(b) and 2(c). The activity of preprocessing of the spectra data is an integral part of modelling to eliminate background information and noise from the useful properties of the scanned samples [3, 20]. In this research, two spectra preprocessing techniques (mean centering (MC) and multiplicative scatter correction (MSC)) were applied because the models developed using the raw spectra data did not give the desired results.

MC is normally used as resolution enhancement method, and it is known to work by simply adjusting a dataset to reposition the centroid of the data to the origin of the coordinate system [21].

$$S_i = S_{ij} - S_{im}. \quad (1)$$

MSC on the other hand is a useful preprocessing technique for the correction of scattered light and inclination of baseline variation. For more information, refer to other authors [21, 22]. For spectra with additive effect (x_i) and multiplicative effect (y_i), it can be represented as

$$S_i = x_i + y_i S_m + e_i, \quad (2)$$

where S_m is the mean of the set of spectra. The residual error vector e_i gives information about the random noise. Hence, the MSC-corrected spectra are calculated as

$$S_{i\,\text{corrected}} = \frac{(S_i - x_i)}{y_i}, \quad (3)$$

where for a measured data containing a set of spectra S_i ($i = 1, 2, \ldots, K$) with data points S_{ij} ($j = 1, 2, \ldots, N$), the mean of these data could be represented as S_{im}, which is the mean of that spectrum.

2.7. Principal Component Analysis (PCA). PCA is an unsupervised data description and dimension reduction techniques which is mostly used to deal with large spectra data [23]. It normally involves the first step of data analysis in order to detect patterns from the data matrix as it brings out visualized data trends in dimensional space [24]. For more information, refer to these authors [25].

2.8. Multivariate Models. The development of computers and software programmes is making chemometric techniques a very powerful tool for processing NIR spectra data as it overcomes the difficulty of multicollinearity and gives scientific statistical inferences for meaningful conclusions to experimental results [26, 27]. Choosing the best method is the next challenge as it is quite a cumbersome process since there existed quite a lot of types. In this research, *K*-nearest neighbor (KNN) and linear discriminant analysis (LDA) were employed comparatively.

K-nearest neighbor is a linear and nonparametric classification method which works based on a distance function that measures the difference or similarity between two stances [28, 29]. For KNN, parameter *K* influences the results of the classification model; hence, the choice of *K* is normally optimized by calculating its potential with several *K* values (normally small *K* values of 3 or 5). It must be known that KNN cannot work well if large differences are present in the number of samples in each class [25]. This therefore makes KNN tool a more suitable technique for modelling similar class groupings.

Linear discriminant analysis is a linear and parametric supervised pattern recognition technique which has found it useful for analyzing spectra data. It works by finding linear combination of features which brings out clearly the ratio of between-class variance and reduces the ratio of within-class variance [30]. For more information, refer to [25]. It is important to note that the performance of LDA is based on the number of principal component factors.

2.9. Partial Least Squares (PLS). Partial least squares (PLS) is a well-known linear multivariate method used for spectra data processing and it can analyze data with strong collinear, noisy, and redundant variables. For more information, refer to [31, 32]. The results of the PLS model are normally evaluated by using three main parameters, namely, the root mean square error of cross-validation (RMSECV), the root mean square error of prediction (RMSEP), and the correlation coefficient (R), among others [33]. These parameters were calculated by the following equations:

$$\text{RMSECV} = \sqrt{\frac{\sum_{i=1}^{n} \left(\widehat{y}_{\backslash i} - y_i\right)^2}{n}}, \quad (4)$$

$$\text{RMSEP} = \sqrt{\frac{\sum_{i}^{n} \left(y_i - \widehat{y}_i\right)^2}{n}}, \quad (5)$$

$$R = \sqrt{1 - \frac{\sum_{i=1}^{n} \left(\widehat{y}_i - y_i\right)^2}{\sum_{i=1}^{n} \left(y_i - \overline{y}\right)^2}}, \quad (6)$$

where n = the number of samples, y_i = the reference measurement results for sample i, $\widehat{y}_{\backslash i}$ = the estimated result for

Table 1: Reference measurement of TSS (°Brix) of total samples used.

Subsets	Number of samples	Total soluble solids (TSS) content		
		Range (°Brix)	Mean (°Brix)	Std (°Brix)
Total samples	90	11.90–18.60	14.81	1.82
Calibration	68	11.90–18.59	14.86	1.88
Prediction	22	12.00–17.50	14.66	1.65

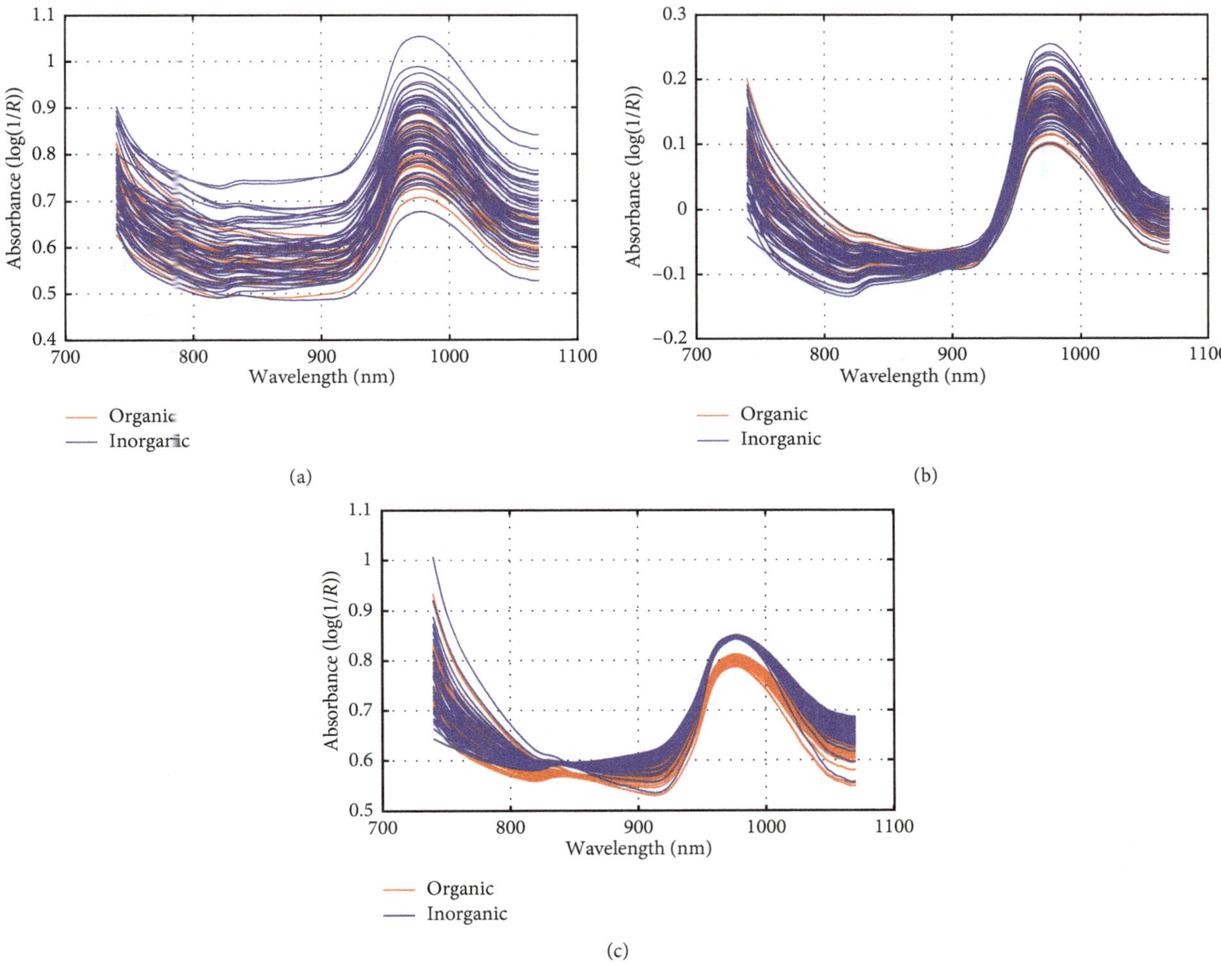

Figure 2: NIR spectra of pineapple fruits. (a) Raw, (b) pretreatment with MC, and (c) pretreatment with MSC.

sample i when the model is constructed with sample i removed, $\hat{y}_i$ = the estimated results of the model for the sample i, and $\overline{y}$ = the mean of the reference measurement results for all samples.

3. Results and Discussion

3.1. Spectra Presentation/Analysis. The fingerprint from the spectra dataset was used to create the statistical models. It could be observed from the spectra profile in Figures 2(a)–2(c) that the major peaks are around 960–1050 nm. The wavelength range corresponds to O-H 2nd overtone and N-H 2nd overtone, which represents H_2O, ROH, ArOH (OH bond on the aromatic group), and NH_2 functional groups [34]. These groups are familiar with major constituents of water, glucose, sucrose, and cellulose of pineapples. TSS is an organic molecule that contains C-H, O-H, C-O, and C-C bonds, and NIR spectroscopy could be used to nondestructively measure this molecule [35, 36]. After preprocessing of the spectra dataset with MC and MSC, a clear separation between organic and inorganic pineapple fruits appeared in MSC pretreatment spectra profile as shown in Figure 2(c). This suggests that organic and inorganic pineapple fruits could be differentiated within 800 nm and 1070 nm range using MSC pretreatment technique. The MSC is therefore a useful tool for correcting baseline shift and light scattering problems related to the spectral dataset as mentioned by other authors [22].

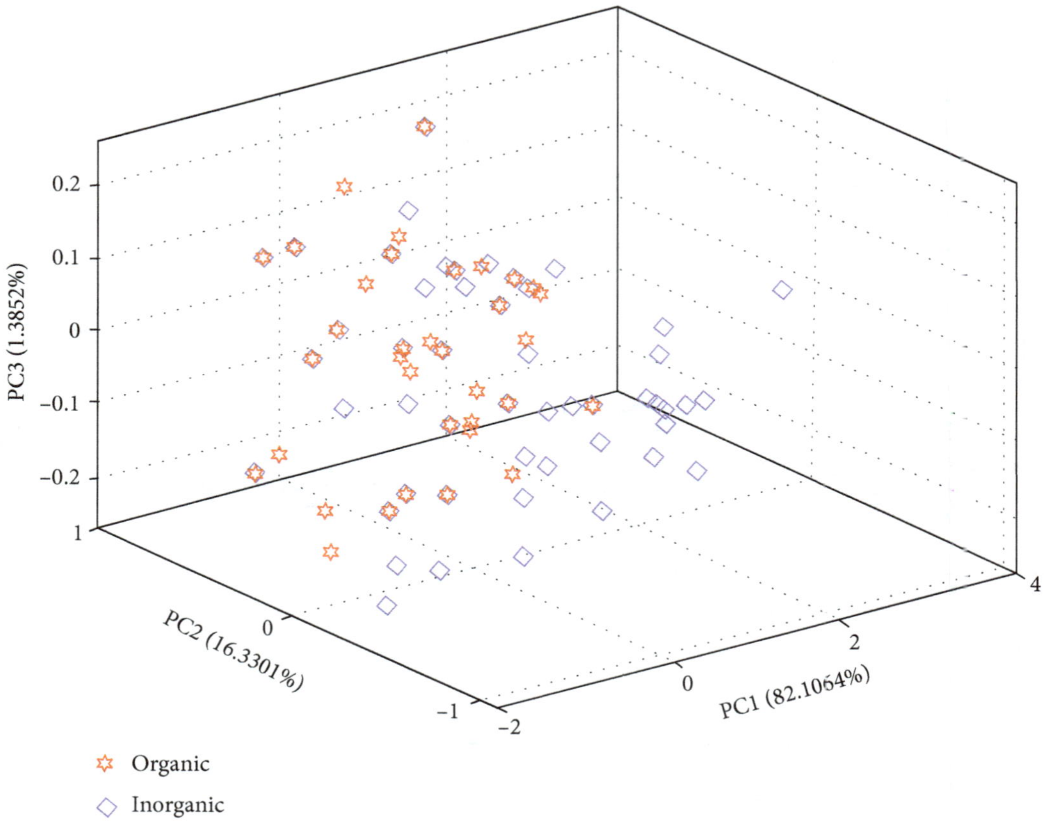

(a)

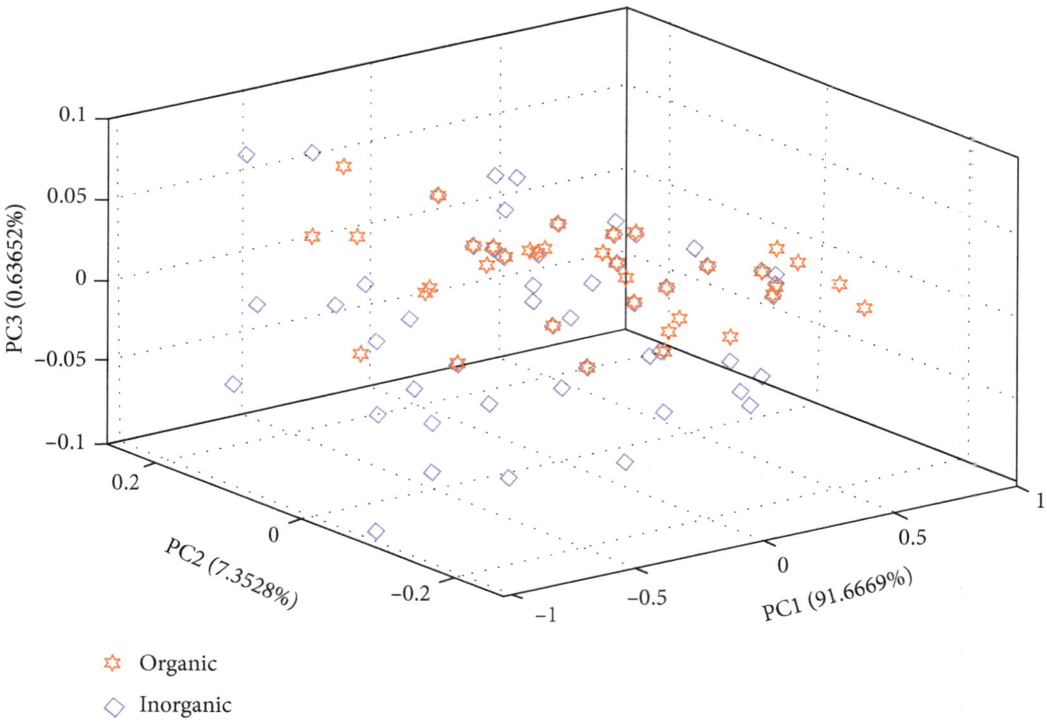

(b)

FIGURE 3: Continued.

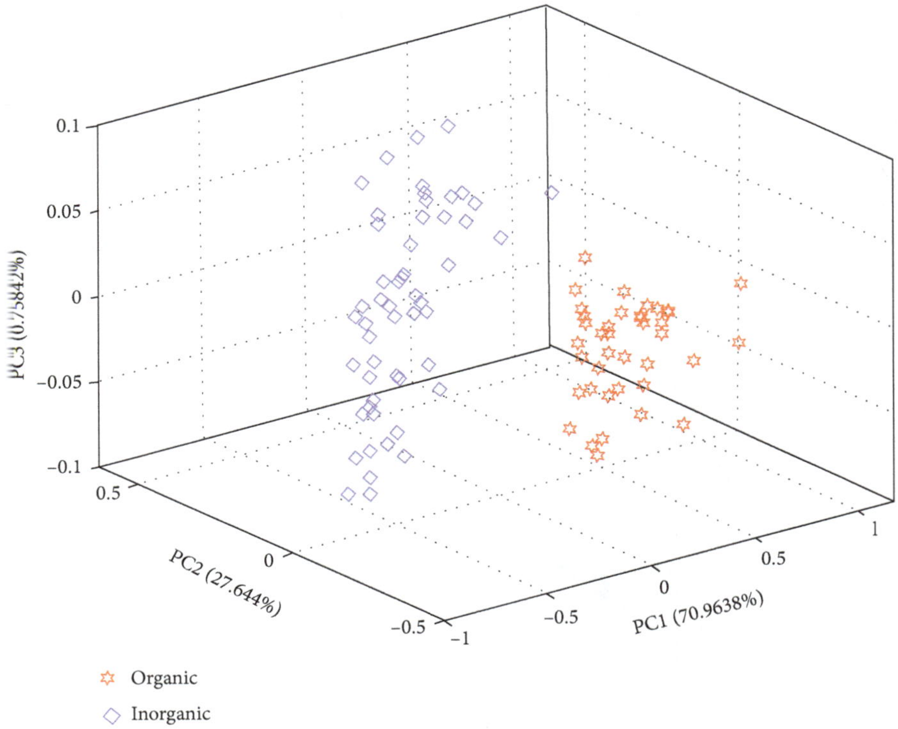

(c)

Figure 3: PCA score plot of spectra data. (a) Raw, (b) preprocessed with MC, and (c) preprocessed with MSC. MSC with PCA technique gave a separation with clear cluster trend.

3.2. Principal Component Analysis (PCA). Principal component analysis was used to identify cluster trend in the spectra data. From the results obtained, it was observed that PCA done on raw and MC preprocessed spectra data did not give any clear cluster trends or separations as shown in Figures 3(a) and 3(b). However, MSC-PCA technique gave a separation with clear cluster trend as shown in Figure 2(c). Again this further proves the unique characteristics of MSC as an effective technique for baseline light scattering corrections as proposed by Geladi and coworkers [22]. PCA was able to identify the most important directions of the variability in the multivariate data space (*X* matrix) and to determine the primary phenomena in the spectra dataset [19]. The PCs (PC1, PC2, and PC3) contain spectra information and its corresponding chemical compositional information hence accounted for 99.36% of the total variance that existed for the 90 pineapple samples used in this study. Pineapples have a considerable difference in their chemical properties according to their preharvest activities and postharvest practices that categorize them as either organic or inorganic.

3.3. Classification Models. In this study, KNN and LDA were attempted for developing classification models for classifying organic and inorganic pineapple fruits.

The results of the classification models are shown in Table 2. The two classification models used performed well using MSC-PCA dataset. KNN and LDA had classification rate above 98% in both the calibration set and prediction set at optimal principal components (PCs) = 3, respectively. This means MSC-PCA preprocessing enhanced the performance of both KNN and LDA as compared to raw and MC dataset. On the other hand, LDA was slightly superior to KNN in the training set. This means LDA was well able to find the linear combination of features and the resulting combination used was a better linear classifier. More so, it could be explained that the good accuracy obtained by the model could be as a result of complex distinct organoleptic and nutritional properties between organically grown pineapple fruits as against conventionally grown ones. This phenomenon is further supported by the evidence that organic production excludes the use of synthetic fertilizer, pesticides, growth regulators, and other chemicals [6] and impact of the fruit's quality and safety due to chemical residues. Other studies have also revealed that organically produced pineapple fruits comparatively have high vitamin C and moderate acidity [4] as well as highest total soluble solids contents [37].

3.4. Quantification Model. Partial least squares model was used for the determination of TSS (°Brix) in both organic and inorganic pineapple fruits. From Figure 4, the measured values correlated linearly with NIR predicted measurements. However, there were some few outliers which subsequently

Table 2: The performance of classification models influenced by preprocessing techniques.

Model		Number of principal components	Correct classification rate (%)	
			Calibration set (68)	Prediction set (22)
KNN	Raw	3	45.56	50.00
	MC	3	52.22	60.00
	MSC	3	98.89	100.00
LDA	Raw	3	61.11	66.67
	MC	3	55.56	50.00
	MSC	3	100.00	100.00

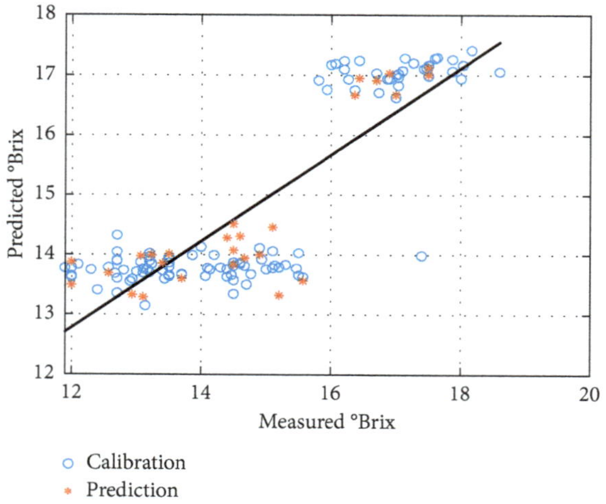

Figure 4: Reference measured versus NIR prediction of TSS in the calibration set and prediction set.

affected the PLS model. From Table 3, it could be seen that the MSC-PLS model was the best comparatively with parameters of Rc = 0.851, and RMSEC = 0.950 °Brix in the calibration set while Rp = 0.854 and RMSEP = 0. 842 °Brix in the prediction set. This result indicates that proper preprocessing technique is an efficient way to improve the accuracy of the PLS model [33]. For a good model, *R* value should be close to unity while RMSEC and RMSEP should be close to zero. The weaknesses of this MSC-PLS model could be attributed to the characteristics of the PLS model, because classic PLS model was built using full spectrum range, which normally consists of useful and irrelevant or redundant information (noise). The noisy spectrum normally reduces the performance of the model. Hence, to improve this model for intact pineapple quality evaluation, other known PLS types should be investigated while comparing with other nonlinear algorithms. Notwithstanding, the results compare favourable with those found by other authors for using VIS-SWNIR spectroscopy for predicting soluble solids content in pineapple fruits [15]. More so, it must be stated that through favourable statistical correlations, the NIR multivariate models predicted the °Brix values.

3.5. Selection of Vital Wavelengths. When developing the PLS model, there is the need to consider how much each wavelength contributes to the final outcome. Figure 5 shows PLS loading weights of the best model and this explains how

Table 3: Prediction results of PLS models under different pretreatment methods.

Pretreatment	Factors	Calibration		Prediction	
		R	RMSEC (°Brix)	*R*	RMSEP (°Brix)
Raw	5	0.559	1.499	0.617	1.274
MC	6	0.504	1.563	0.539	1.363
MSC	5	0.851	0.950	0.854	0.842

the complexity of the PLS model was developed. The loadings show how well the wavelength was taken into consideration by the model components. It is used to understand how much each *x*-variable (wavelengths) contributes to the meaningful variation in the data and to interpret variable relationships as well as interpret the meaning of each model component [38]. The loading weights were normalised so that the length and directions could be made meaningful. From Figure 5, we can see several peaks at certain wavelengths (754, 760, 823, 850, 884, 901, 910, 950, and 960 nm) which are considered to be more useful for the developed PLS model used for determining total soluble solids (TSS, °Brix) content in pineapples. These vital observed wavelengths in this study are closely related to the chemical composition in pineapple fruits. Specifically, these wavelengths are related to the third overtone region comprising OH and CH stretching vibrations of sucrose solutions [39], an important component of TSS. For

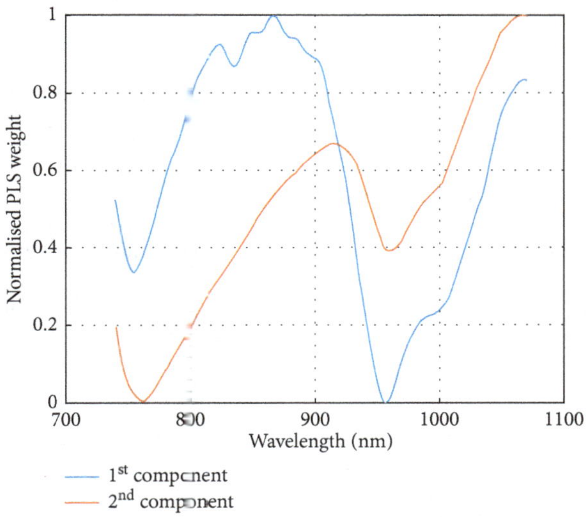

FIGURE 5: PLS loading weights versus wavelength of the top two latent variables.

example, the wavelengths at 910, 950, and 960 nm are related to the chemical group of C-H and O-H which are attributed to TSS, while the region of 750–820 nm reveals sucrose, glucose, and fructose [9, 39].

4. Conclusion

The research has shown the potential of handheld NIR spectroscopic technique for rapid nondestructive measurements of pineapple quality. MSC gave the best PCA cluster trend with clear separation in the first three PCs. The overall results showed that handheld spectrometer coupled with MSC-PCA+LDA model could be used to identify organically and conventionally grown intact pineapple fruits with 100% identification rate in both the training set and prediction set, respectively. On the other hand, PLS regression model could be used for predicting TSS (°Brix) with RMSEC = 0.95 and RMSEP = 0.84 at 5 factors with an accuracy of 85% in both the calibration set and prediction set, respectively. There is a potential of these models to be imported into mobile phone technology for effective all-round application.

Conflicts of Interest

The authors declare that there are no conflicts of interest.

Acknowledgments

The authors highly appreciate the support provided by the University of Cape Coast, Agilent Foundation, and Mars. The supply of fresh pineapple fruits by ACOPPS and AMOPPA pineapple farmers in the Central Region is highly acknowledged. The authors appreciate proofreading assistance provided by Mrs. Winifred Akpene Teye.

References

[1] F. Hossain, "World pineapple production: an overview," *African Journal of Food, Agriculture, Nutrition and Development*, vol. 16, no. 4, pp. 11443–11456, 2016.

[2] Y. Peng and R. Lu, "Analysis of spatially resolved hyperspectral scattering images for assessing apple fruit firmness and soluble solids content," *Postharvest Biology and Technology*, vol. 48, no. 1, pp. 52–62, 2008.

[3] V. Cortés, C. Ortiz, N. Aleixos, J. Blasco, S. Cubero, and P. Talens, "A new internal quality index for mango and its prediction by external visible and near-infrared reflection spectroscopy," *Postharvest Biology and Technology*, vol. 118, pp. 148–158, 2016.

[4] M. Owureku-Asare, J. Agyei-Amponsah, S. W. K. Agbemavor et al., "Effect of organic fertilizers on physical and chemical quality of sugar loaf pineapple (*Ananas comosus* L) grown in two ecological sites in Ghana," *African Journal of Food, Agriculture, Nutrition and Development*, vol. 15, no. 2, pp. 9982–9995, 2015.

[5] S. Bolwig, P. Gibbon, and S. Jones, "The economics of smallholder organic contract farming in tropical Africa," *World Development*, vol. 37, no. 6, pp. 1094–1104, 2009.

[6] D. B. Singh, "Comparative study of organic and conventional farming of pineapple," *Asian Journal of Horticulture*, vol. 4, no. 1, pp. 1–4, 2009.

[7] J. Li, Q. Wang, L. Xu, X. Tian, Y. Xia, and S. Fan, "Comparison and optimization of models for determination of sugar content in pear by portable Vis-NIR spectroscopy coupled with wavelength selection algorithm," *Food Analytical Methods*, vol. 12, no. 1, pp. 12–22, 2019.

[8] B. M. Nicolaï, K. Beullens, E. Bobelyn et al., "Nondestructive measurement of fruit and vegetable quality by means of NIR spectroscopy: a review," *Postharvest Biology and Technology*, vol. 46, no. 2, pp. 99–118, 2007.

[9] H. Wang, J. Peng, C. Xie, Y. Bao, and Y. He, "Fruit quality evaluation using spectroscopy technology: a review," *Sensors*, vol. 15, no. 5, pp. 11889–11927, 2015.

[10] S. Srivichien, A. Terdwongworakul, and S. Teerachaichayut, "Quantitative prediction of nitrate level in intact pineapple using Vis–NIRS," *Journal of Food Engineering*, vol. 150, pp. 29–34, 2015.

[11] A. Moghimi, M. H. Aghkhani, A. Sazgarnia, and M. Sarmad, "Vis/NIR spectroscopy and chemometrics for the prediction of soluble solids content and acidity (pH) of kiwifruit," *Biosystems Engineering*, vol. 106, no. 3, pp. 295–302, 2010.

[12] Y. Liu, X. Chen, and A. Ouyang, "Nondestructive determination of pear internal quality indices by visible and near-infrared spectrometry," *LWT—Food Science and Technology*, vol. 41, no. 9, pp. 1720–1725, 2008.

[13] P. M. Santos, E. R. Pereira-Filho, and L. E. Rodriguez-Saona, "Application of hand-held and portable infrared spectrometers in bovine milk analysis," *Journal of Agricultural and Food Chemistry*, vol. 61, no. 6, pp. 1205–1211, 2013.

[14] L.-M. Yuan, L. Sun, J.-R. Cai, and H. Lin, "A preliminary study on whether the soluble solid content and acidity of oranges predicted by near infrared spectroscopy meet the sensory degustation," *Journal of Food Process Engineering*, vol. 38, no. 4, pp. 309–319, 2015.

[15] K. S. Chia, H. Abdul Rahim, and R. Abdul Rahim, "Prediction of soluble solids content of pineapple via non-invasive low

cost visible and shortwave near infrared spectroscopy and artificial neural network," *Biosystems Engineering*, vol. 113, no. 2, pp. 158–165, 2012.
[16] E. J. N. Marques, S. T. de Freitas, M. F. Pimentel, and C. Pasquini, "Rapid and non-destructive determination of quality parameters in the 'Tommy Atkins' mango using a novel handheld near infrared spectrometer," *Food Chemistry*, vol. 197, pp. 1207–1214, 2016.
[17] J. A. Cayuela and C. Weiland, "Intact orange quality prediction with two portable NIR spectrometers," *Postharvest Biology and Technology*, vol. 58, no. 2, pp. 113–120, 2010.
[18] M.-T. Sánchez, M.-J. De la Haba, J.-E. Guerrero, A. Garrido-Varo, and D. Pérez-Marín, "Testing of a local approach for the prediction of quality parameters in intact nectarines using a portable NIRS instrument," *Postharvest Biology and Technology*, vol. 60, no. 2, pp. 130–135, 2011.
[19] M. Sun, D. Zhang, L. Liu, and Z. Wang, "How to predict the sugariness and hardness of melons: a near-infrared hyperspectral imaging method," *Food Chemistry*, vol. 218, pp. 413–421, 2017.
[20] M. Blanco and I. Villarroya, "NIR spectroscopy: a rapid-response analytical tool," *TrAC Trends in Analytical Chemistry*, vol. 21, no. 4, pp. 240–250, 2002.
[21] Y. Ozaki, W. F. McClure, and A. A. Christy, *Near-Infrared Spectroscopy in Food Science and Technology*, John Wiley & Sons, Hoboken, NJ, USA, 2006.
[22] P. Geladi, D. MacDougall, and H. Martens, "Linearization and scatter-correction for near-infrared reflectance spectra of meat," *Applied Spectroscopy*, vol. 39, no. 3, pp. 491–500, 1985.
[23] W. Kong, C. Zhang, F. Liu, P. Nie, and Y. He, "Rice seed cultivar identification using near-infrared hyperspectral imaging and multivariate data analysis," *Sensors*, vol. 13, no. 7, pp. 8916–8927, 2013.
[24] B. K. Lavine and N. Mirjankar, "Clustering and classification of analytical data," in *Encyclopedia of Analytical Chemistry: Applications, Theory and Instrumentation*, John Wiley & Sons, Hoboken, NJ, USA, 2006.
[25] L. A. Berrueta, R. M. Alonso-Salces, and K. Héberger, "Supervised pattern recognition in food analysis," *Journal of Chromatography A*, vol. 1158, no. 1-2, pp. 196–214, 2007.
[26] Y. Roggo, P. Chalus, L. Maurer, C. Lema-Martinez, A. Edmond, and N. Jent, "A review of near infrared spectroscopy and chemometrics in pharmaceutical technologies," *Journal of Pharmaceutical and Biomedical Analysis*, vol. 44, no. 3, pp. 683–700, 2007.
[27] E. Teye, X. Huang, L. K. Sam-Amoah et al., "Estimating cocoa bean parameters by FT-NIRS and chemometrics analysis," *Food Chemistry*, vol. 176, pp. 403–410, 2015.
[28] R. O. Duda, P. E. Hart, and D. G. Stork, *Pattern Classification*, John Wiley & Sons, Hoboken, NJ, USA, 2012.
[29] L. Jiang, Z. Cai, D. Wang, and S. Jiang, "Survey of improving K-nearest-neighbor for classification," in *Proceedings of the Fourth International Conference on Fuzzy Systems and Knowledge Discovery FSKD 2007*, IEEE, Haikou, China, August 2007.
[30] Q. Chen, J. Cai, X. Wan, and J. Zhao, "Application of linear/non-linear classification algorithms in discrimination of pork storage time using Fourier transform near infrared (FT-NIR) spectroscopy," *LWT–Food Science and Technology*, vol. 44, no. 10, pp. 2053–2058, 2011.
[31] E. Teye and X. Huang, "Novel prediction of total fat content in cocoa beans by FT-NIR spectroscopy based on effective spectral selection multivariate regression," *Food Analytical Methods*, vol. 8, no. 4, pp. 945–953, 2015.
[32] S. Wold, M. Sjöström, and L. Eriksson, "PLS-regression: a basic tool of chemometrics," *Chemometrics and Intelligent Laboratory Systems*, vol. 58, no. 2, pp. 109–130, 2001.
[33] X. Guan, J. Liu, K. Huang, J. Kuang, and D. Liu, "Evaluation of moisture content in processed apple chips using NIRS and wavelength selection techniques," *Infrared Physics & Technology*, vol. 98, pp. 305–310, 2019.
[34] H. Cen and Y. He, "Theory and application of near infrared reflectance spectroscopy in determination of food quality," *Trends in Food Science & Technology*, vol. 18, no. 2, pp. 72–83, 2007.
[35] G. Giovanelli, N. Sinelli, R. Beghi, R. Guidetti, and E. Casiraghi, "NIR spectroscopy for the optimization of postharvest apple management," *Postharvest Biology and Technology*, vol. 87, pp. 13–20, 2014.
[36] T. Ma, X. Li, T. Inagaki, H. Yang, and S. Tsuchikawa, "Noncontact evaluation of soluble solids content in apples by near-infrared hyperspectral imaging," *Journal of Food Engineering*, vol. 224, pp. 53–61, 2018.
[37] M. Darnaudery, P. Fournier, and M. Léchaudel, "Low-input pineapple crops with high quality fruit: promising impacts of locally integrated and organic fertilisation compared to chemical fertilisers," *Experimental Agriculture*, vol. 54, no. 2, pp. 286–302, 2018.
[38] D. Suhandy and M. Yulia, "The use of partial least square regression and spectral data in UV-visible region for quantification of adulteration in Indonesian palm civet coffee," *International Journal of Food Science*, vol. 2017, Article ID 6274178, 7 pages, 2017.
[39] M. Golic, K. Walsh, and P. Lawson, "Short-wavelength near-infrared spectra of sucrose, glucose, and fructose with respect to sugar concentration and temperature," *Applied Spectroscopy*, vol. 57, no. 2, pp. 139–145, 2003.

8

Near-Infrared Spectroscopy and Chemometrics for the Routine Detection of Bilberry Extract Adulteration and Quantitative Determination of the Anthocyanins

Claudio Gardana,[1] **Antonio Scialpi**,[2] **Christian Fachechi**,[2] **and Paolo Simonetti**[1]

[1]*Department of Food, Environmental and Nutritional Sciences (DeFENS), Università Degli Studi di Milano, Via Celoria 2, 20133 Milano, Italy*
[2]*Specchiasol R&D Section, Via Rizzi 1/3, 37012 Bussolengo, Italy*

Correspondence should be addressed to Claudio Gardana; claudio.gardana@unimi.it

Academic Editor: Maria Carmen Yebra-Biurrun

Consumers must be assured that bought food supplements contain both bilberry extract and the anthocyanin amounts that match the declared levels. Therefore, a Fourier transform near-infrared (FT-NIR) spectroscopic method was validated based on principal component scores for the prediction of bilberry extract adulteration and partial least squares regression model for total anthocyanin evaluation. Anthocyanins have been quantified individually in 71 commercial bilberry extracts by HPLC-DAD, and 6 of them were counterfeit. The anthocyanin content in bilberry extracts was in the range 18–34%. Authentic bilberry extracts ($n = 65$) were divided into two parts: one for calibration ($n = 38$) and the other for the validation set ($n = 27$). Spectra were recorded in the range of 4000–12500 cm^{1}, and a good prediction model was obtained in the range of 9400–6096 and 5456–4248 cm^{1} with r^2 of 99.5% and a root-mean-square error of 0.3%. The adulterated extracts subjected to NIR analysis were recognized as noncompliant, thus confirming the results obtained by chromatography. The FT-NIR spectroscopy is an economic, powerful, and fast methodology for the detection of adulteration and quantification of the total anthocyanin in bilberry extracts; above all, it is a rapid, low cost, and nondestructive technique for routine analysis.

1. Introduction

Anthocyanins (ACNs) are one of the most important pigments of vascular plants and are responsible for the shiny orange, pink, red, violet, and blue colours in the flowers and fruits. Anthocyanins are harmless and water soluble, which makes them interesting for use as natural colorants. In addition, in the past decades, there has been increased interest in anthocyanins because of their potential impact on human health [1]. Bilberry (*Vaccinium myrtillus* L.) is one of the richest dietary sources of ACNs, with 15 different major ACNs detected and with a content that varies considerably due to their production by different factories [2]. In traditional European medicine, the bilberry fruit has been used for nearly one thousand years as an astringent for the digestive tract and to maintain vascular integrity [3]. Afterward, several studies have demonstrated the benefits of bilberries in the inhibition of cancer cell growth [4], in the management of visual disorders [5], and in prevention of the onset of metabolic and degenerative diseases [4]. Moreover, bilberry extracts have demonstrated both hypoglycaemic [6] and atheroprotective properties [7]. The ACNs content of fresh bilberries ranges from 0.1 to 1.3% [8, 9], whereas commercial extracts are available in concentrations up to 36% ACNs. The most widely used dietary supplements are generally standardised to 25% ACNs, and this level of concentration (approximately 100 : 1), coupled with the process used to manufacture a standardised high-quality extract, have made bilberry one of the most expensive natural extracts (600–1300 $\$\,\text{kg}^{-1}$). Anthocyanins have also been found in cheaper raw materials such as elderberry, blackcurrant, chokeberry, blackberry, mulberry, black beans [10], black rice husks, and black soy hull [11]. This has led to the possibility of deliberate adulteration of the bilberry extract with less costly (approximately 50\$ kg^{-1} or less) plant species, by so-called "economic adulteration." Artaria et al. [12] found that only 50% of the commercial supplements analyzed containing bilberry extracts were noncompliant under the labelling-compliance terms for ACN content. Likewise, Cassinese et al. [13] reported that only 6 of the 40 finished bilberry products met their specifically stated label claims for ACN content, and approximately 4 were bereft of ACNs. Moreover, 10 extracts differed from that of the typical bilberry, and some of these exhibited a higher content of anthocyanidins. The latter was an index of ACNs degradation due to incorrect processing or storage conditions. More recently, Gardana et al. [9] analyzed 14 bilberry extracts and 12 finished products and found that approximately 50% of the extracts differed significantly from the reference bilberry, suggesting possible adulteration with extracts of mulberry and chokeberry. Moreover, approximately 60% of the extracts and 33% of the food supplements presented a lower anthocyanin content than declared. Besides counterfeiting with other berries and legumes, Pennman et al. [14] found a synthetic dark red-blue dye, Amaranth, in commercial bilberry extracts. Thus, the qualitative and quantitative ACN analysis is pivotal in evaluating the quality of the bilberry extract and dietary supplements. The spectral behaviour of ACN provides very useful information, and spectroscopy is the main technique used in quality control laboratories to determine ACNs [15]. Unfortunately, this method is unspecific, does not provide typical ACN fingerprinting, and does not detect dyes such as Amaranth. To overstep these limitations, Lee et al. [16] developed a pH-differential method to quantify ACNs and detect dyes and compounds with colours that are independent of pH. Regrettably, even this fails to identify extracts produced with plants different from bilberry. Other analytical methods for bilberry ACN determination have been reported, such as thin-layer chromatography, gas chromatography, capillary electrophoresis, nuclear magnetic resonance, and liquid chromatography. The latter is the preferred method for the separation and quantification of 15 ACNs in bilberry using a detection system based on photodiode array detector and/or mass spectrometry [9]. Currently, the European Pharmacopeia [17] adopted a validated liquid chromatography method as the official analytical method for bilberry fruit dry extract quality evaluation. Unfortunately, the pharmacopoeia analytical method may not be sufficient to distinguish genuine bilberry extracts from adulterated material because the detector used, spectrophotometer or DAD, could not be sufficiently specific. Furthermore, the chromatographic method used to determine ACNs in bilberry is destructive and above all requires long analysis time. On the contrary, the near-infrared (NIR) spectroscopy technique is fast and nondestructive, does not require reagents, and could provide additional information regarding sugars [18], acidity [19], antioxidant content, activity [20], moisture, and micronutrients content such as anthocyanins [21]. Additionally, NIR spectroscopy has been revealed as a powerful technique for quality assessment of food [22], natural products [23], and feedstuff [24] and for the analysis of biological materials [25]. Harnly et al. [26] reported that NIR spectrometry failed to distinguish between the authentic and adulterated *G. biloba* powder and stated that the failure was due to the presence of binders or excipients. In fact, the analysis conducted on hydroalcoholic extracts allowed the identification of counterfeit products. Ferrari et al. [27] attempted to detect adulteration of the anthocyanins content of red wines, and recently, Inácio et al. [28] have quantified anthocyanins in *Euterpe oleracea* by NIR spectroscopy. To the best of our knowledge, there has been no report regarding the evaluation of the bilberry extract adulteration by FT-NIR spectroscopy and partial linear square regression (PLSR). Thus, the aims of this study were to establish a routine FT-NIR methodology to assess the possible adulteration of commercial extracts containing *V. myrtillus* and their ACN content. The anthocyanin profiles and their content in the 71 commercial bilberry extracts were preliminarily determined by a validated chromatographic method, the primary method, and afterward, the performances of the FT-NIR established prediction model using PLS regression were compared with those of the primary method. In order to exclude the samples that had a negative impact on the model, the Mahalanobis distance method was used to eliminate the outlier samples in this study. Different preprocessing methods were also compared to obtain the best identification protocol. This study provides a valuable solution for fast and economic identification of adulterated bilberry extracts so as to protect consumers.

2. Materials and Methods

2.1. Chemicals. Standards of cyanidin (Cy-), delphinidin (D-), petunidin (Pt-), peonidin (Pe-), and malvidin (Mv-) and their 3-O-glucoside (glc), Cy-3-O-galactoside (Cy-gal), Pt-gal, Pe-gal, Mv-gal, Cy-arabinoside (Cy-ara), and Cy-rutinoside (Cy-rut) were purchased from Polyphenols Laboratory (Sandnes, Norway). Methanol, acetonitrile hydrochloric, and phosphoric acid were from Merck (Darmstadt). Water was obtained from arium pro apparatus (Sartorius, Milan). Refined and standardised dry extract from the bilberry fruit (Sref) (CRS 1602, code Y0001059) was provided by the European Pharmacopoeia Reference Standard (EDQM, Strasbourg), and bilberry extracts $n = 64$ were provided from different manufacturers over the period 2014–17.

2.2. Sample Preparation for Liquid Chromatography-Diode Array Analysis. Approximately 100 mg of powder was dissolved in approximately 20 ml of a solution of methanol : H_3PO_4 1% in water (10 : 90, v/v). The suspension was sonicated for 10 min at room temperature and centrifuged at 1000 ×g for 5 min, and the supernatant was recovered. The residue, if present, was extracted and treated as described above. The supernatants were combined, and then the final volume was adjusted to 50 ml with 1% H_3PO_4 in water.

2.3. Anthocyanin Determination by Liquid Chromatography-Diode Array. The HPLC system was an Alliance 2695 (Waters, Milford, MA, USA) equipped with a model 2998 photodiode array detector (Waters). A 2.6 μm Kinetex C_{18} column (150 × 4.6 mm, Phenomenex, Torrance, CA) maintained at 45°C carried out the separation. The flow rate was 1.7 ml/min, and the eluents were (A) 1% H_3PO_4 in water and (B) 35% CH_3CN in 1% H_3PO_4. The elution gradient was linear as follows: 0–15 min 14% B; 15–25 min from 14 to 20% B; 25–35 min from 20 to 32% B; 25–45 min from 32 to 50% B; 45–50 min 50% B; 51 min 90% B; and 51–60 min 90% B. Chromatographic data were acquired from 200 to 700 nm and integrated at 520 nm. Anthocyanin stock solutions (1 mg/ml) were prepared in 0.1 M HCl in water and stored at −20°C. Their concentration was evaluated spectrophotometrically by the molar extinction coefficient reported in the literature [29]. Working solutions ($n = 5$) were prepared in the range of 2–50 μg/ml, and twenty microliters was injected into the chromatographic system. Each analysis was carried out in duplicate.

2.4. FT-NIR Reflectance Spectra Acquisition and Model Development. Fourier transform near-infrared (FT-NIR) spectra were recorded in the reflectance mode using a model Tango spectrophotometer (Bruker Optics, Ettlingen) equipped with a gold integrating sphere. Two aliquots for each sample were analyzed, recording spectra in duplicate in order to account for the instrumental or sampling variability. Spectra were recorded in the range of 4000–12500 cm^{-1}, from an average of 64 scans and with a resolution of 8 cm^{-1}. Approximately 20 g of dry extract powder was put in the sample cup, and the data were collected three times for each sample. Principal component analysis (PCA), partial least-squares regression (PLS) modelling, and Mahalanobis distance were performed using OPUS Quant 7.5 (Bruker). Due to the limited number of samples in the data set, cross validation (leave-one-out method) was applied. Thus, authentic bilberry extract samples with known anthocyanin content ($n = 65$) were divided into two parts: one for calibration ($n = 38$) and the other for validation set ($n = 27$). Spectra were not averaged in order to detect any outlier that may arise in the cross-validation/prediction process. Outlier detection was executed to improve model accuracy.

NIR spectra contain large quantities of data that require a combination of statistical and mathematical sciences for their understanding. Therefore, preprocessing is needed to remove noise and background information. Spectral preprocessing was performed including no spectral data preprocessing, smoothing by the Savitzky–Golay (SG) method, multiplicative scatter correction (MSC), first derivative (1stDer) and second derivative (2ndDer) by the SG method, vector normalization (VN), straight line subtraction (SLS), minimum maximum normalization (MMN), subtraction of a constant offset (CO), rank optimization, 1stDer + SLS, 1stDer + VN, and 1stDer + MSC.

In brief, smoothing improves the quality of the spectra by removing noise, mainly consisting of moving average filters and applying the SG algorithm. MSC is used to diminish effects in the spectra caused by artifacts or imperfections such as undesirable scatter effect. This method is often used in diffusive reflection measurements. First and second derivatives eliminate baseline drifts, and small spectral differences are enhanced. To avoid enhancing the noise, which is a consequence of the derivative, spectra are first smoothed by the SG algorithm. VN is used to normalize the spectrum by first calculating the average intensity value and subsequent subtraction of this value from the spectrum. Basically, in diffusive reflection, the interferences from different material densities or particle sizes can often be minimized. MMN is used to transform the data into a desired range by subtracting the minimum value from each individual spectrum and then dividing the range of this spectrum. In SLS, preprocessing a straight line is fitted to the spectrum, using the PLS method, and then subtracted from the respective spectrum. In this way, a linear tilt of the baseline shift is eliminated. In the CO, the spectra are shifted in order to set the y minimum to zero. As an outcome, linear baseline shifts are eliminated.

The rank value, which defines the optimal number of principal components chosen for the analysis, was calculated by plotting the root-mean-square error of calibration (RMSEE) and prediction (RMSEP) values against the correspondent's r^2. The criteria for deleting outliers were (a) samples with residuals higher than 2 and (b) samples with leverage higher than 3 times the average leverage. The capability of the method was identifiable by the root-mean-square error of prediction (RMSEP) value, coefficient of determination (r^2), the bias, and the residual prediction deviation (RPD) value. The last was defined as the ratio between the standard deviation of the population's reference values and the standard error of performance (e.g., RMSEE or RMSEP) bias corrected. The most capable method was the one with the lowest RMSEP value, the highest RPD value, r^2 close to one, and the bias value close to zero. For the qualitative analysis (the identity test), the spectra of the authentic bilberry extracts were included in a library containing different classes of extracts (data not shown). The separation between the groups was assessed by comparison of two spectral classes at a time using the values of selectivity (S) and threshold (T), joined together by the following relationship:

$$S = \frac{D_{ab}}{(T_a + T_b)}, \quad (1)$$

where D_{ab} is the distance between the centres of the two groups, T_a is the threshold value of group a, and T_b is the threshold value of group b.

If $S > 1$, the groups in graphics in two or three dimensions appear to be well spaced, and during validation, they are considered distinct. The method was validated when all the groups were separated from each other with S greater than or equal to 1. The threshold value represents the maximum distance from the centre of the group, and it is defined by the following formula: T = maximum hit + $N \times$ SD, where maximum hit is the distance of the farthest extract from the centre of the group, SD is the standard deviation, and N is the coefficient (between 0 and 1). For the calculation of the threshold value, N was set to 0.25 with a confidence level of 99.99%.

After the group separation, the identity test was performed to determine the *Hit qual* value, which represents the distance of the sample from the centre of the bilberry group.

If *Hit qual* value > threshold, the extract is outside the cluster of bilberry, and therefore, it is no compliant. Moreover, to find outliers in multivariate data, the spectral and anthocyanin concentration information was used to determine Mahalanobis distance (MD). This value could be appropriate to detect adulteration because it can determine whether unknown samples belong to the bilberry group or not according to the spectral residual calculation.

2.5. Statistical Analysis. Statistical analyses were performed with Statistica software (StatSoft Inc., Tulsa, OK, USA). Accuracy was determined by the Wilcoxon test considering significant a level of $p > 0.05$. The normal distribution of residuals was evaluated by the Shapiro–Wilk test considering significant a level of $W > 0.6$.

3. Results and Discussion

3.1. ACNs Determination in Bilberry by Liquid Chromatography-Diode Array Detection. The chromatogram relating to the reference bilberry extract, obtained at 520 nm, showed the presence of 15 main anthocyanins, and the respective aglycones were lower than 0.1%. Quantification of the ACNs was based on authentic standards and for D-gal, D-ara, Pt-ara, and Mv-ara by the Cy-glc calibration curve because pure compounds were not available. The monomeric anthocyanin pigment content of the analyzed bilberry extract samples ranged approximately from 18 to 34%. The precision of the method was tested by both repeatability ($n = 5$) and intermediate precision ($n = 5$), and the coefficient of variation was below 1.6%. The qualitative analysis of the bilberry extracts showed marked differences among them. Indeed, six tested extracts (Sa–Sf) showed a chromatographic profile different from that of the reference extract (Figure S1 in the Supplementary Materials). In particular, the extracts Sa and Sb showed the presence of two main ACNs identified as Cy-glc and Cy-rutinoside. The latter compound is not normally present in the bilberry fruit extracts but has been found in berries such as black mulberry. The results suggest that Sa and Sb were bilberry extracts adulterated with mulberry. On the contrary, the chromatographic profiles of the Sc and Sd extracts did not contain the 15 typical anthocyanins, instead contained only four of them, corresponding to Cy-gal, Cy-glc, Cy-ara, and Cy-xyloside. These ACNs were present in a reference extracts of chokeberry and have been found in chokeberries by different authors [30]. Thus, samples Sc and Sd were extracts of chokeberries and not bilberries. The chromatographic profile of the sample Sf contained mainly Cy-glc and lower amounts of Cy-rut. Jakobek et al. [31] reported a chromatographic profile of the blackberry very similar to that of extract Sf. Therefore, this sample was not bilberry, but a blackberry extract. Lastly, the extract Se did not contain ACNs. Based on the results obtained, the samples Sref and 64 authentic bilberry extracts were included in the database, bilberry group, for the qualitative and quantitative analysis by NIR spectroscopy.

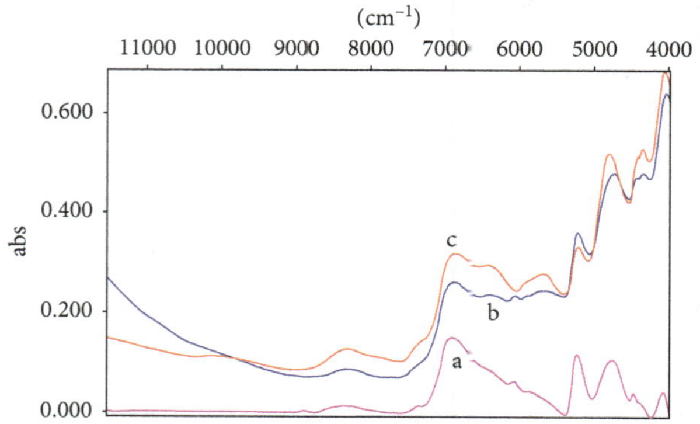

FIGURE 1: FT-NIR mean spectrum from the "outlier" (a), authentic (b), and adulterated (c) bilberry sample acquired from 4000 to 12500 cm^{-1}.

3.2. FT-NIR Spectral Characteristics of Bilberry Extracts. A total of 260 FT-NIR spectra of 65 samples of authentic bilberry extract were recorded in the range 4000–12500 cm^{-1} at room temperature (23 ± 1°C). Spectral data were analyzed by PCA carried out with a validation to search for linear combinations of variables, which best explain the obtained data without taking into account external information. The PC1 and PC2 were responsible for about 61 and 26%, respectively, of the total variance among the examined samples. The scores plot of these PCs indicated that the bilberry group had different spectral patterns, and it could be distinguished from other groups present in library. In particular, the threshold for the group of the bilberry extracts was 0.353. After validation of the library, the adulterated bilberry extracts (Sa–Sf) were subjected to the identity test, and the *Hit qual* values were higher than the threshold of the bilberry group. Only for Sa, the *Hit qual* value (0.357) was close to the threshold. Thus, to better representing the bilberry group, the spectra very different from the average spectrum were removed from the library, and the threshold was recalculated. After removing the spectrum of the "outlier," the threshold was 0.312, and Sa was more distant from the bilberry cluster. Extracts Sa–Sf were identified as non-compliant, thus confirming the results obtained by chromatographic analysis. The spectrum of "outlier," authentic, and adulterated bilberry extract is shown in Figure 1. Xiaowei et al. [21] stated that wavelength in the ranges 4600–4780 and 5780–5990 cm^{-1} corresponded to the UV/Visible absorption bands of anthocyanins. This statement comes from the following considerations. Anthocyanins exhibit two major absorption bands in the regions 270–300 and 520–540 nm. Near-infrared spectroscopy is based on molecular overtone and combination vibrations. Thus, the frequencies within the 270–300 nm range include frequencies of six times of the first selected region (5963–5770 cm^{-1}), whereas the frequencies of the 520–550 nm range (19231–18519 cm^{-1}) include frequencies of four times of the second selected region (4782–4608 cm^{-1}). The spectra of the bilberry extract also shows peaks at 5174 (-OH combination), 6836 (first -OH overtone), and 8330 cm^{-1}, mainly due to water, while the colour and particle size of the product caused the shift observed at wavelengths lower than 9000 cm^{-1}. The water content in the bilberry extracts was in the range of 3–7% (data not shown), and therefore, the intensity of those peaks was not particularly high. The amount of sugar in the extracts was in the range of 25–35% (data not shown). Therefore, the additional signals in the spectral ranges 7500–7200, 5950–5600, and 4600–4000 cm^{-1} were probably due to this component [32]. In particular, the absorption band at 7500–7200 cm^{-1} could be related to frequencies of first overtones of O–H stretching modes and C–H combination vibrations. Bands in the range 5950–5600 and 4600–4000 cm^{-1} could be assigned to first overtones of C–H stretching modes and to combinations of O–H bend/hydrogen-bonded O–H stretch, respectively.

3.3. FT-NIR Quantitative Analysis. Generally, the spectra obtained with the NIR analysis require optimization of the width of the interval of wavelengths considered for the large number of test samples, which introduces a high number of variables. In this way, the signal extracted from the spectra is decomposed by means of PCA to select and eliminate nonrelevant variables (principal component with low eigenvalues) and improve the quality of the calibration model. Thus, spectral components of the signal representing the conditions of minimum error are selected. The choice made in this way allows the optimal bands for characterisation of the samples to be identified. Mathematical pretreatments of the NIR data were carried out to enhance the prediction ability of the models and the qualitative interpretation of the spectra. The best preprocessing strategies chose for the spectra to develop the NIR model were obtained by smoothing (9 points) and straight line subtraction (SLS). Smoothing to remove the noise of the data and SLS fits a straight line to the spectrum and subtracts it. Spectral and chemical data are acquired in the form of matrices, in which each row represents a sample spectrum and then reduced to a few latent variables. Not all principal components are relevant to describe the spectral features, so only the most relevant ones should be used to perform the regression model. In this way, "overfitting" model can be avoided. On the contrary, less latent variables give a lower adaptability because of lacking enough information. Their number in the chemometric model is termed the "rank," and a value below 10 is desirable. In our chemometric model, automatically created after forming the spectral and concentration data and after choosing the optimal preprocessing method, the optimal rank obtained by calibration and prediction was 8 and 7, respectively (Figure S2 in the Supplementary Materials). Thus, all of the spectral processing was performed using a value of rank equal to eight. The calibration curve (n = 38), obtained by cross validation, was constructed based on the data obtained with the PLS (Figure 2): for each sample, reading with the NIR was associated with the concentration value obtained by the validated method. The calibration curve is shown by placing the reference value for ACN % obtained by the chromatographic method on the abscissa and the predicted data in the ordinate. The calibration curve showed good linearity ($r^2 = 99.63$), precision (RMSEE = 0.28), and a RPD value of 12.5. It should be noted that a value of RPD higher than eight is considered, in general, suitable for every application [33]. Table 1 shows the descriptive statistics and number of samples for calibration used in this study to predict the total amount of ACNs in commercial extracts of bilberry. Based on these values, the curve is considered acceptable in terms of linearity and precision (RSD < 1.8%), and the residues had normal distribution ($W = 0.9687$) (Figure S3 in the Supplementary Materials). About accuracy, the results indicated that there was no significant difference (p = 0.203) between the predicted ACN percentage and that determined by chromatography. Moreover, the residues of the calibration curve were normal distributed. The linearity of a model is a reliable indicator of its goodness because it shows that the model can accurately quantify not only the tested samples with an actual content but also all those with a content that is different from the nominal one within the tolerated range or even slightly above it. Then, the calibration model was validated by a validation set (n = 27, Figure 2) and the obtained results for r^2, RMSEP, and RPD were 99.51%, 0.303, and 15.4, respectively (Table 2). The parameters reported in Table 2 allow to state that the curve obtained is validated and can be used to predict the amount of anthocyanins in the bilberry extract in the range 18–34%. For detection of adulteration, the Mahalanobis distance was determined after the calibration and the obtained value was 0.91. In this regard, the NIR analysis suggests that the extracts Sa–Sf must be considered as adulterated because their Mahalanobis distance value was higher than 0.91 (Figure S4 in the Supplementary Materials). The results obtained in this study showed the potential of FT-NIR spectroscopy with PLS regression and chemometric software to discriminate genuine bilberry extracts from those adulterated with anthocyanins extracted from other berries. The results also showed a relationship between near-infrared spectra and the amount of anthocyanins in the bilberry extract. Moreover, this technique allowed the rapid, accurate, and nondestructive quantitation of total anthocyanins in commercial bilberry extracts used for the production of food supplements. Thus, FT-NIR spectroscopy could be applied in a quality control laboratory to monitor adulteration and/or contamination and assess anthocyanin content in the bilberry extract.

4. Conclusions

The results obtained in this study showed the potential of FT-NIR spectroscopy with chemometric techniques to discriminate genuine bilberry extracts from those adulterated with anthocyanins extracted from other berries. The Mahalanobis distance method was successfully used to exclude the outliers, and the differences were removed by preprocessing procedures. The results also showed a relationship between near-infrared spectra and the amount of anthocyanins in the bilberry extract. Moreover, this technique allowed the rapid, accurate, and nondestructive quantification of total anthocyanins in commercial bilberry extracts that are commercially used for the production of food supplements. Thus, FT-NIR spectroscopy could be

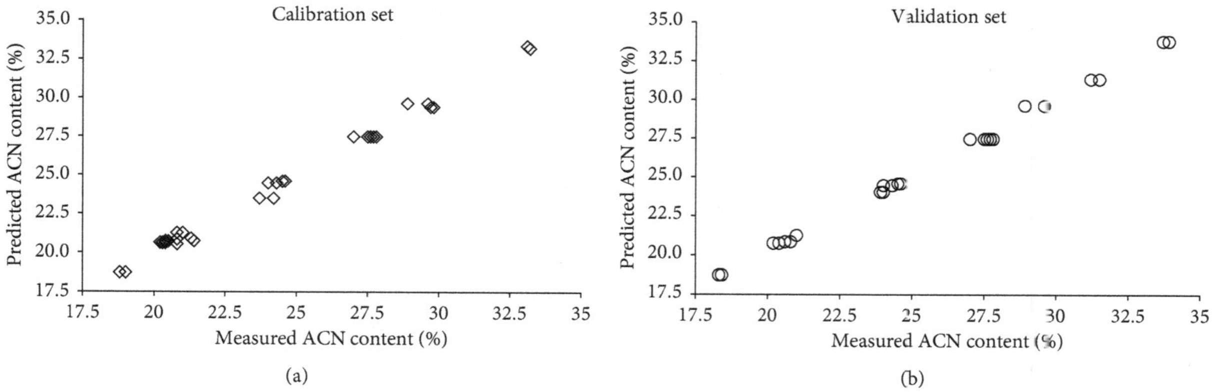

FIGURE 2: Reference measured versus predicted value of calibration (a) and validation (b) samples for total anthocyanins (%) in bilberry extracts using the PLS model.

TABLE 1: Summary of the NIRS calibration model statistics.

Parameters	ACN content in bilberry extracts
ACN content	% dried matter
SEL	0.31
N	38
Outliers	0
Min	18.8
Mean	24.2
Max	33.2
SD	4.1
RMSEE	0.28
r^2	99.63
RPD	12.5
RANK	8
Segments/range (cm^{-1})	2/9400–6096, 5455–4248
Acquisition range (cm^{-1})/step	4000–12500/8
Pretreatments tested	NSDPP, CO, MSC, 1stDer, 2ndDer, VN, SLS, MMN, RO, 1stDer + SLS, 1stDer + VN, and 1stDer + MSC
Pretreatments used	SLS
Regression method	PLS
Threshold	0.353
Mahalanobis distance	0.91

N: number of samples; NSDPP: no spectral data preprocessing; CO: constant offset elimination; MSC: multiplicative scatter correction; 1stDer: first derivative; 2ndDer: second derivative; VN: vector normalization; SLS: straight line subtraction; MMN: minimum maximum normalization; RO: rank optimization.

TABLE 2: Summary of the NIRS validation statistics.

Parameters	Total ACNs in bilberry extracts
ACN content	% dried matter
N	27
Outliers	0
Min	18.3
Mean	25.7
Max	33.9
SD	4.4
RMSEP	0.303
r^2	99.51
SEP	0.28
Rank	7
RPD	15.4
Bias	−0.114
Slope	0.9704
Offset	0.8755

applied in a quality control laboratory to monitor adulteration and/or contamination and assess anthocyanin content in bilberry extract.

Conflicts of Interest

The authors declare that there are no conflicts of interest regarding the publication of this paper.

References

[1] M. J. R. Howes and M. S. J. Simmonds, "The role of phytochemicals as micronutrients in health and disease," *Current Opinion in Clinical Nutrition and Metabolic Care*, vol. 17, no. 6, pp. 558–566, 2014.

[2] A. K. Latti, K. R. Riihinen, and P. S. Kainulainen, "Analysis of anthocyanin variation in wild populations of bilberry (*Vac-*

cinium myrtillus L.) in Finland," *Journal of Agricultural and Food Chemistry*, vol. 56, no. 1, pp. 190–196, 2008.

[3] P. Morazzoni and E. Bombardelli, "*Vaccinium myrtillus* L.," *Fitoterapia*, vol. 67, p. 329, 1996.

[4] N. P. Seeram, L. S. Adams, Y. Zhang et al., "Blackberry, black raspberry, blueberry, cranberry, red raspberry, and strawberry extracts inhibit growth and stimulate apoptosis of human cancer cells in vitro," *Journal of Agricultural and Food Chemistry*, vol. 54, no. 25, pp. 9329–9339, 2006.

[5] R. D. Steigerwalt, B. Gianni, M. Paolo et al., "Effect of Mirtogenol® on ocular blood flow and intraocular hypertension in asymptomatic subjects," *Molecular Vision*, vol. 14, pp. 1288–1292, 2008.

[6] M. Takikawa, S. Inoue, F. Horio et al., "Dietary anthocyanin-rich bilberry extract ameliorates hyperglycemia and insulin sensitivity via activation of AMP-activated protein kinase in diabetic mice," *Journal of Nutrition*, vol. 140, no. 3, pp. 527–533, 2010.

[7] A. Mauray, C. Felgines, C. Morand et al., "Bilberry anthocyanin-rich extract alters expression of genes related to atherosclerosis development in aorta of apo E-deficient mice," *Nutrition Metabolism and Cardiovascular Disease*, vol. 22, no. 1, pp. 72–80, 2012.

[8] D. Muller, M. Schantz, and E. Richling, "High performance liquid chromatography analysis of anthocyanins in bilberries (*Vaccinium myrtillus* L.), blueberries (*Vaccinium corymbosum* L.), and corresponding juices," *Journal of Food Science*, vol. 77, no. 4, pp. 340–344, 2012.

[9] C. Gardana, S. Ciappellano, L. Marinoni et al., "Bilberry adulteration: identification and chemical profiling of anthocyanins by different analytical methods," *Journal of Agricultural and Food Chemistry*, vol. 62, no. 45, pp. 10998–11004, 2014.

[10] G. R. Takeoka, L. T. Dao, G. H. Full et al., "Characterization of black bean (*Phaseolus vulgaris* L.) anthocyanins," *Journal of Agricultural and Food Chemistry*, vol. 45, no. 9, pp. 3395–3400, 1997.

[11] J. H. Lee, N. S. Kang, S. O. Shin et al., "Characterization of anthocyanins in the black soybean (*Glycine max* L.) by HPLC-DAD-ESI/MS analysis," *Food Chemistry*, vol. 112, no. 1, pp. 226–231, 2009.

[12] C. Artaria, R. Pace, G. Maramaldi et al., "Different brands of bilberry extract. A comparison of selected components," *Nutrafood*, vol. 6, pp. 13–18, 2007.

[13] C. Cassinese, E. de Combarieu, M. Falzoni et al., "New liquid chromatography method with ultraviolet detection for analysis of anthocyanins and anthocyanidins in *Vaccinium myrtillus* fruit dry extracts and commercial preparations," *Journal of AOAC International*, vol. 90, pp. 911–919, 2007.

[14] K. G. Pennman, C. W. Halstead, U. Stuhe, K. M. Bone, and R. P. Lehmann, "Bilberry adulteration using the food dye Amaranth," *Journal of Agricultural and Food Chemistry*, vol. 54, no. 19, pp. 7378–7382, 2006.

[15] R. E. Wrolstad, R. W. Durst, and J. Lee, "Tracking color and pigment changes in anthocyanin products," *Trends in Food Science and Technology*, vol. 16, no. 9, pp. 423–428, 2005.

[16] J. Lee, R. W. Durst, and R. E. Wrolstad, "Determination of total monomeric anthocyanin pigment content of fruit juices, beverages, natural colorants, and wines by the pH differential method: collaborative study," *Journal of AOAC International*, vol. 88, pp. 1269–1278, 2005.

[17] European Pharmacopeia 8.0, "Fresh bilberry fruit dry extract, refined and standardized," *Generic Drugs*, vol. 1, pp. 1250–1252, 2014.

[18] J. Y. Chen, H. Zhang, Y. Mia, and M. Asakura, "Non-destructive determination of sugar content in potato tubers using visible and near infrared spectroscopy," *Japan Journal of Food Engineering*, vol. 11, pp. 59–64, 2010.

[19] P. C. Williams, "Implementation of near-infrared technology," in *Near-Infrared Technology in the Agricultural and Food Industries*, P. Williams and K. Norris, Eds., pp. 145–169, American Association of Cereal Chemists, St. Paul, MN, USA, 2nd edition, 2001.

[20] X. Lu and B. A. Rasco, "Determination of antioxidant content and antioxidant activity in food using infrared spectroscopy and chemometrics: a review," *Critical Review in Food Science and Nutrition*, vol. 52, no. 10, pp. 853–875, 2012.

[21] H. Xiaowei, Z. Xiaobo, Z. Jiewen et al., "Measurement of total anthocyanins content in flowering tea using near infrared spectroscopy combined with ant colony optimization models," *Food Chemistry*, vol. 164, pp. 536–543, 2014.

[22] J. H. Qu, D. Liu, J. H. Cheng et al., "Application of near-infrared spectroscopy in food safety evaluation and control: a review of recent research advances," *Critical Review in Food Science and Nutrition*, vol. 53, no. 13, pp. 1939–1954, 2015.

[23] D. Cozzolino, "Near infrared spectroscopy in natural products analysis," *Planta Medica*, vol. 75, no. 7, pp. 746–756, 2009.

[24] F. Cheli, D. Battaglia, L. Puinotti, and A. Baldi, "State of the art in feedstuff analysis: a technique-oriented perspective," *Journal of Agricultural and Food Chemistry*, vol. 60, no. 38, pp. 9529–9542, 2012.

[25] M. Manley, "Near-infrared spectroscopy and hyperspectral imaging: non-destructive analysis of biological materials," *Chemical Society Reviews*, vol. 43, no. 24, pp. 8200–8214, 2014.

[26] J. M. Harnly, D. Luthria, and C. Pei, "Detection of adulterated *Ginkgo biloba* supplements using chromatographic and spectral fingerprints," *Journal of AOAC International*, vol. 95, no. 6, pp. 1579–1587, 2012.

[27] E. Ferrari, G. Foca, M. Vignali, L. Tassi, and A. Ulrici, "Adulteration of the anthocyanin content of red wines: perspectives for authentication by Fourier transform-near infrared and 1H NMR spectroscopies," *Analytica Chimica Acta*, vol. 701, no. 2, pp. 139–151, 2011.

[28] M. R. Inácio, K. M. G. de Lima, V. G. Lopes, J. D. C. Pessoa, and G. H. de Almeida Teixeira, "Total anthocyanin content determination in intact açaí (*Euterpe oleracea* Mart.) and palmitero-juçara (*Euterpe edulis* Mart.) fruit using near infrared spectroscopy (NIR) and multivariate calibration," *Food Chemistry*, vol. 136, no. 3-4, pp. 1160–1164, 2013.

[29] M. M. Giusti, L. E. Rodrıguez-Saona, and R. E. Wrolstad, "Molar absorptivity and color characteristics of acylated and non-acylated pelargonidin-based anthocyanins," *Journal of Agricultural and Food Chemistry*, vol. 47, no. 11, pp. 4631–4637, 1999.

[30] M. Bräunlich, R. Slimestad, H. Wangensteen et al., "Extracts, anthocyanins and procyanidins from *Aronia melanocarpa* as radical scavengers and enzyme inhibitors," *Nutrients*, vol. 5, no. 3, pp. 663–678, 2013.

[31] L. Jakobek, M. Seruga, M. Medvidovic-Kosanovic, and I. Novak, "Anthocyanin content and antioxidant activity of various red fruit juices," *Deutsche Lebensmittel Rundschau*, vol. 103, pp. 58–64, 2007.

[32] L. Xie, X. Ye, D. Liu, and Y. Ying, "Quantification of glucose, fructose and sucrose in bayberry juice by NIR and PLS," *Food Chemistry*, vol. 114, no. 3, pp. 1135–1140, 2009.

[33] N. Yang and G. Ren, "Application of near infrared reflectance spectroscopy to the evaluation of rutin and D-chiro-inositol contents in tartary buckwheat," *Journal of Agricultural and Food Chemistry*, vol. 56, no. 3, pp. 761–764, 2008.

9

Application of VIS/NIR Spectroscopy and SDAE-NN Algorithm for Predicting the Cold Storage Time of Salmon

Ting Wu,[1,2,3,4] **Nan Zhong**[1,2,3] **and Ling Yang**[4]

[1]College of Engineering, South China Agricultural University, Guangzhou, China
[2]Key Laboratory of Key Technology on Agricultural Machine and Equipment, South China Agricultural University, Ministry of Education, Guangzhou, China
[3]Guangdong Provincial Key Laboratory of Food Quality and Safety, Guangzhou 510642, China
[4]School of Information Science and Technology, Zhongkai University of Agriculture and Engineering, Guangzhou 510225, China

Correspondence should be addressed to Nan Zhong; zhongnan@scau.edu.cn

Academic Editor: Jose M. Pedrosa

The cold storage time of salmon has a significant impact on its freshness, which is an important factor for consumers to evaluate the quality of salmon. The efficient, accurate, and convenient protocol is urgent to appraise the freshness for quality checking. In this paper, the ability of visible/near-infrared (VIS/NIR) spectroscopy was evaluated to predict the cold storage time of salmon meat and skin, which were stored at low-temperature box for 0~12 days. Meanwhile, a double-layer stacked denoising autoencoder neural network (SDAE-NN) algorithm was introduced to establish the prediction model without spectral pre-preprocessing. The results showed that, compared with the common methods such as partial least squares regression (PLSR) and back propagation neural network (BP-NN), the SDAE-NN method had a better performance due to its high efficiency in decreasing noise and optimizing the initial weights. The determination coefficient of test sets (R^2_{test}) and root mean square error of test sets (RMSEP) have been calculated based on SDAE-NN, for the salmon meat (skin), the R^2_{test} can reach 0.98 (0.92), and the RMSEP can reach 0.93 (1.75), respectively. It is highlighted that the algorithm is efficient and accurate and that the salmon meat would be more suitable for predicting freshness than the salmon skin. VIS/NIR spectroscopy combined with the SDAE-NN algorithm can be widely used to predict the freshness of various agricultural products.

1. Introduction

Atlantic salmon *(Salmo salar)* are farmed in cold sea water, which is delicious and rich in protein, amino acids, and unsaturated fatty acids. The fish are slaughtered and stored in a freezer under ice when consumers buy salmon in the market. The storage time is an urgent aspect of selling, due to the decrease in the freshness or even deterioration with increasing the storage time. However, it is difficult for consumers to identify the freshness of the salmon unless the salmon is thoroughly spoiled, such as smell or color change. Traditional detection methods, such as quantization index modulation (QIM) [1], mainly depend on artificial detection which may be low efficient and primarily rely on experience. On the other hand, the salmon will have a series of microbial changes and biochemical changes during storage [2, 3]. Some microorganism and chemical detection methods such as total viable count (TVC) [4], total volatile basic nitrogen (TVB-N) [5], and *K* value detection [6] have been used to detect the freshness or storage time of meat. However, the above methods have some disadvantages, such as QIM method which requires trained people and cannot grade the salmon without the head. The microorganism and chemical methods are destructive, not only complicated and time-consuming, but also the waste of raw materials.

Spectral technology is currently considered to be an effective detection technology and has a series of advantages, such as low sample amount, environmentally friendly, and reusable. This technology has been applied to autolytic change detecting [7], fat predicting [8], fish oil predicting [9], adulteration detecting [10, 11], freshness [12, 13], and so on. In the previous reports of spectroscopy technology, chemometric methods such as partial least squares regression (PLSR) [14–16] and back propagation neural network (BP-NN) [17] were used as the main modeling methods. However, the above methods have some disadvantages; for example, before PLSR modeling, the spectra need to be pretreated to eliminate the noise such as scattering and distortion. The BP-NN modeling has some disadvantages such as gradient vanishing [18] and local minima [19], which make it difficult to carry out multilayer training, and the selection of parameters should be careful to avoid overfitting and local optimal.

Therefore, to solve the above problems, stacked denoising autoencoder neural network (SDAE-NN) was used to predict the storage time of salmon. This work combines the stacked autoencoder (SAE) algorithm and denoising technology [20, 21] to improve the neural network model and has the following advantages: (1) The SDAE-NN could effectively avoid the vanishing gradient problem with the use of the layer-wise method [18]. (2) The SDAE-NN could effectively avoid the local optimal problems through unsupervised pretraining [22] and supervised fine tune. (3) The robustness and antinoise ability of SDAE-NN were better than BP-NN and other traditional NN algorithms.

2. Materials and Methods

2.1. Salmon Samples. Norwegian salmon was collected from Junhui Import and Export Company Ltd. (Guangzhou, China). After being caught in the Atlantic, salmon was hit in the head, cut tail, gutted, and rapidly frozen then shipped to China from Norway at a constant temperature of −40°C. In China, 90 salmon meat fillets were collected from the inside middle back parts of different salmons. 66 salmon skin fillets were collected from different salmons back parts without scales. The samples were then quickly packed to the laboratory with ice-packed polyethylene, located with ice under the salmon package, and stored at the constant temperature box (~2°C), and the storage condition was kept unchanged. During the storage period, the ice was added every day to achieve the similar storage conditions in the market.

2.2. VIS/NIR Spectroscopy Measurement. Ocean optics USB4000 (Ocean Optics, USA) Fourier transform fiber spectrometer and 38 mm integrating sphere (Jingyi, China) were used to collect the VIS/NIR reflection spectroscopy. The overall measurement range was from 400 to 1000 nm with a resolution of 1 nm. 15 spectra of salmon meat fillets and 11 spectra of skin fillets were acquired at the storage time of 0, 2, 5, 8, 11, and 16 days. In each measurement day, the samples were rapidly taken to the spectroscopic laboratory and measured at 3 to 5 p.m. Before measuring the salmon spectra, the spectrometer was allowed to warm up for 20 minutes. Then, spectrometer was calibrated through measuring a standard whiteboard as 100% reflectance (W) (the exposure intensity about 55000), a dark background as 0% reflectance (D). The salmon spectrum could then be collected through the following formula.

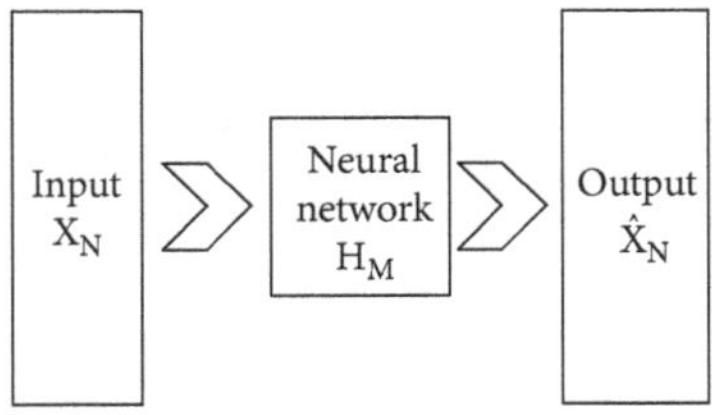

FIGURE 1: Schematic illustration of autoencoder neural network.

$$S_{\text{calibrated}} = \frac{S_{\text{raw}} - D}{W - D} \times 100. \quad (1)$$

Each salmon spectrum was acquired with 64 spectra being scanned successively and averaged. To exclude the effects of the dark current on the instrument, the spectrometer was calibrated once after measuring five salmon spectra. The laboratory environment was kept unchanged at the temperature of 20°C and humidity 30 ± 5%, and a total of 90 salmon meat spectra and 66 salmon skin spectra were collected.

2.3. Algorithm Description

2.3.1. Autoencoder Neural Network. The spectral data always have hundreds or even thousands of dimensions, and many dimensions have a certain correlation. In the past, PCA algorithm was often used for dimensionality reduction, which was a linear reduction method that the new dimensions were the transformation combination of the original dimensions [23]. However, it could not be applied for the nonlinear conditions. The autoencoder is considered as an efficient method for nonlinear dimensionality reduction. In the neural network model, autoencoder [24] is an unsupervised learning algorithm, which can learn the characteristics of the internal structure of data and extract the characteristics more efficiently. The process of the autoencoder neural network was shown in Figure 1.

When training the neural network H_M, if the input of neural network was equal to output $X_N = \widehat{X}_N$, the network could be considered as an autoencoder neural network. When the number of hidden layer neurons M is less than the input spectral data dimensions N, the process from the input X_N to the hidden layer H_M can be considered as a data compression and dimensionality reduction. In this process, the autoencoder neural network can obtain the appropriate initial weights layer by layer through using unsupervised spectroscopic data. During the training process, the hidden layer data information should be lost as few as possible (information-preserving encoding), and the output data reconstructed after training should restore the original data characteristics as much as possible. Through the autoencoder technique, the original data were effectively compressed and the neural network could get better initial weights.

2.3.2. Stacked Denoising Autoencoder. Stacked autoencoder [21] was composed of multilayer autoencoder, and the

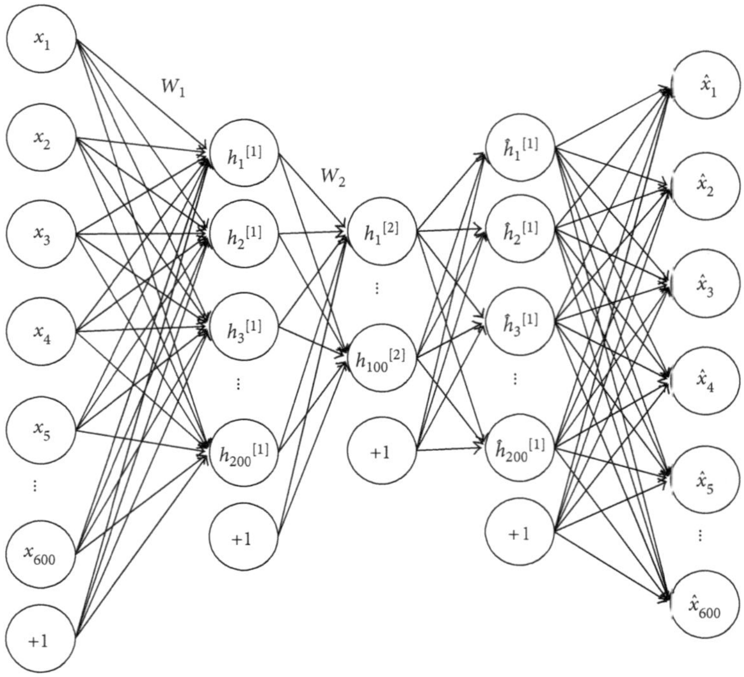

FIGURE 2: Schematic illustration of the stacked autoencoder.

output from the previous layer of autoencoder was used as the input of the subsequent layer. Stacked autoencoder adopts the layer-wise [18] method to train each layer of a neural network by unsupervised training method. A deep-learning network can be established through unsupervised pretraining and supervised output layer regression or classification.

In this work, salmon spectral data were trained by a double-layer autoencoder model. The structure was set up as shown in Figure 2. The input layer ($x_1 - x_{600}$) represents spectral data which have 600 dimensions. The first hidden layer $H^{[1]}$ was compressed to 200 dimensions, and the second hidden layer $H^{[2]}$ was compressed to 100 dimensions. The training steps were as follows:

(1) X»$H^{[1]}$»X, $X = x_1, \ldots, x_{600}$.

(2) $H^{[1]}$»$H^{[2]}$»$H^{[1]}$.

(3) X»$H^{[1]}$»$H^{[2]}$»$\widehat{X}$.

$H^{[1]}$ and $H^{[2]}$ could be separately gotten by the autoencoder training steps (1) and (2). $H^{[2]}$ had 100 dimensions, so it could be considered that the spectral data was compressed from 600 dimensions to 100 dimensions and the 100 dimensions data contained the information of 600 dimensions of input. After that, the model needs to reconstruct the spectral data by making $\widehat{X}_N \approx X_N$. By the stacked autoencoder, the weights of the hidden layer have been trained well and could be used as the initial weights of the neural network.

When noise was added to the input signal, the learning process was more robust by reconstructing the original signal by adding noise data. Stacked denoising autoencoder algorithm (SDAE) was implemented by adding noise to the input data of the stacked autoencoder [2[illegible]]. The stacked autoencoder must learn to remove the noise and get real input that was not contaminated by the noise. Therefore, this forced neural network gives more robust expression of input signals and its adaptability was stronger than the general neural network.

3. Results and Discussion

3.1. Spectral Analysis. When collecting the salmon spectra, after repeated measurement and continuous calibration of the spectrometer, the spectra in the NIR region still had a lot of noise and the spectral analysis of the region was affected. Therefore, in order to avoid the negative effect of noise on the NIR region, only the visible region (400–700 nm) was analyzed and discussed in the paper. The spectra of the salmon meat and skin of different storage days were averaged, as shown in Figure 3. The main difference between the salmon meat and skin at visible light region (400–700 nm) was the color of the skin was different from the salmon meat. When the salmon meat and skin were observed separately, we could see that the spectra of different storage times were not obvious. To better observe the spectral differences in different storage times, a second derivative and 21 points quadratic polynomial Savitzky-Golay smoothing method was used in the spectra of salmon meat (Figure 4). The wavelength range in 400–700 nm has a correlation with freshness because of the change in heme proteins [13, 25], and it could be seen that there are some differences at the peak position of 432, 550, 574, 606, and 621 nm during different storage times. The differences around 432 nm were

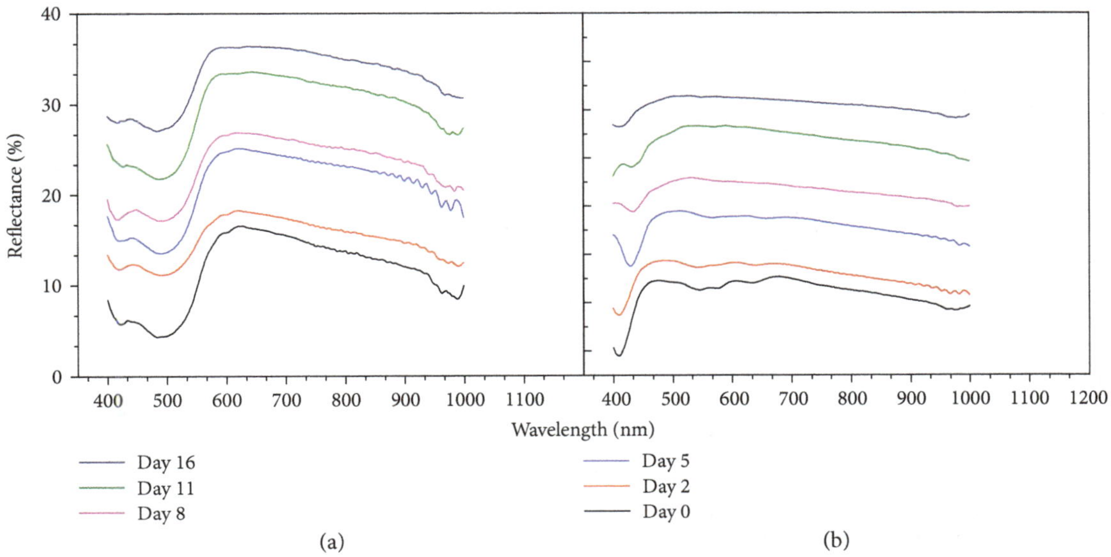

FIGURE 3: The average spectra of the salmon meat (a) and skin (b) at different storage days.

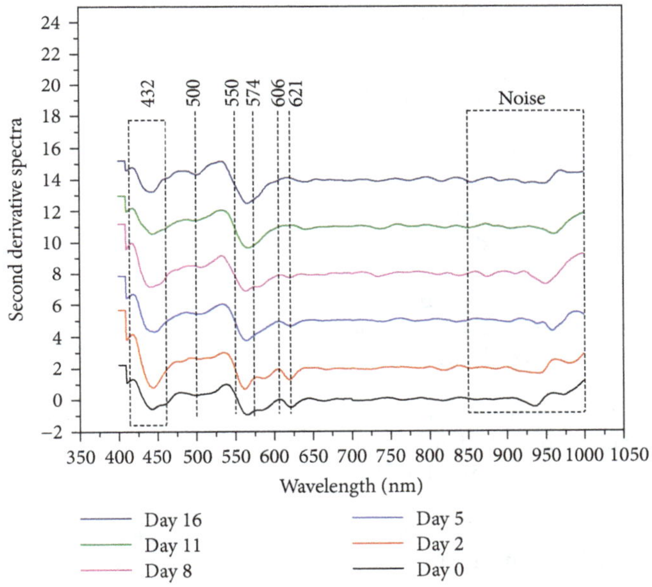

FIGURE 4: The second derivative of the average spectra (salmon meat) at different storage days.

probably due to the absorption by met Hb and/or met Mb, the peak at 550 nm may be mainly caused by deoxymyoglobin, the differences around 574 nm are caused by the content of oxymyoglobin changing during different storage times [26], and the peak around 621 nm may be caused by hemin chloride [27]. Meanwhile, the amount of HbO_2 is different between the fresh and not fresh fish which can be reflected at 606 nm [13]. Another peak at 503 nm may have an influence on freshness because of metmyoglobin, but it is covered by a wide absorption peak at 500 nm (Figure 3) which is assigned to carotenoids, such as astaxanthin and canthaxanthin [28], and has a negative impact to the freshness. Further, the principal component regression- (PCR-) weighted regression coefficient (BW) was analyzed to observe the important spectral peaks (Figure 5), and it could also be observed that 432, 496, 557, and 621 nm had important effects on distinguishing different storage times. The peak at 557 nm is due to deoxymyoglobin [26]; therefore, it can be concluded that the differences in the storage time of the salmon are mainly caused by heme protein changes such as oxymyoglobin, deoxymyoglobin, met Hb, and hemin chloride.

3.2. SDAE Modeling and Analysis. In this work, SDAE neural network algorithm was used to predict the storage time of the salmon; meanwhile, partial least squares regression (PLSR)

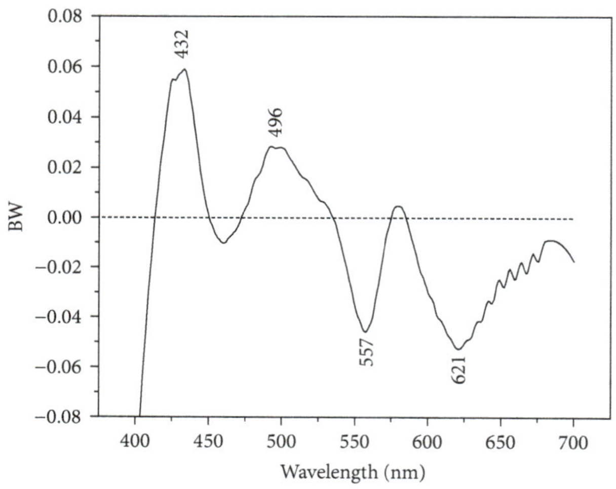

FIGURE 5: Principal component regression-weighted regression coefficient (BW) for salmon meat spectra at different storage days.

and back propagation neural network (BPNN) models also were established to compare the performance of SDAE neural network. In the process of modeling, the salmon meat and skin spectra were used as input data, and the number of storage days was used as the output source. In the salmon meat and skin spectra, 3/4 of the samples were randomly selected for modeling, and the remaining 1/4 samples were used as external prediction samples. Model determination coefficient, prediction mean square error of calibration sets (RMSEC), root mean square error of cross-validation sets (RMSECV), and prediction accuracy were used to evaluate the performance of the model.

Firstly, the salmon meat spectral model was established. In the process of PLSR modeling, when the number of principal components was 6, the model could achieve the best performance with determination coefficient of calibration sets (R^2_{CAL}) 0.97, determination coefficient of cross-validation sets (R^2_{CV}) 0.96, determination coefficient of test sets (R^2_{test}) 0.97, RMSEC 0.99, RMSECV 1.18, and RMSEP 1.0317. A threshold was set in the middle of every two storage days, and when the predicted days exceed the threshold, it could be judged as a prediction error. According to the statistics, when PLSR method was used, the prediction accuracy of calibration sets was 94.7%, and the prediction accuracy of cross-validation sets was 90.8%. When the 25 external test samples were predicted as shown in Figure 6, there were 5 samples misclassified and the prediction accuracy was 80%.

When back propagation (BP) neural network model was built, we used three-layer neural networks and the number of neurons was 200–100–1. The hidden layer activation function was the hyperbolic function (tanh), and the output layer was the linear function "$y=x$." The number of iterations was 1000, and the learning rate was 0.001. When the gradient descent method [29] was used to iterate, the result was shown in Figure 7. Four samples were wrongly predicted with R^2_{CAL} 0.95, R^2_{CV} 0.93, R^2_{test} 0.96, RMSEC 1.4, RMSECV 1.55, and RMSEP 0.99. The accuracy of prediction test sets was 84.6%.

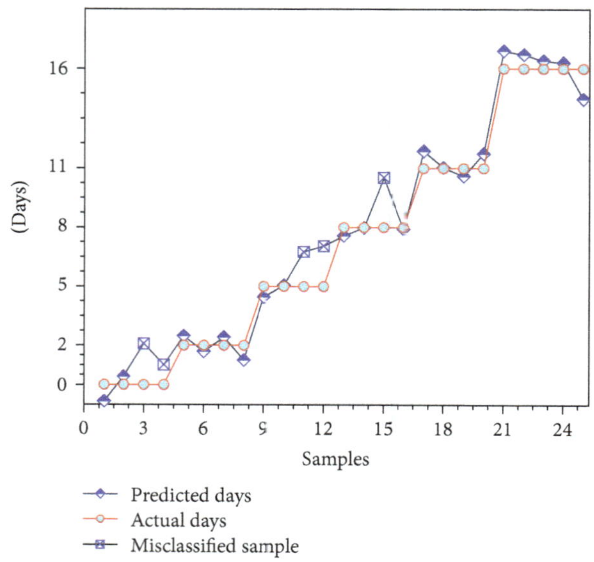

FIGURE 6: Predicted results of PLSR model for test sets of the salmon meat.

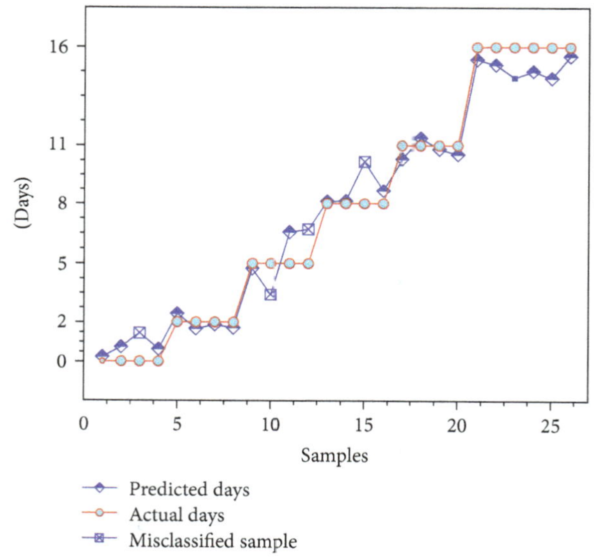

FIGURE 7: The predicted results of BP-NN model for test sets of the salmon meat.

The SDAE-NN was a double autoencoder structure, the number of the first autoencoder neurons was 600–200–600, and the second layer was 200–100–200. The activation function of both layers was the hyperbolic function (tanh), and the output layer activation function was the linear function. Through this algorithm, there were three samples which were wrongly predicted (Figure 8) with R^2_{CAL} 0.95, R^2_{CV} 0.93, R^2_{test} 0.98, RMSEC 1.39, RMSECV 1.54, and RMSEP 0.93. The accuracy of prediction test sets was 88.5%.

PLSR, BP-NN, and SDAE-NN algorithms were also used to model and predict the storage days of the salmon skin. When 9 principal components were used in the

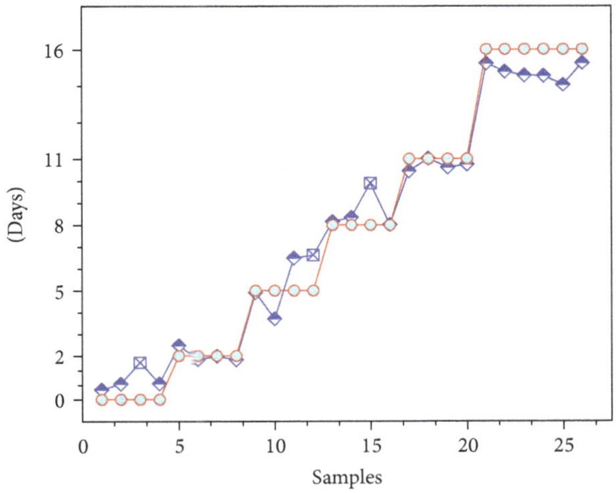

Figure 8: The predicted results of SDAE-NN model for test sets of the salmon meat.

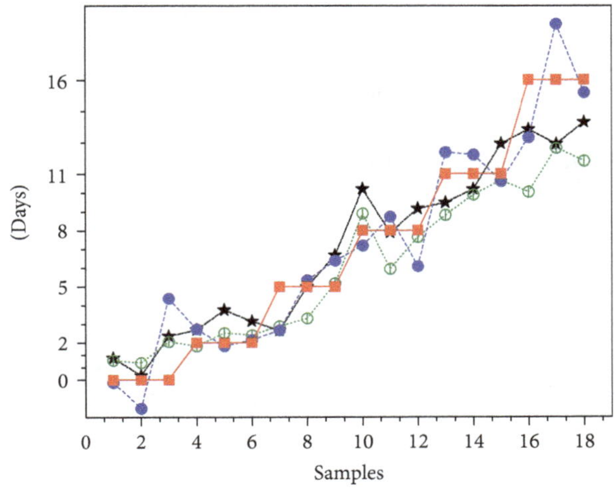

Figure 9: The predicted results of SDAE-NN, BP-NN, and PLSR models for test sets of the salmon skin.

PLSR model, the model performance was best with R^2_{test} 0.90 and RMSEP 2.42. The BP-NN and SDAE-NN methods use the same parameters as salmon meat modeling, When BP-NN was used to predict the test sets, the R^2_{test} was 0.91 and RMSEP was 2.26. When SDAE-NN was used to predict the test sets, the R^2_{test} was 0.92 and RMSEP was 1.75. As shown in Figure 9, the three methods all had a good performance in predicting the storage days of the salmon skin. However, PLSR had the maximum root mean square error and the least determination coefficient. In contrast, the SDAE-NN method had the smallest predicted error and the largest determination coefficient.

Furthermore, we have compared the prediction performance of the salmon skin and salmon meat using the three methods, as shown in Table 1. It can be seen that the performance using the salmon meat was better than salmon skin unrelated to the as-used method. When the salmon skin was used to predict, the best result of R^2_{test} was 0.92 and RMSEP was 1.75. While when the salmon meat was used to predict, the best result of R^2_{test} was 0.98 and RMSEP was 0.93.

The denoising ability of the SDAE-NN model was also studied in this work. In order to improve the denoising ability, Gaussian noise with the variance of 0, 0.01, 0.02, 0.03, 0.04, and 0.05 was added during the autoencoder training, respectively. As shown in Figure 10, when Gaussian noise variance was 0.02 the RMSEP of the model was 0.93, the determination coefficient was 0.98, and the model has the best performance with high robustness and denoising ability.

3.3. Discussion. In this paper, the reason why using the salmon meat could get better prediction result than the salmon skin was that the composition of the salmon meat was different from the salmon skin. On the one hand, the salmon meat had a higher moisture content and was more easily rotted and decayed during storage. The differences of the salmon meat in the absorption peak position and strength were more obvious than the salmon skin, due to the dramatic changes of the chemical composition in the meat, such as bacteria breeding, fat oxidation, and protein decomposition [30]. On the other hand, the visible region of the spectrum (400–700 nm) has been proved to have a great influence on the freshness of fish [13, 31]. The oxidation of heme proteins, such as haemoglobin (Hb) and myoglobin (Mb), was detected and explained with most spectral variation in the visible region [13, 32], and the salmon meat contains more haemoglobin (Hb) and myoglobin (Mb) than the skin.

Nilsen et al. [12] also used spectroscopy to predict the storage time of salmon, and the best result of R^2_{test} 0.98 and RMSEP 1.20 was achieved using the PLSR method. However, it was achieved under the precondition of 10 principal components, which might be overfitting. When the PLSR method was used in this paper, the prediction accuracy could reach 1.03 day under 6 principal components, which was higher than the above reference. It is found that the conditions of sample storage, such as the storage temperature and the degree of exposure, might have a certain effect on the biochemical and chemical changes of the meat [4]. At the same time, the noise of the spectra and the parameters of test instruments also have a certain effect on predicted results.

In the comparison of three kinds of algorithms, we could find that using the neural network method (BP-NN and SDAE-NN) could get better results than the PLSR method. The PLSR is better at solving linear regression issues and is susceptible to noise, so when there was more noise in the spectra, the modeling was not effective enough. In contrast, the neural network method, containing multiple neurons and weights, is more suitable to solve the problem of nonlinear regression and classification. Of course, whether the model parameter has been suited, the set would have a great impact on the effect of the neural network. Since this work carefully selected the related parameters of BP-NN, such as

Table 1: The performance of SDAE-NN, BP-NN, and PLSR models for test sets.

Methods	Salmon meat			Salmon skin		
	PLSR	BP-NN	SDAE-NN	PLSR	BP-NN	SDAE-NN
R^2_{val}	0.97	0.96	0.98	0.9	0.91	0.92
RMSEP	1.03	0.99	0.93	2.42	2.26	1.75

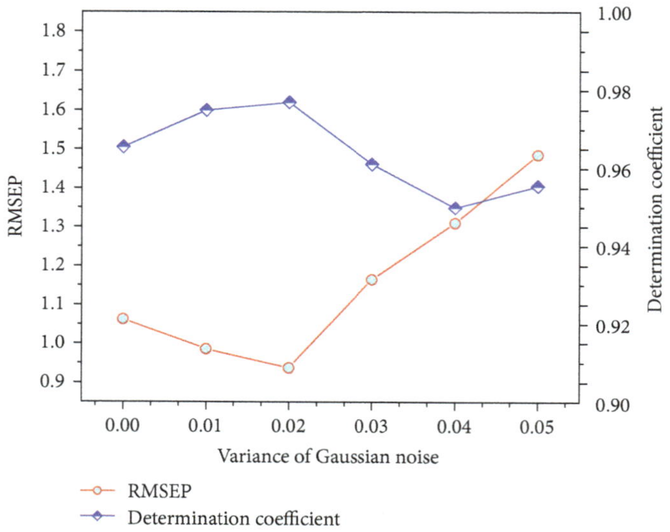

Figure 10: Influence on the model of Gaussian noise with different variance.

the learning rate, the results were better than those of the PLSR method. SDAE-NN algorithm had the best performance of the three algorithms; it is attributed that SDAE-NN is an improved algorithm on the basis of BP-NN. In BP-NN, the weights are initialized randomly and the performance of the BP-NN depends on the selection of the initial value. So it is easily prone to local optimal, and when the number of layers of the neural network increases, gradient descent method used in training networks would have a big probability of creating gradients vanishing and it is difficult for the neural network to select the optimal parameters. In the SDAE-NN algorithm used in this paper, the weights of the neural network were not randomly initialized but obtained through pretraining [22], and it could be considered the optimal weights. In the pretraining, the neural network was trained layer by layer through the autoencoder technology. The hidden features of each layer after training were used as input to the next layer, so that the problem of gradients vanishing could be eliminated as much as possible. At the same time, the process of SAE also reduced the dimensions of spectral data and was considered to be a better dimensionality reduction method than principal component analysis [22].

We have rephrased the highlighted sentence. Please confirm that this is your intended meaning. Due to some noises caused by environment, equipment, samples and baseline drift have a negative impact on the performance of the model, another advantage of SDAE-NN is its denoising ability compared with PLSR.. The PLSR method used in the previous reports [4, 33] needs to perform spectral preprocessing such as multiplicative scatter correction and standard normal variate before modeling to eliminate noise. In this paper, we added some Gaussian noise in the autoencoder process to improve the robustness of the neural network (Figure 10) and to enhance the antinoise ability of the model. Therefore, good prediction results could be obtained in the SDAE-NN method without spectral preprocessing.

It should be pointed out that although SDAE-NN had obtained the best prediction results; however, the prediction accuracy rate of the test sets was only 88.5%. When training the SDAE-NN, it was easy to achieve 100% accuracy of the calibration sets, but the test sets was difficult to achieve this accuracy because the neural network was easy to overfit. At this time, increasing the number of samples is considered to be a better method than adjusting the parameters and structure of neural network itself because the characteristic of neural network is that its performance and accuracy can be significantly improved with the increase of training samples. The prediction accuracy will be closer and closer to 100% when the number of samples is collected enough and the measurements of spectra are accurate.

4. Conclusion

VIS/NIR spectroscopy combined with SDAE-NN algorithm has been used to predict the storage time of the salmon meat and skin. The salmon meat is proved to be a better location that could get better predictive performance than the salmon skin. Compared with PLSR and BPNN algorithm, the SDAE-NN could achieve better determination coefficient and smaller prediction error with $R^2_{test} = 0.98$ and RMSEP = 0.93 day. This technique was nondestructive, low cost, and fast and has no-preprocessing and could be considered as an effective method for predicting the storage time of the salmon. Furthermore, the new technique can be widely adopted to predict the freshness of other agricultural products.

Disclosure

This article does not contain any studies with human participants or animals performed by any of the authors.

Conflicts of Interest

Ting Wu, Ling Yang, and Nan Zhong declare that they have no conflict of interest.

Acknowledgments

The authors thank the anonymous reviewers for their critical and constructive comments and suggestions. This work was partially supported by the National Natural Science Funds for Young Scholar (Grant no. 61501531), Natural Science Foundation of Guangdong Province (Grant no. 2015A030313602), Collaborative Innovation Major Projects of Guangzhou (201508010013 and 201704020030), and Science and Technology Plan of Guangdong Province (Grant nos. 2015A020209173 and 2013B040500013).

References

[1] K. Sveinsdottir, E. Martinsdottir, G. Hyldig, B. Jørgensen, and K. Kristbergsson, "Application of quality index method (QIM) scheme in shelf-life study of farmed Atlantic salmon (*Salmo salar*)," *Journal of Food Science*, vol. 67, no. 4, pp. 1570–1579, 2002.

[2] H. H. Huss, T. Boerresen, P. Dalgaard et al., *Quality and Quality Changes in Fresh Fish*, FAO Fisheries Technical Paper, Rome, Italy, 1995.

[3] V. M. Ocaño-Higuera, A. N. Maeda-Martínez, E. Marquez-Ríos et al., "Freshness assessment of ray fish stored in ice by biochemical, chemical and physical methods," *Food Chemistry*, vol. 125, no. 1, pp. 49–54, 2011.

[4] T. Kimiya, A. H. Sivertsen, and K. Heia, "VIS/NIR spectroscopy for non-destructive freshness assessment of Atlantic salmon (*Salmo salar* L.) fillets," *Journal of Food Engineering*, vol. 116, no. 3, pp. 758–764, 2013.

[5] J. Ortiz, J. P. Vivanco, V. Quitral, M. A. Larraín, G. Concha, and S. P. Aubourg, "Changes in freshness during frozen storage of farmed coho salmon: effect of replacement of synthetic antioxidants by natural ones in fish feeds," *North American Journal of Aquaculture*, vol. 74, no. 2, pp. 224–229, 2012.

[6] K. Fernández, E. Aspe, and M. Roeckel, "Shelf-life extension on fillets of Atlantic salmon (*Salmo salar*) using natural additives, superchilling and modified atmosphere packaging," *Food Control*, vol. 20, no. 11, pp. 1036–1042, 2009.

[7] I. Sone, R. L. Olsen, R. Dahl, and K. Heia, "Visible/near-infrared spectroscopy detects autolytic changes during storage of Atlantic salmon (*Salmo salar* L.)," *Journal of Food Science*, vol. 76, no. 3, pp. S203–S209, 2011.

[8] V. H. Segtnan, M. Høy, F. Lundby, B. Narum, and J. P. Wold, "Fat distribution analysis in salmon fillets using non-contact near infrared interactance imaging: a sampling and calibration strategy," *Journal of Near Infrared Spectroscopy*, vol. 17, no. 5, pp. 247–253, 2009.

[9] W. Klaypradit, S. Kerdpiboon, and R. K. Singh, "Application of artificial neural networks to predict the oxidation of menhaden fish oil obtained from Fourier transform infrared spectroscopy method," *Food and Bioprocess Technology*, vol. 4, no. 3, pp. 475–480, 2011.

[10] M. Kamruzzaman, D. W. Sun, G. ElMasry, and P. Allen, "Fast detection and visualization of minced lamb meat adulteration using NIR hyperspectral imaging and multivariate image analysis," *Talanta*, vol. 103, no. 2, pp. 130–136, 2013.

[11] B. Kuswandi, F. K. Putri, A. A. Gani, and M. Ahmad, "Application of class-modelling techniques to infrared spectra for analysis of pork adulteration in beef jerkys," *Journal of Food Science and Technology*, vol. 52, no. 12, pp. 7655–7668, 2015.

[12] H. Nilsen, M. Esaiassen, K. Heia, and F. Sigernes, "Visible/near-infrared spectroscopy: a new tool for the evaluation of fish freshness?," *Journal of Food Science*, vol. 67, no. 5, pp. 1821–1826, 2002.

[13] A. H. Sivertsen, T. Kimiya, and K. Heia, "Automatic freshness assessment of cod (*Gadus morhua*) fillets by Vis/Nir spectroscopy," *Journal of Food Engineering*, vol. 103, no. 3, pp. 317–323, 2011.

[14] L. W. Mamani-Linares, C. Gallo, and D. Alomar, "Identification of cattle, llama and horse meat by near infrared reflectance or transflectance spectroscopy," *Meat Science*, vol. 90, no. 2, pp. 378–385, 2012.

[15] L. Rodriguez-Saona, M. Giusti, and M. Shotts, *Advances in Infrared Spectroscopy for Food Authenticity Testing*, Woodhead Publishing, Duxford, United Kingdom, 2016.

[16] X. X. Weng, F. Lu, C. X. Wang, and Y. P. Qi, "Discriminating and quantifying potential adulteration in virgin olive oil by near infrared spectroscopy with BP-ANN and PLS," *Spectroscopy and Spectral Analysis*, vol. 29, no. 12, pp. 3283–3287, 2009.

[17] F. G. del Moral, A. Guillén, L. G. del Moral, F. O'Valle, L. Martínez, and R. G. del Moral, "Duroc and Iberian pork neural network classification by visible and near infrared reflectance spectroscopy," *Journal of Food Engineering*, vol. 90, no. 4, pp. 540–547, 2009.

[18] Y. Bengio, P. Lamblin, D. Popovici, and H. Larochelle, "Greedy layer-wise training of deep networks," in *Proceedings of the 19th International Conference on Neural Information Processing Systems*, pp. 153–160, MIT Press, Canada, 2006.

[19] Z. Waszczyszyn, "Fundamentals of artificial neural networks," in *Neural Networks in the Analysis and Design of Structures*, pp. 1–51, Springer Vienna, Vienna, 1999.

[20] Y. LeCun, Y. Bengio, and G. Hinton, "Deep learning," *Nature*, vol. 521, no. 7553, pp. 436–444, 2015.

[21] P. Vincent, H. Larochelle, I. Lajoie, Y. Bengio, and P. A. Manzagol, "Stacked denoising autoencoders: learning useful representations in a deep network with a local denoising criterion," *Journal of Machine Learning Research*, vol. 11, no. 12, pp. 3371–3408, 2010.

[22] G. E. Hinton and R. R. Salakhutdinov, "Reducing the dimensionality of data with neural networks," *Science*, vol. 313, no. 5786, pp. 504–507, 2006.

[23] S. Wold, K. Esbensen, and P. Geladi, "Principal component analysis," *Chemometrics and Intelligent Laboratory Systems*, vol. 2, no. 1–3, pp. 37–52, 1987.

[24] G. E. Hinton and R. S. Zemel, "Autoencoders, minimum description length and Helmholtz free energy," in *Proceedings of the 6th International Conference on Neural Information Processing Systems*, pp. 3–10, Morgan Kaufmann Publishers Inc, Denver, Colorado, USA, 1993.

[25] W. Zijlstra and A. Buursma, "Spectrophotometry of hemoglobin-absorption spectra of bovine oxyhemoglobin, deoxyhemoglobin, carboxyhemoglobin, and methemoglobin," *Comparative Biochemistry and Physiology. Part B: Biochemistry and Molecular Biology*, vol. 118, no. 4, pp. 743–749, 1997.

[26] J. Tang, C. Faustman, and T. A. Hoagland, "Krzywicki revisited: equations for spectrophotometric determination of myoglobin redox forms in aqueous meat extracts," *Journal of Food Science*, vol. 69, no. 9, pp. C717–C720, 2004.

[27] C. Zhou, H. Wang, Y. Chen, and C. Chen, "Effect of L-cysteine and lactose on color stability of porcine red blood cell during freeze-drying and powder storage," *Food Science and Biotechnology*, vol. 21, no. 3, pp. 669–674, 2012.

[28] A. G. Christophersen, P. Knuthsen, and L. H. Skibsted, "Determination of carotenoids in salmonoids," *Zeitschrift für Lebensmittel-Untersuchung und Forschung*, vol. 188, no. 5, pp. 413–418, 1989.

[29] D. E. Rumelhart, G. E. Hinton, and R. J. Williams, "Learning representations by back-propagating errors," *Nature*, vol. 323, no. 6088, pp. 533–536, 1986.

[30] X. L. Zhang, J. Xie, K. Hao, and H. Q. Zhao, "Effects of different cold storage conditions on quality of salmon," *Science and Technology of Food Industry*, vol. 37, no. 17, pp. 316–321, 2016.

[31] I. Sone, R. L. Olsen, A. H. Sivertsen, G. Eilertsen, and K. Heia, "Classification of fresh Atlantic salmon (*Salmo salar* L.) fillets stored under different atmospheres by hyperspectral imaging," *Journal of Food Engineering*, vol. 109, no. 3, pp. 482–489, 2012.

[32] M. P. Richards, A. M. Modra, and R. Li, "Role of deoxyhemoglobin in lipid oxidation of washed cod muscle mediated by trout, poultry and beef hemoglobins," *Meat Science*, vol. 62, no. 2, pp. 157–163, 2002.

[33] K. Ncama, U. L. Opara, S. Z. Tesfay, O. A. Fawole, and L. S. Magwaza, "Application of Vis/NIR spectroscopy for predicting sweetness and flavour parameters of 'Valencia' orange (*Citrus sinensis*) and 'Star Ruby' grapefruit (*Citrus* x *paradisi* Macfad)," *Journal of Food Engineering*, vol. 193, pp. 86–94, 2017.

10

Classification of Geological Samples Based on Soft Independent Modeling of Class Analogy Using Laser-Induced Breakdown Spectroscopy

Ying Zhang,[1] **Ying Li**,[1] **Wendong Li**,[1] **Zigang Sun**,[1] **and Yunfeng Bi**[2]

[1]*Optics and Optoelectronics Laboratory, Ocean University of China, Qingdao, Shandong 266100, China*
[2]*College of Electromechanical and Information Engineering, Shandong University, Weihai, Shandong 264209, China*

Correspondence should be addressed to Ying Li; liying@ouc.edu.cn

Academic Editor: Alessandro De Giacomo

Laser-induced breakdown spectroscopy with soft independent modeling of class analogy is used in the identification of a large number of unprocessed geological samples having similar components in this study. Considering a variety of data from different samples, representative spectral regions representing the major components were extracted. In addition, principal component analysis was applied to remove noninformative variables from the spectrum. The unclassification rate, misclassification rate, and average correct classification rate for 25 types of geological samples were 1.2%, 4.7%, and 94.1%, respectively. These results suggest that laser-induced breakdown spectroscopy using soft independent modeling of class analogy can be used to identify a wide variety of geological samples. Furthermore, we found that this approach can be used to identify spectral differences among similar sample types because of matrix effects and the trace element impurities.

1. Introduction

Laser-induced breakdown spectroscopy (LIBS) [1, 2] is a simple atomic emission technique for multiple elements and provides a semidestructive and efficient analysis, particularly in harsh and dangerous environments. Thus, LIBS has been widely used for various applications, such as industry-oriented analysis [3], archaeological investigation [4], geological and environmental studies [5–7], and jewelry characterization [8]. Geological materials including rocks and minerals convey important information about particular geological environments, and this information can be extremely useful for studies such as determining mineral provenance, reservoir description, prospecting, and geochemical mapping [9–11]. Nevertheless, there is a wide variety of geological materials with overlapping characteristics, thus compromising their proper discrimination. The conventional geological survey depends on the geologist's assessment and subsequent laboratory analyses, which can be time-consuming and complicated. To simplify the analysis, LIBS applications on geological materials have been proposed over the last two decades. Furthermore, multivariate preprocessing methods have been increasingly studied, including approaches based on principal component analysis (PCA) [12, 13], partial least squares discriminant analysis (PLS-DA) [14], graph theory (GT) [15], independent component analysis (ICA) [16], and artificial neural networks (ANNs) [17, 18]. Such methods consider the effect of redundant information and hence increase the efficiency of data analysis and prevent negligible fluctuations resulting from experimental conditions and instrumental instability [19]. Specifically, soft independent modeling of class analogy (SIMCA) is widely used to classify high-dimensional data because it incorporates PCA for dimensionality reduction [20]. It was originally developed to increase the accuracy and speed of classification in near-infrared spectroscopy [21–25] and subsequently applied to the classification of LIBS [26, 27]. Although suitable results have been reported in the classification of some geological materials, it is still challenging to provide a method that suitably classifies a large number of materials, especially when they present similar major elements. In this study, we applied the SIMCA and PCA to LIBS data aiming to classify a wide variety of geological samples.

2. Experiment and Methods

2.1. LIBS Instrument. Figure 1 illustrates the complete experimental system used in this study. The Brilliant B Nd: YAG Laser (Quantel SA, Les Ulis Cedex, France) was operated at a fundamental wavelength of 1064 nm, a repetition rate of 10 Hz, and a pulse width of 10 ns. The laser energy was optimized to maximize the peak intensity without saturating the intensified charge-coupled device (ICCD) camera. The excitation energy from this laser was focused on a target with a long-focus ($f_1 = 100$ mm) lens to prevent contamination from spatter particles generated by the laser shots. Light emitted from the plasma was collected by a pair of matching coaxial fused silica planoconvex lenses ($f_2 = f_3 = 38.1$ mm) and guided into a 230 μm diameter optical fiber for linking with a Mechelle 5000 spectrometer (Andor Technology Ltd., Belfast, UK). Then, the dispersed light from the spectrometer was recorded with an iStar DH734i-18F-03 ICCD camera (Andor Technology Ltd., Belfast, UK) having a wide spectral range (212 nm–1032 nm) with 0.1 nm resolution. The angle between the collection direction and the sample stage surface was approximately 45°. The samples were fixed on a rotating platform and mechanically rotated to different positions following laser ablation. The crater effects were minimized, and the inhomogeneity among samples was partially compensated by collection from different positions.

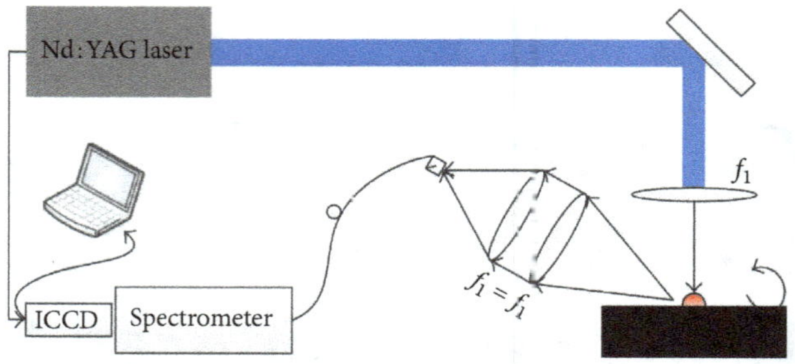

FIGURE 1: Diagram of a typical LIBS experimental setup for studying geological samples.

TABLE 1: Description of geological samples used in this study.

No.	Name
1	Carbuncle
2	K feldspar
3	Albite
4	Tektite
5	Magnetite
6	Demantoid
7	Oolitic hematite
8	Pink pyroxene
9	Idocrase
10	Black hematite
11	Biotite
12	Hornblende
13	Calcite
14	Labradorite
15	Magnesite
16	Rhodonite
17	Reniform hematite
18	Banded siliceous rocks
19	Diopside
20	Selenite
21	Alabaster
22	Anhydrite
23	Moonstone
24	Barite
25	Hypersthene

2.2. Samples and Measurements. This study involved the analysis of 25 types of geological samples representing a mixture of minerals and rocks (carbuncle), which are listed in Table 1. Figure 2 shows photographs of six types of samples used in this study. These samples were obtained from the China Institute of Geology in Qingdao City, Shandong Province. Five different blocks of each sample were collected at different but nearby geographical locations. In addition, several geological samples with a similar chemical composition were purposely considered in this study to verify the robustness of the proposed model. In fact, compositionally similar minerals can exhibit a very high spectral correlation, thus posing a challenge to intersample discrimination. For instance, sample No. 20, 21, and 22 can be considered as a type of gypsum, which is mostly composed of calcium sulfate ($CaSO_4$), whereas sample No. 7, 10, and 17 basically consist of iron oxides (Fe_2O_3), and sample No. 8, 19, and 25 were also considered of the same type. Samples were measured using LIBS without pretreatment to obtain raw data from in-field measurements. The five separate blocks of each geological sample were detected, from which three were assigned for determining the method parameters, whereas the remaining two were used to test the method performance. To partially balance spectral heterogeneity, each spectrum was determined from 5 laser shots and 20 spectra acquired per block at different points on the sample surface. The integration time and delay were 15 μs and 200 ns, respectively, to eliminate continuum emission.

2.3. Model. Multivariate analysis can be applied to reduce or compress spectral data while retaining important spectral information of the samples [28]. In particular, SIMCA is a widely used supervised pattern recognition method to classify sample spectra within specific categories. It consists of a collection of PCA models and can provide independent classification for each category, as detailed in [29, 30]. Specifically, PCA calculations were carried out in order to reduce the dimensionality of the data set, allowing an overview of the samples. The results from PCA are typically analyzed by score and loading plots. The score plots allow the identification of samples, by verifying if there are similarities or not, and the identification of outliers and clusters.

(a) (b) (c) (d) (e) (f)

FIGURE 2: Some geological samples used in this study: (a) selenite, (b) spodumene, (c) barite, (d) hematite, (e) moonstone, and (f) labradorite.

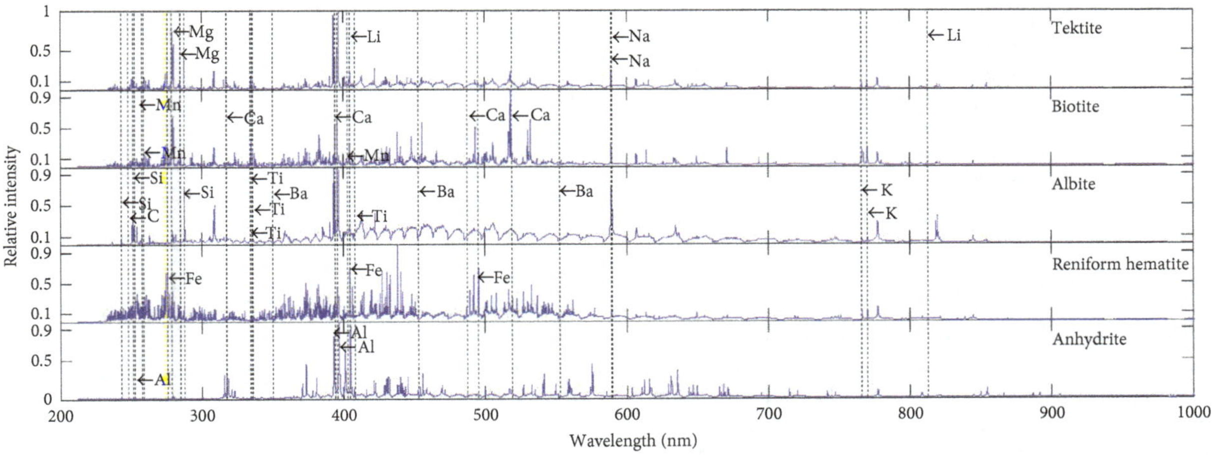

FIGURE 3: Typical spectral lines of five mineral samples: tektite, biotite, albite, reniform hematite, and anhydrite.

Loading plots permit the identification of variables that have greater importance for the sample positions in the score plots. The optimal number of principal components (PCs) to characterize the data set was based on the total value of the principal component retained variance [30, 31]. In addition, we used a toolbox for SIMCA that was developed by the Milano Chemometrics and QSAR Research Group at the University of Milano-Bicocca in Italy [32, 33]. Moreover, we considered a statistical confidence level of 95% ($\alpha = 0.05$) and implemented the calculations using MATLAB version 7.2.

A total of 1000 spectra (40 spectra × 25 samples) from known samples were used to build the SIMCA recognition model, and 500 spectra (20 spectra × 25 samples) were applied for optimizing the parameters based on cross validation. The remaining 1000 spectra (40 spectra × 25 samples) from unknown samples as a test set were used to determine the classification performance.

2.4. Emission Line Selection. The analytical spectral line of an element in the plasma is related to the ejected sample mass and depends on the laser radiation parameters, that is, energy and focusing. Either random or systematic changes of these parameters can strongly affect the analytical precision and accuracy and may introduce nonlinearity in the classification [34]. Moreover, the roughness of the sample surface further increases nonlinearity by the interaction between the laser and samples. However, normalization can compensate some shortcomings and signal variations resulting from experimental conditions and instrumental instability [35]. In this study, the signals corresponding to various elements were normalized with the total spectrum intensity. Figure 3 illustrates the normalized spectral lines of five mineral samples, where the highlighted elements were identified using the NIST Atomic Spectral Database. Emissions in the spectra from the analyzed 5 samples correspond to the elements showing high similarities and differences among distinct LIBS fingerprints.

The recorded spectra consist of more than 20,000 pixels spanning a wide wavelength region from ultraviolet to near-infrared. However, a substantial portion of the feature space may not be relevant for classification. Hence, feature selection must be applied to eliminate spurious correlations, especially when interclass differences are subtle. Feature selection allows to reduce regions of LIBS data that do not convey useful classification information. We found that the most important features correspond to wavelengths from the elemental emission lines of K, Li, Na, Ba, Mn, Ca, Al, Ti, Si, Mg, and Fe. In fact, these elements are the main components of the continental crust and determine unique chemical fingerprints, which are useful for geological study. For classification, a set of spectral regions from the major elements commonly used in spectral analysis were selected [36], totaling 1107 variables. The selected spectral variables are listed in Table 2 and illustrated in Figure 4.

Table 2: Emission lines for classification based on major elements from geological samples.

Element	Analytical line (nm)	Extracted wavelength range (nm)
Fe I	251.810	251.27–252.14
Mg II	279.553	278.75–280.72
Mg II	280.27	
Si I	288.158	288.07–288.33
Ti II	336.123	335.79–336.51
Ca II	393.366	391.48–399.04
Al I	394.401	
Al I	396.152	
Mn I	403.307	402.82–403.80
Ba II	493.408	488.72–495.35
Na I	588.995	588.62–589.93
Na I	589.592	
Li I	670.777	670.55–671.13
Li I	670.793	
K I	766.489	764.26–771.85
K I	769.896	

3. Results and Discussion

3.1. PCA Optimization. Problems of unsatisfied collinearity and high computational cost persisted after selecting the spectral regions. Hence, the selected LIBS regions were projected into lower-dimensional independent variables using PCA, and those having the maximal interclass variance and minimal intraclass variance were iteratively determined. The selection of principal components greatly affects the classification capabilities and can prevent both under- and overfitting problems of classification. The scores and loadings of PCA for the 25 types of samples are shown in Figure 4, where the principal components PC1, PC2, and PC3 contribute 28.84%, 22.17%, and 14.34% of the overall variance, respectively. In addition, high loading indicates that elements corresponding to that wavelength have a high effect on the principal components [37] and also a high concentration of the corresponding element in the samples. The loading and score of PC1 suggest a high correlation with the concentrations of Ca and Al. Similarly, PC2 is clearly related to Ca, Al, Si, and Mg; PC3 is more relevant to Ba, Mn, Ca, and Al. Considering that the first three principal components convey only 65.35% of the original information, other principal components were also introduced into for SIMCA. Figure 5 shows that an increasing number of principal components reduce the root mean square error of cross validation (RMSECV), thus indicating a more accurate selection. The RMSECV converges to a stable minimum after considering approximately 15 principal components. By selecting the most relevant principal components based on the RMSECV results, the classification for the validation data set was as follows: unclassification of 0.4%, misclassification of 1.4%, and correct classification of 98.2%.

3.2. SIMCA Evaluation. Table 3 lists the SIMCA classification results for the test data. The results also demonstrate that misclassification mostly occurs among similar samples,

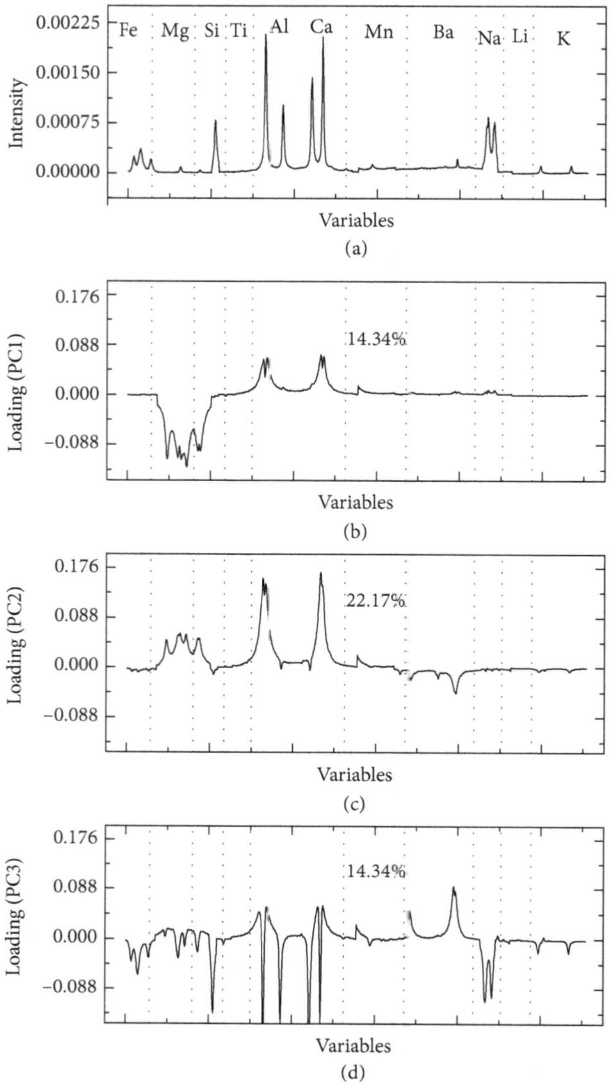

Figure 4: (a) A recombination of selected spectral variables in Table 2 for the tektite sample; (b) PC1 loading plot; (c) PC2 loading plot; (d) PC3 loading plot. Comparison between the PCA loading plots of the first three principal components is shown, the percentages represent the explained variance for each principal component. These three principal components, combined, explain 65.35% of the net variance in the data set. The loadings were compared to the recombined spectrum in order to observe the variables in the spectrum that contribute to classification.

suggesting the difficulty to classify minerals that have the same cations as major constituents, such as gypsum (sample No. 20 (selenite), 21 (alabaster), and 22 (anhydrite)) and hematite (sample No. 7 (oolitic hematite), 10 (black hematite), and 17 (reniform hematite)). Nevertheless, the correct classification rate among similar samples remained acceptable. The correct classification rate among similar samples can be related to the slight differences resulting from physical matrix effects including hardness, structure, and texture. These factors can result in different amounts of ablated mass and in a consequent variation of the spectral

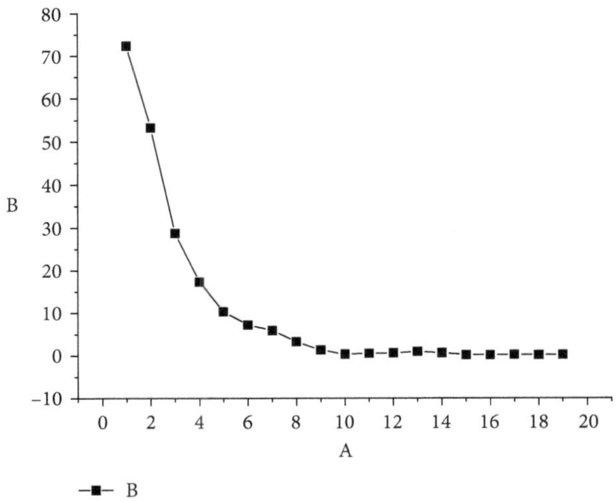

FIGURE 5: Estimated root mean square error of cross validation using SIMCA according to the number of principal components for 25 validation samples.

TABLE 3: Classification performance using SIMCA with PCA for 25 geological samples.

No.	Correct classification	Misclassification	Unclassification
1	60%	40%	0
2	97.5%	2.5%	0
3	100%	0	0
4	100%	0	0
5	97.5%	2.5%	0
6	100%	0	0
7	97.5%	2.5%	0
8	100%	0	0
9	87.5%	12.5%	0
10	90%	10%	0
11	100%	0	0
12	100%	0	0
13	97.5%	2.5%	0
14	92.5%	0	7.5%
15	80%	0	20%
16	92.5%	0	7.5%
17	100%	0	0
18	100%	0	0
19	97.5	0	2.5%
20	100%	0	0
21	80%	20%	0
22	97.5%	2.5%	0
23	100%	0	0
24	100%	0	0
25	85%	15%	0
Average results	94.1% (941/1000)	4.7% (47/1000)	1.2% (12/1000)

lines, even when having a similar chemical composition of the geological samples [38]. Therefore, these influences can be useful in studies for recognizing similar geological samples. Another reason for the ability to distinguish similar samples can be attributed to minor impurities, whose distribution trace in natural minerals provides different spectral features among similar matrices.

The correct classification rate of 60% for carbuncle is lower than that of other minerals. This may be ascribed to the fact that the carbuncle as a type of rock is a combination of mineral grains, porosity, and cement-mixed body. There are also subtle differences between each block of carbuncle because they come from the different locations. The measurements on different points of the carbuncle block surface do not exhibit consistent proportions. In contrast, minerals are a single substance or a compound formed by geological action, presenting a relatively fixed chemical composition.

The remaining unclassified and misclassified samples can be attributed to the target heterogeneity. To obtain representative spectra for calibration/validation in the model from heterogeneous samples, it is necessary to collect LIBS information from a sufficiently large number of analysis spots, but this number could have not been sufficient to obtain representative analyses for all minerals in this study.

Overall, a high performance is achieved for all the samples, with an average correct classification rate of 94.1% and negligible rates of unclassification and misclassification. Therefore, the proposed SIMCA with LIBS was capable of correctly classifying samples in multiple categories, even when presenting similar compositions. Furthermore, the selected wavelength ranges, which reduce the amount of analyzed data to the major elements of the geological minerals, allowed to retrieve a successful and more efficient classification.

4. Conclusions

In this paper, LIBS combined with SIMCA is evaluated for the robust classification of a wide range of mineral types, which can be suitable for real-world applications, an improvement over many previous studies that were limited to several minerals. The LIBS data of different minerals can be used to identify samples based on major constituents. In fact, PCA allows to evaluate the discriminating ability of different elements present in geological samples by the corresponding loadings and scores of principal components. In this study, PCA-optimized SIMCA was employed to classify 25 types of geological samples. Although the correct identification was compromised for some varieties of rock such as carbuncle, which is composed of a variety of minerals having a high degree of trace element variability, the overall correct classification rate of 94.1% was high and the unclassification rate of 1.2% and the misclassification rate of 4.7% were acceptable.

Conflicts of Interest

The authors declare that there are no conflicts of interest regarding the publication of this paper.

Acknowledgments

This study was supported by the National Natural Science Foundation of China (Grant no. 41503063). The authors would like to thank Ye Tian for the helpful discussion and suggestions and Zhao Luo for providing the language assistance.

References

[1] R. Gaudiuso, M. Dell'Aglio, O. D. Pascale, G. S. Senesi, and A. D. Giacomo, "Laser induced breakdown spectroscopy for elemental analysis in environmental, cultural heritage and space applications: a review of methods and results," *Sensors*, vol. 10, no. 8, pp. 7434–7468, 2010.

[2] D. A. Cremers and R. C. Chinni, "Laser-induced breakdown spectroscopy—capabilities and limitations," *Applied Spectroscopy Reviews*, vol. 44, no. 6, pp. 457–506, 2009.

[3] A. Sarkar, V. Karki, S. K. Aggarwal et al., "Evaluation of the prediction precision capability of partial least squares regression approach for analysis of high alloy steel by laser induced breakdown spectroscopy," *Spectrochimica Acta Part B: Atomic Spectroscopy*, vol. 108, pp. 8–14, 2015.

[4] A. Ramil, A. J. López, and A. Yáñez, "Application of artificial neural networks for the rapid classification of archaeological ceramics by means of laser induced breakdown spectroscopy (LIBS)," *Applied Physics A*, vol. 92, no. 1, pp. 197–202, 2008.

[5] J. Rakovský, O. Musset, J. Buoncristiani et al., "Testing a portable laser-induced breakdown spectroscopy system on geological samples," *Spectrochimica Acta Part B: Atomic Spectroscopy*, vol. 74-75, pp. 57–65, 2012.

[6] R. S. Harmon, F. C. DeLucia, C. E. McManus et al., "Laser-induced breakdown spectroscopy-an emerging chemical sensor technology for real-time field-portable, geochemical, mineralogical, and environmental applications," *Applied Geochemistry*, vol. 21, no. 5, pp. 730–747, 2006.

[7] N. L. Lanza, R. C. Wiens, S. M. Clegg et al., "Calibrating the ChemCam laser-induced breakdown spectroscopy instrument for carbonate minerals on Mars," *Applied Optics*, vol. 49, no. 13, pp. C211–C217, 2010.

[8] L. E. Garcıa-Ayuso, J. Amador-Hernández, J. M. Fernández-Romero, and M. L. De Castro, "Characterization of jewellery products by laser-induced breakdown spectroscopy," *Analytica Chimica Acta*, vol. 457, no. 2, pp. 247–256, 2002.

[9] J. M. Anzano, M. A. Villoria, A. Ruíz-Medina, and R. J. Lasheras, "Laser-induced breakdown spectroscopy for quantitative spectrochemical analysis of geological materials: effects of the matrix and simultaneous determination," *Analytica Chimica Acta*, vol. 575, no. 2, pp. 230–235, 2006.

[10] S. L. Lui and A. Koujelev, "Accurate identification of geological samples using artificial neural network processing of laser-induced breakdown spectroscopy data," *Journal of Analytical Atomic Spectrometry*, vol. 26, no. 12, pp. 2419–2427, 2011.

[11] J. L. Gottfried, R. S. Harmon, F. C. De Lucia, and A. W. Miziolek, "Multivariate analysis of laser-induced breakdown spectroscopy chemical signatures for geomaterial classification," *Spectrochimica Acta Part B: Atomic Spectroscopy*, vol. 64, no. 10, pp. 1009–1019, 2009.

[12] V. K. Unnikrishnan, K. S. Choudhari, S. D. Kulkarni, R. Nayak, V. B. Kartha, and C. Santhosh, "Analytical predictive capabilities of laser induced breakdown spectroscopy (LIBS) with principal component analysis (PCA) for plastic classification," *RSC Advances*, vol. 3, no 48, pp. 25872–25880, 2013.

[13] Y. Lee, S. H. Nam, K. S. Ham et al., "Multivariate classification of edible salts: simultaneous laser-induced breakdown spectroscopy and laser-ablation inductively coupled plasma mass spectrometry analysis," *Spectrochimica Acta Part B: Atomic Spectroscopy*, vol. 118, pp. 102–111, 2016.

[14] V. A. Da Silva, M. Talhavini, I. C. Peixoto, J. J. Zacca, A. O. Maldaner, and J. W. B. Braga, "Non-destructive identification of different types and brands of blue pen inks in cursive handwriting by visible spectroscopy and PLS-DA for forensic analysis," *Microchemical Journal*, vol. 116, pp. 235–243, 2014.

[15] E. Grifoni, S. Legnaioli, G. Lorenzetti, S. Pagnotta, and V. Palleschi, "Application of graph theory to unsupervised classification of materials by laser-induced breakdown spectroscopy," *Spectrochimica Acta Part B: Atomic Spectroscopy*, vol. 118, pp. 40–44, 2016.

[16] S. Moncayo, J. D. Rosales, R. Izquierdo-Hornillos, J. Anzano, and J. O. Caceres, "Classification of red wine based on its protected designation of origin (PDO) using laser-induced breakdown spectroscopy (LIBS)," *Talanta*, vol. 158, pp. 185–191, 2016.

[17] G. Vítková, K. Novotný, L. Prokeš et al., "Fast identification of biominerals by means of stand-off laser-induced breakdown spectroscopy using linear discriminant analysis and artificial neural networks," *Spectrochimica Acta Part B: Atomic Spectroscopy*, vol. 73, pp. 1–6, 2012.

[18] J. El Haddad, M. Villot-Kadri, A. Ismael et al., "Artificial neural network for on-site quantitative analysis of soils using laser-induced breakdown spectroscopy," *Spectrochimica Acta Part B: Atomic Spectroscopy*, vol. 79-80, pp. 51–57, 2013.

[19] M. Defernez and E. K. Kemsley, "The use and misuse of chemometrics for treating classification problems," *TrAC Trends in Analytical Chemistry*, vol. 16, no. 4, pp. 216–221, 1997.

[20] S. Bicciato, A. Luchini, and C. Di Bello, "PCA disjoint models for multiclass cancer analysis using gene expression data," *Bioinformatics*, vol. 19, no. 5, pp. 571–578, 2003.

[21] C. Krafft, L. Shapoval, S. B. Sobottka, K. D. Geiger, G. Schackert, and R. Salzer, "Identification of primary tumors of brain metastases by SIMCA classification of IR spectroscopic images," *Biochimica et Biophysica Acta (BBA)-Biomembranes*, vol. 1758, no. 7, pp. 883–891, 2006.

[22] W. J. Dunn III and S. Wold, "An assessment of the carcinogenicity of *N*-nitroso compounds by the SIMCA method of pattern recognition," *Journal of Chemical Information and Computer Sciences*, vol. 21, no. 1, pp. 8–13, 1981.

[23] S. Wold, K. Esbensen, and P. Geladi, "Principal component analysis," *Chemometrics and Intelligent Laboratory Systems*, vol. 2, no. 1–3, pp. 37–52, 1987.

[24] W. Hu, X. X. Guo, X. C. Wang et al., "Rapid discrimination of different grades of white croaker surimi by tri-step infrared spectroscopy combined with soft independent modeling of class analogy (SIMCA)," *Food Analytical Methods*, vol. 9, no. 4, pp. 831–839, 2016.

[25] M. Foot and M. Mulholland, "Classification of chondroitin sulfate A, chondroitin sulfate C, glucosamine hydrochloride and glucosamine 6 sulfate using chemometric techniques," *Journal of Pharmaceutical and Biomedical Analysis*, vol. 38, no. 3, pp. 397–407, 2005.

[26] M. Hoehse, A. Paul, I. Gornushkin, and U. Panne, "Multivariate classification of pigments and inks using combined

Raman spectroscopy and LIBS," *Analytical and Bioanalytical Chemistry*, vol. 402, no. 4, pp. 1443–1450, 2012.

[27] P. Pease and V. Tchakerian, "Source provenance of carbonate grains in the Wahiba Sand Sea, Oman, using a new LIBS method," *Aeolian Research*, vol. 15, pp. 203–216, 2014.

[28] F. C. De Lucia and J. L. Gottfried, "Rapid analysis of energetic and geo-materials using LIBS," *Materials Today*, vol. 14, no. 6, pp. 274–281, 2011.

[29] M. Khanmohammadi, A. B. Garmarudi, K. Ghasemi, H. K. Jaliseh, and A. Kaviani, "Diagnosis of colon cancer by attenuated total reflectance-Fourier transform infrared microspectroscopy and soft independent modeling of class analogy," *Medical Oncology*, vol. 26, no. 3, pp. 292–297, 2009.

[30] O. G. Meza-Márquez, T. Gallardo-Velázquez, and G. Osorio-Revilla, "Application of mid-infrared spectroscopy with multivariate analysis and soft independent modeling of class analogies (SIMCA) for the detection of adulterants in minced beef," *Meat Science*, vol. 86, no. 2, pp. 511–519, 2010.

[31] Q. Godoi, F. O. Leme, L. C. Trevizan et al., "Laser-induced breakdown spectroscopy and chemometrics for classification of toys relying on toxic elements," *Spectrochimica Acta Part B: Atomic Spectroscopy*, vol. 66, no. 2, pp. 138–143, 2011.

[32] D. Ballabio and V. Consonni, "Classification tools in chemistry. Part 1: linear models. PLS-DA," *Analytical Methods*, vol. 5, no. 16, pp. 3790–3798, 2013.

[33] M. Alvarez Guerra, D. Ballabio, J. M. Amigo, J. R. Viguri, and R. A. Bro, "Chemometric approach to the environmental problem of predicting toxicity in contaminated sediments," *Journal of Chemometrics*, vol. 24, no. 7-8, pp. 379–386, 2010.

[34] N. B. Zorov, A. A. Gorbatenko, T. A. Labutin, and A. M. Popov, "A review of normalization techniques in analytical atomic spectrometry with laser sampling: from single to multivariate correction," *Spectrochimica Acta Part B: Atomic Spectroscopy*, vol. 65, no. 8, pp. 642–657, 2010.

[35] P. Pořízka, J. Klus, A. Hrdlička et al., "Impact of laser-induced breakdown spectroscopy data normalization on multivariate classification accuracy," *Journal of Analytical Atomic Spectrometry*, vol. 32, no. 2, pp. 277–288, 2017.

[36] F. Sorrentino, G. Carelli, F. Francesconi et al., "Fast analysis of complex metallic alloys by double-pulse time-integrated laser-induced breakdown spectroscopy," *Spectrochimica Acta Part B: Atomic Spectroscopy*, vol. 64, no. 10, pp. 1068–1072, 2009.

[37] N. J. McMillan, S. Rees, K. Kochelek, and C. McManus, "Geological applications of laser-induced breakdown spectroscopy," *Geostandards and Geoanalytical Research*, vol. 38, no. 3, pp. 329–343, 2014.

[38] G. S. Senesi, G. Tempesta, P. Manzari, and G. Agrosì, "An innovative approach to meteorite analysis by laser-induced breakdown spectroscopy," *Geostandards and Geoanalytical Research*, vol. 40, no. 4, pp. 533–541, 2016.

11

Long-Term Ripening Evaluation of Ewes' Cheeses by Fourier-Transformed Infrared Spectroscopy under Real Industrial Conditions

Jonathan Andrade, Cristina Guimarães Pereira, Thamiris Ranquine, Cosme Antonio Azarias, Maria José Valenzuela Bell, and Virgilio de Carvalho dos Anjos

Grupo de Engenharia e Espectroscopia de Materiais, Departamento de Física, Universidade Federal de Juiz de Fora, Rua José Lourenço Kelmer s/n São Pedro, 36036-900 Juiz de Fora, MG, Brazil

Correspondence should be addressed to Jonathan Andrade; dr.andradej@gmail.com

Academic Editor: Jose S. Camara

The ripening changes over time of special cheeses (Pecorino, ewes' ripe, and Gouda) made with ewes' milk were evaluated using FTIR/ATR spectroscopy during approximately one year. The midinfrared FTIR/ATR analyses were carried out in different ripening times between the cheese varieties and processed by means of multivariate statistical approaches. Overall, during the maturation, we observed a downward trend of the absorbance intensity of the amide group peaks (1700 to 1500 cm^{-1}), which is linked to the breakdown of peptide bonds. Similar behavior was obtained for the lipidic region (3000 to 2800 cm^{-1} and 1765 to 1730 cm^{-1}). Hierarchical cluster analysis and principal component analysis allowed the evaluation of the physicochemical changes of the cheeses. The proteolysis occurs in a fast pace during the first trimester of the ripening process, and the lipids are converted to smaller species as the times goes by. Our results indicate that infrared spectroscopy can be a useful tool in determining optimal temporal parameters in stages involving the development, production, and even a possible estimation of shelf life of cheeses.

1. Introduction

Cheese is the most abundant processed dairy product in the world. The global production is around 23 million tons, representing an increase of 791,000 tons between 2013 and 2014 [1]. The United States, for example, was responsible for approximately a quarter of the world production in 2014 and expected to reach 286,000 tons in export trades by the end of 2017 [1, 2]. Thus, the quality control of cheeses is of great importance, especially when a long period of ripening is required before marketing the product. About 70% of the total milk produced in Brazil is processed into cheeses, where approximately 1100 tons are consumed within the domestic market per year. The special cheese market is growing throughout the years, as well as the research and development of products adapted for the Brazilian market [3].

The physical-chemical and sensorial characteristics of dairy products can be changed by several factors including herd genetics, health, feeding, and environmental conditions, among others [4]. Cheese ripening involves several biochemical reactions, leading to the development of the unique sensorial attributes for each cheese variety [5–10]. The formation of smaller peptides and free amino acids has a direct impact on flavor and texture, due to the breakdown of the protein network [7]. In addition to that, lipolysis generates free fatty acids and organic acids that are further metabolized to volatile and nonvolatile flavor components [11].

Usual methods for analyzing cheese composition and ripening index are time-consuming and expensive. There is a need for new methodologies and instrumentation to observe the main changes during the ripening process. In fact, nowadays, several noninvasive spectroscopic techniques have gained more and more space due to several reasons: (i) they are environmentally friendly due to the absence of reagents; (ii) they are fast due to the absence of sample preparation; and (iii) they are susceptible to miniaturization [12–15]. In particular, instrumental technique such as midinfrared (MIR) spectroscopy has been widely used in the analysis of food products [16]. Encompassing the spectral range of 4000 to 400 cm^{-1}, MIR spectroscopy allows access to the vibrational modes of the samples which may result in bands that are assigned to specific chemical groups. In 2005, AOAC international has approved the Fourier transform infrared (FTIR) method for determination of fat, protein, lactose, and total solid contents in some dairy products [17]. Since then, studies on cheese ripening using spectroscopic methods have been published [18–23]. However, they were all carried out during a relatively short period of time, that is, less than 3 months. A better exploration of the maturation time requires a longer period of analysis, once the development of sensory attributes such as texture and flavor is directly related to the rate of breakdown of the casein matrix over time, especially in hard and semihard-type cheeses [11].

This paper deals with three ewes' cheeses under real industrial conditions and named "Pecorino," "Gouda," and "ewes' ripe." Although present in the European and North American markets, the "Pecorino" and "Sheep Gouda," the information about ewes' cheeses are in general scarce [24]. Concerning Brazil, the three cheeses are promising novelties in the market, especially the "ewes' ripe," as it is a variation of "Minas" cheese (made originally with cow's milk). To evaluate them, attenuated total reflection (ATR) unit was coupled to FTIR spectrometer to follow the changes in the cheeses over a long period (approximately one year). As a result, we show that infrared spectroscopy provides secure hints to determine optimal temporal parameters in ripening stages involving the development of the aforementioned cheeses.

2. Material and Methods

2.1. Cheese Samples. A dairy industry with Brazilian federal inspection seal (SIF), located in the municipality of Soledade de Minas, Minas Gerais, Brazil, supplied three different varieties of cheese (Pecorino, Gouda, and ewes' ripe) made with ewes' milk. The cheeses had different production dates and at aging times (1 to 12 months), so that represents real industrial process conditions.

2.2. Cheese Composition. The compositions of the supplied cheeses were carried out in terms of moisture, fat, and protein content to classify them according to the Brazilian legislation. The measurements were carried out following Lutz [25] and Brasil [26], which establish the physical and chemical official analytical methods for control of milk and dairy products. All samples were tested in duplicate.

2.3. FTIR/ATR Spectroscopy Measurements. Infrared spectroscopy was performed using a FT-MIR spectrophotometer (Bruker, model Vertex 70), with ATR accessory and OPUS 6.5 software. Infrared spectra were recorded in absorbance mode, between 4000 and 400 cm^{-1}, with a resolution of 4 cm^{-1} and 32 scans accumulated by sample. The data were analyzed in the software OriginPro 8.0, Excel, and Minitab.

Table 1: Physicochemical characterization of Pecorino (2nd month), Gouda (2nd month), and ewes' ripe (1st month) cheeses. Values obtained from duplicates and expressed as percentages ± standard deviation.

Cheese	Moisture (%)	Total protein (%)	FDM (%)
Pecorino (2 months*)	27.6 ± 0.1^a	26.886 ± 0.1^a	53.5 ± 0.4^a
Gouda (2 months*)	36.0 ± 0.0^b	19.916 ± 0.1^b	58.2 ± 0.4^b
Ewes' ripe (1 month*)	43.0 ± 0.1^c	21.936 ± 0.1^c	56.2 ± 0.0^c
p value	0.004494**	0.000001**	0.000000**

FDM: fat in dry matter. Superscript letters were used to indicate the group classification. *Ripening time. **Means without a common letter are significantly different ($p \leq 0.05$).

Cheese slices were prepared with approximately 4 cm^2 of area and thickness of 1 cm. The samples were taken from the center of the cheeses to avoid any contamination during the storage time. The measurements were performed in quadruplicate, using two pieces of each cheese.

2.4. Statistical Analysis. The differences in group mean values were submitted to analysis of variance (one-way ANOVA). Tukey HSD test was carried out when significant differences among groups were obtained. The statistical tests were performed with a significance level equals to 5% using Statistica 7.1 software. The spectra obtained between 3000 and 2800 cm^{-1}, 1766 and 1730 cm^{-1}, and 1700 and 1000 cm^{-1} were submitted to multivariate analysis using Minitab 16.1.1 software. Hierarchical cluster analysis (HCA) was performed with different linkage methods (data not shown), where the clustering using the McQuitty linkage provided the best overview of the results. Finally, the principal component analysis (PCA) was applied to the spectral sets to evaluate the evolution during ripening.

3. Results and Discussion

3.1. Physical-Chemical Characterization of the Cheeses. The earliest stage of maturation of each cheese received, that is Pecorino (2nd month), Gouda (2nd month), and ewes' ripe (1st month), was submitted to physical-chemical analysis to estimate values of moisture, total protein, and fat in dry matter (FDM) as presented in Table 1. All cheese samples were classified as "fatty" since their FDM values were between 45.0% and 59.9%. In relation to moisture, the Pecorino cheese revealed being the variety with the lowest value, "hard," while ewes' ripe and Gouda were classified as "average moisture" or "semihard" [27]. Moisture values were within the allowed range according to the local regulation [26]. Analysis of variance revealed that the cheeses presented significant difference between them, regarding the above parameters.

3.2. FTIR/ATR-MIR Spectroscopy. Figure 1 shows the absorption spectra of Pecorino (a), Gouda (b), and ewes' ripe (c)

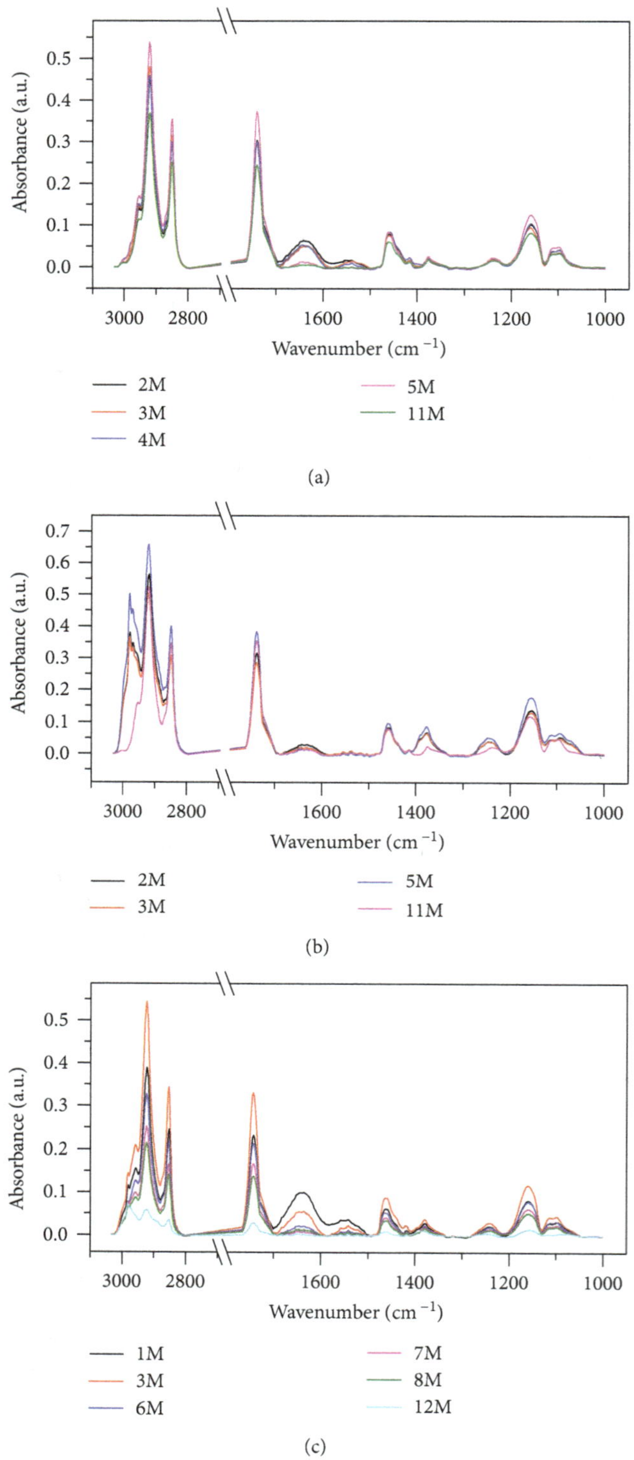

FIGURE 1: Fourier transform infrared spectra (3700 to 1000 cm^{-1}) of Pecorino (a), Gouda (b), and ewes' ripe (c) cheese samples during ripening. M: months.

cheeses over some months (M). In general, for all samples, the spectra presented bands around the same spectral positions which, nevertheless, differ in intensity a priori as well as along the time. The difference in intensity per se is related to the nature of the cheese and is weighted to different functional groups [20, 28, 29]. The differences along the time have their origins linked to the biochemical processes occurring during ripening.

The biochemistry of the cheese ripening is usually evaluated under three general subjects: proteolysis, lipolysis, and glycolysis [11, 30]. Among these processes, the first two are considered the most important for the great majority of the matured cheeses [5–10]. Therefore, the spectra were subdivided in narrow regions, to ease the interpretation of the data over time. These regions were conveniently handled by multivariate statistical analysis.

3.2.1. Amide I and II Region. The area between 1700 and 1500 cm^{-1} are related to peptide bonds (CO–NH), representing two major peaks: amide I in 1640 cm^{-1} (νC=O, νC–N) and amide II in 1550 cm^{-1} (δN–H, νC–N) [29, 31]. The behavior of these peaks during the ripening processes of the three types of cheese can be observed in Figure 2.

In general, all varieties studied have shown a decrease in the absorbance in both bands during the development of the ripening process. This indicates the occurrence of the hydrolysis of proteins by breakdown of the peptide bonds. This conclusion is supported by Fox et al. [6].

Martín-del-Campo et al. [29] have used midinfrared spectroscopy to study the ripening phenomena of two cheese zones (core and under-rind) of Camembert cheeses. Significant changes were recorded for amide I and II bands in the under-rind and only for amide II in the core. The maturation time analyzed was of 27 days where the absorbance values decreased until the tenth day and then tended to increase up to the end of the maturation. Mazerolles et al. [32] studied sixteen experimental semihard cheeses in the range of amide I and II at 1, 21, 51, and 81 days of maturation. Their results have shown a continuous decrease in the amide I band and an increase of amide II.

In Figure 2, one can observe that even though the absorbance intensities have decreased systematically over-ripening time, they do not occur in the same way among the cheese varieties analyzed and not with the same rate in the months of ripening. Regarding the Pecorino cheese (Figure 2(a)), the highest variation in absorbance levels (linked to proteolysis) appeared in the months between 4 and 5 of the maturation process. In ewes' ripe cheeses (Figure 2(c)), this fact could be verified in the months between 1 and 3, while in Gouda cheeses (Figure 2(b)), the proteolysis seemed to happen in a more constant pace, meaning that the variations among the months were similar up to the 5th month. The difference in the intensity decay of the peaks could be explained by the fact that the casein hydrolysis (proteolysis), lipid hydrolysis (lipolysis), and lactose fermentation processes are dependent on the milk composition, the fabrication process, pH, salt content, water activity, storage temperature, and specially, the type of yeast present in the cheese [8, 11, 30].

Studies evaluating the amide peaks in cheeses have stated that changes in the intensity of the amide I peak might be associated with modification in the secondary structure of casein, protein aggregation, and protein/water interaction

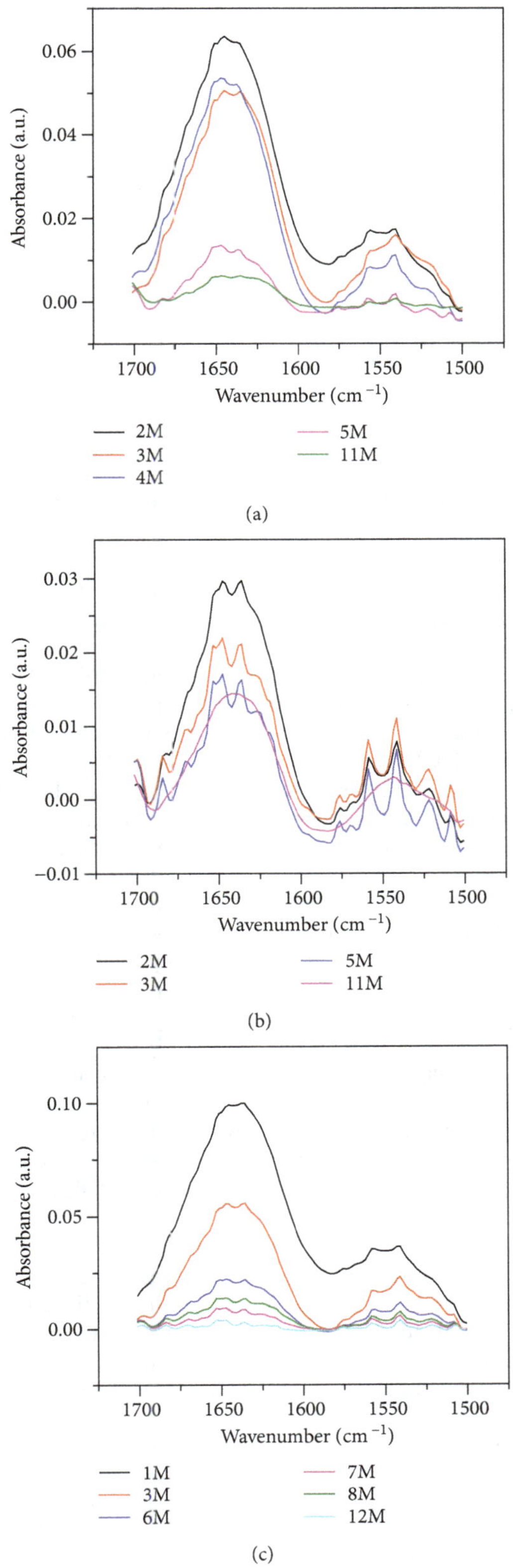

FIGURE 2: Fourier transform infrared spectra (1700 to 1500 cm^{-1}) of Pecorino (a), Gouda (b), and ewes' ripe (c) cheese samples during ripening. M: months.

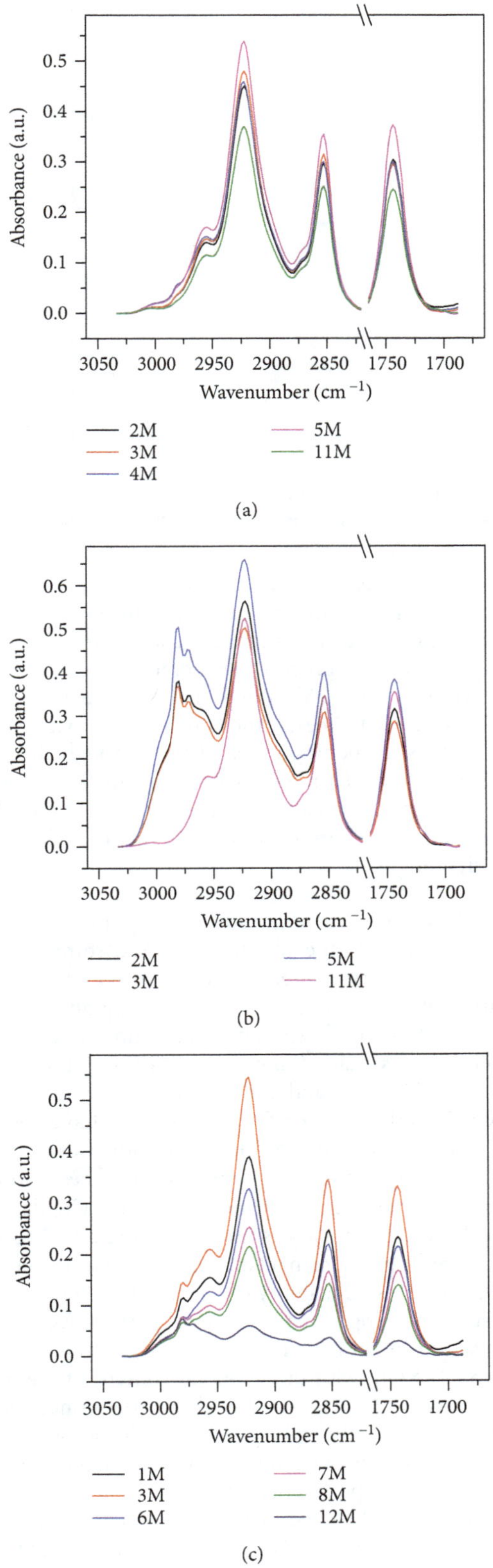

FIGURE 3: Fourier transform infrared spectra (3000 to 2800 cm^{-1} and 1765 to 1730 cm^{-1}) of Pecorino (a), Gouda (b), and ewes' ripe (c) cheese samples during ripening. M: months.

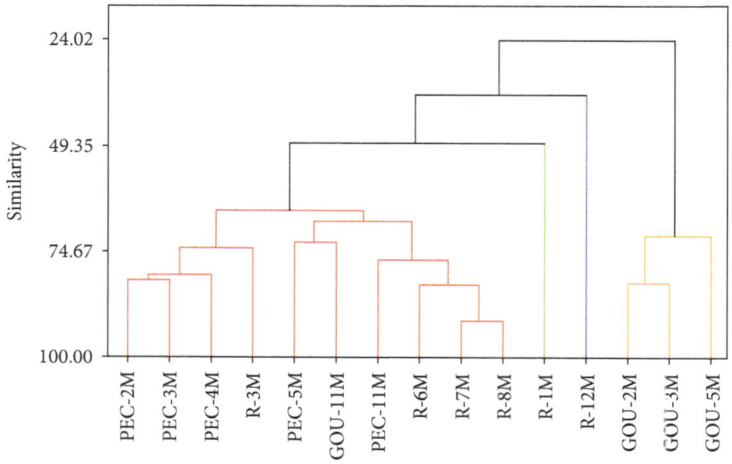

FIGURE 4: Hierarchical cluster analysis of the 15 samples of Pecorino (PEC), Gouda (GOU), and ewes' ripe (R) varieties of cheese. The ripening times are represented by its number followed by M (months).

[29, 32–34]. Therefore, since the cheeses studied here had different maturation times and were produced with different batches of milk, the factors mentioned above may have influenced the proteolysis process in different ways.

Farkye and Fox [35] stated that for the most of the semihard and hard cheeses, the proteolysis process is commonly used as a ripening index. Thus, through the behavior and evolution of the peaks obtained in the FTIR/ATR spectra, the technique can be extremely useful to verify the occurrence of the proteolysis phenomenon during the cheese ripening.

3.2.2. Lipid and Carbohydrate Regions. Figure 3 shows three peaks associated with the lipid fraction. According to Dufour et al. [28], the determination of fat is commonly used by the association of the C = O (~1750 cm^{-1}) group and acyl chain C-H (3000–2800 cm^{-1}) stretch wavenumbers. In Figure 3, in general, the signal of methylene peaks (bands around 2922 and 2852 cm^{-1}) and those associated with ester and organic acid carbonyl (νC=O in 1743 cm^{-1}) have decreased as the cheeses matured. Similar behavior was found by Martín-del-campo et al. [29]. However, the cheeses with ripening time of 5 months (Pecorino), 5 months (Gouda), and 3 months (ewes' ripe) have shown the highest absorbance bands. According to Ticiani et al. [36], at the highest lactation stage, the ewes lose the ability of keeping the fat content stable. A study developed by Brito et al. [37] showed that the ewes raised in Brazil have the peak of lactation occurring in spring (Sept–Nov), with increased fat content of the milk in approximately 2.5%. Once the above-mentioned cheeses were made during the springtime, the deviation observed in the absorbance peaks may rely in the fluctuations of fat content during the peak of the lactation.

Collins et al. [38] stated that lipolysis occurs in most or all cheeses, but its intensity is more extensive in some hard Italian varieties and in Blue cheese. In addition, regarding the ripening of Cheddar, Gouda, and Swiss cheeses, the contribution of lipolysis is relatively low. Thus, among the cheese varieties studied here, the Gouda made with ewes' milk was the one that has had the least influence of lipolysis and smaller decreases in absorption during its maturation process (Figure 3(b)).

The peaks located around 1377 cm^{-1} (δ_s CH_3) are assigned to glucose and galactose. Surrounding 1161 cm^{-1} are located peaks related to the sum of lactose (tertiary alcohol νC–OH) and monosaccharide (endocyclic oxygen νC–O) signals (Figure 1) [29]. The evolution of these peaks during the ripening process has shown a similar behavior as the lipid bands, that is, a decrease in the absorbance values.

3.3. Ripening Time Discrimination by Multivariate Analysis. The different cheeses and the correspondent ripening times were submitted to the multivariate approaches HCA and PCA to observe similarity patterns and possible correlations amongst the samples. The wavenumber range used for multivariate analysis was chosen to include a wide variety of chemical groups. The spectra were subdivided in two parts: one containing information regarding amide groups, alkanes, carbon-oxygen ester bonds (between 1700 and 1000 cm^{-1}), and another with wavenumbers characterized with absorption peaks typical of lipid sources (3000–2800 cm^{-1} and 1765–1730 cm^{-1}) [22, 39].

Figure 4 shows the resulting dendrogram of the cluster analysis that compared the different cheese types and ripening times to obtain the similarities regarding their spectra. It was possible to observe a clear separation of the early stage samples among the 3 cheese types. A small cluster containing the earliest Gouda samples was the one that revealed to be less similar. Regarding the ewes' ripe cheese, the 1st month of maturation and the 12th month were related despite their differences. The other samples of ewes' ripe cheese, in between the 1st and 12th month, were closely related with Pecorino samples and the 11th maturation month of Gouda. These samples were all grouped together in a large cluster. The similarities between the oldest Pecorino (11 months of maturation) and Gouda (11 months of maturation), with the intermediate months of the ewes' ripe variety, suggest that the different ways that the cheeses were processed did

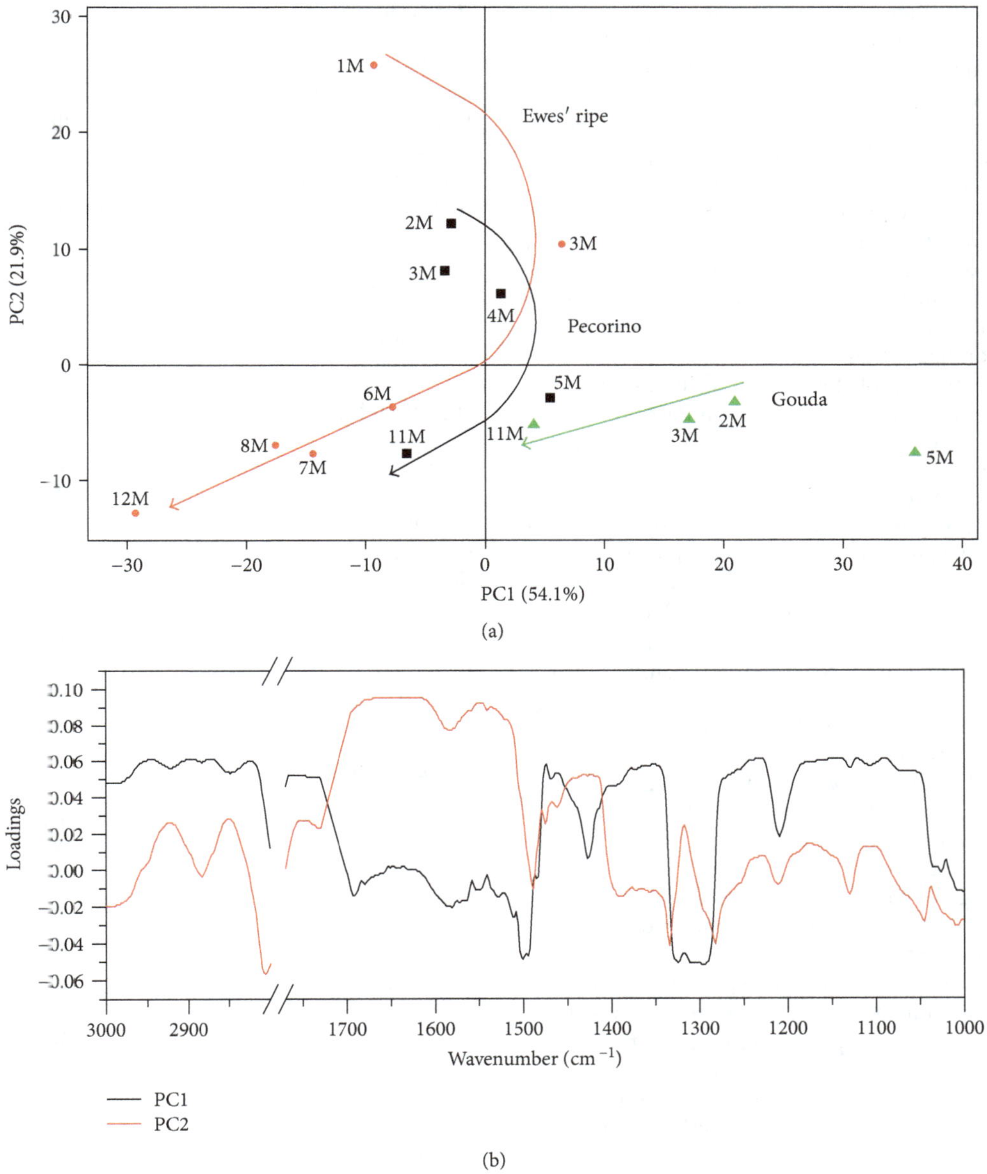

FIGURE 5: Principal component analysis of the selected wavelength spectra of the 15 samples of Pecorino (■), Gouda (▲), and ewes' ripe (●) varieties of cheese. (a) Score plot. (b) Loading plot. M: months.

not affect the physicochemical reactions driving the maturation of the cheeses.

PCA analysis resulted in PC1 and PC2 describing above 75% of total variance of the spectra (Figure 5(a)). A separation between the different cheeses and their ripening times agreed with the pattern obtained with cluster analysis. In relation to PC1 (54.1% of total variance), the main difference was regarding the Gouda and ewes' ripe cheeses. PC2 accounted for 21.9% of the variance of the data. In the loading plot (Figure 5(b)), the spectral region linked to lipids (3000–2800 cm^{-1}) and other chemical groups related to carbohydrates and to the fingerprint of the samples (1280–1000 cm^{-1}) weighted positively to PC1, while the region associated to ester groups (1330–1280 cm^{-1}) and proteins (1700–1500 cm^{-1}) was present in opposition. The PC2 on the other hand seemed to be more related to bonds linked to proteins, where more intense loadings were obtained.

Regarding Pecorino cheese, the samples of the 2nd and 3rd months of maturation were correlated to both PCs. Next to them was also located at the sample of the 4th month. The amide I and II regions were the main responsible for that proximity according to the loading plot (Figure 5(b)). The higher levels of proteolysis during the ripening of this cheese led to a decrease in number of peptide bonds (Figure 2(a)). The samples of the 5th and 11th months of Pecorino cheese were not well correlated in relation to PC1, and there was a significant difference between the earliest samples in relation to PC2. The 5th-month sample revealed high absorbance levels in the peaks assigned to lipids (Figures 3(a) and 5(a)). The altered profile observed in this sample might be related, as already mentioned, to the changes in the milk fat level during the lactation period.

The Gouda cheese samples were all well correlated in relation to the PC2, while on PC1, the most variability between them could be explained. PCA validated the spectral observations, where the protein region was not significantly different as in the other cheeses (Figures 2(b) and 5(a)). In case of the peaks linked to lipids (Figure 3(b)), the Gouda samples showed higher absorbance values compared to the other cheeses (Figures 3(a) and 3(c)). This fact corroborates with the positioning of these samples on the positive scale of PC1. The positioning of the 5th month of Gouda might be related with the same issue observed on Pecorino's.

In the ewes' ripe cheeses (Figure 5(a)), the lipolysis was the main reason for the variability of the samples. The lipid region, which was linked to PC1, showed a reduction pattern, except for the sample of the 3rd month that might have also been affected by the fat content deviation during the lactation stage (Figure 3(c)). Regarding PC2, the sample of the 1st and 3rd months, on the other hand, revealed the highest peaks in the protein region (Figure 2(c)). These wavenumbers, once again, weighted significantly to the earliest samples, just as in Pecorino cheese.

Different authors also have demonstrated discrimination between different cheeses by means of PCA [29, 32, 40]. They have found a link between ripening and absorbance changes of amide I and II bands studying the earliest stages of ripening; however, the changes observed were not as evident as the ones found in this study.

Our results agreed to the fact that proteolysis and lipolysis are the main factors driving the maturation of cheeses. A timeline could be observed in relation to both principal components, and a trendline of the ripening process was drawn in Figure 5(a). It was concluded that the proteolysis was more intense during the earliest stages up to the end of the first trimester. Thus, different cheese types and maturation times resulted in different spectral information which must be analyzed with caution during the association with ripening index of cheeses.

4. Conclusions

In this study, we have interpreted and differentiated cheeses during long maturation times by means of midinfrared spectroscopy technique (FTIR-MIR-ATR). The main factors influencing the ripening process (lipolysis and proteolysis) were stressed by multivariate analysis. The infrared spectroscopy proved to be efficient on following the changes in time occurring in the cheeses regarding proteolysis and lipolysis during the maturation period. Therefore, FTIR shows potential to assist in the quality control of dairy products. While evaluating the maturation degree, the method can be envisaged with the robustness of the chemometric tools. Implemented in dairy plants, the FTIR/ATR will allow automation of the cheese ripening process without the need of sample treatment as well as subsidize a better understanding about their shelf life.

Additional Points

Highlights. (i) Ripening of special ewes' cheeses was evaluated over 1 year via FT-MIR-ATR. (ii) Proteolysis and lipolysis were well characterized throughout ripening process. (iii) HCA and PCA of the spectra clearly separated the ripening times of the samples. (iv) A timeline relationship between major molecular changes and ripening was built

Conflicts of Interest

The authors declare that they have no conflicts of interest.

Acknowledgments

This work was funded by Coordenacão de Aperfeiçoamento de Pessoal de Nível Superior (CAPES) foundation (Ministry of Education of Brazil, Brasília, DF), Fundação de Amparo à Pesquisa de Minas Gerais (FAPEMIG) foundation, Brazil, and Conselho Nacional de Desenvolvimento Científico e Tecnológico (CNPq), Brazil. The authors also thank CAPES foundation regarding the bursaries provided to Jonathan Andrade, Cristina Guimarães Pereira, and Thamiris Ranquine.

References

[1] FAOSTAT, *Statistics Database*, Food and Agriculture Organization of the United Nations, 2014, http://www.fao.org/faostat/en/.

[2] USDA, *Dairy: World Markets and Trade, United States Department of Agriculture*, Washington DC, USA, 2016.

[3] Milkpoint, "Abiq: mercado de queijos tem alto potencial de crescimento no Brasil," 2017, January 2018, https://www.milkpoint.com.br/noticias-e-mercado/giro-noticias/abiq-mercado-de-queijos-tem-alto-potencial-de-crescimento-no-brasil-105515n.aspx.

[4] M. Bergamaschi, E. Aprea, E. Betta et al., "Effects of dairy system, herd within dairy system, and individual cow characteristics on the volatile organic compound profile of ripened model cheeses," *Journal of Dairy Science*, vol. 98, no. 4, pp. 2183–2196, 2015.

[5] P. Mcsweeney and M. Sousa, "Biochemical pathways for the production of flavour compounds in cheeses during ripening: a review," *Le Lait*, vol. 80, pp. 293–324, 2000.

[6] P. F. Fox, T. P. Guinee, T. M. Cogan, and P. L. H. McSweeney, *Fundamentals of Cheese Science*, 2000.

[7] M. J. Sousa, Y. Ardö, and P. L. H. McSweeney, "Advances in the study of proteolysis during cheese ripening," *International Dairy Journal*, vol. 11, no. 4-7, pp. 327–345, 2001.

[8] V. K. Upadhyay, P. L. H. McSweeney, A. A. A. Magboul, and P. F. Fox, "Proteolysis in cheese during ripening BT - starter cultures: general aspects," 2004, http://www.sciencedirect.com/science/article/pii/S1874558X04800769%5Cn(null).

[9] V. Fallico, P. L. H. McSweeney, K. J. Siebert, J. Horne, S. Carpino, and G. Licitra, "Chemometric analysis of proteolysis during ripening of Ragusano cheese," *Journal of Dairy Science*, vol. 87, no. 10, pp. 3138–3152, 2004.

[10] J.-R. Albani, *Fluorescence Spectroscopy in Food Analysis*, 2006.

[11] P. L. H. McSweeney, "Biochemistry of cheese ripening," *International Journal of Dairy Technology*, vol. 57, no. 2-3, pp. 127–144, 2004.

[12] W. W. G. Nascimento, M. P. F. de Souza, A. C. M. M. Valente, V. C. dos Anjos, M. A. M. Furtado, and M. J. V. Bell, "Results from portable and of low cost equipment developed for detection of milk adulterations," *Food Science and Technology*, vol. 37, pp. 38–41, 2017.

[13] M. P. Brandao, M. Gouvea Neto, V. C. dos Anjos, and M. J. V. Bell, "Detection of adulteration of goat milk powder with bovine milk powder by front-face and time resolved fluorescence," *Food Control*, vol. 81, pp. 168–172, 2017.

[14] M. P. Brandao, V. de Carvalho dos Anjos, and M. J. V. Bell, "Time resolved fluorescence of cow and goat milk powder," *Spectrochimica Acta Part A: Molecular and Biomolecular Spectroscopy*, vol. 171, pp. 193–199, 2017.

[15] R. Alves da Rocha, I. M. Paiva, V. Anjos, M. A. M. Furtado, and M. J. V. Bell, "Quantification of whey in fluid milk using confocal Raman microscopy and artificial neural network," *Journal of Dairy Science*, vol. 98, no. 6, pp. 3559–3567, 2015.

[16] M. P. Brandao, A. P. Carmo, M. J. V. Bell, and V. C. Anjos, "Characterization of milk by infrared spectroscopy," *Revista do Instituto de Laticínios Cândido Tostes*, vol. 373, pp. 30–33, 2010.

[17] W. Horwitz and G. W. Latimer, *Official Methods of Analysis of AOAC International*, Association of Official Analytical Chemistry International, Gaithersburg, MD, USA, 18th edition, 2005.

[18] I. González-Martín, J. M. Hernández-Hierro, R. Morón-Sancho, J. Salvador-Esteban, A. Vivar-Quintana, and I. Revilla, "Determination of the percentage of milk (cow's, ewe's and goat's) in cheeses with different ripening times using near infrared spectroscopy technology and a remote reflectance fibre-optic probe," *Analytica Chimica Acta*, vol. 604, no. 2, pp. 191–196, 2007.

[19] I. González-Martín, J. M. Hernández-Hierro, A. Vivar-Quintana, I. Revilla, and C. González-Pérez, "The application of near infrared spectroscopy technology and a remote reflectance fibre-optic probe for the determination of peptides in cheeses (cow's, ewe's and goat's) with different ripening times," *Food Chemistry*, vol. 114, no. 4, pp. 1564–1569, 2009.

[20] M. J. Lerma-García, A. Gori, L. Cerretani, E. F. Simó-Alfonso, and M. F. Caboni, "Classification of Pecorino cheeses produced in Italy according to their ripening time and manufacturing technique using Fourier transform infrared spectroscopy," *Journal of Dairy Science*, vol. 93, no. 10, pp. 4490–4496, 2010.

[21] H. Kraggerud, T. Næs, and R. K. Abrahamsen, "Prediction of sensory quality of cheese during ripening from chemical and spectroscopy measurements," *International Dairy Journal*, vol. 34, no. 1, pp. 6–18, 2014.

[22] L. E. Rodriguez-Saona, N. Koca, W. J. Harper, and V. B. Alvarez, "Rapid determination of Swiss cheese composition by Fourier transform infrared/attenuated total reflectance spectroscopy," *Journal of Dairy Science*, vol. 89, no. 5, pp. 1407–1412, 2006.

[23] A. Subramanian, V. B. Alvarez, W. J. Harper, and L. E. Rodriguez-Saona, "Monitoring amino acids, organic acids, and ripening changes in Cheddar cheese using Fourier-transform infrared spectroscopy," *International Dairy Journal*, vol. 21, no. 6, pp. 434–440, 2011.

[24] J. O. A. do Amarante, *Queijos do Brasil e do Mundo: para iniciantes e apreciadores*, Mescla Editorial, São Paulo, 2015.

[25] I. A. Lutz, *Métodos Físico-Químicos para Análise de Alimentos*, 4th edition4th edition, , 2004.

[26] *BRASIL, Instrução Normativa no 68 de 12/12/2006. Oficializa os Métodos Analíticos Oficiais Físico-Químicos, para Controle de Leite e Produtos Lácteos*, Diário Oficial da República Federativa do Brasil, 2006.

[27] *BRASIL, Portaria No. 146 de 07 de março de 1996. Aprovar os Regulamentos Técnicos de Identidade e Qualidade dos Produtos Lácteos*, Diário Oficial da República Federativa Do Brasil, 1996.

[28] E. Dufour, G. Mazerolles, M. F. Devaux, G. Duboz, M. H. Duployer, and N. Mouhous Riou, "Phase transition of triglycerides during semi-hard cheese ripening," *International Dairy Journal*, vol. 10, no. 1-2, pp. 81–93, 2000.

[29] S. T. Martín-del-Campo, D. Picque, R. Cosío-Ramírez, and G. Corrieu, "Middle infrared spectroscopy characterization of ripening stages of Camembert-type cheeses," *International Dairy Journal*, vol. 17, no. 7, pp. 835–845, 2007.

[30] P. L. H. McSweeney, "Biochemistry of cheese ripening: introduction and overview," *Cheese: Chemistry, Physics and Microbiology*, vol. 1, pp. 347–360, 2004.

[31] A. Subramanian and L. Rodriguez-Saona, "Chapter 7 – fourier transform infrared (FTIR) spectroscopy," in *Infrared Spectroscopy for Food Quality Analysis and Control*, D.-W. Sun, Ed., pp. 145–178, Academic Press, San Diego, CA, USA, 2009.

[32] G. Mazerolles, M.-F. Devaux, G. Duboz, M.-H. Duployer, N. Mouhous Riou, and E. Dufour, "Infrared and fluorescence spectroscopy for monitoring protein structure and interaction changes during cheese ripening," *Le Lait*, vol. 81, no. 4, pp. 509–527, 2001.

[33] A. Kulmyrzaev, É. Dufour, Y. Noël et al., "Investigation at the molecular level of soft cheese quality and ripening by infrared and fluorescence spectroscopies and chemometrics - relationships with rheology properties," *International Dairy Journal*, vol. 15, no. 6-9, pp. 669–678, 2005.

[34] L. Vannini, D. Baldi, and R. Lanciotti, "Use of Fourier transform infrared spectroscopy to evaluate the proteolytic activity of *Yarrowia lipolytica* and its contribution to cheese ripening," *International Journal of Food Microbiology*, vol. 69, no. 1-2, pp. 113–123, 2001.

[35] N. Y. Farkye and P. F. Fox, "Objective indices of cheese ripening," *Trends in Food Science & Technology*, vol. 1, pp. 37–40, 1990.

[36] E. Ticiani, E. C. Sandri, J. De Souza, F. Batistel, and D. E. De Oliveira, "Persistência da lactação e composição do leite em ovelhas leiteiras das raças Lacaune e East Friesian," *Ciência Rural*, vol. 43, no. 9, pp. 1650–1653, 2013.

[37] M. A. Brito, F. D. González, L. A. Ribeiro et al., "Composição do sangue e do leite em ovinos leiteiros do sul do Brasil: variações na gestação e na lactação," *Ciência Rural*, vol. 36, no. 3, pp. 942–948, 2006.

[38] Y. F. Collins, P. L. H. McSweeney, and M. G. Wilkinson, "Lipolysis and free fatty acid catabolism in cheese: a review of current knowledge," *International Dairy Journal*, vol. 13, no. 11, pp. 841–866, 2003.

[39] D. A. Skoog, F. J. Holler, and S. R. Crouch, *Principles of Instrumental Analysis*, vol. 152Thompson Brooks/Cole, 6th edition, 2006.

12

Spectroscopic Characteristics of Treated-Color Natural Diamonds

Meili Wang,[1] Guanghai Shi,[1] Joe C. C. Yuan,[2] Wen Han,[3] and Qing Bai[1]

[1]*State Key Laboratory of Geological Processes and Mineral Resources, China University of Geosciences, Beijing, China*
[2]*Taidiam Technology (Zhengzhou) Co. Ltd., Zhengzhou, China*
[3]*National Gems & Jewelry Technology Administrative Center, Beijing, China*

Correspondence should be addressed to Guanghai Shi; shigh@cugb.edu.cn

Academic Editor: Vincenza Crupi

With the increasing availability of treated-color diamonds on the market, their characterization is becoming more and more critical to the jewelry testers and customers. In this investigation, ten color diamonds treated by irradiation (4 pieces), HPHT (3 pieces), and multiprocess (3 pieces) were examined by spectroscopic methods. These diamonds are classified to be type Ia according to their FTIR characteristics. Using microscope and DiamondView, the internal features (such as distinctive color zoning and graphitized inclusions) and complex natural growth structures were observed, which show that the samples are more likely artificially colored natural diamonds. Through photoluminescence spectroscopy, a combination of optical centers was detected, including N-V^0 at 575 nm, N-V$^-$ at 637 nm, H3 at 503 nm, H2 at 986 nm, and GR1 at 741 and 744 nm. Combining with the previous studies, treatment conditions for the studied diamonds were estimated depending on the presence and/or absence of the optical centers. In addition, the coloration mechanism of the samples (blue, green, and red) during the treatment process was also discussed. It is suggested that a number of techniques should be combined in order to make a reliable identification for such diamonds.

1. Introduction

Diamonds, one of the most precious gemstones, have been very popular on the jewelry market, especially the fancy color gem-quality diamonds. However, most natural diamonds show brown hue in color and have some internal or external defects, and only a few of them are suitable for using as gemstones [1]. Therefore, considerable scientific work has been continuously done to enhance the color of diamonds so as to make them exhibiting fancy color, in order to offset the scarcity of natural diamonds and to satisfy the demand for color gem-quality diamonds of the jewelry market.

Generally, the gemologists have begun to enhance the color of diamond since the 1930s [2] and diamond treatments in the modern era mostly involve high-energy irradiation, high pressure and high temperature (HPHT) treatment, low pressure and high temperature (LPHT) annealing, and combination of these treatments [3–11]. Such as Vins [12] reported that natural diamonds with brown hue could be altered to display purple to red color, through multiprocess treatments including HPHT treatment (6-7 GPa at over 2150°C), high-energy irradiation, and LPHT annealing (lower than 1100°C).

With the increasing availability of the treated-color diamonds in the jewelry market, characterization of them becomes more and more critical to the gem testers and customers. However, by far, there are not enough works for systematically studying the treated-color diamonds and most of them are focused on examining the diamonds treated by the same method or that from the same company [13–15]. This study attempts to characterize the natural diamonds treated by different process (irradiation, HPHT, and multiprocess) using spectroscopic methods, to present the coloration mechanism of them during treatment, and to provide more data of the identification and the treatment method for the gemologists and the jewelry market.

2. Materials and Methods

Ten treated-color natural diamonds were collected for this investigation (4 irradiation-treated, 3 HPHT-treated, and 3 multiprocess-treated diamonds, see Table 1 and Figure 1).

TABLE 1: Description of the treated-color natural diamonds used in this study.

	Sample number	Color	Weight (ct)	Clarity	Fluorescence to long-wave UV	Fluorescence to short-wave UV
Treated-irradiated	TR-1	Blue	0.4	SI_2	Very weak blue, chalky	Invert
	TR-2	Dark green	0.24	SI_1	Moderate green, chalky	Invert
	TR-3	Blue	0.24	VVS_2	Invert	Invert
	TR-4	Dark green	0.36	VS_1	Invert	Invert
Treated-HPHT	TR-5	Yellowish green	0.38	VS_1	Greenish yellow	Moderate greenish yellow
	TR-6	Yellow	0.42	SI_1	Weak blue, chalky	Invert
	TR-9	Greenish yellow	0.29	SI_2	Blue, chalky	Greenish yellow
Treated-multiprocess	TR-7	Dark purplish pink	0.39	I_1	Pink, chalky	Moderate orange
	TR-10	Dark purplish pink	0.35	SI_2	Weak pink, chalky	Weak orange
	TR-11	Slight dark red	0.30	SI_1	Very weak blue	Invert

FIGURE 1: These faceted diamonds are treated by irradiation, HPHT, and multiprocess treatments, respectively, and they range in size from 0.24 to 0.42 ct.

All the faceted samples are round brilliants and range in size from 0.24 to 0.42 ct. Using a standard gemological binocular microscope, the internal features of all diamonds were examined. The reactions to ultraviolet (UV) light were checked with a conventional four-watt combination long-wave (365 nm) and short-wave (254 nm) lamp (Table 1). In addition, we recorded the fluorescence and phosphorescence images by the Diamond Trading Company (DTC) DiamondView instrument (illumination wavelengths < 230 nm, Welbourn et al. [16]) in the laboratory of National Gems & Jewelry Technology Administrative Center (NGTC) in Beijing, and which could also reveal the internal growth features of the samples.

Two other spectroscopic analyses were conducted on most of the samples at NGTC laboratory. Infrared absorption spectra in the midinfrared (4000–400 cm^{-1}, 2 cm^{-1} resolution) were measured by a Thermo Nicolet Nexus iN10 Fourier-transform infrared (FTIR) spectrometer, furnished with KBr and quartz beam splitters. A total of 64 scans were collected for each sample. A Renishaw InVia Raman confocal microspectrometer was used to acquire the low temperature photoluminescence (PL) spectra, with the samples cooled by submerging in liquid nitrogen. To activate various types of defects in these treated-color natural diamonds, three different excitation wavelengths (473, 532, and 785 nm) were employed.

3. Results

3.1. Microscopic Characteristics. Viewing with the binocular gemological microscope, natural mineral inclusion and internal growth graining were observed in some of the treated-color diamonds, which indicated that they had grown naturally, as shown in Figures 2(a) and 2(b). Some of them display color zoning, as the most characteristic feature of the treated diamonds (Figure 2(c)). In addition, other evidences of treatment were also noted, such as one diamond exhibited etched facet near to the girdle which was not removed during polishing (Figure 2(d)) and some graphitized inclusions formed during the HPHT treatment were also shown (Figure 2(e)). Besides, under magnification with crossed polarizer, anomalous birefringence with high interference colors caused by residual internal strain was observed in all treated diamonds (Figure 2(f)).

3.2. DiamondView Imaging. When exposed to the high-intensity ultrashort wavelengths of the DTC DiamondView, the treated-color diamonds exhibited strong green, blue, orange, and orangey red fluorescence (Figure 3). In general, one sample could display two or more different fluorescence color. For example, numerous red, orangey red, blue, and slight green fluorescence lines were simultaneously present in the TR-10 sample (Figure 3(h)). The TR-6 sample with blue fluorescence displayed some irregular green regions as shown in Figure 3(e). In addition, the luminescence images of the examined diamonds revealed complex growth patterns, which is a typical reflection of complex growth history and also the natural origin.

3.3. Infrared Absorption Spectroscopy. As shown in Figures 4(a) and 4(b), all the irradiation-treated diamonds are assigned to be type IaAB according to their infrared spectra. The absorption from the H1a defect (the <001> dinitrogen split interstitial N_{2I}; [17]) at 1450 cm^{-1} is observed in all the irradiated diamonds, which points out that they have been irradiated or irradiated with subsequently annealing at relatively low temperatures (~300°C; Clark et al. [18]). The absence of band of 2800 cm^{-1} in the blue diamonds is another evidence for the existence of irradiation. Absorption peaks located at near 1356~1380 cm^{-1} are related to C-C platelets, which represent for the layers of interstitial carbon atoms generated as a byproduct of nitrogen aggregation [19]. In addition, the intensity of band of 1405 cm^{-1} is found to be positive related to that of 3107 cm^{-1}, indicating that both of them may originate from the same VN_3H defect [20]. The peak at 2785 cm^{-1} was reported to be produced by the vibrations of 1405 and 3107 cm^{-1} [21]. Besides, the double peak at ~2926 and ~2854 cm^{-1}, attributing to C-H stretch [22], is observed in all the samples. Unknown weak or sharp bands of the two dark green diamonds are also detected at 3254, 3236, 3069, 3056, 2812, 2785, 1553,1524, 1528, 1499, and 1435 cm^{-1}, and we speculate that some of them may arise from the hydrogen in diamonds.

The FTIR spectra of the HPHT-treated samples display typical features of type Ia, with various aggregation states of nitrogen. As represented in Figures 4(c) and 4(d), the TR-6 sample shows stronger A-aggregate absorption while the other two show stronger B-aggregate features. The occurrence of 1331 cm^{-1}, associated with the positive charge state of the single substitutional nitrogen defect (N^+_s), may indicate that the diamonds have been treated by HPHT process which can create isolated nitrogen [23]. Additionally, the two TR-5 and TR-9 samples also exhibit absorption related to hydrogen impurities at 3107 cm^{-1}.

The multiprocess-treated diamonds are proved to be type IaAB (Figures 4(e) and 4(f)), where the concentration of A-aggregates is stronger than that of B-aggregates. Weak absorptions at 1344 and 1331 cm^{-1} attributed to isolated nitrogen impurities are recorded in all the samples. In addition, the H1a defect due to irradiation is observed in this group of samples. Other weak and broad features also occurred in some of them, including peaks at 3107, 1502, and 1405 cm^{-1}, where the 1502 cm^{-1} center was suggested to be related to the H1a center according to their positive correlation in intensity [24].

3.4. Photoluminescence Spectroscopy. Low temperature photoluminescence spectra were collected for most of the samples using three laser excitations with wavelengths of 473, 532, and 785 nm. The major PL emission lines are illustrated below based on each laser excitation.

As illustrated in Figures 5(a)–5(c), when the 473 nm laser excitation is employed, all the treated diamonds display the H3 (503.2 nm) defect, with side bands between ~512 and ~553 nm, and with various intensity between different samples. A faint emission line at 496 nm, assigned to be H4 center, occurs in the irradiation-treated and multiprocess-treated diamonds. At the same time, a set of emissions at 484, 486, and 491 nm are present in some of the samples in these two groups; however, it is unknown whether they are related with the H4 defect or not. Besides, the 3H defect (a radiation center of diamond [22, 25]) is not found in the irradiation-treated diamonds, which needs further exploration. Nitrogen-vacancy centers ($NV^0/^-$), with zero-phonon lines at 575 and 637 nm, are obvious in the multiprocess-treated diamonds. However, they are absent in the HPHT-treated diamonds and only one of the irradiation-treated diamonds shows very weak NV^0 center. In addition, one spectrum of the HPHT-treated diamonds displays an intense peak at 478.9 nm (not shown), which was suggested to be an indicator of possible HPHT treatment [26]. The 588 nm emission, considered as an evidence of multiprocess treatment [26], is observed in the multiprocess-treated diamonds. These spectra also reveal peaks at 634.4 nm in the HPHT-treated diamonds, at 489.1, 558.1, 613.5, 647.1, and 657.9 nm in the irradiation-treated diamonds, and at 600.2 nm in the multiprocess-treated diamonds.

The PL spectra taken with 532 nm laser excitation reveal sharp NV center emission systems in most of the treated samples, with the intensity of the 637 nm ZPL more intense than that of 575 nm ZPL. Another band is detected at ~658 nm in most of the samples, which was suggested to be created at high temperatures and then destructed by HPHT treatment [27]. Strong doublet emissions at 741.4 and 744.7 nm, attributed to GR1 center, are observed in the irradiation-treated diamonds. In addition, numerous emission lines at 536.0, 539.7, 542.9, 551.0, 558.2, 588.1, 563.6, 603.5, 609.3, 612.5, 640.0, 647.3, ~680, ~715, ~723, ~757, ~787, and ~800 nm are detected in the spectra of the irradiation-treated diamond. Other features are also recorded

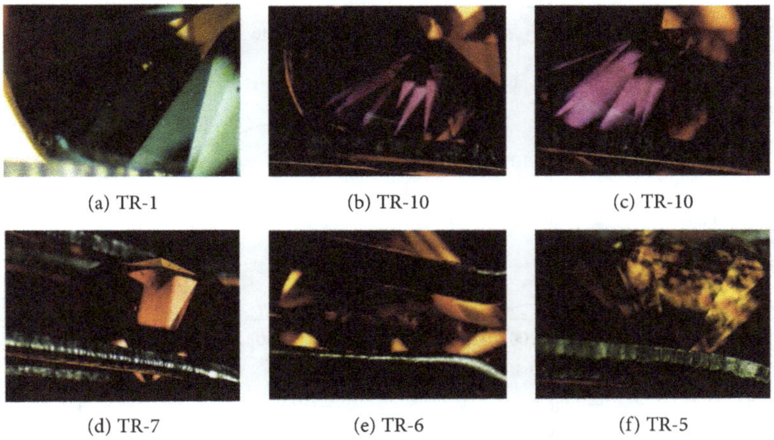

FIGURE 2: A broad range of internal characteristics were noted with magnification, including natural mineral inclusions and some graphitized inclusions, as well as color zoning and etched facet near to the girdle. In addition, anomalous birefringence with high interference colors was also observed.

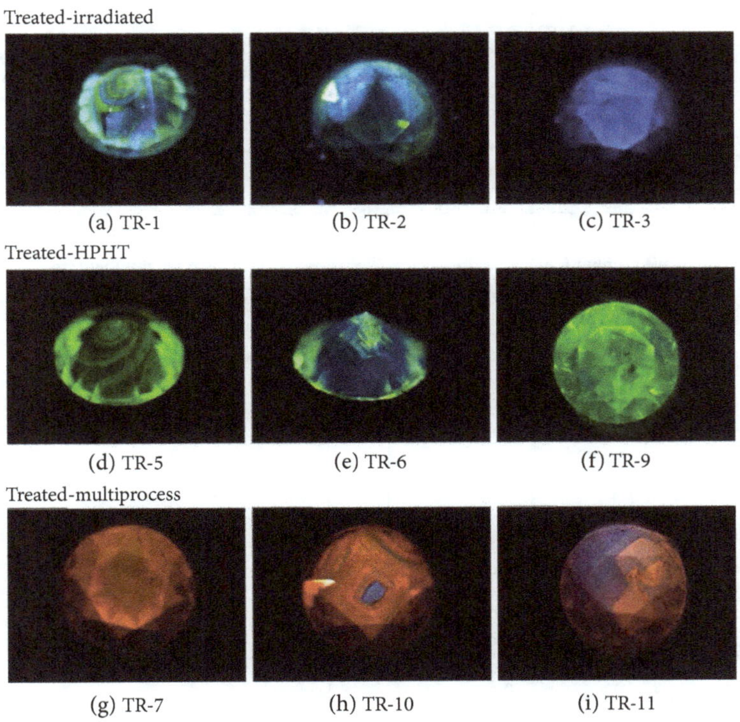

FIGURE 3: DiamondView images showed complex growth patterns of the treated-color diamonds, which revealed the natural origin. The dominant luminescence was strong green, blue, orange, and orangey red, and while one sample usually displayed two or more different fluorescence.

in the HPHT-treated diamond, including peaks at 536.1, 539.8, 555.3, 565.3, 566.2, and 601 nm. The defects responsible for most of these emission lines have not been identified.

Infrared laser excitation at 785 nm produces two emission features (876.4 nm, the first order diamond Raman line, and 986.2 nm, the H2 center) in the HPHT-treated diamonds. We also observe another weak peak at 959.7 nm in one of these samples. In the irradiation-treated diamonds, many weak and sharp peaks are recorded, including those at 795.3, 799.2, 802.5, 805.4, 806.4, 839.7, 890.5, 900.6, 926.3, 936.7, 948.6, 960.3, and 986.4 nm. The spectra of the multiprocess-treated diamonds reveal several features at 789.9, 805.1, 815.9,

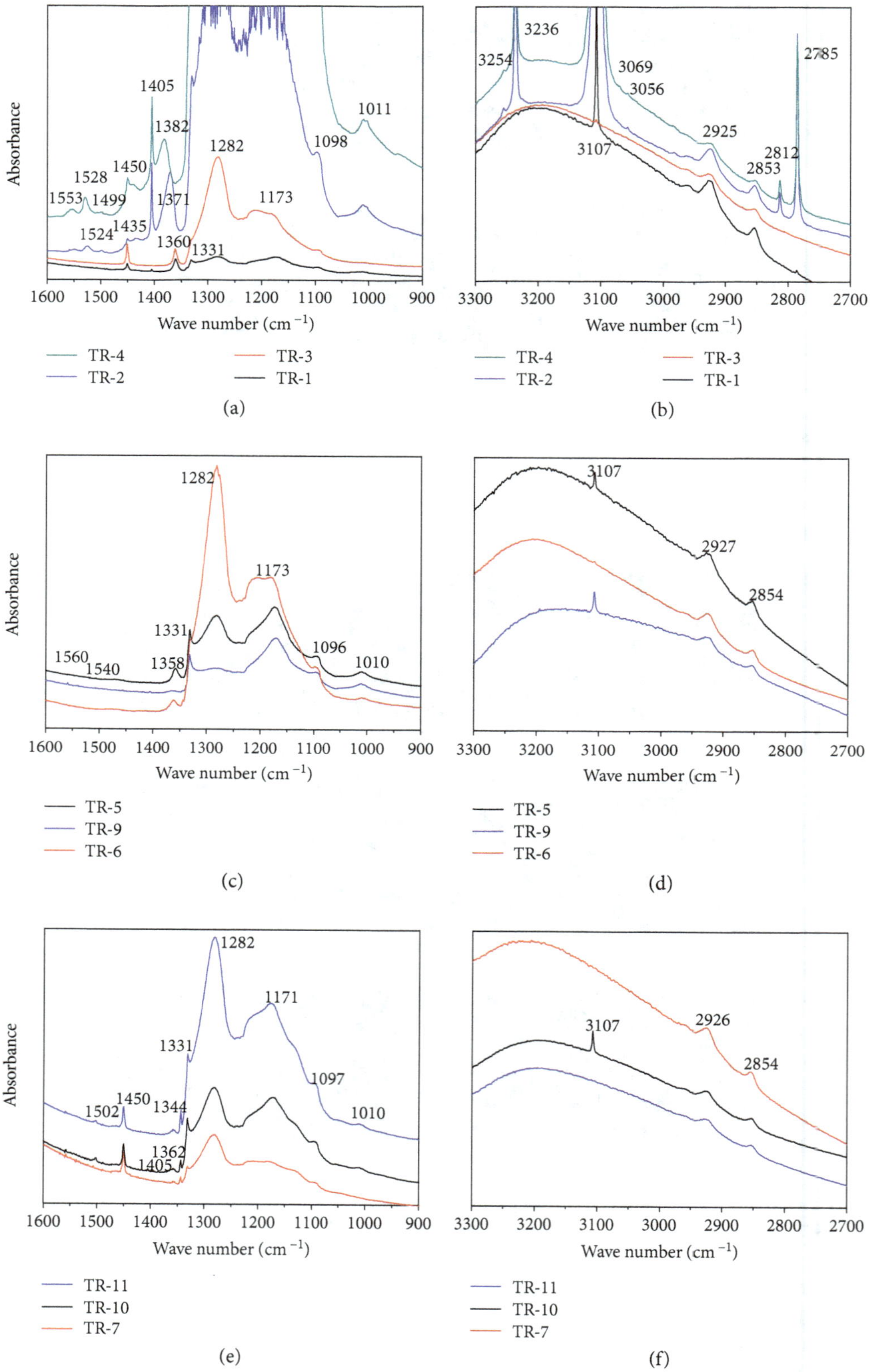

FIGURE 4: The FTIR spectra of the treated-color diamond collected in the midinfrared range reveal typical features of type Ia diamonds, with various aggregation states of nitrogen. The H1a absorption at 1450 cm $^{-1}$ and the isolated nitrogen-related peaks (1344 and 1331 cm $^{-1}$) were also observed in some of them.

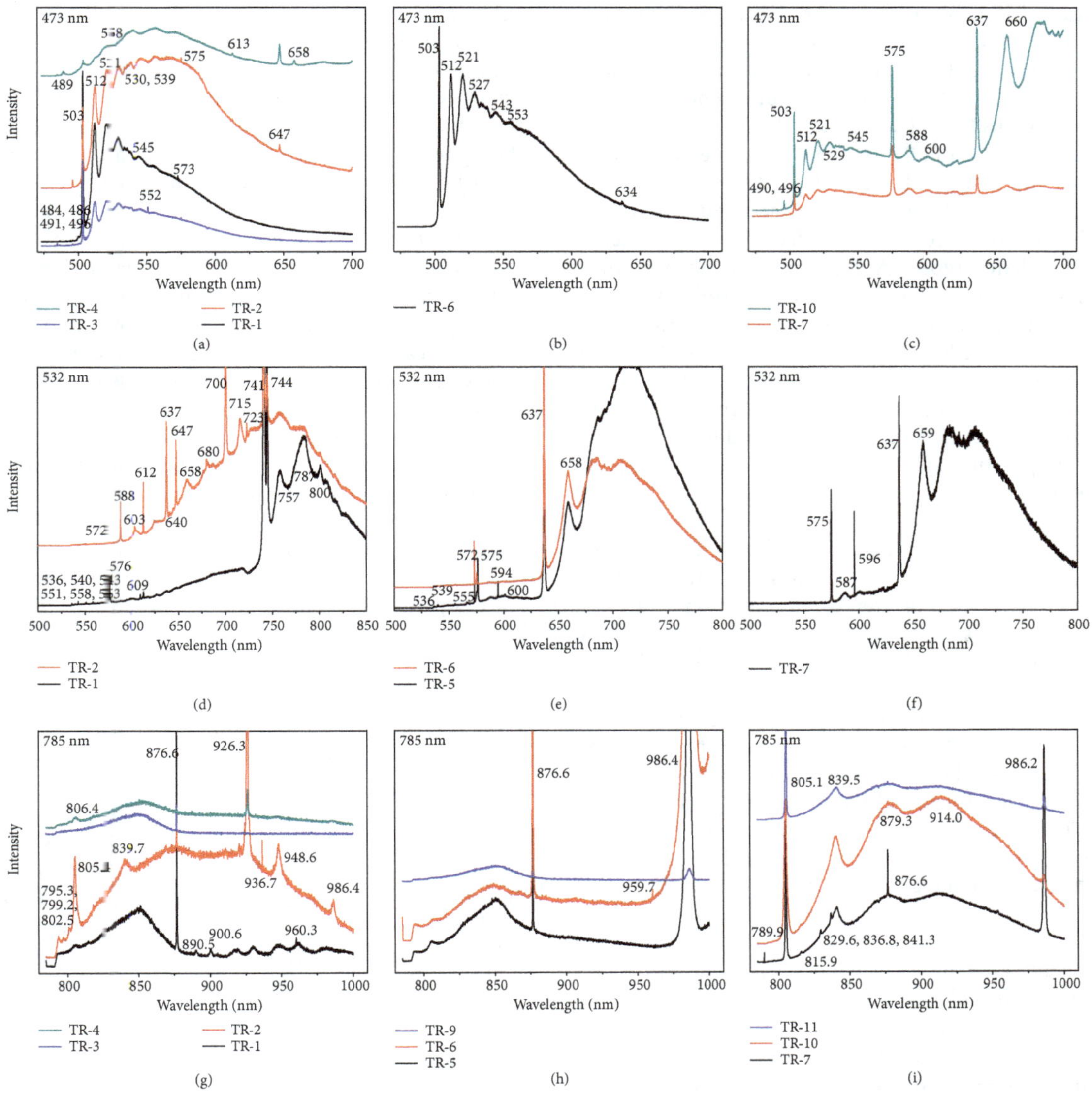

Figure 5: Photoluminescence spectra were taken at liquid-nitrogen temperatures for most of the samples using three laser excitation wavelengths of 473, 532, and 785 nm. Strong H3 center, NV center, GR1 center, and H2 center were observed in some of these treated-color diamonds. The presence and/or absence of these centers have an indicative role to determine the type of treatment and the process of the treatment.

829.6, 836.8, 839.5, 841.3, and 986.2 nm. Unfortunately, the assignments for most of them are not sure.

4. Discussions

4.1. Spectroscopic Features and Related Treatment. Usually, the condition of treatment correlates with the presence and/or absence of the optical centers, and thus analysis of the optical centers may contribute to reveal the condition of treatment. The variations of some optical centers with different treatment processes and conditions in previous works are reviewed as followed.

4.1.1. H3 and H2 Centers. The H3 (503.2 nm) and H2 (986.2 nm) centers are related to the nitrogen-vacancy-nitrogen (NVN) defect in neutral and negative charge state, respectively. Commonly, the H3 center has been observed in diamonds involving with irradiation and high temperature annealing (or HPHT annealing) [28–30] or been directly created in plastically deformed brown diamonds through heating at high temperature [31, 32], while various concentrations of the H3 defect can also occur in some natural diamonds [33–36]. In nitrogen-containing diamonds, high-energy irradiation followed by annealing at temperatures over 500°C could result in the formation of H3 center [26]. Collins [37] reported that the H3 center had maximum intensity at approximately 800°C and annealed out at temperatures above 1500°C in irradiated diamonds. In terms of the HPHT annealing treatment for the irradiated type Ia diamonds, the H3 absorption just began to decrease after 1650°C

[28] or reduced largely or disappeared after high pressure annealing at 2300°C [29]. While in most of the HPHT-treated diamonds, the H3 center may form in brown plastically deformed type IaA and IaB diamonds when annealed at a temperature 1800°C at 7.0 GPa and then its concentration increased up to about 2150°C [32, 38] or decreased considerably at temperatures above 2100°C [39, 40].

The H2 center, with negative charged state, is usually observed in diamonds containing defects which can act as donors. In irradiated type IaA diamonds, the absorption of the H2 center started to increase after annealing at around 1250°C and reached a maximum at 1600°C or disappeared when annealed at 2300°C and high pressure [28, 29]. Through HPHT annealing with temperatures above 1800/1900°C, in type IaA/IaB diamond with plastic deformation, the absorption of the H2 center appeared and increased with increasing temperature and then began to decrease at 2150/2200°C [38, 39].

4.1.2. GR1 Center. The GR1 center is attributed to the neutral isolated vacancy (V^o), located at 741 and 744 nm. Weak line at 741 nm has been found in a few natural diamonds [41, 42]. Through high-energy irradiation (electrons, gamma-rays, neutrons, and ions), the GR1 center could be artificially created in most of diamonds [22, 26]. Kiflawi et al. [43] reported that there was an initial reduction in the strength of the GR1 absorption at 390–410°C followed by sharp decrease beyond 700°C in the irradiated diamonds. In addition, the GR1 center is completely destroyed in high-nitrogen diamonds when heated at 800°C for 1 h, whereas heating above 900°C was required in low-nitrogen specimens [28, 44]. The GR1 center was absent in irradiated diamonds upon HTHT treatment at 1000°C [26].

4.1.3. NV Center. One of the nitrogen-vacancy defects locates at 575 nm in neutral charge state (NV^0), and the other one locates at 637 nm in negative charge state (NV^-). In irradiated diamonds, the 575 nm center annealed out at approximately 1400°C; however, in some cases, it was still stable at temperatures above 1600°C and it could survive to about 2000°C when the diamonds were heated under high pressure [22, 25]. Vins and Eliseev [38] suggested that the A defects began to dissociate and some C defects (isolated nitrogen impurities) were formed in brown type IaA diamond after HPHT annealing at 2100°C, and a weak absorption of the NV defects simultaneously appeared.

The 637 nm center in electron irradiation type Ia diamonds began to develop after annealing at 800°C [5] and may anneal out at about 1500°C [25]. In addition, when the irradiated type Ia diamonds are subjected to 2300°C at 5.0 GPa, the 637 nm absorption is extremely weak or undetectable [29]. In type Ia brown diamonds, the NV^- center was produced by HPHT treatment at 2000°C and above [4, 34] and could be rather strong through a few minutes of HPHT treatment at a temperature of 2300°C [45].

4.1.4. Other Features. The 3107 cm^{-1} center is attributed to H-related vibration and is assigned to be mode of VN_3H [20, 46]. In general, the behavior of the 3107 cm^{-1} during treatment is complex. After electron irradiation and subsequent prolonged annealing at 1200°C for 24 h, the intensity of the 3107 cm^{-1} absorption in type Ia diamonds can be considerably reduced or even disappeared [47]. The study of De Weerdt and Kupriyanov [48] showed that the 3107 cm^{-1} absorption was stable in type IaA/B diamonds when annealing at 1900°C and 7.0 GPa. As summarized by Dobrinets et al. [26] and Vins and Yelisseyev [39], HPHT treatment could strongly reduce the intensity of the 3107 cm^{-1} in 80% of cases. The C center is related to the isolated nitrogen, located at 1344 and 1131 cm^{-1}. Usually, the presence of C center in the FTIR spectrum of type IaAB is an indication of possible HPHT treatment. After HPHT treatment performed at 2000°C and above, the C defects with detectable concentration appeared [38, 45, 49].

Combining the above data with the presence and/or absence of the optical centers in the diamonds of this study, it can be inferred that the irradiation-treated diamonds were irradiated without annealing or the annealing temperature was very low, the minimum annealing temperature was ~2000°C for the HPHT-treated diamonds and the annealing temperature for the multiprocess-treated diamond was between 2000 and 2300°C.

4.2. Cause of Color

4.2.1. Blue. In natural diamonds, the blue color is rare. The type IIb diamonds owe their blue color to the absorption of the boron acceptors [22, 50, 51]. Some natural type Ia diamonds can also show the blue-gray color, which is caused by the high concentration of hydrogen [52, 53]. Moreover, the GR1 center produced by irradiation is another origin of the blue color. A few diamonds are naturally irradiated to display blue color [22], and a similar color can also be induced by artificial irradiation [28, 54]. The blue diamonds in this study were artificially irradiated, which give rise to strong GR1 center.

4.2.2. Yellow to Green. In natural diamonds, the yellow color is usually caused by the absorption of the N3 and N2 optical centers ("cape-yellow") and/or the C defects (canary-yellow). The green hue of some natural diamonds is due to the GR1 center produced by natural irradiation [26, 54]. HPHT treatment is a common method to induce yellow/green color to the diamonds. After HPHT annealing at 1800°C, a yellow hue due to the C center was created in brown type Ia diamonds [49, 55]. With increasing temperature, the H2 center—additional green color modifier—was produced [25, 26, 56]. In general, the resulting green/yellow color is a combination of the absorptions of H3, H2, and C defect in the spectrum and also the same as the HPHT-treated diamonds in this study. Strong GR1 center created by artificial irradiation can also give rise to dark green color, such as the sample TR-2 and TR-4.

4.2.3. Red. Pure red diamond is extremely rare in nature, and some pink diamonds with moderate brightness and high saturation can display red color [54]. After multiprocess treatment, the diamonds could also show red color. Usually, the type IaAB natural diamonds are chosen to be treated to induce the red color [9, 12]: firstly, HPHT annealing is performed at over 2150°C to achieve high concentration of C center and then the diamonds are irradiated with 2–4 MeV energy to create enough isolated vacancies; finally, the diamonds are annealing at a temperature exceeding 1100°C in order to combine the C defects with the vacancies into NV centers; after these, the purple to red color could be obtained. The main reason for the red color is the absorption of N-V centers at 575 and 637 nm, with some influence from other absorption features [15, 26, 35]. It is obvious that the red color of the diamonds in this study has been induced by a similar process, because the N-V centers at 575 and 637 nm were also observed in these multiprocess-treated diamonds.

4.3. Identification Features. Generally, a number of techniques should be combined to identify a certain diamond, as it is difficult to be identified just by means of a single feature [57]. Combination of the presence of internal growth graining and the absence of growth structures typical of HPHT- and CVD-grown diamonds is a common feature of the diamonds in this work, which could confirm that they have grown in nature. Usually, after HPHT treatment, some of the diamonds could display obvious color zoning (especially around the culet or girdle), etched facet near to the girdle which was not repolished, and some graphitized inclusions along fracture surfaces.

The absorption caused by H1a at 1450 cm^{-1} in the spectra of FTIR is observed in all the irradiated samples of this study (Figure 4). Besides, the GR1 center related to the isolated neutral vacancies is another common center in irradiated diamonds; however, it easily disappears after high temperature annealing (Figure 5).

Fisher [23] suggested that aggregation of nitrogen in natural diamonds tended to follow a progression: firstly, the single nitrogen aggregates to form pairs, and only after this transformation is completed, the A centers will aggregate further to form B centers. It is rare to find a normal color untreated diamond simultaneously exhibiting strong absorption from both single nitrogen and B centers, and therefore the occurrence of isolated nitrogen associated with the B centers could be considered as a possible indicator of HPHT treatment. It is true that various aggregation states of nitrogen (C, A, and B centers) are observed in the FTIR spectra of the HPHT-treated samples in this work. Similarly, the formation of H2 center requires for the existence of the donors (e.g., the C center), and therefore the existence of isolated nitrogen together with H3 and H2 centers is another possible property related to HPHT treatment [55, 58]. Although the intensity of the 637 nm is found to be stronger than that of 575 nm for the HPHT-treated diamonds in this work, which is considered to be a typical characteristic of HPHT treatment in previous works [4, 49, 59, 60], however, Collins [34] suggested that it is just a rough criterion and works well for approximately 75% of the HPHT-treated diamonds.

5. Conclusions

Recently, the color diamonds have become highly popular and gotten a great value in the jewelry market. Through high-energy irradiation, HPHT treatment, LPHT annealing, and combination of these treatments, many industries and researchers have continuously focused on enhancing the color of diamonds so as to achieve a fancy color. Therefore, the study of these treated-color diamonds appears particularly important for the jewelry testers and customers. In this work, we collected ten treated-color diamonds with different treatment process and characterized them through the most widely used optical gemological and spectroscopic techniques (IR absorption spectrum, PL spectra, and DiamondView). The possible conditions of the treatment and the coloration mechanism of these samples (blue, green, and red) were discussed. It is suggested that the conclusive identification of these treated-color diamonds should be made by a combination of gemological and spectroscopic properties, such as their distinctive growth patterns, color zoning, and absorption features in the IR and PL spectra.

Conflicts of Interest

The authors declare that there is no conflict of interest regarding the publication of this paper.

Acknowledgments

This work was financially supported by the National Natural Science Foundation of China (Grant no. 41603064) and China Postdoctoral Science Foundation (Grant no. 2016M591222).

References

[1] G. E. Harlow, "What is diamond?," in *The Nature of Diamond*, G. E. Harlow, Ed., Cambridge University Press, Cambridge, 1998.

[2] K. Nassau, *Gemstone Enhancement History, Science and State of the Art*, Butterworth-Heinemann Ltd, Oxford, 2nd edition, 1994.

[3] A. T. Collins and I. Kiflawi, "The annealing of radiation damage in type Ia diamond," *Journal of Physics: Condensed Matter*, vol. 21, no. 36, article 364209, 2009.

[4] T. Hainschwang, A. Katrusha, and H. Vollstaedt, "HPHT treatment of different classes of type I brown diamonds," *Journal of Gemmology*, vol. 29, no. 5-6, pp. 61–73, 2005.

[5] T. Hainschwang, A. Respinger, F. Notari, H. J. Hartmann, and C. Günthard, "A comparison of diamonds irradiated by high fluence neutrons or electrons, before and after annealing," *Diamond and Related Materials*, vol. 18, no. 10, pp. 1223–1234, 2009.

[6] M. Hall and T. M. Moses, "Gem trade lab notes: diamond-blue and pink, HPHT annealed," *Gems & Gemology*, vol. 36, no. 3, pp. 254–259, 2000.

[7] Y. F. Meng, C. S. Yan, J. Lai et al., "Enhanced optical properties of chemical vapor deposited single crystal diamond by low-pressure/high-temperature annealing," *Proceedings of the National Academy of Sciences of the United States of America*, vol. 105, no. 46, pp. 17620–17625, 2008.

[8] S. S. Vagarali, S. W. Webb, W. E. Jackson, W. F. Banholzer, T. R. Anthony, and G. R. Kaplan, "High pressure/high

temperature production of colorless and fancy-colored diamonds," U.S. patent 6692714, 2004.

[9] V. G. Vins, "The technique of production of fancy red diamonds," Russian Federation patent 2237113, 2004.

[10] V. G. Vins, A. P. Yelisseyev, S. S. Lobanov, D. V. Afonin, A. Y. Maksimov, and A. Y. Blinkov, "APHT treatment of brown type Ia natural diamonds: dislocation movement or vacancy cluster destruction?," *Diamond and Related Materials*, vol. 19, no. 7-9, pp. 829–832, 2010.

[11] V. G. Vins and O. V. Kononov, "A model of HPHT color enhancement mechanism in natural gray diamonds," *Diamond and Related Materials*, vol. 12, no. 3-7, pp. 542–545, 2003.

[12] V. G. Vins, "Technique of production of fancy red diamonds," US Patent 2007/0053823A1, 2007.

[13] I. M. Reinitz, P. R. Buerki, J. E. Shigley, S. F. McClure, and T. M. Moses, "Identification of HPHT-treated yellow to green diamonds," *Gems & Gemology*, vol. 36, no. 2, pp. 128–137, 2000.

[14] J. E. Shigley, "High-pressure-high-temperature treatment of gem diamonds," *Elements*, vol. 1, no. 2, pp. 101–104, 2005.

[15] W. Wang, C. P. Smith, M. S. Hall, C. M. Breeding, and T. M. Moses, "Treated-color pink-to-red diamonds from Lucent Diamonds Inc," *Gems & Gemology*, vol. 41, no. 1, pp. 6–19, 2005.

[16] C. M. Welbourn, M. Cooper, and P. M. Spear, "De beers natural versus synthetic diamond verification instruments," *Gems & Gemology*, vol. 32, no. 3, pp. 156–169, 1996.

[17] S. Liggins, M. E. Newton, J. P. Goss, P. R. Briddon, and D. Fisher, "Identification of the dinitrogen <001> split interstitial H1a in diamond," *Physical Review B*, vol. 81, no. 8, article 0852145, 2010.

[18] C. D. Clark, R. W. Ditchburn, and H. B. Dyer, "Absorption spectra of natural and irradiated diamonds," *Proceedings of the Royal Society A: Mathematical, Physical and Engineering Sciences*, vol. 234, no. 1198, pp. 363–381, 1956.

[19] G. S. Woods, "Platelets and the infrared absorption of type Ia diamonds," *Proceedings of the Royal Society A: Mathematical, Physical and Engineering Sciences*, vol. 407, no. 1832, pp. 219–238, 1986.

[20] F. S. Gentile, S. Salustro, M. Causa, A. Erba, P. Carbonniere, and R. Dovesi, "The VN3H defect in diamond: a quantum-mechanical characterization," *Physical Chemistry Chemical Physics*, vol. 19, no. 33, pp. 22221–22229, 2017.

[21] A. Mainwood, "Nitrogen and nitrogen-vacancy complexes and their formation in diamond," *Physical Review B*, vol. 49, no. 12, pp. 7934–7940, 1994.

[22] A. M. Zaitsev, *Optical Properties of Diamond: A Data Handbook*, Springer-Verlag, Berlin, 2001.

[23] D. Fisher, "Brown diamonds and high pressure high temperature treatment," *Lithos*, vol. 112, pp. 619–624, 2009.

[24] W. Wang, P. Doering, J. L. R. Tower et al., "Strongly colored pink CVD lab-grown diamonds," *Gems & Gemology*, vol. 46, no. 1, pp. 4–17, 2010.

[25] L. Tretiakova, "Spectroscopic methods for the identification of natural yellow gem-quality diamonds," *European Journal of Mineralogy*, vol. 21, no. 1, pp. 43–50, 2009.

[26] I. A. Dobrinets, V. G. Vins, and A. M. Zaitsev, *HPHT-Treated Diamonds, Springer Series in Material Science*, Berlin, 2013.

[27] J. E. Shigley, E. Fritsch, J. I. Koivula, N. V. Sobolev, I. Y. Malinovsky, and Y. N. Pal'yanov, "The gemological properties of Russian gem-quality synthetic yellow diamonds," *Gems & Gemology*, vol. 29, no. 4, pp. 228–248, 1993.

[28] A. T. Collins, "Optical centres produced in diamond by radiation damage," *New Diamond and Frontier Carbon Technology*, vol. 17, no. 2, p. 47, 2007.

[29] A. T. Collins, A. Connor, C. Ly, A. Shareef, and P. M. Spear, "High-temperature annealing of optical centers in type-I diamond," *Journal of Applied Physics*, vol. 97, no. 8, article 083517, 2005.

[30] V. G. Vins and A. P. Yelisseyev, "Effect of high pressure, high temperature annealing on impurity-defect structure of natural diamonds," *Perspectives Materials*, vol. 1, pp. 49–57, 2010, (in Russian).

[31] E. J. Brookes, A. T. Collins, and G. S. Woods, "Cathodoluminescence at indentations in diamonds," *Journal of Hazardous Materials*, vol. 4, pp. 98–105, 1993.

[32] F. De Weerdt, R. Galloway, and A. Anthonis, "Defect aggregation and dissociation in brown type-Ia diamonds by annealing at high pressure and high temperature (HPHT)," *In Defect and Diffusion Forum*, vol. 226, pp. 49–60, 2004.

[33] A. T. Collins, "Colour centers in diamond," *Journal of Gemmology*, vol. 18, no. 1, pp. 37–75, 1982.

[34] A. T. Collins, "The detection of colour-enhanced and synthetic gem diamonds by optical spectroscopy," *Diamond and Related Materials*, vol. 12, no. 10-11, pp. 1976–1983, 2003.

[35] A. T. Collins, H. Kanda, and H. Kitawaki, "Colour changes produced in natural brown diamonds by high-pressure, high-temperature treatment," *Diamond and Related Materials*, vol. 9, no. 2, pp. 113–122, 2000.

[36] F. De Weerdt and J. Van Royen, "Defects in coloured natural diamonds," *Diamond and Related Materials*, vol. 10, no. 3-7, pp. 474–479, 2001.

[37] A. T. Collins, "Investigating artificially colored diamond," *Nature*, vol. 273, no. 5664, pp. 654-655, 1978.

[38] V. G. Vins and A. P. Eliseev, "Effect of annealing at high pressures and temperatures on the defect-admixture structure of natural diamonds," *Inorganic Materials: Applied Research*, vol. 1, no. 4, pp. 303–310, 2010.

[39] V. G. Vins and A. P. Yelisseyev, "Physical fundamentals behind modern techniques of natural diamond enhancement," *Australian Gemmologist*, vol. 24, pp. 219–221, 2008.

[40] V. G. Vins, A. P. Yelisseyev, and V. Sarin, "Physics behind the modern methods of enhancement of natural diamonds," *Precision Metals Precision Stones*, vol. 12, no. 180, pp. 155–163, 2008.

[41] D. Fisher, D. J. F. Evans, C. Glover, C. J. Kelly, M. J. Sheehy, and G. C. Summerton, "The vacancy as a probe of the strain in type IIa diamonds," *Diamond and Related Materials*, vol. 15, no. 10, pp. 1636–1642, 2006.

[42] H. Lim, S. Park, H. Cheong, H. M. Choi, and Y. C. Kim, "Discrimination between natural and HPHT-treated type IIa diamonds using photoluminescence spectroscopy," *Diamond and Related Materials*, vol. 19, no. 10, pp. 1254–1258, 2010.

[43] I. Kiflawi, A. T. Collins, K. Iakoubovskii, and D. Fisher, "Electron irradiation and the formation of vacancy-interstitial pairs in diamond," *Journal of Physics: Condensed Matter*, vol. 19, no. 4, article 046216, 2007.

[44] G. Davies, "Charge states of the vacancy in diamond," *Nature*, vol. 269, no. 5628, pp. 498–500, 1977.

[45] F. De Weerdt and A. T. Collins, "HPHT annealing of natural diamond," *New Diamond and Frontier Carbon Technology*, vol. 17, no. 2, 2007.

[46] J. P. Goss, P. R. Briddon, V. Hill, R. Jones, and M. J. Rayson, "Identification of the structure of the 3107 cm^{-1} H-related defect in diamond," *Journal of Physics: Condensed Matter*, vol. 26, no. 14, article 145801, 2014.

[47] V. G. Vins, "Changing diamond colour," *Best Gemology*, vol. 1, no. 3, pp. 19–30, 2001.

[48] F. De Weerdt and I. N. Kupriyanov, "Report on the influence of HPHT annealing on the 3107 cm^{-1} hydrogen related absorption peak in natural type Ia diamonds," *Diamond and Related Materials*, vol. 11, no. 3-6, pp. 714-715, 2002.

[49] A. T. Collins, "Colour of diamond and how it may be changed," *Journal of Gemmology*, vol. 27, no. 6, pp. 341–359, 2001.

[50] A. T. Collins and P. M. Spear, "Optically active nickel in synthetic diamond," *Journal of Physics D: Applied Physics*, vol. 15, no. 12, pp. L183–L187, 1982.

[51] S. C. Lawson, H. Kanda, and M. Sekita, "New nickel-related optical absorption in high-pressure synthetic diamond," *Philosophical Magazine B*, vol. 68, no. 1, pp. 39–46, 1993.

[52] J. Darley and J. M. King, "Natural color hydrogen-rich blue-gray diamond," *Gems & Gemology*, vol. 43, no. 2, pp. 155-156, 2007.

[53] E. Fritsch and K. Scarratt, "Natural-color nonconductive gray-to-blue diamonds," *Gems & Gemology*, vol. 28, no. 1, pp. 35–42, 1992.

[54] H. Kitawaki, "Gem diamonds: causes of colors," *New Diamond and Frontier Carbon Technology*, vol. 17, no. 3, pp. 119–126, 2007.

[55] Y. C. Kim and H. M. Choi, "A study on the HPHT-processed NOUV diamonds by means of their gemological and spectroscopic properties," *Journal of the Korean Crystal Growth and Crystal Technology*, vol. 15, no. 3, pp. 114–119, 2005.

[56] W. Wang and M. Hall, "HPHT-treated type Ia diamond with a green component caused by the H_2 defect," *Gems & Gemology*, vol. 43, no. 2, pp. 156–158, 2007.

[57] M. E. Newton, "Treated diamond: a physicist's perspective," *Gems & Gemology*, vol. 42, no. 3, pp. 84-85, 2006.

[58] F. De Weerdt and J. Van Royen, "Investigation of seven diamonds, HPHT treated by NovaDiamond," *Journal of Gemmology*, vol. 27, no. 4, pp. 201–208, 2000.

[59] J. P. Chalain, E. Fritsch, and H. A. Haenni, "Diamants de type IIa et traitement HPHT: Identificatio," *Revue de Gemmologie*, vol. 141, pp. 50–53, 2001.

[60] L. Tretiakova and Y. Tretyakova, "In significance of spectroscopic methods for the identification defects in diamond," in *9th International Kimberlite Conference, Extended Abstract No. 9IKC-A-00042*, 2008.

13

Automation of the Peak Fitting Method in Bone FTIR Microspectroscopy Spectrum Analysis: Human and Mice Bone Study

Marc Gardegaront, Delphine Farlay, Olivier Peyruchaud, and Hélène Follet

INSERM, Université de Lyon, UMR1033, F-69008 Lyon, France

Correspondence should be addressed to Hélène Follet; helene.follet@inserm.fr

Academic Editor: Arnaud Cuisset

FTIR microspectroscopy (FTIRM) is a commonly used nondestructive method to characterise thin bone sections. However, spectrum analysis methods are often highly sensitive to small variations (e.g., boundary limits), thus implying a time-consuming and redundant analysis process. To solve this issue, software has been developed based on several algorithms to automate the analysis. Furthermore, a rigorous framework has been established concerning the peak fitting method to obtain the systematic best potential solution. Validation of the automatic method has been performed by comparison with the manual method. Results and validation proved the reliability of the automatic process. The developed algorithms provide the means necessary to fully compare the results between bone FTIRM studies and between different laboratories.

1. Introduction

Bone analysis methods are quite varied [1–3], ranging from the noninvasive, e.g., X-ray scanning, to the destructive, e.g., mechanical rupture tests. FTIR microspectroscopy (FTIRM) is a promising field which characterises bone samples, as it is able to provide information on the chemical composition and structural organisation of the analysed matter without damaging it on a local level of a thin section [4–9]. This method combines optic, quantum mechanics, chemistry, and biology knowledge, which leads to a complex understanding that might have an impact on the conclusion of an FTIR study [10]. In addition, the spectrum analysis is laborious if done "manually."

Several different methods are currently used to analyse FTIRM spectra (peak fitting method [11–13], second derivative method [13, 14], etc.), and each has its advantages and limitations; moreover, none are considered to be a gold standard. We chose to stick with the peak fitting method used in [11, 12]. The second derivative method, although quite promising due to its objective and purely mathematical analysis, is too sensitive to noise (producing peaks that are not there initially) and quite inadequate in the case of peaks that are too close and with different full width at half maximum (FWHM) (the wider peak is then masked by the thinner one). This is why the second derivative method is not chosen (for more explanation, cf. Appendix C).

Traditionally, the peak fitting of a bone spectrum is done by selecting a region of interest (ROI) in the bone spectrum, followed by fitting the curve with a sum of Gaussians. Previously, this fitting was done by assuming both position and characteristics of the peaks (height and FWHM), giving a starting state for the least-squares fitting method. The result of the fitting process was then left to the operator for visual verification. As one can easily imagine, this technique proved to be fastidious, as correct results were typically obtained through several tries. Indeed, the regression may sometimes be highly sensitive and be executed multiple times for one ROI in order to find the correct fitting result.

In a bone spectrum (Figure 1), six distinct regions are identified as the vibrational response of different constituents:

(i) The amide (I, II, and III) areas (1300, 1700) cm^{-1} corresponding to the response of the organic matrix in the bone tissue (with also a ν_3CO_3 contribution at around 1400 cm^{-1})

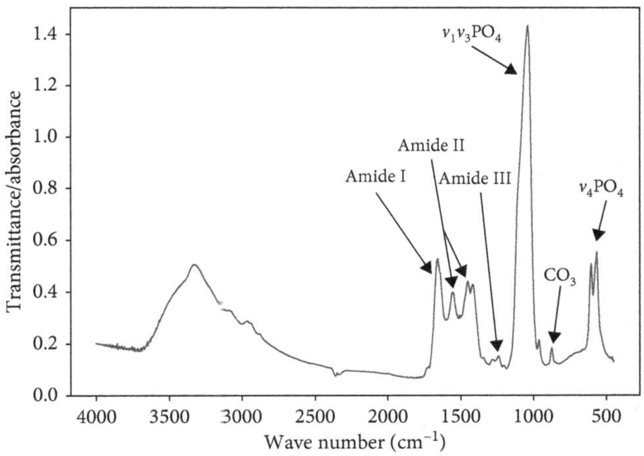

FIGURE 1: Typical FTIRM spectrum of human bone between the $450\,cm^{-1}$ and $4000\,cm^{-1}$ boundary. Amides I, II, and III represent the organic matrix of the bone, while the $\nu_1\nu_3PO_4$, ν_4PO_4, and ν_2CO_3 are part of the mineral matrix.

(ii) The $\nu_1\nu_3PO_4$ area (900, 1200) cm^{-1} corresponding to the symmetric and antisymmetric stretching of the phosphates

(iii) The ν_2CO_3 area (800, 900) cm^{-1} corresponding to the symmetrical stretching of the carbonates (CO_3), which also contains a large HPO_4 contribution, which is removed before analysis by using a wisely placed baseline correction

(iv) The ν_4PO_4 area (500, 650) cm^{-1} corresponding to the antisymmetric deformation of the phosphates (PO_4).

We automated the methods used in [9, 11, 14, 15], which are biologically relevant methods (not meant to be spectroscopically accurate), to obtain the five parameters within these regions to characterise the sample:

Directly measured on the spectra are as follows:

(i) Mineral-to-organic ratio (mineral/matrix): ratio of the $\nu_1\nu_3PO_4$ area (910, 1184) cm^{-1} over the amide I area (1592, 1730) cm^{-1} [14]

(ii) Carbonation ratio of the ν_2CO_3 area (862, 894) cm^{-1} over the $\nu_1\nu_3PO_4$ area (910, 1184) cm^{-1} [14].

Postpeak fitting is given as follows

(i) Crystallinity Index: inverse of the FWHM of the ~$604\,cm^{-1}$ peak [15]

(ii) Mineral Maturity (MM): ratio of the apatitic (~$1030\,cm^{-1}$ peak) over nonapatitic (~$1110\,cm^{-1}$ peak) part of the $\nu_1\nu_3PO_4$ area [15]

(iii) Collagen Maturity (Col Mat): ~$1660\,cm^{-1}$ peak over ~$1690\,cm^{-1}$ peak (part of the amide I area) [5, 11].

The manual peak fitting method might have induced subjectivity and imprecision, as there were no strict means to analyse bone spectra. Also, these methods are not always spectroscopically accurate (i.e., there are not always the right number of peaks corresponding to the exact composition of the studied material); however, these methods have been validated and are sufficient to conclude the proportion of compounds in the analysed bone.

These methods might be improved by the careful study and translation of the analysis part into an algorithm implemented into custom software.

The aim of this technical note was to offer a solution to the automation process of the peak fitting method used in FTIR spectrum analysis of the bone. Software, i.e., LYOS spectrum analysis, written in Python, has been developed to fully automate the analysis process of healthy human and thin mouse bone section's infrared spectra, even though the software can be used to analyse other types of spectrum.

2. Materials and Methods

The validation process has been done using the following:

(1) Fifty-one trabecular thin sections of human L2 vertebra from 51 donors (French body donation to science program, declaration number: DC-2015-2357; Laboratory of Anatomy, Faculty of Medicine Lyon Est, University of Lyon, France) (20 spectra per sample at $30 \times 100\,\mu m^2$ spatial resolution)

(2) Four trabecular and cortical thin sections of human iliac crest from four donors (Laboratoire d'anatomie, Université Claude Bernard, Lyon, France) (78 spectra at $50 \times 50\,\mu m^2$ spatial resolution)

(3) Twelve cortical thin sections of human femur from 12 subjects (MULTIPS: ANR-13-BS09-0006-01) (12 spectra at $150 \times 150\,\mu m^2$ spatial resolution)

(4) One trabecular and cortical thin section of mouse tibia from one mouse (Balb/C E2A-55 #DR2015-58 MENESR: #APAFIS# 39132015121 0903612v2, Lys-bone: ANR-15- CE14-0010) (10 spectra at $30 \times 30\,\mu m^2$ spatial resolution).

After dissection, samples were fixed in 70% ethanol for two weeks, dehydrated for 48 h in 100% ethanol, substituted for 48 h in methylcyclohexane (solvent that removes lipids in the samples), and then embedded in methyl methacrylate (MMA). The sections of $2\,\mu m$ thickness were obtained with a microtome Polycut E (Reichert-Jung, Leica, Germany) and stored between two slides at room temperature.

The raw spectra were collected at $1\,cm^{-1}$ resolution and averaged by 40 scans per scanned region in transmission mode with a Perkin–Elmer GXII AutoIMAGE microscope (Norwalk, CT, USA) equipped with a wide-band detector (mercury-cadmium-telluride) (7800–400 cm^{-1}). The instrument used a Cassegrain objective of numerical aperture 0.6. The system has a spatial resolution of $10\,\mu m$ at typical midinfrared wavelengths. Contribution from air was subtracted from the original spectrum.

The commercial software Spectrum (v6.2.0.0055) provided by Perkin–Elmer and Grams/AI 8.0 Spectroscopy Software provided by the Thermo Electron Corporation offer various preprocessing options and customisation for the automatic peak fitting process. The mandatory methods used in these software have been implemented and improved for automation in the custom Python software and are described below, such as a transformation from transmittance to absorbance values, a spectrum component removal (to extract the components of interest from a raw spectrum), and a baseline correction method (to remove the high-scale effects of a bone FTIRM measure).

2.1. Transmittance to Absorbance. The FTIRM spectrum of a thin section of the bone is obtained in transmittance. However, the spectrum analysis is conventionally done on an absorbance spectrum. The transmittance spectrum (T, in percent) is then transformed into its absorbance (A) form by the following function:

$$A = -\log_{10}\left(\frac{T}{100}\right). \quad (1)$$

2.2. Spectrum Component Removal. Thin bone sections could not be realised without an embedding medium (methyl methacrylate, MMA). This medium creates a potential infrared-reactive component in the spectrum that must be removed in order to obtain the bone-only spectrum. This component presents a high absorption peak around 1730 cm^{-1} (Figure 2). The first step is to scan the MMA to have its pure spectrum in the same range as the bone spectrum (450, 4000) cm^{-1}. Then, a custom algorithm of optimisation subtracts the MMA spectrum from the bone spectrum (cf. Appendix A). This process is done by selecting the boundaries of the highest MMA peak (1826, 1706) cm^{-1} between which is computed with the second derivative of the bone spectrum (corresponding to the curvature of the spectrum). An iterative loop consecutively guesses a coefficient by which the whole MMA spectrum is multiplied, and the subtraction result is evaluated on the convexity of the shape of the spectrum. The convexity of a curve is mathematically defined by the positivity of the second derivative. The best result is the one that has the minimum negative part. Each loop will check the negative part of the second derivative of the bone-embedded spectrum and try to minimise it by subtracting the weighted MMA-only spectrum (Figure 3). The algorithm can be summed up as follows:

Find y_i which $\int f(y_i)$ is minimum

with y_i: the spectrum subtracted by the ith weighted MMA-only spectrum

and $f(y_i) = \begin{cases} 0, & \text{if } y_i > 0, \\ y_i, & \text{if } y_i \le 0. \end{cases}$

2.3. Baseline Correction. The baseline of the bone in the (450, 4000) cm^{-1} is convex. The purpose is to realign this line horizontally. A multilinear baseline is chosen as an approximation of the effects of the interferences (beam spreading, sample heating, etc). Based on the convexity hypothesis, the algorithm used to find the baseline will first detect the local minima using a Savitzky–Golay smoothing [16, 17] (keeping intensity and position of the local extrema of the raw spectrum) and draw the segments between two minima containing the most minima above it while forbidding the presence of minima beneath it (Figure 4).

2.4. Peak Fitting. The peak fitting method is one of the methods used in the conventional FTIRM spectrum analysis [18] (least-squares method with trust region reflective algorithm to find the best suited parameters). The curve fitting function used is curve_fit from the scipy.optimize module version 1.0.0. Usually, only a first guess state for each Gaussian peak is given for the manual method, allowing the peaks to place everywhere with any characteristics (position, FWHM, and height). This first guess-only method often induces inaccurate results and forces the operator to execute the method multiple times in order to find a valid fit. The automation process is implemented by adding boundaries to find each peak's parameter. These boundaries are carefully chosen for the solution to be unique (restricting the plurality of solutions induced by the manual method with first state guessing only). Ideally, boundaries should be chosen considering the characteristics of each sample (temperature, internal constraints, crystal properties, etc.). However, which characteristic and how each influences a peak's parameter cannot be answered precisely with our current knowledge. Thus, a more empirical approach has been chosen to determine the boundaries of each peak. Based on a set of more than a thousand spectra considered to be well peak fitted, some schemes have been observed. Eight peaks were found in the amide area, five peaks in the $\nu_1\nu_3PO_4$ area, and four peaks in the ν_4PO_4 area. Boundaries for each peak's parameters have been established and are registered in Table 1. For each area, the first guess of each peak located at a specific wave number is set at 90% of the absorbance in height and at 5 cm^{-1} in FWHM. The ranges for the peak fitting of each area are chosen to give mathematically correct results. They are related to the biologically relevant range given in the introduction part but generally need to be wider, as mathematical information is present beyond the biologically relevant range (for example, the amide area range is (1300, 1700) cm^{-1} and the mathematical range used is (1300, 1787) cm^{-1}, the same for the $\nu_1\nu_3PO_4$ and ν_4PO_4). The algorithm can be summed up as follows:

Find the jth Gaussians parameters G_j, x_{0j}, ω_j

for which $f(x)$ minimizes $\sum_i (y_i - f(x_i))^2$

with $f(x) = \sum_j G_j * e^{-(x-x_{0j}/\omega_j)^2}$

and G_j: amplitude of the jth Gaussian,

x_{0j}: position of the jth Gaussian,

ω_j: $FWHM/2\sqrt{2}$ of the jth Gaussian

with respect to the constraints to each Gaussian's parameters explicited in Table 1.

2.4.1. The Amide Area (1300, 1787) cm^{-1}. The area of interest inside the amide part of the spectrum lies around the (1600, 1730) cm^{-1} boundary (Figure 5). Even if (1300, 1600) cm^{-1} is not used in the measured parameters, it was kept to accurately fit the (1300, 1730) cm^{-1} area. The main peaks are the dashed-blue peak (~1690 cm^{-1}) and the dashed-purple peak (~1660 cm^{-1}). The 1660 cm^{-1} peak is highly influenced by the 1633 cm^{-1} (dashed orange), which is as well influenced by the 1600 cm^{-1} peak (full green peak).

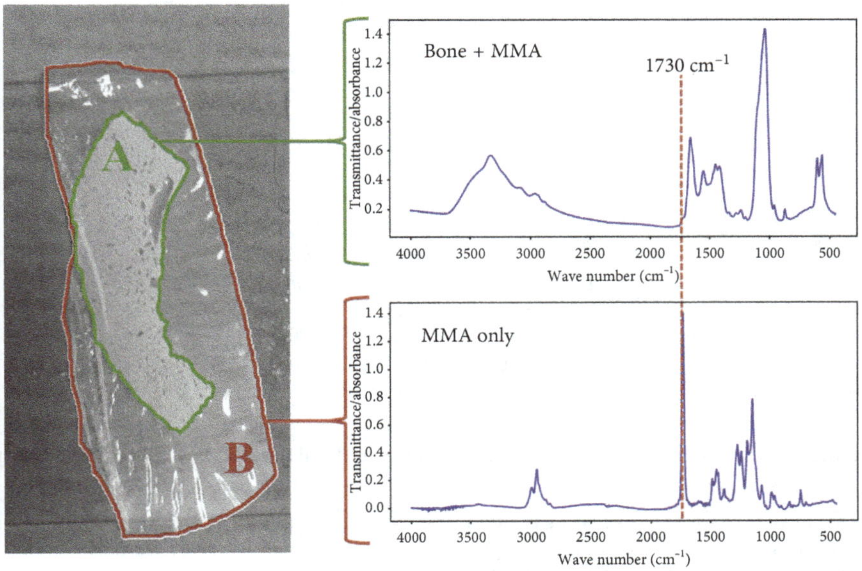

FIGURE 2: Typical sample of bone embedded in MMA (left) with its respective spectrum (top right) and the pure MMA spectrum (bottom right). A: cortical bone (human femur) embedded in MMA. B: methyl methacrylate (MMA). The MMA spectrum is taken outside the bone region of the sample.

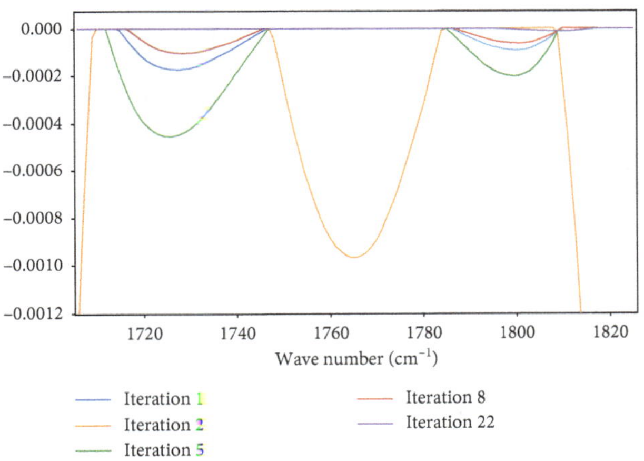

FIGURE 3: Visual representation of the algorithm. Evolution of the negative part of the second derivative of the bone spectrum between the MMA peak of interest boundaries (1706 cm^{-1} and 1826 cm^{-1}) through the iterations. Each iteration corresponds to the subtraction of the MMA-embedded bone spectrum minus the weighted MMA spectrum (the weight changes with each iteration). *Note.* An iteration is not always better than the previous one (e.g., iterations 1 and 2), but it tends nonetheless to converge to the expected result (e.g., iteration 22), which is the convexity of the spectrum between the 1706 cm^{-1} and 1826 cm^{-1} boundaries. The final result of the MMA removal can be observed in Figure 1 (the MMA contribution, including the 1730 cm^{-1} peak, in the bone spectrum is removed).

In bone, the 1633 cm^{-1} peak must be higher than the local minima at 1600 cm^{-1}, and the height of the 1600 cm^{-1} peak greatly influences the height of the 1633 cm^{-1}. All the parameters are constrained with static values (unchanged regarding the spectrum) (cf. Table 1). It may happen that the MMA peak is not correctly removed due to small differences of the shape of the MMA peak between the pure MMA spectrum and the MMA component in the bone + MMA spectrum. In order to minimise the implications of the residual MMA peak, an additional peak is included for the peak fitting method, which is located at 1745 cm^{-1}. This is an automatic process realised when the algorithm detects a significant peak between 1730 cm^{-1} and 1760 cm^{-1} (i.e., when the positive area of the derivate between those values is superior to the half of the negative area of the derivate).

2.4.2. The $\nu_1\nu_3PO_4$ Area (910, 1184) cm^{-1}. The two peaks of interest here are the 1030 and 1110 cm^{-1} peak (full purple and green) (Figure 6). These two peaks are mainly influenced by the position of the peaks between 1060 and 1070 cm^{-1}. Every parameter for this area determination is a static value (cf. Table 1).

2.4.3. The ν_4PO_4 Area (531, 659) cm^{-1}. Only the 604 cm^{-1} peak (full yellow) is used to characterise the bone in the ν_4PO_4 area (Figure 7). This peak is mainly influenced by the 575 cm^{-1} peak. In bone, the 575 cm^{-1} peak's height must be inferior to the 565 cm^{-1} peak to avoid the 575 cm^{-1} peak to spread too much on the 604 cm^{-1} peak area. Every parameter for this area is a static value (cf. Table 1).

2.4.4. Statistical Analysis. The validation of the spectrum analysis methods has been done using the root mean square of the coefficient of variation to assess the variation in manual methods, with the Bland–Altman method to compare manual and automatic results.

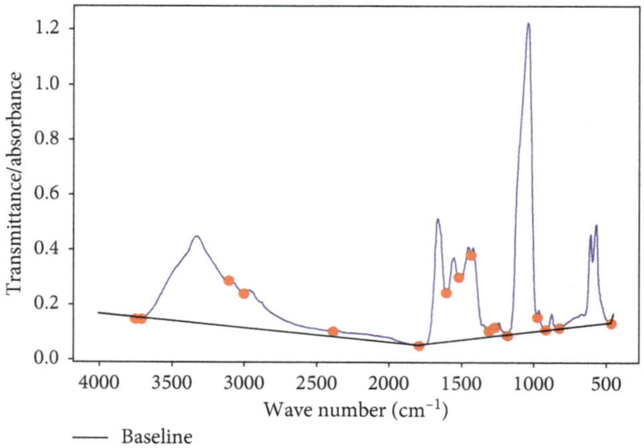

FIGURE 4: Graphical representation of the baseline creation. Red dotes are local minima, and the black line is the created baseline.

TABLE 1: Boundaries of each peak used for the peak fitting method. The ν_4PO_4 area contains 4 peaks, the $\nu_1\nu_3PO_4$ area contains 5 peaks, and the amide area contains 8 peaks. Each peak's parameter (height, FWHM, and position) is searched between the defined minimum and maximum boundaries, expressed in the table, by the fitting algorithm.

		ν_4PO_4 area (531, 659) cm^{-1}				$\nu_1\nu_3PO_4$ area (910, 1184) cm^{-1}					Amide area (1300, 1787) cm^{-1}							
First guess	Position	540	565	575	604	962	1030	1060	1070	1110	1340	1420	1450	1550	1600	1633	1660	1690
Height	Minimum	0	0	0	0	0	0	0	0	0	0	0	0	0	0	0	0	0
	Maximum	∞	∞	∞	∞	∞	∞	∞	∞	∞	∞	∞	∞	∞	∞	∞	∞	∞
FWHM (cm^{-1})	Minimum	0	0	0	0	0	0	0	0	0	0	0	0	0	0	0	0	0
	Maximum	60	50	40	∞	70	70	40	40	70	∞	∞	∞	∞	30	∞	∞	∞
Position (cm^{-1})	Minimum	530	555	565	−∞	932	1000	1045	1055	1080	1310	1390	1420	1520	1585	1633	1631	1690
	Maximum	550	575	585	+∞	992	1060	1075	1085	1140	1370	1450	1480	1580	1615	1633	1689	1690

∞: infinity.

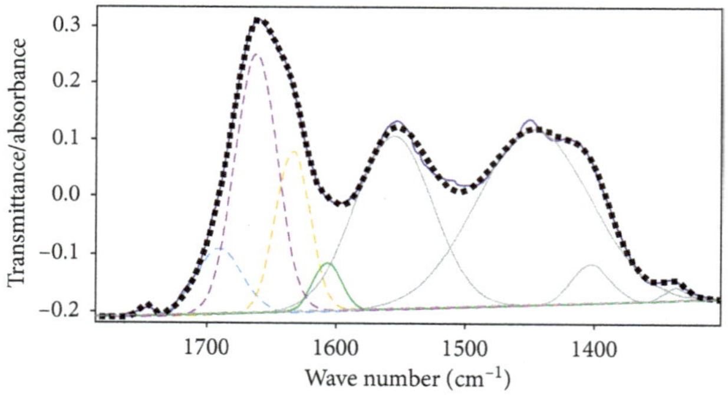

FIGURE 5: Original spectrum in the amide area (dark blue line) recovered by the Gaussian sum fitting (dotted-black line) and the 8 individual peaks beneath (dashed and full lines) + 1 peak for the MMA (1730 cm^{-1}). The two peaks directly used to compute the Collagen Maturity parameter are the 1660 cm^{-1} peak (dashed-purple line) and the 1690 cm^{-1} peak (dashed-blue one).

3. Results and Discussion

3.1. Validation. The first step of the validation process is done by analysing the precision of the manual peak fitting method. Two types of precision are measured: (1) the intraoperator precision that measures the ability of one operator to obtain the same result multiple times on the same spectrum and (2) the interoperator precision that measures the ability to get the same result from two different operators on the same spectrum.

The first operator (operator 1) is experienced with the bone FTIRM peak fitting, and is considered the reference. The second (operator 2) is an inexperienced operator who has been given the instructions to execute the peak fitting analysis.

Five different bone spectra have been peak fitted with the manual method two times by each operator (cf. Table 2). For the manual peak fitting, the root mean square of the coefficient of variation shows that results are operator-dependant.

Then, we compared both automatic and manual methods based on the spectra extracted from the human samples and

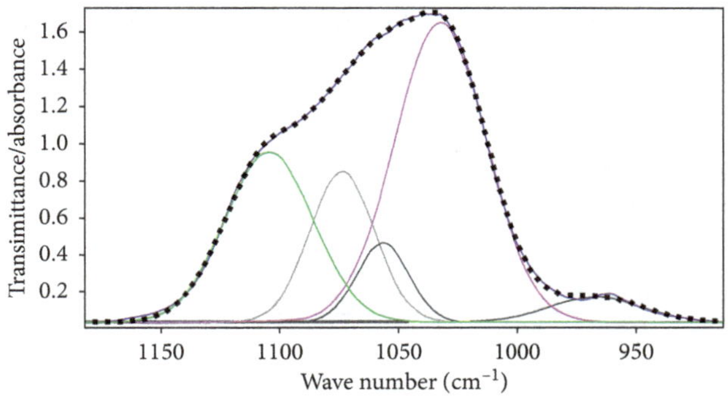

FIGURE 6: Original spectrum in the $\nu 1 \nu 3PO_4$ area (dark blue line) recovered by the Gaussian sum fitting (dotted-black line) and the 5 individual peaks beneath (full lines). The two peaks directly used to compute the Mineral Maturity parameter are the 1030 cm^{-1} peak (purple line) and the 1110 cm^{-1} peak (dashed-green one).

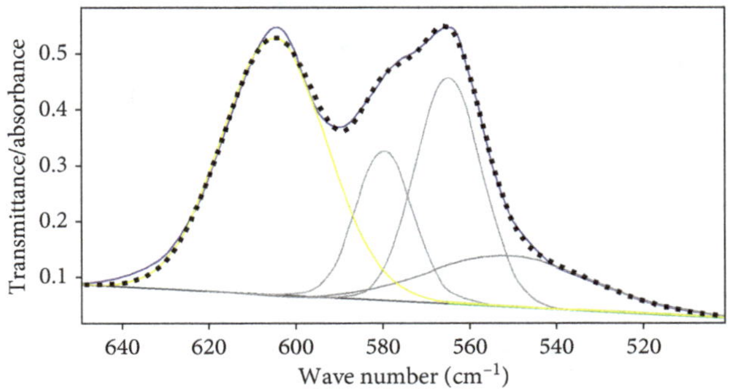

FIGURE 7: Original spectrum in the $\nu 4PO_4$ area (dark blue line) recovered by the Gaussian sum fitting (dotted-black line) and the 4 individual peaks beneath (full lines). The peak used to compute the Crystallinity parameter is 604 cm^{-1} (yellow line).

mouse sample. The Bland–Altman graph plots the means and the differences between automatic and manual methods measured by a human operator. For the human bones, Figure 8 shows, at first sight, an almost null shift for every parameter. The ratio of valid results (i.e., results between the mean ± 1.96 ∗ SD) is 92% for the Crystallinity Index, 93% for the Mineral Maturity, and 92% for the Collagen Maturity.

For the mice tibia (Figure 9), the shift is also almost null for every parameter. The ratio of valid results under the aforementioned hypothesis is 90% of valid results for the Crystallinity Index, 100% for the Mineral Maturity, and 100% for the Collagen Maturity.

3.2. Discussion. Gaussians have been long used to establish methods and peaks of interests in human bone, meaning changing the fitting function would end in a change of the peaks of interest. Thus, to maintain consistency with previous studies, Gaussians are kept for the fitting method. However, it should be noted that Lorentzians or Voigt functions would be more accurate regarding the quantum reality of the spectra [19].

The development of the automatic method not only allowed a significant gain of time but also suppressed the interoperator and intraoperator variability.

Currently, there is no gold standard to check the exactitude of the results obtained. After careful examination of the automatic peak fitting of these spectra, it has been found that the results obtained with the automatic method are valid compared with the analysis done by the expert using the manual method.

The observed shifts are mainly due to the choice of boundaries within which the peak fitting method is applied. Indeed, there are no strict boundaries required for the manual peak fitting of the ν_4PO_4 area. This induces the variation of the shift, as shown in Appendix B. This problem is solved by the automatic method that fixes those boundaries, chosen after careful analysis of several sets of boundaries.

We could argue that it is not the most exact method in terms of localisation of the peaks, compared with the manual method (which is not a gold standard method and can be different from one laboratory to another). However, the main advantage provided by the automatic peak fitting is the absence of intraoperator and interoperator variation and reproducibility. The second advantage of the automatic process is the normalisation of its preprocessing methods (MMA removal and baseline subtraction), removing the influence a human operator has in the spectrum preparation. Finally, the automatic method reduces drastically the time needed to analyse a spectrum (10 s is required for the automatic process, while the manual process takes 5 to 10 min per spectra).

4. Conclusions

The algorithms described in this paper offer the possibility to analyse quickly an infrared spectrum of the thin bone section.

Table 2: Differences of the manual peak fitting analysis intraoperator, interoperator, maximum absolute error, and variation of results using the automatic method for each parameter, based on the analysis of 5 bone spectra. Operator 1 is an experienced operator, and operator 2 is an inexperienced operator to whom has been given the instructions to analyse the spectra.

	Crystallinity Index	Mineral Maturity	Collagen Maturity
RMS of coefficient of variation (operator 1) (%)	0.7	1.4	0.7
RMS of coefficient of variation (operator 2) (%)	1.7	0.4	1.0
RMS of coefficient of variation (interoperator) (%)	2.0	1.2	0.7
Maximum absolute error	0.0013 cm	0.07	0.11
Variation of results using automatic method	No variation	No variation	No variation

RMS = root mean square.

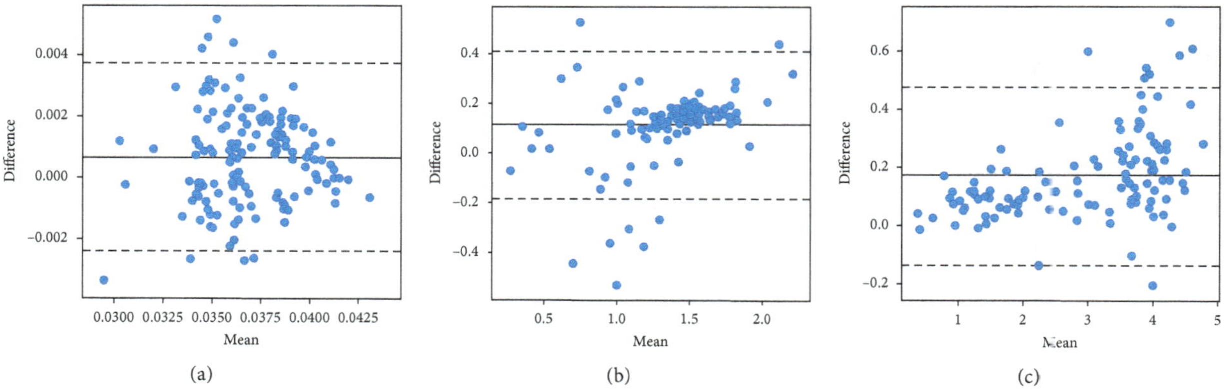

Figure 8: The Bland–Altman plot of the human bone spectra (manual method-automatic method values). The straight line is the mean of the differences. The dashed line is the mean of the differences plus or minus 1.96 ∗ SD, representing the limit of agreement. 92% of Crystallinity Index values, 93% of Mineral Maturity values, and 92% of Collagen Maturity values are within the dashed lines. Shifts of 0.0007 cm for the Crystallinity Index, 0.11 for the Mineral Maturity, and 0.17 for the Collagen Maturity are observed.

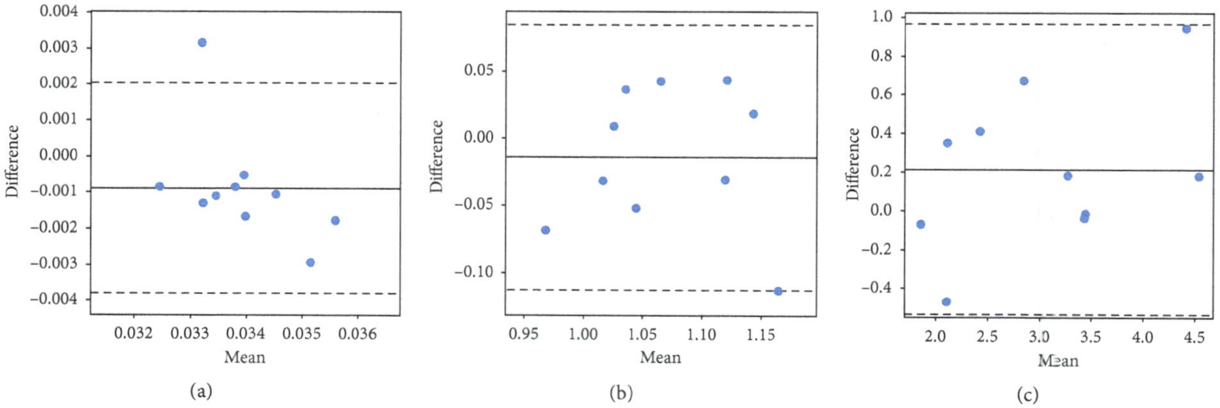

Figure 9: The Bland–Altman plot of the mice bone spectra (manual method-automatic method values). The straight line is the mean of the differences. The dashed line is the mean of the differences plus or minus 1.96 ∗ SD, representing the limit of agreement. 90% of Crystallinity Index values, 100% of Mineral Maturity values, and 100% of Collagen Maturity values are within the dashed lines. Shifts of −0.0009 cm for the Crystallinity Index, −0.01 for the Mineral Maturity, and 0.21 for the Collagen Maturity are observed.

This automatic method will allow the analysis of large sets of samples, which confer robust results.

By using machine learning methods on large data of deconvolved FTIRM spectra, the automatic method could be improved. This could bring specificity of peak fitting parameters regarding the type of the sample analysed, thus improving the exactitude and reliability of the results.

The software will be available under GNU GPL license.

Appendix

A. Simple Algorithm of Spectrum Component Removal

Variables used in the algorithm (Figure 10):

OrigSpectrumData: raw MMA-embedded bone spectrum.

RemSpectrumData: MMA-only spectrum.

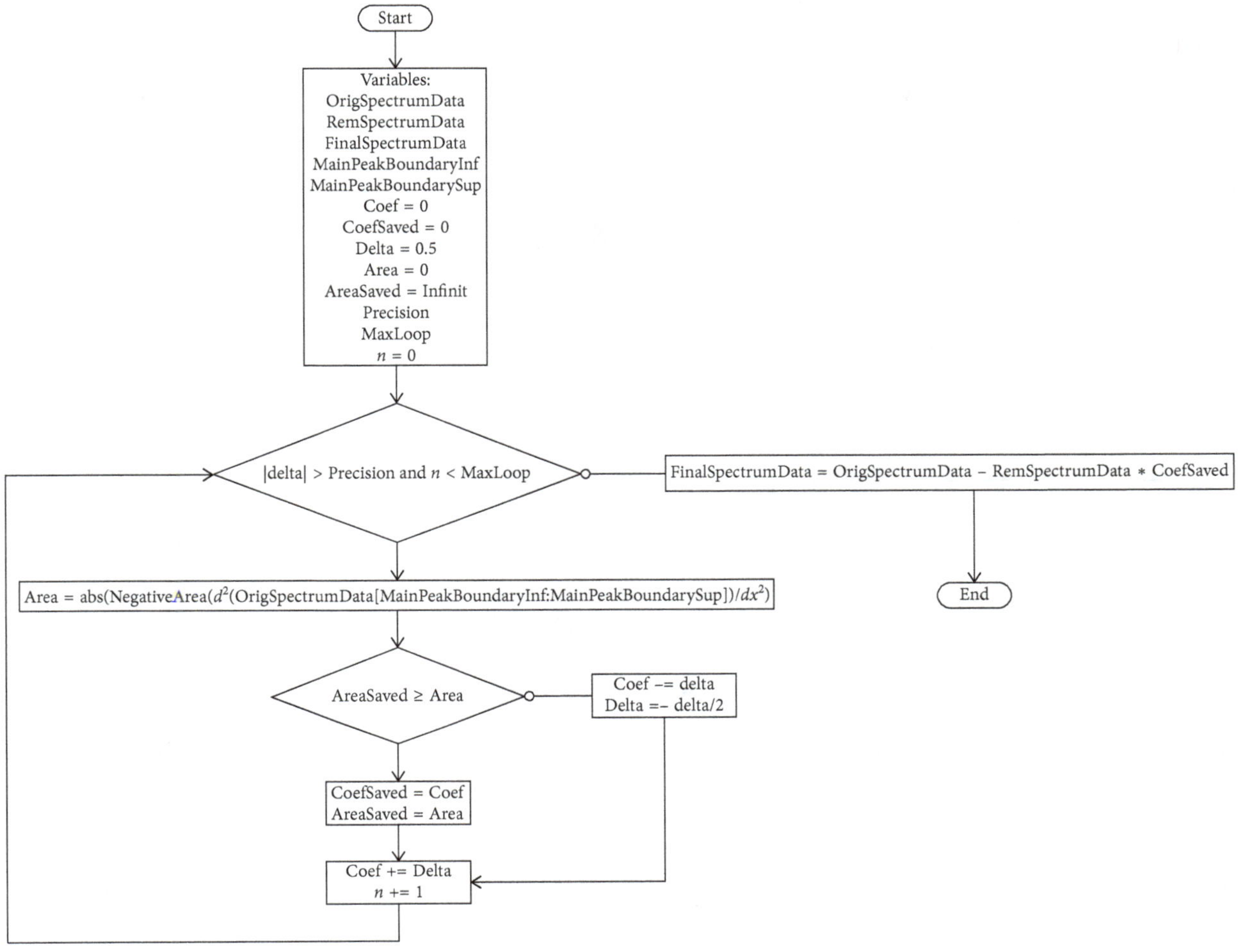

FIGURE 10

FinalSpectrumData: bone spectrum (removed from the MMA component).

MainPeakBoundaryInf: inferior boundary of the position of the main MMA peak.

MainPeakBoundarySup: superior boundary of the position of the main MMA peak.

Coef: coefficient used to weight the MMA-only spectrum.

CoefSaved: the current best coefficient used to weight the MMA-only spectrum.

Delta: increment for the coefficient.

Area: area of the negative part of the second derivative of OrigSpectrumData − RemSpectrumData ∗ coef.

AreaSaved: current minimum area.

Precision: maximum precision for the coefficient.

MaxLoop: maximum number of loop.

n: current number of iteration.

B. An Example of the Choice of the Region of Interest Boundaries for the Same Spectrum for the Crystallinity Index

In Figure 11(a), if we choose the (536, 634) cm^{-1} boundaries, compared to the (531, 659) cm^{-1} (Figure 11(b)), we can see the Crystallinity Index varies from 0.040 to 0.035. The chosen boundaries for the automatic deconvolution are those of the bottom figure (531, 659) cm^{-1}.

C. Searching the Position of Underlying Peaks in a Spectrum: Commentary on the Method of Second Derivative

The purpose of this study is to balance advantage/disadvantage of the method using the second derivative for the analysis of spectra.

The second derivative of the Gaussian function $(d^2/dx^2\,(a * e^{-(x-b)^2/2c^2}))$ gives a peak located at the same abscissa for the maximal amplitude, but its pseudofull width at half maximum (FWHM) is around $0.53\omega0$ ($\omega0$ is the FWHM of the original Gaussian).

The main advantage is that peaks are well discerned using the second derivate compared to the original Gaussian, and then, this value ($0.53\ \omega0$) permits to better discriminate peaks in spectra. An example is given in Figure 12. Taking a Gaussian (blue) with FWHM = 1, after calculation of the second derivative of this function (green), we can see that the peaks are located at the same abscissa, and the second

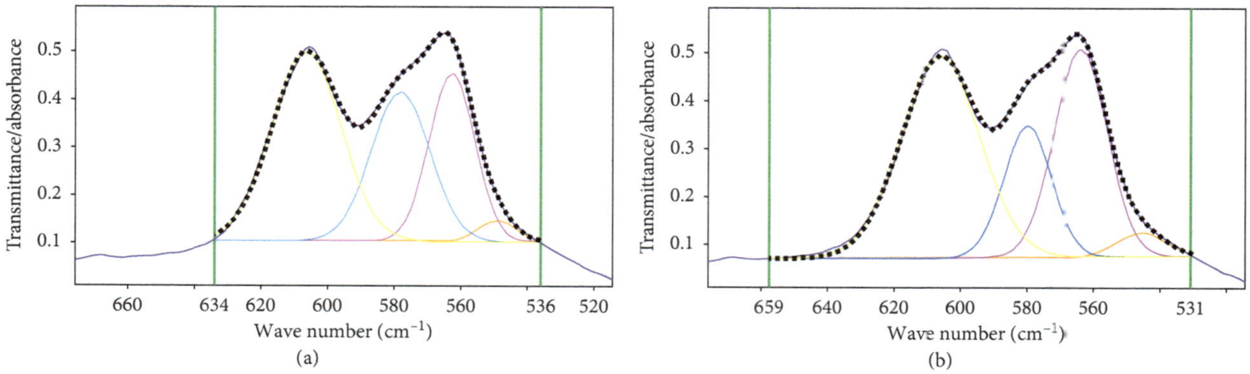

Figure 11

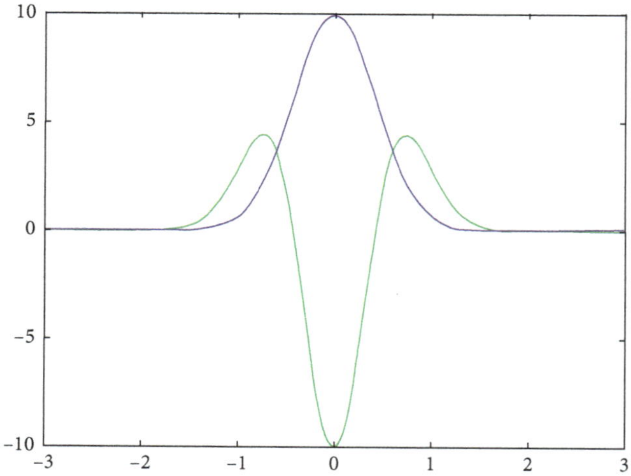

Figure 12: Gaussian with FWHM = 1 (blue) and its second derivative (green).

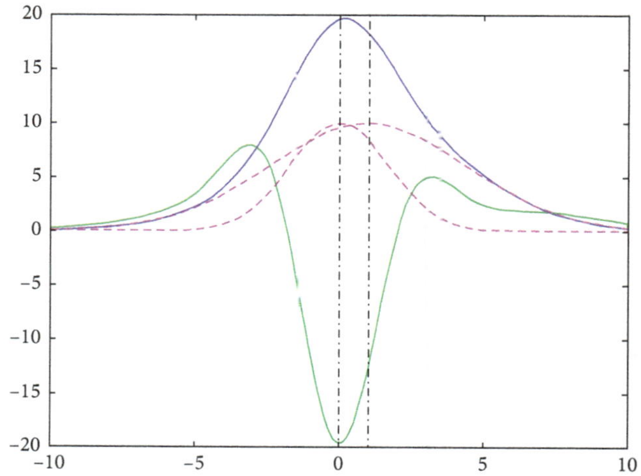

Figure 13: Sum of two Gaussians (blue), its second derivative (green), and the individual peaks (pink). It shows that only one peak is found. The wider peak is hidden in the right part of the second derivative.

derivative also shows a thinner peak (Figure 12, an example of a Gaussian with its FWHM = 1 and, for visualization, its second derivative normalized by the maximum of the Gaussian).

However, the second derivative brings up two major disadvantages:

- The emergence of two perturbations from either side of the main peak (cf. the 2 small green peaks around the main one).
- The maximal amplitude of the second derivative (A'') depends not only on the maximal amplitude of the original Gaussian A, but also of its FWHM ($A'' \approx 2.77 * A/\omega 0^2$). In other words, thinner peaks in the original spectrum will be more present and visible in the second derivative than wider peaks. In some cases, it may happen that the wider peak in the original spectrum will be hidden (in the second derivative) by the thinner one (Figure 13).

If we use more complex spectra, the discrimination of peaks becomes more difficult. An example is given in Figure 14. As previously mentioned, the 5 original Gaussian curves (pink) are summed into one spectrum (blue). The second derivative (green) curve shows only 3 minima, which corresponds to the thinner or isolated original peaks. The position of the peak is also slightly shifted. In an ideal analysis, we should get 5 minima corresponding to the 5 original peaks (in abscissa).

A simple example is given in Figure 13. Two simple original Gaussian curves (dotted pink, vertical dotted line showing the position of the maximal amplitude for each) are summed into one spectrum (plain blue line). The second derivative of this spectrum is shown in green. As we can see, this method does not allow the obtaining of the 2 initial pink peaks. One of the wide pink peaks has been "hidden" after using the second derivative method.

And finally, if we introduce noise and a Savitzky–Golay smoothing (commonly find in the spectra treatment), other peaks start to appear and the position of peaks slightly shifts (Figure 15).

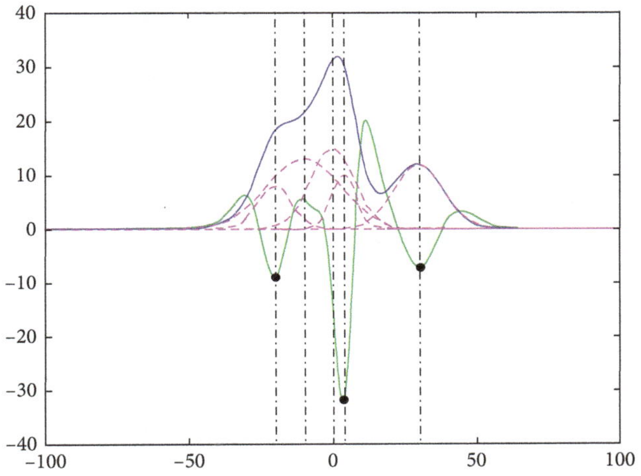

FIGURE 14: Second derivative method applied to a 5-peak spectrum (in dashed-black line, the true position of peaks. The found peaks are highlighted with a black dot).

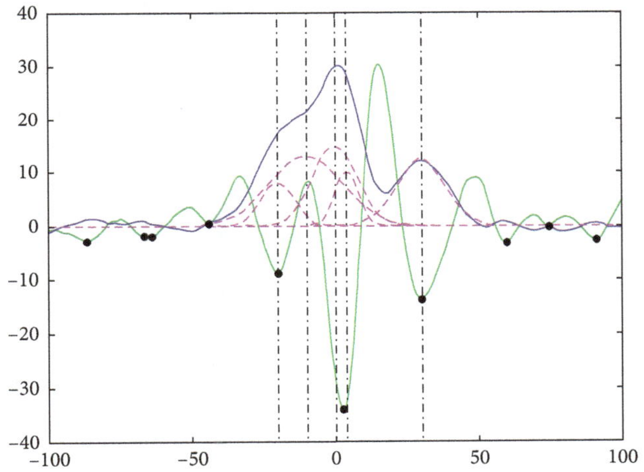

FIGURE 15: Second derivative applied to the same spectrum as Figure 14, with the introduction of noise, and then, a Savitzky–Golay smoothing is applied.

An example is given in Figure 15. As previously mentioned, the 5 original Gaussian curves (pink) are summed into one spectrum. This curve is degraded by adding noise, and then, a smoothing is applied (Savitzky–Golay smoothing), to correspond to the "classical" spectra we obtain in the bone (blue curve). The second derivative (green) curve shows more minima (9) than the original Gaussian curves. The position of the peak is also slightly shifted.

The method second derivative is applied to our bone sample which is shown in Figure 16.

One of the domains of the bone spectrum of the bone is shown between 900 and 1200 cm^{-1} (phosphate), with smoothing and baseline correction. The second derivative (green) is calculated, and outcomes enounced previously are visible on the graph. There are 5 minima. One appears (the

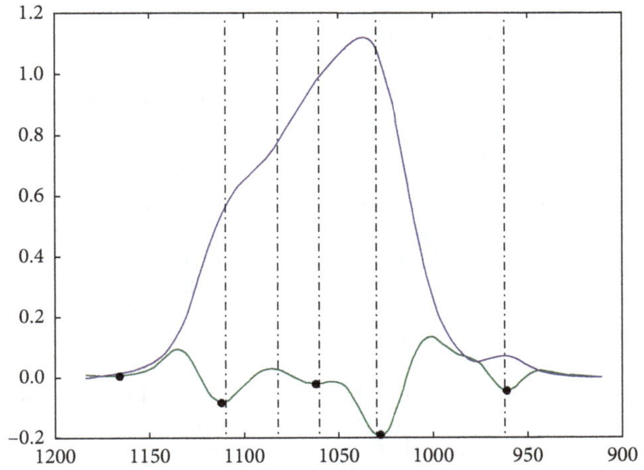

FIGURE 16: Location of peaks by the second derivative method on a bone spectrum between 900 and 1200 cm^{-1}. The approximate position of peaks is marked in dashed-black lines. A peak is not detected by this method, and another one is created on the left part of the spectrum.

first one) on the left and does not correspond to the presence of a real peak, and one disappears around (1100–1050 cm^{-1}).

In conclusion, the second derivative method shows great advantages (peak thinning, easier discrimination, and localization of maxima in the case of a simple curve). However, in the case of complex spectra, the "purely mathematical" aspect of this method does not take into account the real composition of the analyzed samples and may give partially false results. In our lab, we decided to use another method: the peak fitting using a fix number of peaks, previously defined in [11, 12, 15], and based on the composition of the material (bone).

Conflicts of Interest

There are no conflicts to declare.

Acknowledgments

This study was partly funded by the ANR project MULTIPS (ANR-13-BS09-0006) and by the ANR project LYSBONE (ANR-15-CE14-0010), which provided the mice. The authors would like to thank Pr. Gérard Panczer from Université Claude Bernard Lyon 1 for his expert advice in FTIR analysis and his scientific review of the manuscript.

References

[1] S. C. Cowin, *Bone Mechanics Handbook*, CRC Press, Boca Raton, FL, USA, 2001.

[2] D. Chappard, M. F. Baslé, E. Legrand, and M. Audran, "New laboratory tools in the assessment of bone quality," *Osteoporosis International*, vol. 22, no. 8, pp. 2225–2240, 2011.

[3] L. Imbert and A. Boskey, "Effects of drugs on bone quality," *Clinical Reviews in Bone and Mineral Metabolism*, vol. 14, no. 3, pp. 167–196, 2016.

[4] D. Farlay and G. Boivin, "Bone mineral quality," in *Osteoporosis, Book*, Y. Dionyssiotis, Ed., IntechOpen Limited, London, UK, 2012.

[5] E. P. Paschalis, K. Verdelis, S. B. Doty, A. L. Boskey, R. Mendelsohn, and M. Yamauchi, "Spectroscopic characterization of collagen cross-links in bone," *Journal of Bone and Mineral Research*, vol. 16, no. 10, pp. 1821–1828, 2001.

[6] A. L. Boskey, L. Spevak, and R. S. Weinstein, "Spectroscopic markers of bone quality in alendronate-treated postmenopausal women," *Osteoporosis International*, vol. 20, no. 5, pp. 793–800, 2009.

[7] A. L. Boskey, E. Donnelly, E. Boskey et al., "Examining the relationships between bone tissue composition, compositional heterogeneity, and fragility fracture: a matched case-controlled FTIRI study," *Journal of Bone and Mineral Research*, vol. 31, no. 5, pp. 1070–1081, 2015.

[8] E. Durchschlag, E. P. Paschalis, R. Zoehrer et al., "Bone material properties in trabecular bone from human iliac crest biopsies after 3- and 5-year treatment with risedronate," *Journal of Bone and Mineral Research*, vol. 21, no. 10, pp. 1581–1590, 2006.

[9] R. Y. Huang, L. M. Miller, C. S. Carlson, and M. R. Chance, "In situ chemistry of osteoporosis revealed by synchrotron infrared microspectroscopy," *Bone*, vol. 33, no. 4, pp. 514–521, 2003.

[10] C. Brolly, J. Parnell, and S. Bowden, "Raman spectroscopy: caution when interpreting organic carbon from oxidising environments," *Planetary and Space Science*, vol. 121, pp. 53–59, 2016.

[11] D. Farlay, M.-E. Duclos, E. Gineyts et al., "The ratio 1660/1690 cm^{-1} measured by infrared microspectroscopy is not specific of enzymatic collagen cross-links in bone tissue," *PLoS One*, vol. 6, no. 12, Article ID e28736, 2011.

[12] Y. Bala, D. Farlay, P. D. Delmas, P. J. Meunier, and G. Boivin, "Time sequence of secondary mineralization and microhardness in cortical and cancellous bone from ewes," *Bone*, vol. 46, no. 4, pp. 1204–1212, 2010.

[13] W. F. Maddams, "The scope and limitations of curve fitting," *Applied Spectroscopy*, vol. 34, no. 3, pp. 245–267, 1980.

[14] E. P. Paschalis, E. DiCarlo, F. Betts, P. Sherman, R. Mendelsohn, and A. L. Boskey, "FTIR microspectroscopic analysis of human osteonal bone," *Calcified Tissue International*, vol. 59, no. 6, pp. 480–487, 1996.

[15] D. Farlay, G. Panczer, C. Rey, P. Delmas, and G. Boivin, "Mineral maturity and crystallinity index are distinct characteristics of bone mineral," *Journal of Bone and Mineral Metabolism*, vol. 28, no. 4, pp. 433–445, 2010.

[16] A. Savitzky and M. J. E. Golay, "Smoothing and differentiation of data by simplified least squares procedures," *Analytical Chemistry*, vol. 36, no. 8, pp. 1627–1639, 1964.

[17] J. Steinier, Y. Termonia, and J. Deltour, "Smoothing and Differentiation of Data by Simplified Least Square Procedure," *Analytical Chemistry*, vol. 44, no. 11, pp. 1906–1909, 1972.

[18] B. Stuart, *"Spectral Analysis", Infrared Spectroscopy: Fundamentals and Applications*, Wiley, Hoboken, NJ, USA, 2005.

[19] R. J. Meier, "On art and science in curve-fitting vibrational spectra," *Vibrational Spectroscopy*, vol. 39, no. 2, pp. 266–269, 2005.

14

Characterization and Identification of Coal and Carbonaceous Shale Using Visible and Near-Infrared Reflectance Spectroscopy

En Yang, Shirong Ge, and Shibo Wang

School of Mechatronic Engineering, China University of Mining and Technology, Xuzhou 221116, China

Correspondence should be addressed to Shibo Wang; wangshb@cumt.edu.cn

Academic Editor: Vincenza Crupi

Because of the high organic carbon concentration in carbonaceous shale, a large proportion of carbonaceous shales are often misclassified into coals using visible and near-infrared (VIS-NIR) reflectance spectroscopy in the field of coal-gangue identification of hyperspectral remote sensing of coal mine. In order to study spectral characterization of coal and carbonaceous shale, three bituminite samples and three carbonaceous shales were collected from a coal mine of China, and their spectral reflectance curves were obtained by a field spectrometer in the wavelength range of 350–2500 nm. Only one carbonaceous shale could be easily identified from the three bituminite samples according to obvious absorption valleys near 1400 nm, 1900 nm, and 2200 nm of its reflectance curve while the other two carbonaceous shales have similar reflectance curves to the three bituminite samples. The effect of carbon concentration on reflectance curve was simulated by the mixed powder of ultralow ash bituminite and clay in 0.5 mm grain size under various mixing ratios. It was found that absorption valleys near 1400 nm, 1900 nm, and 2200 nm of the mixed powder become not obvious when the bituminite content is more than 30%. In order to establish an effective identification method of coal and carbonaceous shale, 250 other samples collected from the same coal mine were divided into 150 training samples and 100 prediction samples. Principal component analysis (PCA) and Gauss radial basis kernel principal component analysis (GRB-KPCA) were employed to extract principal components (PCs) of continuum removed (CR) spectra of the training samples in eight selected wavelength regions which are related to the main mineral and organic compositions. Two support vector machine- (SVM-) based models PCA-SVM and GRB-KPCA-SVM were established. The results showed that the GRB-KPCA-SVM model had better identification accuracies of 94% and 92% for powder and nature block prediction samples, respectively.

1. Introduction

Coal is the main energy source in China and still the irreplaceable main energy in the foreseeable future. Carbonaceous shale which is sandwiched in coal seam and was formed during coal formation is mainly composed of minerals and organic carbon. During the coal production, lots of carbonaceous shales which are the main part of gangues are often cut together with coals and transported to the ground. These gangues which occupy a large amount of land on the ground can cause damages to the environment, and the ones which are mixed in coals increase the ineffective transports [1, 2].

Due to the huge coal demand and coal production capacity requirement, lots of advanced spectral technologies have been employed in coal mine investigation, coal-gangue identification, and coal property determination [3–6]. Reflectance spectrum is easily acquired using cheap instrument and suitable for online analysis in the visible and near-infrared (VIS-NIR) region [7]. However, in coal mine areas, most carbonaceous shales have similar black color to coals. Meanwhile, due to the high carbon concentration in carbonaceous shale, there is no obviously different spectral absorption feature between coals and many carbonaceous shales, and they may have similar spectral characterization in the VIS-NIR region. Therefore, a large proportion of carbonaceous shales are often misclassified into coals using VIS-NIR reflectance spectroscopy in the field of coal-gangue identification of hyperspectral remote sensing of coal mine [8–10].

Under the same measurement conditions, the difference between reflectance spectra of coal and carbonaceous shale mainly depends on the material compositions. Coal which is mainly composed of complicated organic matters including aliphatic structures, aromatic structures and oxygen-containing functional group structures often contains many minerals while carbonaceous shale which is mainly composed of clay minerals and quartz often contains many organic matters [11]. The spectral properties of coal and carbonaceous shale had been studied by some scholars in the VIS-NIR region [12, 13]. However, the effect of carbon concentration on the VIS-NIR reflectance spectra of coal and carbonaceous shale had seldom been investigated.

Some scholars have focused on various experiments, algorithms, and models to improve the spectral identification accuracies of coals and gangues including a large number of carbonaceous shales in coal mine areas, such as Mao's normalized difference coal index (NDCI) model [8], Le's improved distribution maps model, and Song's classification method based on combination of reflectance spectra in the VIS-NIR region and spectral emissivity in the thermal infrared (TIR) region [9, 10]. However, all the above-mentioned scholars have not accurately discriminated coals and carbonaceous shales by the reflectance spectra in the VIS-NIR region, and carbonaceous shale is the main type of gangue that reduces the classification accuracies of coals and gangues. For instance, in Song's experiment, based on the reflectance spectra in the VIS-NIR region, all carbonaceous shales were misclassified into coals in all samples including 12 coals and 9 carbonaceous shales [10]. Therefore, an effective identification method based on the reflectance spectra of coal and carbonaceous shale in the VIS-NIR region is very necessary to be studied.

In this paper, spectral reflectance characterization (350–2500 nm) of typical bituminite and carbonaceous shale samples of China was studied. The effect of carbon concentration on spectral reflectance curves of coal and carbonaceous shale in the VIS-NIR region was investigated through an experiment. An effective coal-carbonaceous shale identification method based on VIS-NIR reflectance spectroscopy was established.

2. Materials and Methods

2.1. Spectra and Compositions Acquisition of Typical Samples. Figure 1 shows six typical samples including three bituminite samples and three carbonaceous shales collected from Malan coal mine area in Shanxi province of China. It is obvious that the appearances of carbonaceous shales and bituminite samples in Figure 1 are similar.

Reflection spectra of rough or mat solid surfaces of remote samples are closer to reflection spectra of powder samples in larger grain size (0.25–1.6 mm) than smaller grain size to determine the compositions of the surfaces [14]. For powder samples of sedimentary rock, larger particle size (0.25–1.2 mm) displays lower reflectance, but shows more obvious spectral features in their spectral bands [12]. The ancillary files of sedimentary rocks in ASTER spectral library version 2.0 of JPL pointed out that reflectance spectra of fresh (rough) solid surfaces of sedimentary or metamorphic rocks could be approximately simulated by spectra of the very coarse particulate (0.5–1.5 mm) samples [15]. Some scholars had used reflectance spectra of powdered carbonate rock (a type of sedimentary rock) samples with grain size fractions between 0.125 mm and 0.5 mm and known chemical compositions to simulate reflectance spectra of fresh solid surfaces of block carbonate rock samples [16]. Coal and carbonaceous shale belong to sedimentary rock. Therefore, in this study, the six typical samples were crushed, air-dried, and sieved to achieve size fraction of 0.5 mm. These powder samples are divided into three fractions which were used for spectra measurement, X-ray fluorescence (XRF) analysis, and proximate analysis, respectively.

The sample powder in 0.5 mm grain size used for spectra measurement was placed in a culture dish, and the powder surface was smoothed with the culture dish cover. Then the powder surface was measured by an ASD FieldSpec field spectrometer (Analytical Spectral Devices Inc., USA) in a dark laboratory. The spectral range is 350–2500 nm, and the spectral resolution is 3 nm for 350–1000 nm, 8.5 nm for 1000–1800 nm, and 6.5 nm for 1800–2500 nm. As is shown in Figure 2, each sample was horizontal, and the fiber optic cable with field of view (FOV) of 25° of the spectrometer was 15 cm vertical above the surface of the sample. With the distance of 15 cm to the sample, the recorded spectrum was an average spectral reflectance of the surface area of 35 cm^2, which equals a circle of 6.5 cm in diameter. A halogen lamp at a distance of 15 cm above the surface of the sample was the light source, and the incidence angle was 45°. Ten spectral reflectance curves of each sample were recorded in the first direction, then the sample was rotated 90° in the horizontal plane and measured to record the second 10 curves and so on, obtaining 40 curves of the sample in all four directions. Mean of the 40 curves of each sample was calculated, and a total of 6 spectral curves of the samples in Figure 1 were obtained with a spectral dimension of 2151 in the wavelength range of 350–2500 nm.

Through XRF by an S8 Tiger spectrometer (Bruker Inc., Germany) and proximate analysis by a VMF10/6 volatile muffle furnace (Carbolite Inc., United Kingdom), an AAF12/18 ash muffle furnace (Carbolite Inc., United Kingdom), and a BS124S electronic analytical balance (Sartorius Inc., Germany), the contents of major mineral elements, volatile, and ash of these six samples were acquired. Categories (coal or rock) of these samples were determined by their ash yields according to ISO 11760 that ash yield of coal is less than or equal to 50% comparing with rock.

2.2. Spectra Acquisition of Bituminite and Clay Mixtures. Ultralow ash bituminite powder (2% ash yield, 8% volatile) with a grain size of 0.5 mm was mixed evenly with clay powder (62% kaolinite, 19% quartz, 10% hydromuscovite, 4% montmorillonite, 3% limonite, and 1.2% illite) with a grain size of 0.5 mm under various mixing ratios to simulate the effect of carbon concentration on reflectance

Figure 1: Six typical bituminite (a–c) and carbonaceous shale (d–f) samples of China.

curve. The weight mixing ratios of bituminite to clay were 0 : 100, 5 : 95, 10 : 90, . . ., 95 : 5, and 100 : 0 varying at the interval of 5%, and a total of 21 mixed powder samples were made. Reflectance spectra of these 21 samples were obtained under the same condition as the sample in Figure 2.

2.3. Spectra and Categories Acquisition of Modeling Samples. Two hundred other bituminite and carbonaceous shale samples collected from Malan coal mine were randomly divided into 150 training samples and 50 prediction samples in order to establish an effective identification model of coal and carbonaceous shale. All these 200 modeling samples were crushed, air-dried, and sieved to achieve a size fraction of 0.5 mm. Each powder sample was divided into two fractions which were used for spectra measurement and proximate analysis, respectively. The training set included 56 bituminite samples and 94 carbonaceous shales, and the prediction set included 22 bituminite samples and 28 carbonaceous shales according to ISO 11760 and the ash yields acquired through proximate analysis of all these 200 samples. Reflectance spectra of these 200 powder samples in 0.5 mm grain size were obtained under the same condition as the sample in Figure 2. The preprocessing method of

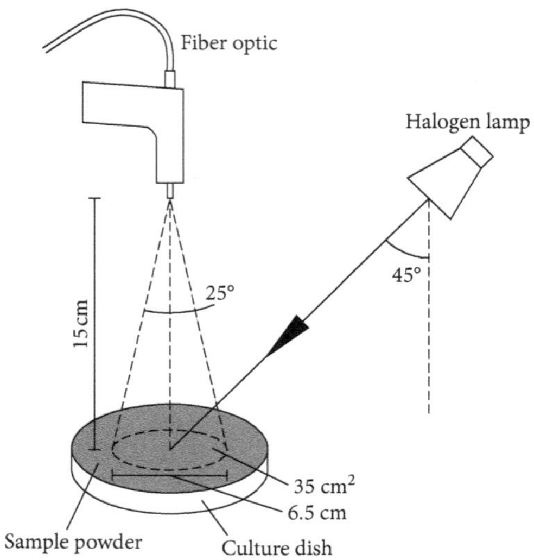

FIGURE 2: Reflectance spectra acquisition of powder sample in 0.5 mm grain size using a field spectrometer and a halogen lamp light source.

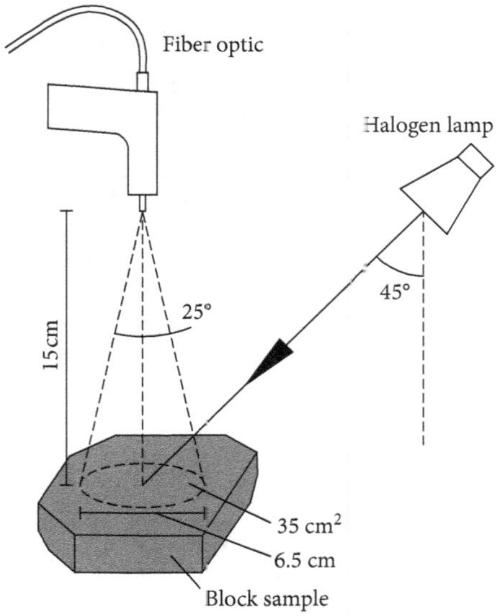

FIGURE 3: Reflectance spectra acquisition of the top solid surface of each block sample using a field spectrometer and a halogen lamp light source.

continuum removal (CR) was employed to remove background noise of the reflectance curves and isolate particular absorption features [17].

2.4. Spectra and Categories Acquisition of Block Samples. Fifty extra nature block samples collected from the same coal mine as above 200 samples were air-dried and measured using the same field spectrometer and halogen lamp in the same dark laboratory. Reflectance spectra of these 50 block samples were used to verify the practicability of the identification model established by the training samples discussed above. As is shown in Figure 3, each block sample was horizontal, and the fiber optic cable of the spectrometer was vertical above the top solid surface of the sample. The recorded spectrum was the average spectral reflectance of the circular area of about 35 cm^2. As the sample in Figure 2, mean of 40 reflectance curves in all four directions of each sample was acquired in the wavelength range of 350–2500 nm. The spectral reflectance curves of top solid surfaces of the 50 block samples were preprocessed by CR.

After reflectance spectra acquisition, the small sample pieces were cut from the measured circular area of the top solid surface of each sample and crushed into powder for proximate analysis. The 50 block samples included 19 bituminite samples and 31 carbonaceous shales according to ISO 11760 and the ash yields acquired through proximate analysis of these 50 cut samples.

2.5. Kernel Principal Component Analysis. Principal component analysis (PCA) is used for spectral feature extraction, and good results can be achieved when dealing with linear problems [18–20]. However, this method has some limitations when dealing with many spectral analyses, which are often not simple linear problems. Kernel principal component analysis (KPCA) is based on the PCA [21–23]. With the kernel function introduced, the data are mapped into a high-dimensional space, and then, the nonlinear features are extracted by the PCA method, thus the unsatisfactory results of PCA under nonlinear data distribution are improved.

The procedures of KPCA and PCA are similar, but in KPCA, the kernel function is used instead of the original data [22]. For the linearly inseparable spectral data set $\mathbf{X}_{l\times p}$ ($\mathbf{X}_{l\times p} \subset \mathbf{R}^p$) of the samples, where $\mathbf{X}_{l\times p} = [\mathbf{x}_1, \mathbf{x}_2, \ldots, \mathbf{x}_l]^T$, $\mathbf{x}_i = [\mathbf{x}_{i,1}, \mathbf{x}_{i,2}, \ldots, \mathbf{x}_{i,p}]$, $i = 1, 2, \ldots, l$, l is the number of samples and p is the spectral dimension, and they could be mapped into a higher dimension space F through nonlinear transformation Φ.

$$\begin{cases} \Phi : \mathbf{R}^p \longrightarrow \mathbf{F} \\ \mathbf{x}_i \longrightarrow \Phi(\mathbf{x}_i) \end{cases} \quad (1)$$

$\Phi(\mathbf{X}_{l\times p})$ is defined as

$$\Phi(\mathbf{X}_{l\times p}) = [\Phi(\mathbf{x}_1), \Phi(\mathbf{x}_2), \ldots, \Phi(\mathbf{x}_l)]^T, \quad (2)$$

and kernel function matrix is defined as

$$\mathbf{K} = K(\mathbf{X}_{l\times p}) = \Phi(\mathbf{X}_{l\times p})\Phi(\mathbf{X}_{l\times p})^T = \begin{bmatrix} k(\mathbf{x}_1, \mathbf{x}_1) & \cdots & k(\mathbf{x}_1, \mathbf{x}_l) \\ \vdots & \ddots & \vdots \\ k(\mathbf{x}_l, \mathbf{x}_1) & \cdots & k(\mathbf{x}_l, \mathbf{x}_l) \end{bmatrix}, \quad (3)$$

where kernel function $k(\mathbf{x}_i, \mathbf{x}_j) = \Phi(\mathbf{x}_i)^T \Phi(\mathbf{x}_j)$ $i, j = 1, 2, 3 \ldots l$. The covariance matrix $\mathbf{C}$ in F is

$$\mathbf{C} = \frac{1}{l}\Phi(\mathbf{X}_{l\times p})^T \Phi(\mathbf{X}_{l\times p}) = \frac{1}{l}\sum_{i=1}^{l} \Phi(\mathbf{x}_i)^T \Phi(\mathbf{x}_i). \quad (4)$$

Eigenvector $\mathbf{V}$ and eigenvalue λ of $\mathbf{C}$ satisfy the relationship $\mathbf{CV} = \lambda\ \mathbf{V}$ and $\boldsymbol{\alpha} = (\alpha_1, \alpha_2, \ldots, \alpha_l)$ exists, making

$$V = \sum_{i=1}^{l} \alpha_i \Phi(\mathbf{x}_i). \quad (5)$$

Based on Equation (5), the relationship $\mathbf{K\alpha} = l\lambda\ \boldsymbol{\alpha}$ can be concluded and $\boldsymbol{\alpha}$ can be calculated. Based on Equation (5) and $\boldsymbol{\alpha}$, the nonlinear features of any inputted spectral data $\mathbf{x}$ in the higher dimension space $\mathbf{F}$ can be expressed by the following equation:

$$(\mathbf{V}, \Phi(\mathbf{x})) = \sum_{i=1}^{l} \alpha_i k(\mathbf{x}_i, \mathbf{x}). \quad (6)$$

The above process is derived when the following equation,

$$\sum_{i=1}^{l} \Phi(\mathbf{x}_i) = 0, \quad (7)$$

is assumed, but in actual spectral data, K' is used instead of $\mathbf{K}$ by

$$K' = \mathbf{K} - \mathbf{LK} - \mathbf{KL} + \mathbf{LKL}, \quad (8)$$

where

$$\mathrm{L} = \left(\frac{1}{l}\right)_{l\times l}. \quad (9)$$

In this study, Gauss radial basis (GRB) kernel function $k(\mathbf{x}_i, \mathbf{x}_j)$ is chosen and expressed by

$$k(\mathbf{x}_i, \mathbf{x}_j) = \exp\left(-\frac{\|\mathbf{x}_i - \mathbf{x}_j\|^2}{2\sigma^2}\right), \quad (10)$$

where $\mathbf{x}_i$ and $\mathbf{x}_j$ are the spectral data of the i-th and j-th sample and σ is the width parameter.

2.6. Support Vector Machine. Support vector machine (SVM) is a machine-learning algorithm based on statistic learning theory (SLT) established by Vapnik et al. in 1990s, in which structural risk minimization (SRM) criterion is used to reduce the upper bound of the model generalization error while minimizing the sample error in order to improve the generalization ability of the model [24]. SVM can solve the problem of pattern recognition in small samples and nonlinear and high-dimensional data space [25, 26]. When dealing with nonlinear problems, the support vector classification (SVC) maps data in original space into high-dimensional space and constructs an optimal classification plane in the high-dimensional space so that the distances of all samples from the plane are maximized. Radial basis function (RBF) is chosen as the kernel function of SVM in this study.

2.7. The Proposed Procedure for Establishment of Coal-Carbonaceous Shale Prediction Models. Principal components (PCs) were obtained through PCA and KPCA of CR spectra of the training samples in eight selected wavelength regions which are related to the main mineral and organic compositions. With the PCs combined with the categories (coal and rock) as input parameters, the SVM model was trained.

The optimal width parameter σ in formula (10) of Gauss radial basis kernel principal component analysis (GRB-KPCA) was determined by cross validation. For each width parameter σ from 1 to 100 in step length of 1, the maximum average prediction accuracy (PA) of validation set of the training set was obtained through 5-fold cross validation in SVM. At the same time, based on the maximum average PA, the optimal penalty factor c and RBF variance parameter γ of SVM for each σ were obtained. Then the optimal width parameter σ was determined by the maximum value of all maximum average PAs of all width parameters σ. GRB-KPCA-SVM model was established according to the abovementioned procedure. Figure 4 shows flowcharts of procedures of the two prediction models: PCA-SVM and GRB-KPCA-SVM.

Prediction accuracies (PAs) of categories of the 50 powder and 50 nature block prediction samples were employed to assess the prediction models. Establishment of prediction models and assessment of the models were all performed in Matlab R2013b software (MathWorks Inc., USA).

3. Results and Discussion

3.1. Spectral Reflectance Curves and Compositions of Typical Samples. Spectral reflectance curves of the six representative samples in Figure 1 in the wavelength range of 350–2500 nm are shown in Figure 5. Eight spectral feature bands are marked using No. 1–8 in Figure 5, and interactive processes of the related major compositions are shown in Table 1. In Figure 5, absorption valleys of 1400 nm (5) (5 is the spectral feature band number, and so on), 1900 nm (6), and 2200 nm (7) bands of the reflectance curve of carbonaceous shale (f) are more obvious than those of the other five samples. Carbonaceous shale (f) could be easily identified from the three bituminite samples according to the above three obvious absorption valleys, while the other two carbonaceous shales have similar reflectance curves to the three bituminite samples. The reflectance curves of carbonaceous shale (d) and bituminite (c) have similar obvious absorption valleys of 700 nm (2) and 870 nm (3) bands. The frequent absorption valleys of 400–550 nm (1) and 1000–1100 nm (4) bands appear in all reflectance curves, and the typical strong absorption bands from 2300 nm to 2500 nm (8) resulting from organic matters appear in all reflectance curves, too. There is another obvious different feature between reflectance curve of carbonaceous shale (f) and that of the other five samples that the overall shape of the reflectance curve of carbonaceous shale (f) is straight while the other five are obviously concave.

Table 2 shows the contents of major mineral elements (SiO_2, Al_2O_3, and Fe_2O_3) through XRF analysis and the contents of volatile and ash through proximate analysis of these six samples in Figure 1. Matrix of XRF refers to C, H, O, and N in organic matters and matrix of carbonaceous shale (d) and (e) (47.80% and 41.25%) is much more than that of carbonaceous shale (f) (20.24%). This may be the

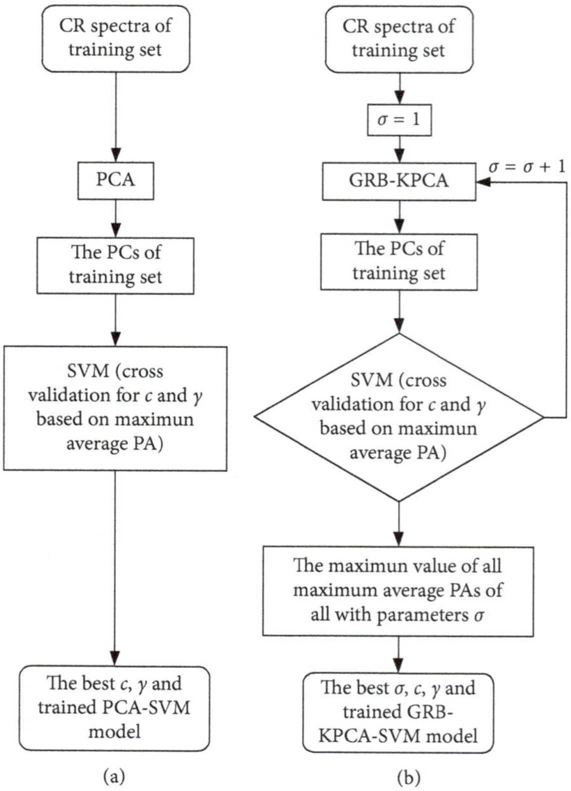

FIGURE 4: Flowcharts of procedures of PCA-SVM (a) and GRB-KPCA-SVM (b) models.

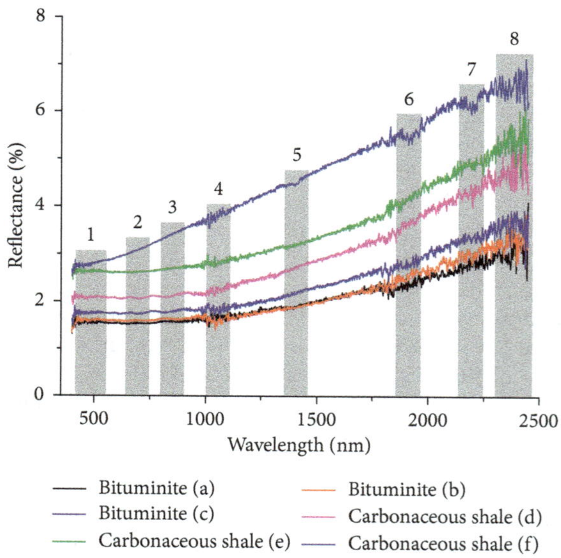

FIGURE 5: Spectral reflectance curves of the six representative samples in Figure 1 in the wavelength range of 350–2500 nm.

main reason why reflectance curves of carbonaceous shale (d) and (e) are similar to that of the three bituminite samples [12, 31, 32]. The obvious absorption valleys of 1400 nm, 1900 nm, and 2200 nm bands of the reflectance curve of carbonaceous shale (f) depends on the low matrix content (20.24%) and clay and quartz minerals from which the high ash yield (78.94%) originates [27, 28]. The top two highest ferriferous mineral element contents (3.39% and 3.14% Fe_2O_3) in Table 2 are the main cause of obvious absorption valleys of 700 nm and 870 nm bands of carbonaceous shale (d) and bituminite (c) [27, 28].

According to the main compositions in Table 2 and the major related interactive processes in Table 1, the eight selected wavelength regions in Table 1 can reflect the main composition information of a bituminite or carbonaceous shale sample. From the waveforms of reflectance curves in Figure 5 and composition contents in Table 2, it can easily be deduced that matrix content and ash yield which are correlated to organic matters and minerals of a sample are the key factors to determine the waveform of reflectance curve of the sample. In other words, the effect of carbon concentration on the VIS-NIR reflectance spectra of coal and carbonaceous shale should be systemically investigated.

3.2. Spectral Reflectance Characterization of Bituminite and Clay Mixtures. Figure 6 shows spectral reflectance curves of the 21 mixed powder samples of dry ultralow ash bituminite and clay in 0.5 mm grain size in the wavelength range of 350–2500 nm. The bituminite content increases from 0% to 100% at the interval of 5%. The eight spectral feature bands of Table 1 are also marked in Figure 6. The pure clay powder has the highest spectral reflectance curve. However, spectral reflectance curve of the mixed powder of 5% bituminite

Table 1: Spectral regions related to the major compositions [27–30].

No.	Absorption band (nm)	Major related interactive process	Selected wavelength region (nm)
1	400, 430, 450, 490, 510, 520, 550	Fe^{2+} and Fe^{3+} in ferriferous minerals	400–550
2	700	Fe^{3+} in ferriferous minerals	650–750
3	870	Fe^{3+} in ferriferous minerals	820–920
4	1000–1100	Fe^{2+} in ferriferous minerals	1000–1100
5	1400	OH^- vibrations in clay minerals	1350–1450
6	1900	H_2O vibrations in milky quartz	1850–1950
7	2200	Al–OH vibrations in clay minerals	2150–2250
8	2300–2500	Overtones and combinations of organic absorption groups	2300–2500

Table 2: X-ray fluorescence (XRF) and proximate analysis of samples in Figure 1.

Sample	XRF analysis (wt. %)				Proximate analysis (wt. %)	
	SiO_2	Al_2O_3	Fe_2O_3	Matrix	Volatile	Ash
Bituminite (a)	8.63	5.01	0.87	82.60	13.74	16.05
Bituminite (b)	10.61	6.17	0.81	79.12	16.14	18.22
Bituminite (c)	10.68	8.40	3.39	73.94	21.57	23.95
Carbonaceous shale (d)	26.46	19.98	3.14	47.80	7.86	51.80
Carbonaceous shale (e)	32.13	21.20	1.04	41.25	4.31	57.55
Carbonaceous shale (f)	45.65	30.14	0.92	20.24	1.51	78.94

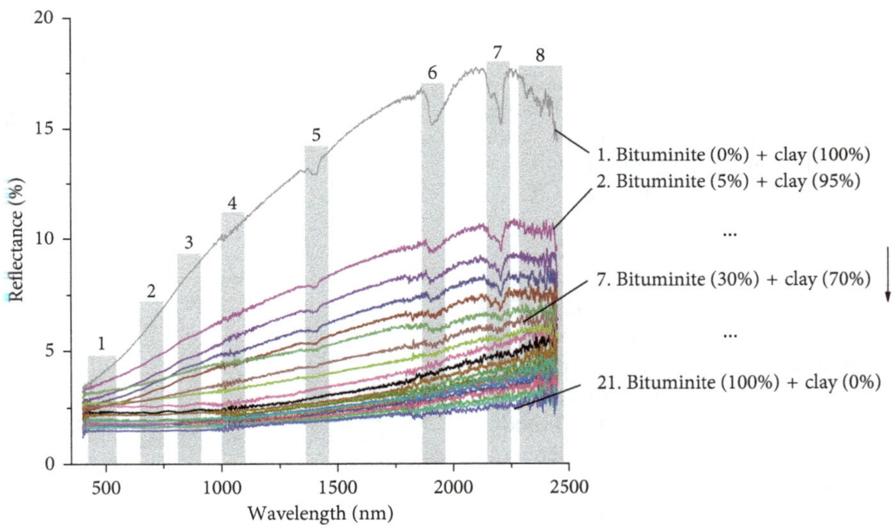

Figure 6: Spectral reflectance curves of 21 mixed powder samples of dry bituminite and clay in 0.5 mm grain size under different mixing ratios in the wavelength range of 350–2500 nm.

content dramatically falls comparing with the curve of pure clay powder especially in the near-infrared (NIR) region (780–2500 nm). When bituminite content is more than 5%, with increasing bituminite content, spectral reflectance curve of the mixed powder falls at a smaller interval. With increasing bituminite content, the depths of absorption valleys of 1400 nm, 1900 nm, and 2200 nm bands of the reflectance curve all decrease. When the bituminite content is more than 30%, absorption valleys of 1400 nm, 1900 nm, and 2200 nm bands of the reflectance curve become not obvious, and the overall shape of the reflectance curve changes from convex to concave. The overall shapes of the curves are more dense, similar, and horizontal when the bituminite contents are more than 40%. The pure bituminite powder has the lowest spectral reflectance curve, and the overall shape of the curve is close to the horizontal line.

The experimental results of mixed powder of bituminite and clay explain the spectral reflectance curves of the six representative samples in Figure 5. It could be deduced that it is difficult to identify carbonaceous shale from bituminite according to the waveform of spectral reflectance curve of the sample in the VIS-NIR region when carbon concentration of the carbonaceous shale is more than about 30%. However, the material composition information contained in the reflectance spectra in the eight selected wavelength regions in Table 1 might provide reference for identifying carbonaceous shale with high carbon concentration from bituminite. And an effective identification method based on the reflectance spectra of coal and carbonaceous shale in the VIS-NIR region should be studied.

3.3. Spectra of Modeling Samples and Establishment of Identification Models. In order to establish an effective identification model of coal and carbonaceous shale, spectral reflectance curves of 200 powder samples in 0.5 mm grain size including 150 training samples and 50 prediction samples in the wavelength range of 350–2500 nm were obtained and shown in Figure 7(a). Figure 7(b) shows the 200 spectral curves preprocessed by CR. It is obvious that less reflectance curves (12 samples) with overall convex shapes could be easily identified from others according to the CR method in Figure 7(b). But it is difficult to identify most of the samples because of many overlappings, crossovers, and similarities among these spectral curves.

Figure 7(b) indicates that the absorption features are amplified. CR spectra of the training set in the eight selected wavelength regions in Table 1 were used to establish identification models. PCA was first employed to eliminate irrelevant information and reduce dimensionality of preprocessed spectra in the eight regions. Table 3 lists cumulative contribution rates of the first ten PCs of CR spectra of the training set in the eight regions through PCA. The cumulative contribution rate of the first three PCs is 90.0403%, and the 3D spatial distribution of the first three PCs of the 150 training samples through PCA is exhibited in Figure 8(a). Figure 8(a) indicates that a large part of rock (carbonaceous shale) balls are mixed in coal balls.

The first ten PCs of the training set through PCA combining with the categories (0 stands for coal and 1 stands for rock) were taken as input variables to train the SVM model. 5-fold cross validation was employed to determine the optimal penalty parameter c and radial basis function variance parameter γ. The search ranges of c and γ were both from 2^{-10} to 2^{10}, and step length of the exponent was 0.5. The process of determining the optimal parameter c and γ is shown in Figure 8(b). With $c = 2^{1.5}$ and $\gamma = 2^{-2}$ as optimal parameters, the maximum average PA of categories of the validation set is 90.67%.

KPCA which has more abilities to deal with the nonlinear data distribution might get a better result when employed to reduce dimensionality of preprocessed spectra of the training set in the eight regions. When the width parameter σ in Formula (10) of GRB-KPCA increases from 1 to 100 in step length of 1, Figure 9 shows the variation trend of the maximum average PA of categories of the validation set through GRB-KPCA-SVM. As is shown in Figure 9, the optimal width parameter σ is 63 for GRB-KPCA-SVM. Table 3 lists the cumulative contribution rates of the first ten PCs of CR spectra of the training set in the eight regions through GRB-KPCA (σ = 63). The cumulative contribution rate of the first three PCs

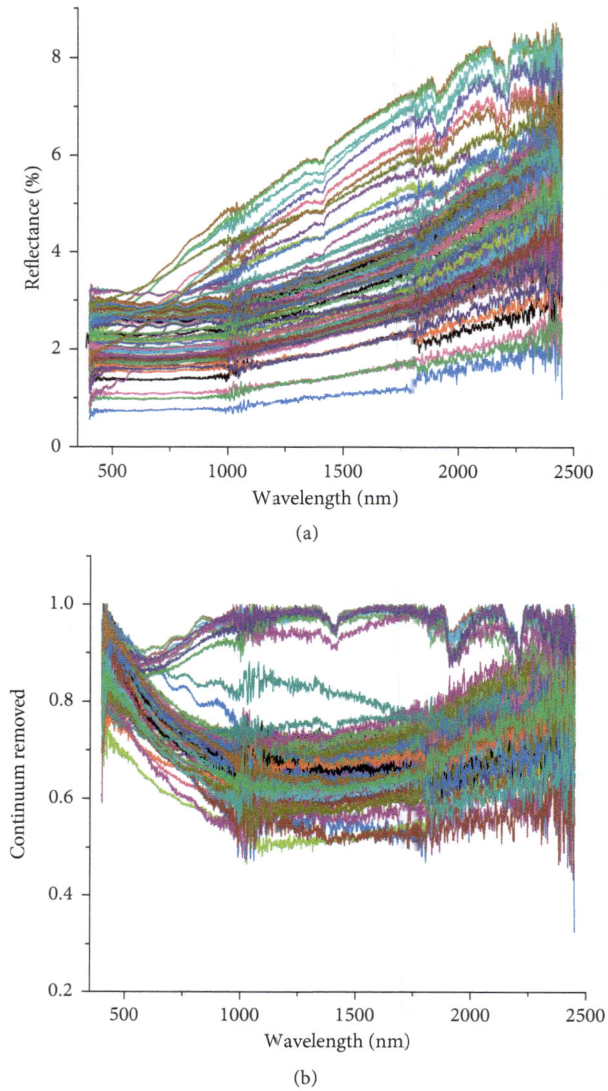

Figure 7: Spectral reflectance curves (a) and continuum-removed spectral curves (b) of 200 modeling samples in the wavelength range of 350–2500 nm.

Table 3: Cumulative contribution rates of the first ten PCs of continuum-removed spectra of the 150 training samples in the eight selected wavelength range in Table 1.

PC	PCA (%)	GRB-KPCA (σ = 63) (%)
PC1	50.0150	62.1243
PC2	82.8114	92.8212
PC3	90.0403	98.1378
PC4	93.9476	98.2543
PC5	96.6978	98.3647
PC6	97.0984	98.7834
PC7	97.2688	99.0579
PC8	98.3690	99.1521
PC9	98.6629	99.3829
PC10	98.9098	99.9251

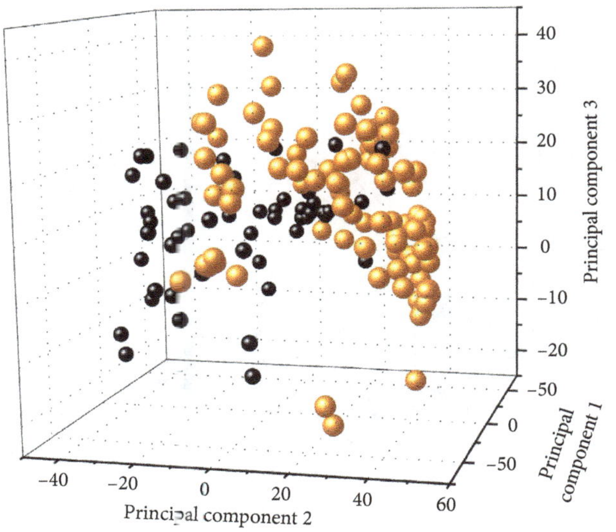

(a)

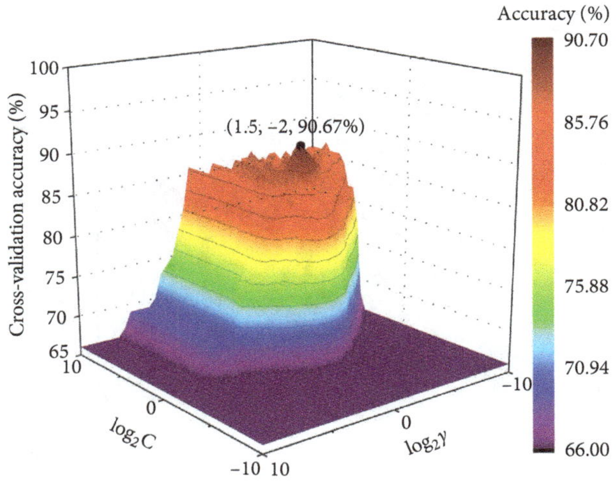

(b)

FIGURE 8: (a) 3D spatial distribution of the first three PCs of CR spectra of the 150 training samples in the eight selected regions in Table 1 through PCA; (b) process of searching for the optimal penalty parameter c and the radial basis function variance parameter γ of PCA-SVM model through cross validation with the first ten PCs as input variables.

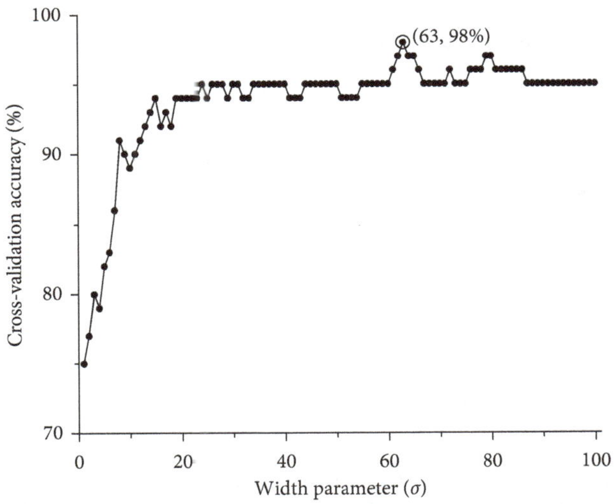

FIGURE 9: Process of determining the optimal parameter σ of GRB-KPCA: the variation trend of the maximum average PA of categories of the validation set through GRB-KPCA-SVM with the change of σ.

is 98.1378%, and the 3D spatial distribution of the first three PCs of the 150 training samples through GRB-KPCA ($\sigma = 63$) is exhibited in Figure 10(a). Figure 10(a) indicates that the 3D spatial cluster of coal and rock (carbonaceous shale) balls is better than that of PCA, and the mixed part of balls decreases.

When the width parameter σ in Formula (10) of GRB-KPCA is 63, the first ten PCs of the training set through GRB-KPCA combining with the categories (0 stands for coal and 1 stands for rock) were taken as input variables to train the SVM model. 5-fold cross validation was employed to determine the optimal penalty parameter c and radial basis function variance parameter γ. The search ranges of c and γ were also both from 2^{-10} to 2^{10} and step length of the exponent was also 0.5. The process of determining the optimal parameter c and γ is shown in Figure 10(b). The optimal parameters are $c = 2^4$ and $\gamma = 2^0$ based on GRB-KPCA-SVM, and the maximum average PA of categories of the validation set is 98%.

From the results of cross validation of the validation set, the dimension reduction method using KPCA is better than that of PCA for CR spectra of the training set in the eight selected wavelength regions in Table 1. Two coal-carbonaceous shale prediction models, PCA-SVM and GRB-KPCA-SVM, were established according to above processes.

3.4. Identification of Samples by the Prediction Models

3.4.1. Identification of Powder Prediction Samples. CR spectra of the 50 prediction samples in 0.5 mm grain size in the eight selected wavelength regions in Table 1 were input into the above PCA-SVM ($c = 2^{1.5}$, and $\gamma = 2^{-2}$) and GRB-KPCA-SVM ($\sigma = 63$, $c = 2^4$, and $\gamma = 2^0$) models, and the results predicted by the two models are shown in Figure 11.

3.4.2. Spectra and Identification of Block Samples. Spectral reflectance curves and CR curves of top solid surfaces of the 50 extra block samples in the wavelength

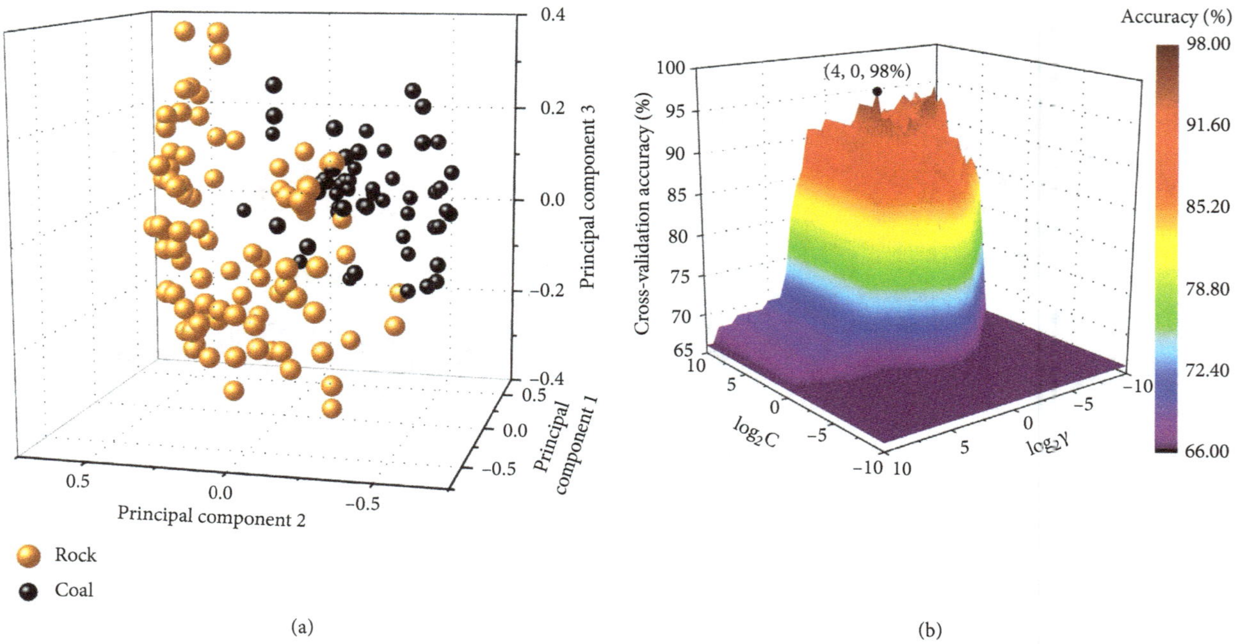

FIGURE 10: (a) 3D spatial distribution of the first three PCs of CR spectra of the 150 training samples in the eight regions in Table 1 through GRB-KPCA ($\sigma = 63$); (b) process of searching for the optimal penalty parameter c and radial basis function variance parameter γ of GRB-KPCA-SVM model through cross validation with the first ten PCs as input variables.

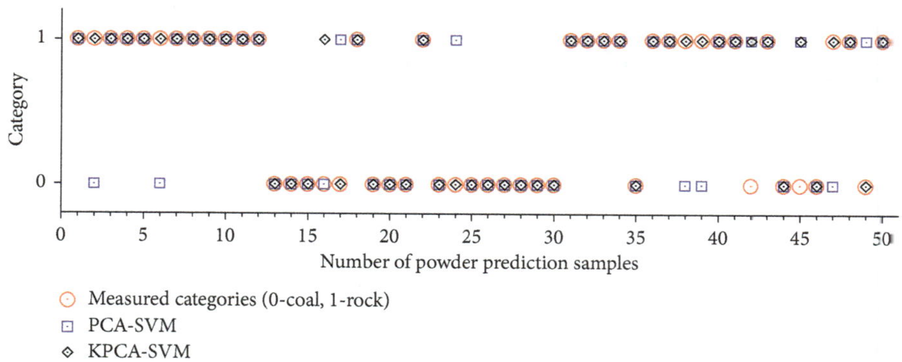

FIGURE 11: Prediction results of 50 powder prediction samples by PCA-SVM and GRB-KPCA-SVM models.

range of 350–2500 nm are shown in Figure 12(a) and Figure 12(b), respectively. Figure 12 indicates that spectral curves of block samples are more similar to spectral curves of powder samples in 0.5 mm grain size in Figure 7. Likewise, it is difficult to identify most of these block samples using the observed spectral curves in Figure 12.

In order to further verify the practicability of the above two trained models, categories of the 50 extra nature block samples were predicted. CR spectra of the 50 block samples in the eight selected wavelength regions in Table 1 were input into the above PCA-SVM ($c = 2^{1.5}$, and $\gamma = 2^{-2}$) and GRB-KPCA-SVM ($\sigma = 63$, $c = 2^4$, and $\gamma = 2^0$) models. The identification results of categories of the 50 block samples by the two models are shown in Figure 13.

3.4.3. Performance of Prediction Models. Identification accuracies of categories of the 50 powder prediction samples and the 50 extra block samples are listed in Table 4. The 150 training samples were also predicted by the two models and are listed in Table 4.

According to the results in Table 4, it can be seen that the categories of coal and carbonaceous shale samples based on GRB-KPCA-SVM have higher PAs comparing with PAs of PCA-SVM. The PA of 92% for nature block samples is acceptable, and GRB-KPCA-SVM is a more practicable and ideal coal-carbonaceous shale identification method. And the GRB-KPCA-SVM model provides a valid reference for coal-carbonaceous shale identification using VIS-NIR reflectance spectroscopy.

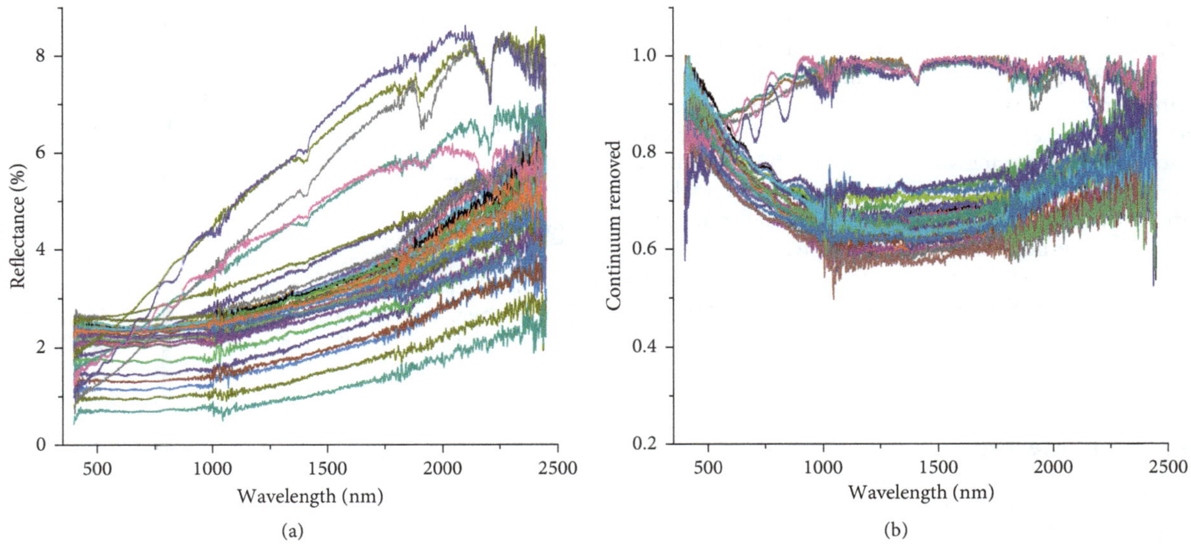

FIGURE 12: Spectral reflectance curves (a) and continuum-removed spectral curves (b) of 50 nature block samples in the wavelength range of 350–2500 nm.

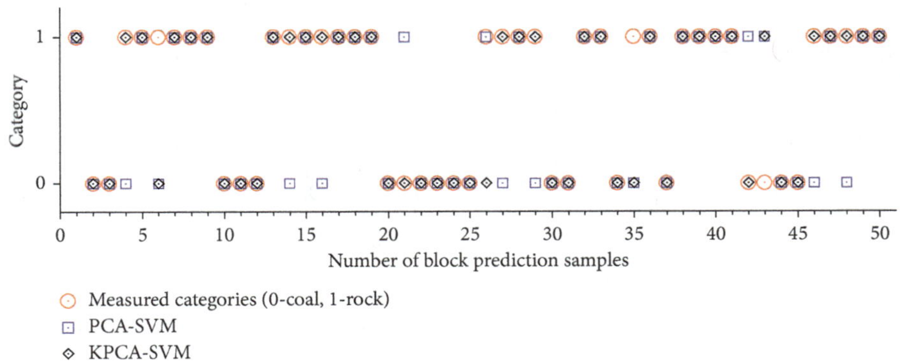

FIGURE 13: Prediction results of 50 nature block samples by PCA-SVM and GRB-KPCA-SVM models.

TABLE 4: Identification accuracies of categories of all bituminite and carbonaceous shale samples collected from Malan coal mine area in Shanxi province of China by the prediction models.

Model	150 training samples	50 prediction samples	50 block samples
PCA-SVM ($c = 2^{1.5}$, $\gamma = 2^{-2}$)	92%	80%	76%
GRB-KPCA ($\sigma = 63$)-SVM ($c = 2^4$, $\gamma = 2^0$)	100%	94%	92%

4. Conclusions

In order to solve the problem of misclassification between coal and carbonaceous shale using VIS-NIR reflectance spectroscopy in the field of coal-gangue identification of hyperspectral remote sensing of coal mine, spectral reflectance characterization of six typical bituminite, and carbonaceous shale samples of China in the wavelength range of 350–2500 nm were studied. In the six samples, two carbonaceous shales without obvious absorption valleys have similar reflectance curves to the three bituminite samples. Through simulation experiment of the effect of bituminite content on reflectance curve of mixed powder of ultralow ash bituminite and clay, it was found that absorption valleys near 1400 nm, 1900 nm, and 2200 nm of the mixed powder become not obvious when bituminite content is more than 30%. It could be deduced that when carbon concentration of the carbonaceous shale is more than about 30%, it is difficult to identify carbonaceous shale from bituminite according to the waveform of spectral reflectance curve of the sample in the VIS-NIR region.

Eight wavelength regions including 400–550 nm, 650–750 nm, 820–920 nm, 1000–1100 nm, 1350–1450 nm, 1850–1950 nm, 2150–2250 nm, and 2300–2500 nm which are related to the main mineral and organic compositions of the samples were selected. CR spectra of 150 other coal and carbonaceous shale samples of China in the eight selected wavelength regions were used to establish two coal-carbonaceous shale prediction models: PCA-SVM and GRB-KPCA-SVM. The two models were verified through predicting categories of 50 powder samples and 50 nature block samples. The GRB-KPCA-SVM model with prediction accuracies of 94% and 92% for powder and nature block prediction samples, respectively, is a more practicable and ideal coal-carbonaceous shale identification method. The basic principles and models in this paper provide references for coal-carbonaceous shale identification using VIS-NIR reflectance spectroscopy in coal mines.

Conflicts of Interest

The authors declare that there are no conflicts of interest regarding the publication of this paper.

Acknowledgments

This work was supported by the Joint Funds of the National Natural Science Foundation of China (grant numbers U1610251 and U1510116), the National Key Research and Development Program of China (grant number 2018YFC0604503), and the Priority Academic Program Development (PAPD) of Jiangsu Higher Education Institutions.

References

[1] Y. Z. Sun, J. S. Fan, P. Qin, and H. Y. Niu, "Pollution extents of organic substances from a coal gangue dump of Jiulong Coal Mine, China," *Environmental Geochemistry and Health*, vol. 31, no. 1, pp. 81–89, 2009.

[2] J. Fan, Y. Sun, X. Li et al., "Pollution of organic compounds and heavy metals in a coal gangue dump of the Gequan Coal Mine, China," *Chinese Journal of Geochemistry*, vol. 32, no. 3, pp. 241–247, 2013.

[3] D. Xiao, B. T. Le, Y. Mao et al., "Research on coal exploration technology based on satellite remote sensing," *Journal of Sensors*, vol. 2016, Article ID 8214801, 9 pages, 2016.

[4] Z. Song and C. Kuenzer, "Spectral reflectance (400–2500 nm) properties of coals, adjacent sediments, metamorphic and pyrometamorphic rocks in coal-fire areas: a case study of Wuda coalfield and its surrounding areas, northern China," *International Journal of Coal Geology*, vol. 171, pp. 142–152, 2017.

[5] X. Wang, K. Hu, L. Zhang, X. Yu, and E. Ding, "Characterization and classification of coals and rocks using terahertz time-domain spectroscopy," *Journal of Infrared Millimeter and Terahertz Waves*, vol. 38, no. 2, pp. 1–13, 2017.

[6] M. T. Bona and J. M. Andrés, "Reflection and transmission mid-infrared spectroscopy for rapid determination of coal properties by multivariate analysis," *Talanta*, vol. 74, no. 4, pp. 998–1007, 2008.

[7] D. W. Kim, J. M. Lee, and J. S. Kim "Application of near infrared diffuse reflectance spectroscopy for on-line measurement of coal properties," *Korean Journal of Chemical Engineering*, vol. 26, no. 2, pp. 489–495, 2009.

[8] Y. Mao, B. Ma, S. Liu et al., "Study and validation of a remote sensing model for coal extraction based on reflectance spectrum features," *Canadian Journal of Remote Sensing*, vol. 40, no. 5, pp. 327–335, 2014.

[9] B. T. Le, D. Xiao, D. Okello et al., "Coal exploration technology based on visible-infrared spectra and remote sensing data," *Spectroscopy Letters*, vol. 50, no. 8, pp. 440–450, 2017.

[10] L. Song, S. Liu, M. Yu, Y. Mao, and L. Wu, "A classification method based on the combination of visible, near-Infrared and thermal infrared spectrum for coal and gangue distinguishment," *Spectroscopy and Spectral Analysis*, vol. 37, no. 2, pp. 416–422, 2017.

[11] A. K. Permana, C. R. Ward, Z. Li, and L. W. Gurba, "Distribution and origin of minerals in high-rank coals of the south Walker Creek area, Bowen basin, Australia," *International Journal of Coal Geology*, vol. 116, no. 9, pp. 185–207, 2013.

[12] G. R. Hunt and J. W. Salisbury, "Visible and near-infrared spectra of minerals and rocks: XI. Sedimentary rocks," *Modern Geology*, vol. 5, pp. 211–217, 1976.

[13] E. A. Cloutis, "Quantitative characterization of coal properties using bidirectional diffuse reflectance spectroscopy," *Fuel*, vol. 82, no. 18, pp. 2239–2254, 2003.

[14] G. R. Hunt and R. K. Vincent, "The behavior of spectral features in the infrared emission from particulate surfaces of various grain sizes," *Journal of Geophysical Research*, vol. 73, no. 18, pp. 6039–6046, 1968.

[15] A. M. Baldridge, S. J. Hook, C. I. Grove, and G. Rivera, "The ASTER spectral library version 2.0," *Remote Sensing of Environment*, vol. 113, no. 4, pp. 711–715, 2009.

[16] N. Zaini, F. van der Meer, and H. van der Werff, "Determination of carbonate rock chemistry using laboratory-based hyperspectral imagery," *Remote sensing*, vol. 6, no. 5, pp. 4149–4172, 2014.

[17] Q. Fang, H. Hong, L. Zhao et al., "Visible and near-infrared reflectance spectroscopy for investigating soil mineralogy: a review," *Journal of Spectroscopy*, vol. 2018, Article ID 3168974, 14 pages, 2018.

[18] M. Blanco and A. Peguero, "Analysis of pharmaceuticals by NIR spectroscopy without a reference method," *Trends in Analytical Chemistry*, vol. 29, no. 10, pp. 1127–1136, 2010.

[19] D. Barbin, G. Elmasry, D. Sun, and P. Allen, "Near-infrared hyperspectral imaging for grading and classification of pork," *Meat Science*, vol. 90, no. 1, pp. 259–268, 2012.

[20] Y. Liu, X. Sun, and A. Ouyang, "Nondestructive measurement of soluble solid content of navel orange fruit by visible-NIR spectrometric technique with PLSR and PCA-BPNN," *LWT-Food Science and Technology*, vol. 43, no. 4, pp. 602–607, 2010.

[21] J. Wu, J. Wang, and L. Liu, "Feature extraction via KPCA for classification of gait patterns," *Human Movement Science*, vol. 26, no. 3, pp. 393–411, 2007.

[22] M. B. Richman and I. Adrianto, "Classification and regionalization through kernel principal component analysis," *Physics and Chemistry of the Earth*, vol. 35, no. 9, pp. 316–328, 2010.

[23] G. Fu, F. Y. Shih, and H. Wang, "A kernel-based parametric method for conditional density estimation," *Pattern Recognition*, vol. 44, no. 2, pp. 284–294, 2011.

[24] V. N. Vapnik, *The Nature of Statistical Learning Theory*, Springer, Berlin, Germany, 1995.

[25] T. Oommen, D. Misra, N. K. C. Twarakavi et al., "An objective analysis of support vector machine based classification for remote sensing," *Mathematical Geosciences*, vol. 40, no. 4, pp. 409–424, 2008.

[26] C. Chang and C. Lin, "LIBSVM: a library for support vector machines," *ACM transactions on Intelligent Systems and Technology*, vol. 2, no. 3, pp. 1–27, 2011.

[27] G. R. Hunt and J. W. Salisbury, "Visible and near-infrared spectra of minerals and rocks: I silicate minerals," *Modern Geology*, vol. 1, pp. 283–300, 1970.

[28] G. R. Hunt, "Spectral signatures of particulate minerals in the visible and near infrared," *Geophysics*, vol. 42, no. 3, pp. 501–513, 1977.

[29] J. M. Andrés and M. T. Bona, "Analysis of coal by diffuse reflectance near-infrared spectroscopy," *Analytica Chimica Acta*, vol. 535, no. 1, pp. 123–132, 2005.

[30] X. Xie, Y. Zhao, P. Qiu et al., "Investigation of the relationship between infrared structure and pyrolysis reactivity of coals with different ranks," *Fuel*, vol. 216, pp. 521–530, 2018.

[31] R. N. Clark, "Spectral properties of mixtures of montmorillonite and dark carbon grains: implications for remote sensing minerals containing chemically and physically adsorbed water," *Journal of Geophysical Research: Solid Earth*, vol. 88, no. B12, pp. 10635–10644, 1983.

[32] R. N. Clark, 'Spectroscopy of rocks and minerals, and principles of spectroscopy," *Manual of Remote Sensing*, vol. 3, pp. 3–58, 1999.

15

Evaluation of Quality Parameters of Apple Juices Using Near-Infrared Spectroscopy and Chemometrics

Katarzyna Włodarska[1] **Igor Khmelinskii**[2] **and Ewa Sikorska**[1]

[1]*Faculty of Commodity Science, Poznań University of Economics and Business, al. Niepodległości 10, 61-875 Poznań, Poland*
[2]*Universidade do Algarve, FCT, DQB and CEOT, Campus de Gambelas, 8005-139 Faro, Portugal*

Correspondence should be addressed to Ewa Sikorska; ewa.sikorska@ue.poznan.pl

Academic Editor: Jose S. Camara

Near-infrared (NIR) spectra were recorded for commercial apple juices. Analysis of these spectra using partial least squares (PLS) regression revealed quantitative relations between the spectra and quality- and taste-related properties of juices: soluble solids content (SSC), titratable acidity (TA), and the ratio of soluble solids content to titratable acidity (SSC/TA). Various spectral preprocessing methods were used for model optimization. The optimal spectral variables were chosen using the jack-knife-based method and different variants of the interval PLS (iPLS) method. The models were cross-validated and evaluated based on the determination coefficients (R^2), root-mean-square error of cross-validation (RMSECV), and relative error (RE). The best model for the prediction of SSC ($R^2 = 0.881$, RMSECV = 0.277 °Brix, and RE = 2.37%) was obtained for the first-derivative preprocessed spectra and jack-knife variable selection. The optimal model for TA ($R^2 = 0.761$, RMSECV = 0.239 g/L, and RE = 4.55%) was obtained for smoothed spectra in the range of 6224–5350 cm^{-1}. The best model for the SSC/TA ($R^2 = 0.843$, RMSECV = 0.113, and RE = 5.04%) was obtained for the spectra without preprocessing in the range of 6224–5350 cm^{-1}. The present results show the potential of the NIR spectroscopy for screening the important quality parameters of apple juices.

1. Introduction

Over the past years, the application of the near-infrared (NIR) spectroscopy coupled with chemometrics has gained wide acceptance in different fields, including food and agricultural products [1–6].

NIR spectroscopy is based on the absorption of electromagnetic radiation in the range of 12,500–4000 cm^{-1} [2, 7]. The NIR spectra consist of broad overlapping bands arising from overtones and combination tones of the fundamental vibrations involving C-H, O-H, and N-H chemical bonds. These bonds are the primary structural components of organic molecules; thus, NIR is very useful for measurements of biological and organic systems, including foods. Due to the wealth of chemical information provided by the NIR spectra, they allow simultaneous determination of several constituents and/or of diverse sample properties [4, 7].

One of the main advantages of the NIR technique is its nondestructive character and simple and rapid measurements. Different measurement modes enable direct analysis of both liquid and solid samples without any preparation. Due to its advantages, the NIR technique coupled with chemometrics provides a rapid, effective, and cost-saving alternative to the conventional methods in routine, high-throughput analysis of foods. NIR has been used to assess both the properties and concentrations of the food components, being also a well-established tool for process monitoring.

Using NIR for quality control requires chemometric methods to extract useful information out of complex spectra of the products studied [8]. Practical applications usually require development of multivariate calibration, which define the relationships between the measured spectra and the content of the compound or property of interest, obtained by the respective reference methods. Multivariate regression methods are used for developing quantitative models, with partial least squares (PLS) regression being the most widely used. A lot of factors impact the performance of the calibration models, one of the important issues being an appropriate choice and application of chemometric methods. The collected spectra are usually preprocessed mathematically to reduce noise and enhance the analytical information. This improves the results of the subsequent data analysis and leads to better calibration models [9]. The regression analysis may be performed using the entire NIR spectra. However, many studies showed improvements when calibrations were developed in a selected spectral region as compared to the full-spectrum model [10]. Several methods have been developed to objectively identify the important variables (spectral regions), being more efficient than the traditional approach based on the knowledge of the spectroscopic properties of the sample and/or analysis of the regression results performed on the entire spectra [10, 11].

An important area of NIR application is the analysis of fruit and vegetables and products of their processing [5, 12–14]. Considerable attention has been devoted to studies of the apple properties using NIR [12, 15]. Apples are very popular due to their pleasant flavour and beneficial health effects, being a relevant dietary source of phytochemicals, including phenolics [16].

NIR spectroscopy has been successfully used to evaluate a range of intact apple quality attributes such as the soluble solids content, titratable acidity, sugar content, vitamin C, total polyphenols, starch index, chlorophyll content, firmness, and mealiness [17–19]. The feasibility of using variable selection methods for determination of the apple quality parameters such as soluble solids content was also demonstrated [17, 20, 21].

Despite the amount of research carried out to date on using NIR to evaluate properties of the intact apples, the number of published papers that study the apple juice is rather limited. Spectroscopy in the NIR range was used to predict sugar content in the apple juice [22], detect adulteration [23], and differentiate between the apple juices on the basis of apple variety [24]. The combination of NIR spectroscopy and fluorescence enabled detection of quality deterioration of the apple juice during storage and heating [25]. Application of this method for determination of the quality parameters of apple wine was also reported recently [26].

The important characteristics of apple juices related directly to their quality are soluble solids content (SSC) and titratable acidity (TA). The limits for these parameters in marketed apple juices are defined by the Code of Practice developed by the European Fruit Juice Association, which provides reference for the control of juice quality on the EU market. SSC is one of the major characteristics used to indicate sweetness of fresh and processed fruit products [13]. Titratable acidity is related to the organic acid contents; these compounds contribute to the sour taste and also stabilize colour and extend the shelf life of fresh fruit and their processed products. The overall taste of fruit is more closely related to the ratio of SSC and TA than to the individual parameters; therefore, this ratio is used as an index of sensory acceptability of the fruit taste [27].

The aim of the present study was to test feasibility of the NIR spectroscopy in developing the calibration models for predicting the main quality parameters of the apple juices: SSC, TA, and SSC/TA. We also explored the possibilities to optimize the models using jack-knife variable selection, different variants of the interval PLS variable selection, and preprocessing methods.

2. Materials and Methods

2.1. Apple Juices. Apple juices that are available on the market were evaluated in this study. The samples included clear and cloudy juices reconstituted from the concentrate, direct juices that were pasteurized, and freshly squeezed juices. The total of thirty juices from 15 different producers was studied; all of these samples were studied in duplicate, using two different production batches.

2.2. NIR Measurements. The spectra were collected using an FT-NIR spectrophotometer (MPA; Bruker Optics, Ettlingen, Germany). The instrument performance was validated before measurements by running automatic tests according to the manufacturer's procedure. Spectral acquisition and instrument control were performed using OPUS software (v. 5; Bruker Optics, Ettlingen, Germany). The spectra were acquired in the range of 12,500–4000 cm^{-1} with the resolution of 8 cm^{-1} and with 64 scans coadded to obtain the averaged spectrum. The measurements were performed using transmittance techniques in cuvettes with the optical pathlength of 2 mm. The cuvettes were placed into a temperature-controlled cell holder, and measurements were conducted at a constant temperature of 35°C, controlled by the OPUS software. The spectra were recorded after centrifugation (15,000 rpm for 5 min), with six replicated spectra collected for each of the juices.

2.3. Determination of the Chemical Parameters. The soluble solids contents (SSC) of the juices were determined using an Abbe refractometer (model DR-A1's Conbest) at 20°C, calibrated with distilled water. The SSC was expressed as Brix degrees (°Brix), with all of the measurements carried out in triplicate.

Titratable acidity (TA) was measured using a pH meter (S220 SevenCompact™; Mettler Toledo), by titrating 25 ml of the juice sample with 0.1 M NaOH to the pH endpoint of 8.1. The results were expressed as grams of malic acid per litre of the juice (g/L). These measurements were performed in triplicate.

2.4. Data Analysis

2.4.1. Regression Methods. Partial least squares (PLS) regression was used to establish the calibration models between the NIR spectra (the **X** matrix) and the quality parameters of the apple juices (the **Y** matrix). The PLS method models both the **X**- and **Y**-matrices simultaneously, finding the latent variables in **X** that best predict the latent variables in **Y** [28]. We used all thirty juice samples for developing and optimizing the calibration models. The average spectra were used in the analysis.

2.4.2. Validation of the Regression Models. Full leave-one-out (LOO) cross-validation was applied to all of the regression models. The regression models were evaluated using the determination coefficient (R^2), the root-mean-square error of cross-validation (RMSECV), and the relative error (RE), calculated as the percentage ratio of RMSECV to the average value of the studied parameter in the calibration set. The optimal number of components was chosen as the minimum on the plot of the RMSECV as a function of the number of components.

2.4.3. Spectral Preprocessing. We used different preprocessing methods in order to remove noise, baseline, and scattering effects from the spectra. Savitzky–Golay smoothing with the filter width of 15 data points was used to remove spectral noise, while the baseline was corrected using the baseline offset and the first and second derivatives. The baseline offset involved linear offset subtraction, which shifted the spectra in order to set the minimum value to zero. The first-order derivative is normally used to eliminate constant baseline shifts, and the second-order derivative also eliminates the baseline slope [9]. The derivatives were calculated using the Savitzky–Golay algorithm, with the filter width of 15 data points. Multiplicative scatter correction (MSC) and standard normal variate (SNV) were applied for the correction of the light-scattering effects [9]. The MSC estimates the correction coefficients for additive and multiplicative scattering effects by regressing the spectrum to be corrected on a reference spectrum [9]. The average spectrum of the calibration set was used as a reference. The SNV corrects the spectra by first calculating the mean spectrum and subsequently subtracting this mean from the spectrum to be corrected. Then, that value is divided by the standard deviation of the spectrum [9]. The spectra were preprocessed using each of the single methods and/or their following combinations: smoothing and baseline, smoothing and SNV, smoothing and MSC, MSC and the first-order derivative, MSC and the second-order derivative, SNV and the first-order derivative, and SNV and the second-order derivative. The order of application of the different preprocessing methods was as indicated in the preceding description.

The preprocessing was performed on the average NIR spectra. Prior to PLS analysis, all of the spectra were mean-centred.

2.4.4. Variable Selection. The variable selection methods applied in this work include the jack-knife method and different variants of the interval PLS (iPLS) [29].

The jack-knife is a method used for calculating the standard errors of the regression coefficient estimated in the PLS regression model [30]. The regression coefficients are then divided by their estimated standard errors, giving the *t*-test values to be used for testing the significance of the variables used in the model [11]. These calculations were carried out using The Unscrambler v. 9 8 software (CAMO, Norway).

The iPLS method subdivides the data into non-overlapping sections, obtaining a local PLS model in each section, in order to determine the most useful variable range. The comparison between all of the local models is usually based on the RMSECV values, obtained from the validation [11]. An optimal data range may be found by reducing or increasing the existing trial ranges, or by removing or adding new variables [20]. Presently, we used different variants of the iPLS method as implemented in the OPUS software for selection of the optimal variable ranges [31].

The iPLS (NIR) variant used an NIR spectrum (with the 12,500–11,263 cm^{-1} and 5349–4779 cm^{-1} ranges excluded) that was divided into five frequency ranges, each corresponding to specific absorption bands. The local PLS models were tested in each of the selected ranges on their own and in all of their possible combinations. This procedure coincides with the synergy interval PLS (SiPLS) [10].

The iPLS (A) and iPLS (B) variants used the entire NIR spectrum (in the 12,500–4000 cm^{-1} range, with the 5349–4779 cm^{-1} range excluded) divided into ten subranges. The iPLS (A) started the calculation with all of the 10 subranges and next successively excluded one of the subranges. This procedure continued until the RMSECV value did not improve any further. This procedure coincides with the backward iPLS (BiPLS) [10].

The iPLS (B) starts the calculation to find the optimum spectral range with one of the subranges. After finding the best subrange, a second subrange is added. After the best combination of the two subranges is found, a third subrange is added, and so on. The best combination of the subranges was thus searched by adding and leaving out further subranges. This procedure coincides with the forward iPLS (FiPLS) [10].

The selection of variables was performed on differently preprocessed spectra. The algorithm implemented in the OPUS software enables automatic searching for the optimal combinations of the preprocessing method with the spectral range based on the minimum value of the RMSECV criterion. The 5349–4779 cm^{-1} spectral range was excluded from the calculations due to the high absorbance values, clearly exceeding the useful range of the instrument.

Finally, all of the PLS models with different combinations of the preprocessing methods and the variable ranges were calculated using The Unscrambler v. 9.8 (CAMO, Norway).

3. Results and Discussion

3.1. NIR Spectra of Apple Juices. The thirty apple juice samples studied included different juice categories available on the market. They included juices reconstituted from the concentrate, both clear and with added fruit pulp, and direct juices, pasteurized and freshly squeezed.

Figure 1 shows the NIR absorbance spectra collected for the apple juices studied.

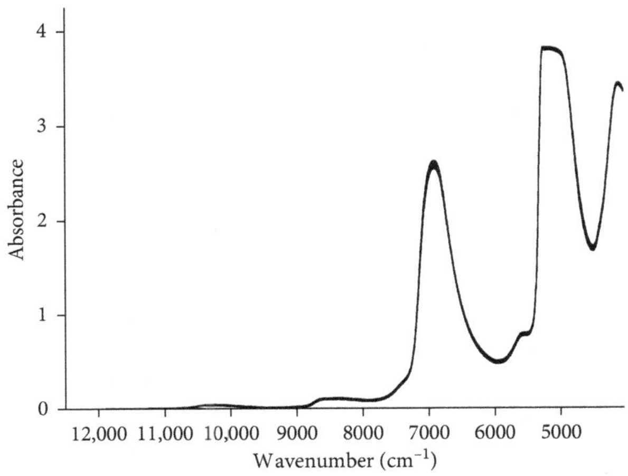

FIGURE 1: The NIR spectra of the apple juices under study.

Very similar characteristic spectral patterns were observed in all of the measured spectra, which were visually indistinguishable. Generally, the positions of the main absorption bands coincided with those obtained for intact apples [32] and other fruit juices [33].

The absorbance spectra are dominated by water absorption, which is the main component of the apple juices. The absorption bands for water were reported at 10,309 cm^{-1} (the second overtone of the O-H stretching band), 8403 cm^{-1} (the combination of the first overtone of the O-H stretching and the O-H bending bands), 6896 cm^{-1} (the first overtone of the O-H stretching band and a combination band), 5154 cm^{-1} (combination of the O-H stretching band and the O-H bending band), and 4444 cm^{-1} [12, 13, 34].

Sugars and organic acids are the main constituents of apple juices, besides water. The most dominant sugar in the apple fruit is fructose, followed by glucose and sucrose. Malic acid is the principal organic acid found in apples. Other components of the apple juice include polyphenolic compounds, vitamins, and some amino acids [35]. All of these components should contribute to the spectra in different NIR ranges; however, their bands are largely suppressed by the dominant water absorption bands. The absorption bands in fruit juices at 6896, 5587, and 4413 cm^{-1} were attributed to sucrose, fructose, and glucose [2]. In fact, the absorption spectra of glucose, fructose, and sucrose are very similar to each other in aqueous solutions, with characteristic bands at 6301–6317, 4716–4710, and 4403–4397 cm^{-1} [36].

The first, second, and third overtones of the C-H stretching vibrations (CH group) are observed, respectively, in the ranges of 5550–6250 cm^{-1}, 8100–9100 cm^{-1}, and 11,000 cm^{-1}. The bands arising from the overtones of OH, CH, and CH_2 deformation vibrations are observed below 5400 cm^{-1} [13]. The combination band of the C-H bond in sugars and organic acids was reported at 4323 cm^{-1} [32].

The absorption bands characteristic for the carboxylic acids appear at 6222 cm^{-1} (C-O from COOH), 8873 cm^{-1} (O-H from carboxylic acids), and 6959 cm^{-1} (C=O from saturated and unsaturated carboxylic acids) [15].

TABLE 1: The soluble solids content (SSC), the titratable acidity (TA), and the ratio of the soluble solids to the titratable acidity (SSC/TA) of the apple juices in the calibration set (n = 30 samples).

Parameter	Range	Mean	SD
SSC (°Brix)	11.0–13.6	11.68	0.79
TA (g/L)	4.51–6.09	5.25	0.48
SSC/TA	1.84–2.71	2.24	0.28

3.2. Multivariate Calibration

3.2.1. Chemical Characteristics of the Calibration Set. All of the thirty apple juice samples were used as the calibration set for developing and optimizing the calibration models. Chemical characteristics of the calibration set including the mean values, ranges, and standard deviations of the soluble solids contents (SSC), titratable acidity (TA), and the SSC/TA ratio are presented in Table 1.

The solids content in the studied apple juices was in the range of 11.0 to 13.6 °Brix. The titratable acidity was 5.25 g/L on average and ranged from 4.51 g/L to 6.09 g/L. These values are within the limits established for apple juices by the Code of Practice [37].

The ratio of SSC and TA fell in a narrow range of 1.84–2.71 in all of the juices studied, being the key parameter determining the taste of fruit products.

3.2.2. Development and Optimization of the Calibration Models. Multivariate PLS regression was used to model the relations between the NIR spectra and the properties of the juices (SSC, TA, and SSC/TA). Different methods of preprocessing and variable selection were tested. The preprocessing methods included smoothing, multiplicative scatter correction (MSC), standard normal variate (SNV), and baseline correction techniques, and the latter included baseline offset and calculation of the first and second spectral derivatives; both single methods and some of their combinations were tested.

The optimal variable ranges for the raw and differently preprocessed spectra were determined using the jack-knife method and three variants of the iPLS method. The jack-knife method was applied to the regression coefficient of the PLS regression models obtained for the analysis of the entire NIR spectra. The iPLS models were developed on ten spectral subranges of equal width, or on five subranges, selected to include specific absorption bands. These five spectral intervals were 11,262–9407 cm^{-1}, 9406–7498 cm^{-1}, 7497–6225 cm^{-1}, 6224–5350 cm^{-1}, and 4778–4000 cm^{-1}. The idea of variable selection is to identify a subset of the data that produces the lowest prediction error for the parameter of interest. Different combinations of the preprocessing and variable selection methods were evaluated in order to find the optimal procedure. We compared the prediction performance of these local models with that of the global full-spectrum model. We evaluated the models on the basis of cross-validation, R^2, the RMSECV, and the RE value [12].

Table 2 presents the optimal calibration models obtained using each of the tested variable selection methods for each

Table 2: Characteristics of the optimal regression models for the prediction of the soluble solids content (SSC), titratable acidity (TA), and ratio of the soluble solids content to the titratable acidity (SSC/TA) of the apple juices under study.

Parameter	Variable selection	Spectral range (cm^{-1})	Preprocessing	LV	R^2	Calibration RMSECV	RE (%)	R^2	Cross-validation RMSECV	RE (%)
SSC	None	Full range	None	6	0.891	0.257	2.20	0.800	0.360	3.08
	Jack-knife		*1st derivative*	*3*	*0.911*	*0.232*	*1.99*	*0.881*	*0.277*	*2.37*
	NIR	9406–7498; 6224–5350	2nd derivative	10	0.987	0.088	0.75	0.853	0.309	2.65
	A, B	10,109–8516; 6137–5334	SNV + 1st derivative	4	0.904	0.242	2.07	0.838	0.324	2.77
TA	None	Full range	None	7	0.804	0.209	3.98	0.512	0.341	6.50
	Jack-knife		Smooth + SNV	7	0.862	0.175	3.33	0.733	0.252	4.80
	NIR	*6224–5350*	*Smooth*	*9*	*0.889*	*0.157*	*2.99*	*0.761*	*0.239*	*4.55*
	A	10,904–10,106; 9314–7722; 6137–5334	Smooth + SNV	6	0.829	0.195	3.71	0.717	0.260	4.95
	B	10,109–8516; 6137–5334	Smooth + MSC	10	0.893	0.155	2.95	0.713	0.262	4.99
SSC/TA	None	Full range	None	7	0.902	0.086	3.84	0.707	0.154	6.88
	Jack-knife		Smooth	8	0.931	0.073	3.26	0.835	0.116	5.18
	NIR	*6224–5350*	*None*	*8*	0.940	*0.068*	*3.04*	*0.843*	0.113	*5.04*
	A, B	6137–5334	None	10	0.941	0.067	2.99	0.828	0.118	5.27

The best model for each of the parameters is italicized; none: spectra without preprocessing; 1st derivative: first-order derivative; 2nd derivative: second-order derivative; smooth: smoothing; SNV: standard normal variate; MSC: multiplicative scatter correction; NIR, A, and B are different versions of the iPLS method.

of the parameters studied. The characteristics of the models developed using the full raw spectra are also presented for comparison.

Finally, for each of the parameters studied, we identified a combination of the preprocessing method and the spectral range, which provided the model with the best prediction performance. The predicted versus measured plots and the regression coefficient plots for these models with the best performance for each of the parameters studied are shown in Figure 2.

(1) SSC Calibration Models. The parameters listed in Table 2 demonstrate good capacity of the NIR spectroscopy to predict the SSC of the apple juices. Indeed, a relatively good model for SSC was obtained for the analysis of the entire NIR spectra without any preprocessing. Preprocessing and variable selection improved the model parameters. Thus, the best model for SSC prediction was obtained for the first spectral derivative and variables selected by the jack-knife method; these variables are shown in Figure 2(a). The respective model was characterized by the R^2 of 0.881 and the RE value of 2.37%.

The variable selection by iPLS also led to model improvement as compared to the full-spectra models. The optimal intervals selected using iPLS (NIR) were 9406–7498 cm^{-1} and 6224–5350 cm^{-1}, which combined with the second-derivative preprocessing gave the model with a slightly higher value of RE of 2.65%. The optimal intervals selected using both iPLS (A) and iPLS (B) were the same (10,109–8516 cm^{-1} and 6137–5334 cm^{-1}), which combined with the SNV and the first-derivative spectral preprocessing gave the models with slightly higher errors than the other variable selection methods tested (RE equal to 2.77%).

The results obtained for SSC modeling are comparable with the literature data; typical values of RMSEP for intact apples were around 0.5 °Brix or even higher (1–1.5 °Brix), when the external validation was performed using fruit test sets collected in different seasons and orchards [12].

(2) TA Calibration Models. The calibration model developed using raw data and the full-spectral range for TA showed poor performance with a low R^2 value and high RMSECV (Table 2). Spectral preprocessing combined with variable selection markedly increased the prediction ability of these models. However, it should be noted that even the optimized models were characterized by rather low R^2 values in the range between 0.713 and 0.761.

The best model for the TA prediction was obtained for the spectral range selected by the iPLS (NIR) method (6224–5350 cm^{-1}) for smoothed spectra (Figure 2(b)). This model was characterized by the R^2 of 0.761 and RE value of 4.55%. The application of iPLS (A) and iPLS (B) methods led to the selection of a wider spectral range as compared to the iPLS (NIR). In addition to the range of 6137–5334 cm^{-1} selected by both iPLS (A) and iPLS (B) methods, the 10,904–10,106 cm^{-1} and 9314–7722 cm^{-1} regions were selected by iPLS (A) and the 10,109–8516 cm^{-1} region was selected by iPLS (B). Models with a similar predictive ability resulted from the combination of iPLS (A) with smoothing and SNV, and of iPLS (B) with smoothing and MSC. These models had a lower prediction ability as compared to the iPLS (NIR) models. A model with an intermediate predictive ability (RE of 4.80%) was obtained for the analysis of the smoothed and SNV-corrected spectra with the variables selected by jack-knife.

The lower predictive ability obtained for the TA models as compared to the SSC models is in accordance with the literature data. This result may be explained by the lower concentration of acids compared to that of sugars [12], and/or lower NIR spectral sensitivity to acids, due to the lower number of functional groups per molecule.

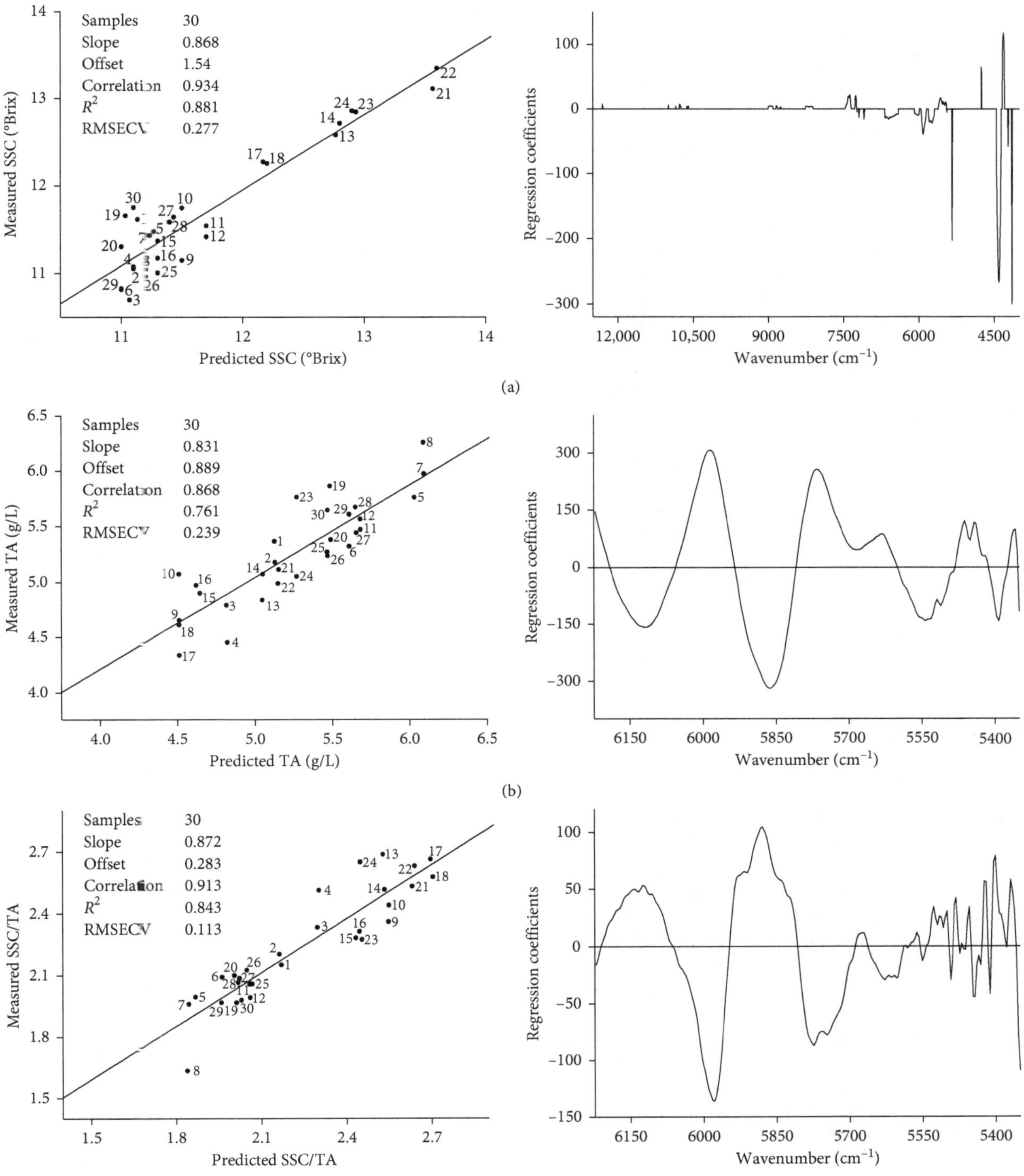

Figure 2: The results of PLS regression analysis for (a) SSC, (b) TA, and (c) SSC/TA. Left panel: predicted versus measured plots for the cross-validation. Right panel: regression coefficients.

(3) SSC/TA Calibration Models. The regression analysis for SSC/TA performed on raw spectra in the full-spectral range gave a model with the R^2 equal to 0.707 and RE equal to 6.88% (Table 2). Also, in this case, PLS models were significantly improved by applying an appropriate combination of spectral preprocessing and variable selection methods. The performances of optimized models for the SSC/TA prediction were intermediate as compared to those of the SSC and TA models.

The best model was obtained for the analysis performed on spectra without preprocessing, using the variables selected by the iPLS (NIR) method, in the range of 6224–5350 cm^{-1} (Figure 2(c)). This model was characterized by the RE of 5.04%. A slightly inferior performance was produced by the models that used spectra without any preprocessing and variables selected using the iPLS (A) or iPLS (B) method in the range of 6137–5334 cm^{-1}. The combination of smoothing and variable selection using the jack-knife method provided a model with intermediate performance (R^2 of 0.835 and RE of 5.18%).

Summing up, preprocessing and variable selection had a marked effect on the model performance. The two variants of the iPLS method, versions (A) and (B), each based on the same ten intervals, selected similar spectral ranges and provided PLS models with a similar performance. On the contrary, for the parameters studied, using the intervals based on the chemical knowledge of the NIR spectrum of the iPLS (NIR) variant produced better performing models as compared to iPLS (A) or iPLS (B). Application of the jack-knife method enabled selection of variables that gave models with a similar or better performance as compared to the iPLS method.

The iPLS-based models with the best performance for each of the chemical parameters studied used the 6224–5350 cm^{-1} range (or a similar 6137–5334 cm^{-1} range), indicating that spectral bands containing chemically significant information on the parameters studied are present in this spectral region. The models for TA and SSC/TA using this range only gave good calibration results, while the calibration model for SSC required additional spectral ranges.

4. Conclusions

In the present study, we developed and optimized the calibration models for the prediction of characteristic parameters in apple juices. We demonstrated that NIR coupled with multivariate calibration is a suitable method for determination of the parameters, which are crucial for quality assessment (SSC and TA) and additionally for sweet-sour taste (SSC/TA) evaluation of apple juices. An optimal combination of the mathematical preprocessing of the spectra and selection of the variable range had to be found individually for each of the parameters studied, leading to a significant improvement of the model performance. The usage of an objective variable selection method may speed up the process of model optimization, identifying the spectral ranges with significant chemical information. The identification of the important spectral variables may contribute to the development of NIR screening sensors for the quality and sensory-related properties of apple juices. Such applications require further studies on extended sample sets.

Conflicts of Interest

The authors declare that they have no conflicts of interest.

Acknowledgments

Grant 2016/23/B/NZ9/03591 from the National Science Centre, Poland, is gratefully acknowledged.

References

[1] J. T. Alander, V. Bochko, B. Martinkauppi, S. Saranwong, and T. Mantere, "A review of optical nondestructive visual and near-infrared methods for food quality and safety," *International Journal of Spectroscopy*, vol. 2013, Article ID 341402, 36 pages, 2013.

[2] H. Cen and Y. He, "Theory and application of near infrared reflectance spectroscopy in determination of food quality," *Trends in Food Science and Technology*, vol. 18, no. 2, pp. 72–83, 2007.

[3] J. U. Porep, D. R. Kammerer, and R. Carle, "On-line application of near infrared (NIR) spectroscopy in food production," *Trends in Food Science and Technology*, vol. 46, no. 2, pp. 211–230, 2015.

[4] H. Huang, H. Yu, H. Xu, and Y. Ying, "Near infrared spectroscopy for on/in-line monitoring of quality in foods and beverages: a review," *Journal of Food Engineering*, vol. 87, no. 3, pp. 303–313, 2008.

[5] L. Wang, D.-W. Sun, H. Pu, and J.-H. Cheng, "Quality analysis and classification and authentication of liquid foods by near-infrared spectroscopy: a review of recent research developments," *Critical Reviews in Food Science and Nutrition*, vol. 57, no. 7, pp. 1524–1538, 2017.

[6] T. Woodcock, C. O'Donnell, and G. Downey, "Review: better quality food and beverages: the role of near infrared spectroscopy," *Journal of Near Infrared Spectroscopy*, vol. 16, no. 1, pp. 1–29, 2008.

[7] M. Manley, "Near-infrared spectroscopy and hyperspectral imaging: non-destructive analysis of biological materials," *Chemical Society Reviews*, vol. 43, no. 24, pp. 8200–8214, 2014.

[8] D. Cozzolino, W. U. Cynkar, N. Shah, and P. Smith, "Multivariate data analysis applied to spectroscopy: potential application to juice and fruit quality," *Food Research International*, vol. 44, no. 7, pp. 1888–1896, 2011.

[9] J. Engel, J. Gerretzen, E. Szymańska et al., "Breaking with trends in pre-processing?," *TrAC Trends in Analytical Chemistry*, vol. 50, pp. 96–106, 2013.

[10] Z. Xiaobo, Z. Jiewen, M. J. W. Povey, M. Holmes, and M. Hanpin, "Variables selection methods in near-infrared spectroscopy," *Analytica Chimica Acta*, vol. 667, no. 1-2, pp. 14–32, 2010.

[11] C. M. Andersen and R. Bro, "Variable selection in regression—a tutorial," *Journal of Chemometrics*, vol. 24, no. 11-12, pp. 728–737, 2010.

[12] B. M. Nicolaï, K. Beullens, E. Bobelyn et al., "Nondestructive measurement of fruit and vegetable quality by means of NIR spectroscopy: a review," *Postharvest Biology and Technology*, vol. 46, no. 2, pp. 99–118, 2007.

[13] L. S. Magwaza, U. L. Opara, H. Nieuwoudt, P. J. R. Cronje, W. Saeys, and B. Nicolaï, "NIR spectroscopy applications for internal and external quality analysis of citrus fruit—a review," *Food and Bioprocess Technology*, vol. 5, no. 2, pp. 425–444, 2011.

[14] J.-L. Li, D.-W. Sun, and J.-H. Cheng, "Recent advances in nondestructive analytical techniques for determining the total soluble solids in fruits: a review," *Comprehensive Reviews in Food Science and Food Safety*, vol. 15, no. 5, pp. 897–911, 2016.

[15] H. Wang, J. Peng, C. Xie, Y. Bao, and Y. He, "Fruit quality evaluation using spectroscopy technology: a review," *Sensors*, vol. 15, no. 5, pp. 11889–11927, 2015.

[16] V. I. Candrawinata, J. B. Golding, P. D. Roach, and C. E. Stathopoulos, "From apple to juice—the fate of polyphenolic compounds," *Food Reviews International*, vol. 29, no. 3, pp. 276–293, 2013.

[17] S. Fan, B. Zhang, J. Li, W. Huang, and C. Wang, "Effect of

spectrum measurement position variation on the robustness of NIR spectroscopy models for soluble solids content of apple," *Biosystems Engineering*, vol. 143, pp. 9–19, 2016.

[18] M. Schmutzler and C. W. Huck, "Simultaneous detection of total antioxidant capacity and total soluble solids content by Fourier transform near-infrared (FT-NIR) spectroscopy: a quick and sensitive method for on-site analyses of apples," *Food Control*, vol. 66, pp. 27–37, 2016.

[19] A. Peirs, N. Scheerlinck, K. Touchant, and B. M. Nicolaï, "Comparison of Fourier transform and dispersive near-infrared reflectance spectroscopy for apple quality measurements," *Biosystems Engineering*, vol. 81, no. 3, pp. 305–311, 2002.

[20] Z. Xiaobo, Z. Jiewen, H. Xingyi, and L. Yanxiao, "Use of FT-NIR spectrometry in non-invasive measurements of soluble solid contents (SSC) of 'Fuji' apple based on different PLS models," *Chemometrics and Intelligent Laboratory Systems*, vol. 87, no. 1, pp. 43–51, 2007.

[21] X. Zou, J. Zhao, and Y. Li, "Selection of the efficient wavelength regions in FT-NIR spectroscopy for determination of SSC of 'Fuji' apple based on BiPLS and FiPLS models," *Vibrational Spectroscopy*, vol. 44, no. 2, pp. 220–227, 2007.

[22] T. Temma, K. Hanamatsu, and F. Shinoki, "Measuring the sugar content of apples and apple juice by near infrared spectroscopy," *Optical Review*, vol. 9, no. 2, pp. 40–44, 2002.

[23] L. León, J. D. Kelly, and G. Downey, "Detection of apple juice adulteration using near-infrared transflectance spectroscopy," *Applied Spectroscopy*, vol. 59, no. 5, pp. 593–599, 2005.

[24] L. M. Reid, T. Woodcock, C. P. O'Donnell, J. D. Kelly, and G. Downey, "Differentiation of apple juice samples on the basis of heat treatment and variety using chemometric analysis of MIR and NIR data," *Food Research International*, vol. 38, no. 10, pp. 1109–1115, 2005.

[25] D. Zhu, B. Ji, Z. Qing, C. Wang, and M. Zude, "The detection of quality deterioration of apple juice by near infrared and fluorescence spectroscopy," in *Computer and Computing Technologies in Agriculture IV: 4th IFIP TC 12 Conference, CCTA 2010, Nanchang, China, Selected Papers, Part III*, D. Li, Y. Liu, and Y. Chen, Eds., pp. 84–91, Springer, Berlin, Germany, 2011.

[26] M. Ye, T. Yue, Y. Yuan, and Z. Li, "Application of FT-NIR spectroscopy to apple wine for rapid simultaneous determination of soluble solids content, pH, total acidity, and total ester content," *Food and Bioprocess Technology*, vol. 7, no. 10, pp. 3055–3062, 2014.

[27] N. Abu-Khalaf and B. S. Bennedsen, "Near infrared (NIR) technology and multivariate data analysis for sensing taste attributes of apples," *International Agrophysics*, vol. 18, pp. 203–211, 2004.

[28] S. Wold, M. Sjöström, and L. Eriksson, "PLS-regression: a basic tool of chemometrics," *Chemometrics and Intelligent Laboratory Systems*, vol. 58, no. 2, pp. 109–130, 2001.

[29] L. Nørgaard, A. Saudland, J. Wagner, J. P. Nielsen, L. Munck, and S. B. Engelsen, "Interval partial least-squares regression (iPLS): a comparative chemometric study with an example from near-infrared spectroscopy," *Applied Spectroscopy*, vol. 54, no. 3, pp. 413–419, 2000.

[30] F. Westad and H. Martens, "Variable selection in near infrared spectroscopy based on significance testing in partial least squares regression," *Journal of Near Infrared Spectroscopy*, vol. 8, no. 2, pp. 117–124, 2000.

[31] Opus. Spectroscopy Software, *Version 6. User Manual. Quant*, Bruker Optik GmbH, Ettlingen, Germany, 2006.

[32] G. Giovanelli, N. Sinelli, R. Beghi, R. Guidetti, and E. Casiraghi, "NIR spectroscopy for the optimization of postharvest apple management," *Postharvest Biology and Technology*, vol. 87, pp. 13–20, 2014.

[33] L. Xie, X. Ye, D. Liu, and Y. Ying, "Quantification of glucose, fructose and sucrose in bayberry juice by NIR and PLS," *Food Chemistry*, vol. 114, no. 3, pp. 1135–1140, 2009.

[34] A. F. Omar, H. Atan, and M. Z. MatJafri, "Peak response identification through near-infrared spectroscopy analysis on aqueous sucrose, glucose, and fructose solution," *Spectroscopy Letters*, vol. 45, no. 3, pp. 190–201, 2012.

[35] J. Wu, H. Gao, L. Zhao et al., "Chemical compositional characterization of some apple cultivars," *Food Chemistry*, vol. 103, no. 1, pp. 88–93, 2007.

[36] F. J. Rambla, S. Garrigues, and M. de la Guardia, "PLS-NIR determination of total sugar, glucose, fructose and sucrose in aqueous solutions of fruit juices," *Analytica Chimica Acta*, vol. 344, no. 1, pp. 41–53, 1997.

[37] European Fruit Juice Association, *AIJN Code of Practice for the Evaluation of Fruit and Vegetable Juice*, European Fruit Juice Association, Brussels, Belgium, 2012, in Polish.

16

Use of LSPR Spectroscopy Biosensing for *In Situ* Identification of Arsenic from Bioleaching of Realgar by *Acidithiobacillus ferrooxidans*

Ruixiang Xu,[1] Wenbin Zhao,[1] Shenghui Xu,[2] Lei Yan,[3] Zhengrong Wu,[1] Yuheng Liu,[1] Wantong Ma,[1] and Peng Chen[1]

[1]*School of Pharmacy, Lanzhou University, No. 199 Donggang West Road, Lanzhou 730020, China*
[2]*College of Earth and Environmental Sciences, Lanzhou University, Key Laboratory of Western China's Environmental Systems, Ministry of Education, Lanzhou 730000, China*
[3]*College of Life Science and Technology, Heilongjiang Bayi Agricultural University, Daqing 163319, China*

Correspondence should be addressed to Peng Chen; chenpeng@lzu.edu.cn

Academic Editor: Alessandro Longo

Localized surface plasmon resonance (LSPR) spectroscopy has received extensive attention as a new method in chemical and biological analysis that can be integrated with nanotechnology. As a new analytical method, LSPR possesses various advantages over the traditional bioanalysis method of enzyme-linked immunosorbent assay (ELISA), including a label-free procedure, low cost, high response speed, simple operation and structure, and ease of storage and transport. Additionally, *in situ* and high-throughput measurements can be achieved. This study aims to solve the problem of the lack of *in situ* and highly efficient monitoring methods for current realgar bioleaching processes. An LSPR chip is made to monitor the changes in arsenic content in the process of realgar bioleaching. A convenient, fast, sensitive, and high-throughput method of bioleaching process investigation based on the LSPR spectroscopic *in situ* monitoring technique is proposed. The LSPR chip provided a highly specific selectivity and a linear detection of arsenic content in the range of 1.0–100.0 μM with detection limit (LOD) 0.898 μM. The developed chip was applied to the quantification of realgar bioleaching sample with satisfactory results.

1. Introduction

Bioleaching technology is a newly developed hydrometallurgical technology [1]. This technique is cross-fused with the preparation of medicinal realgar and uses natural acidophilic microorganisms to perform bioleaching [2]. Recent studies have confirmed that the realgar leachate obtained using *Acidithiobacillus ferrooxidans* (*A. ferrooxidans*) showed strong anticancer activity in both *in vitro* and *in vivo* experiments [3, 4]. Compared to traditional preparatory methods, the bioleaching of realgar can significantly increase its solubility and bioavailability. In addition, this method is highly efficient, ecofriendly, and of low cost [5]. In the bioleaching of realgar, an important aspect is monitoring the changes in the concentration of arsenic during the process [2]. Traditionally, the molybdenum blue spectrophotometric method is utilized to measure the arsenic content. This method is tedious, requires a specific experimental setup, and needs to be airtight. The final gaseous arsenic product of the reaction is toxic and difficult to collect, affecting both the experimental results and the experimenters. Additionally, the experimental setup does not allow high-throughput processing in the case of high amounts. Another commonly implemented method utilizes instrumental analysis by inductively coupled plasma atomic emission spectroscopy (ICP-AES). ICP-AES is safer and more sensitive, but the instrumentation is relatively expensive, and, more importantly, it is difficult to conduct *in situ* analysis.

As a new analysis method, localized surface plasmon resonance (LSPR) provides various advantages compared to the traditional bioanalysis technique enzyme-linked immunosorbent assay (ELISA) [6]. These advantages include label-free procedures, low cost, high response speed, simple operation and structure, and ease of storage and transport [7]. In addition, *in situ* and high-throughput measurements can be achieved. In recent years, there have been extensive advances in synthetic nanotechnology [8]. It is possible to control and synthesize nanoparticles with specific morphologies, sizes, and properties according to the need. For precious metal nanoparticles (NPs) that are specifically utilized in LSPR analysis, the refractive index sensitivities for the dielectric environment of various types of NPs are in the following order: nanostars > nanoprisms > bipyramids > nanorods > nanocubes nanospheres [9]. The LSPR sensitivity of non-spheroidal particles cannot be described analytically, but it has been shown experimentally and in electrodynamics simulations that particle shape plays a large role in determining the sensitivity [10]. In particular, particles with sharp tips produce much higher refractive index sensitivities than would be predicted from their aspect ratios alone. Over the past decade, a myriad of new nanoparticle shapes with ever-increasing refractive index sensitivities have been developed [11]. In this study, we selected the triangular Ag@Au core-shell NP as the nanostructure on which the LSPR analysis is based. Ag@Au core-shell NPs possess the following advantages: (I) for the Ag@Au core-shell NP we synthesized, the refractive index sensitivity to the media environment could reach 570 nm/RIU. This sensitivity is significantly greater than that of the previously reported triangular silver NPs and spherical gold and silver NPs [12]. (II) Conceptually, Ag NPs, due to the stronger and narrower plasmon resonance peak, could be widely adapted for use in chemical sensors, biosensors, and imaging and catalytic applications. However, the structure of silver NPs can change under conditions such as in the presence of salt solutions, halogens, oxidants, heat, and ultraviolet light [13–16].

This property has greatly limited the practical application of silver NPs in various areas. Nevertheless, when a triangular silver NP is coated with an even, thin layer of gold, the stability can be greatly increased, thus increasing the value of silver NPs in practical applications [17]. Gold (III) NPs are easy to modify and exhibit good biocompatibility [18]. Moreover, coating silver NPs with a layer of gold not only effectively protects the silver NPs from environmental influences but also effectively improves the physical and chemical properties of the silver NPs [19]. Therefore, this study provides a new solution to the difficulty of monitoring the change in arsenic content in the process of bioleaching realgar by producing an LSPR chip to investigate the bioleaching process via a convenient, rapid, sensitive, and high-throughput LSPR spectroscopic *in situ* monitoring technique.

In summary, this work is focused on the following tasks: (1) prepare triangular Ag@Au core-shell NPs, based on which a chip is produced using Ag@Au triangular NPs, and perform further characterization; (2) investigate realgar bioleaching by *A. ferrooxidans* and implement the prepared LSPR chip to monitor changes in arsenic content in the bioleaching process.

2. Materials and Methods

2.1. Realgar Bioleaching System. The strain *A. ferrooxidans* BY-3 (CCTCC-M203071) was obtained from the discarded copper mine in Baiyin City, Gansu Province, China, and purified by solid medium streak-plate separation. This strain was selected based on its growth and physical and chemical properties. The strain was entered into GenBank after 16S rRNA nucleic acid sequencing (DQ676505) [20]. The culture medium for strain *A. ferrooxidans* BY-3 (CCTCC-M203071) was 9K culture medium containing 44.69 g/L ferrous sulfate [21]. *A. ferrooxidans* is used to leach out the arsenic within realgar in water solutions. Using a 10% inoculum volume, strain *A. ferrooxidans* BY-3 was inoculated into the culture medium containing 5 g/L of realgar powder. The pH was adjusted to 1.7, the agitation speed was 150 r/min, and the bacteria were cultured in a shaker at a constant temperature of 30°C for 100 h until the logarithmic growth phase.

2.2. Preparation of Triangular Ag@Au Core-Shell Nanoparticles. Fixed amounts of $AgNO_3$ and sodium citrate solutions were added to H_2O_2. Then, triangular silver NPs were synthesized by adding a fixed amount of freshly prepared cold $NaBH_4$ solution while stirring the mixture. The resulting triangular silver NP solution was homogeneously mixed with fixed amounts of 3 mL PVP (5%), 450 μl diethylamine, and 1.2 mL ascorbic acid solutions (0.05 M). The mixture, while stirring for 1 h, reacted with the 0.6 mL $HAuCl_4$ growth solution, 1.2 mL PVP (5%), 240 μl KI, and 9 mL H_2O and produced Ag@Au core-shell triangular NPs (~0.4 nM) [22].

2.3. Preparation of Chip Based on Ag@Au Triangular Nanoparticles

2.3.1. Pretreatment of Glass Substrate. Standard plain microscope glass slides were cut into small pieces. The glass pieces were cleaned and then sonicated in piranha acid solution and ammonia solution for surface hydroxylation. The glass pieces were then rinsed with a large volume of pure water, blow-dried with argon, and soaked in APTES ((3-aminopropyl)triethoxysilane) solution for surface amination. The surface-aminated glasses were further sonicated in methanol, rinsed with a large volume of pure water, blow-dried with argon, and oven-dried.

2.3.2. Assembly of Ag@Au Core-Shell Nanoparticles on Glass Slide Surface. Pretreated glass slides were placed on a watch glass. A certain concentration of Ag@Au NP solution was poured into the watch glass to a certain level above the slide surface. After a period for assembly, the glass slides were removed and washed with a large volume of ultrapure water, blow-dried with argon, and analyzed by UV-Vis extinction spectroscopy.

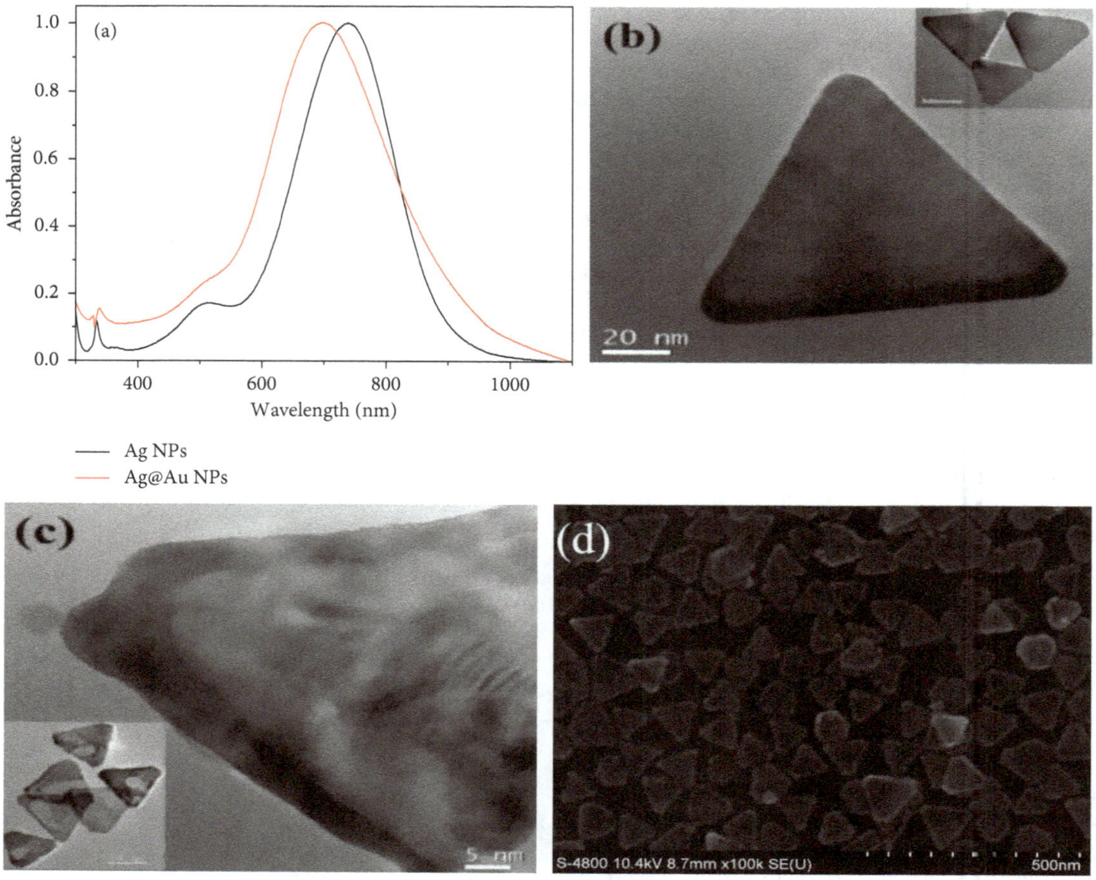

FIGURE 1: (a) UV-Vis extinction spectra of prepared triangular silver NPs (black line) and Ag@Au core-shell NPs (red line). (b) and (c) TEM images of triangular silver NPs and Ag@Au core-shell NPs. (d) A scanning electron microscopy image of Ag@Au core-shell NPs.

3. Results and Discussion

3.1. Preparation and Property Tests of Ag@Au Triangular Nanoparticle Chip

3.1.1. Preparation of Triangular Ag@Au Core-Shell Nanoparticles. Figure 1 displays the UV-Vis extinction spectra of the prepared triangular silver NPs (Figure 1(a), black line) and the Ag@Au core-shell NPs (Figure 1(a), red line). Figures 1(b) and 1(c) show the transmission electron microscopy (TEM) images of the triangular silver NPs and the Ag@Au core-shell NPs. Figure 1(d) shows the scanning electron microscopy image of the Ag@Au core-shell NPs.

3.1.2. Evaluation of the Stability of Ag@Au Core-Shell Nanoparticles. As seen in Figure 2, the stability of the prepared Ag@Au core-shell triangular NPs was fairly good. The NPs were stably present in buffer and salt solutions. Note that the position of the maximum LSPR peaks (λ_{max}) was almost unchanged, but the intensity was decreased. The main cause was the trace coagulation of NPs in high concentrations of PBS buffer or salt solution and a certain amount of loss of NPs during each measurement. In our study, LSPR measurement is mainly based on the position change of the maximum LSPR peaks, and thus, the change in intensity does not influence the measurement results.

3.1.3. Evaluation of the Sensitivity of the Ag@Au Core-Shell Nanoparticles. As illustrated in Figure 3, the detection sensitivity of this type of core-shell NP in solution was high. From a linear regression analysis of the relationship between the LSPR λ_{max} position of the core-shell NPs and the solution refractive index (n), the sensitivity could reach as high as 570 nm/RIU.

3.1.4. Preparation of Ag@Au Triangular Nanoparticle Chip. The core-shell NPs to be assembled onto glass slides still showed localized surface plasmon resonance and could thus result in LSPR peaks. Meanwhile, due to the difference in refractive indexes of the air, glass, and water, a blue shift of LSPR peaks, as illustrated in Figure 4(a), was observed for glass slides loaded with NPs. The peak shape was similar to the peak shape of the monodispersed Ag@Au core-shell NPs in aqueous solution. On the other hand, upon observing the glass slides after assembly, the color of the slides was

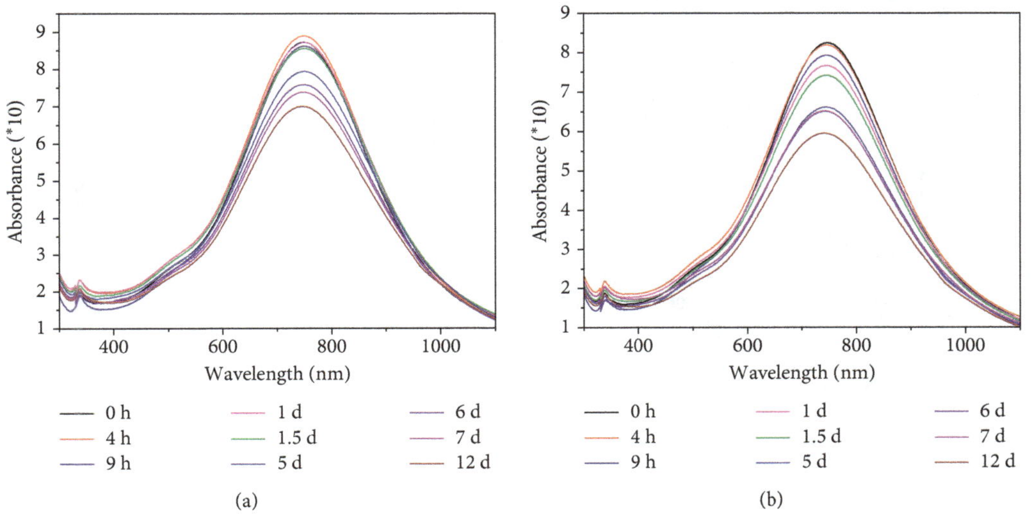

Figure 2: UV-Vis extinction spectra of Ag@Au core-shell NPs stored for a period of time in pure water (a) and PBS buffer (10 mM, pH = 7.4) (b).

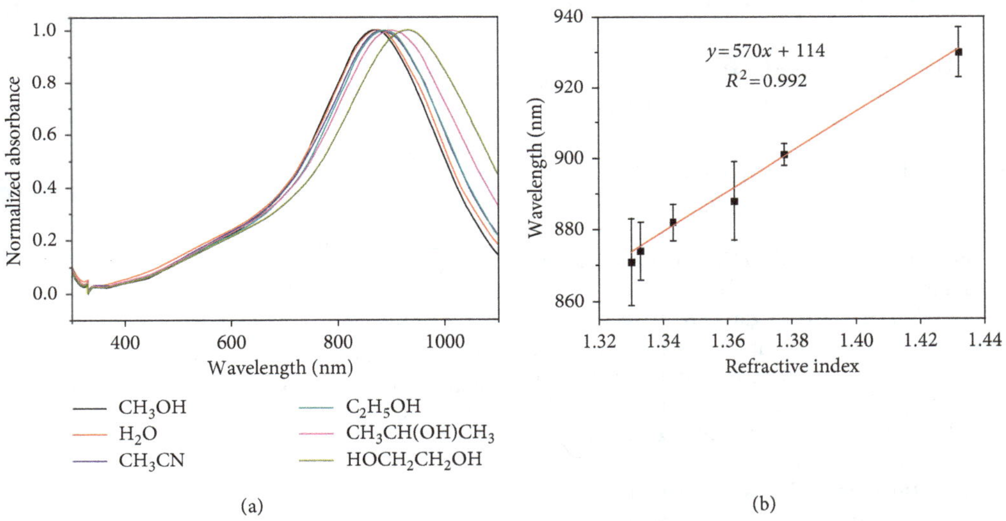

Figure 3: (a) UV-IR extinction spectra of Ag@Au NPs in 6 different solvents: CH_3OH ($n = 1.3301$), H_2O ($n = 1.333$), CH_3CN ($n = 1.343$), C_2H_5OH ($n = 1.362$), $CH_3(OH)CH_3CH$ ($n = 1.3776$), and $HOCH_2CH_2OH$ ($n = 1.4318$). (b) Relationship between the LSPR λ_{max} shift of Ag@Au core-shell NPs and the refractive index (n) of the solvent.

similar to the color of the monodispersed NPs; both were blue, but the glass slide was of lighter color. These two observations suggest that when NPs assembled on the glass surface, they were fixed onto the slides in a nonaggregated state. As shown in Figure 4(b), the full-width half-maximum of the LSPR peak was narrowest for slides soaked for 2 h, so the soaking time was set at 2 h for the subsequent experiments.

3.2. Application of LSPR Chip in Monitoring the Change in Arsenic Content in the Bioleaching Process. Glutathione (GSH) and dithiothreitol (DTT) were treated as recognition units and modified onto the Ag@Au core-shell NP chip through the Au-S chain reaction. Acidophilic bacteria, such as *A. ferrooxidans*, can increase the solubility of realgar, and the main reactions are as follows:

$$As_2S_2 + 14H_2O \xrightarrow{A.ferrooxidans} 2H_3AsO_3 + 2HSO_4^- + 20H^+ + 18e^- \quad (1)$$

$$H_3AsO_3 + H_2O \xrightarrow{A.ferrooxidans} H_3AsO_4 + 2H^+ + 2e^- \quad (2)$$

The sulfhydryl group of GSH or DTT can react with arsenate and reduce As(V) to As(III). As(III) can form the As-$(SG)_3$ complex with GSH in solution, while As(V) can

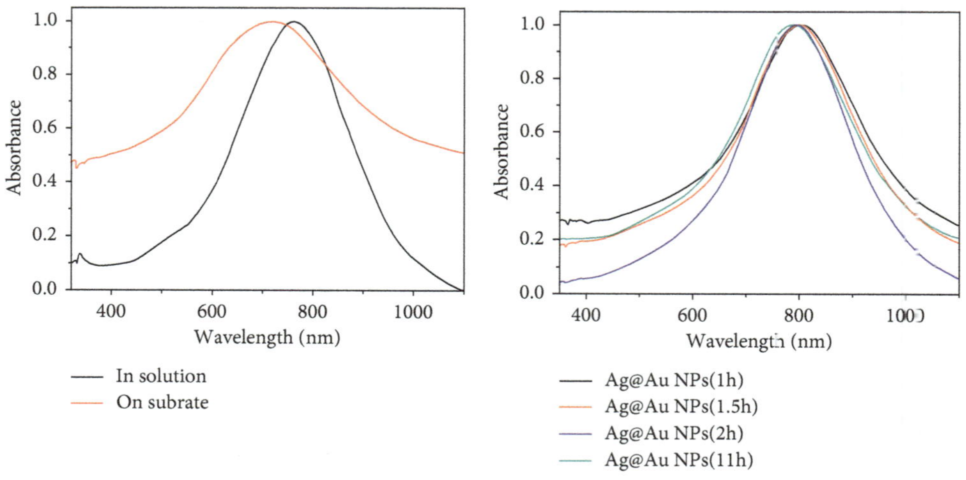

(a) (b)

FIGURE 4: (a) UV-IR extinction spectra of Ag@Au core-shell NPs in aqueous solution and assembled on glass slides. (b) LSPR peaks, before and after normalization, of glass slides soaked in Ag@Au core-shell NP solution for different periods of time.

oxidize GSH to form oxidized GSH, GS-SG. The main mechanisms are as follows:

$$H_2AsO_4^- + 5GSH + H^+ \longrightarrow As-(SG)_3\,(III) + GS-SG + 4H_2O \quad (3)$$

$$HAsO_4^{2-} + 5GSH + 2H^+ \longrightarrow As-(SG)_3\,(III) + GS-SG + 4H_2O \quad (4)$$

As recognition units, GSH and DTT can be modified onto the Ag@Au core-shell NP chip by the Au-S chain reaction. When the GSH or DTT on glass slides reacts with arsenate to form the As-$(SG)_3$ complex and GS-SG, the refractive index of the LSPR chip surface is altered, resulting in changes to the LSPR peak. The goal of analysis can then be achieved by UV-IR absorbance spectroscopic characterization. The analysis mechanism is illustrated in Figure 5. The LSPR chip was used through UV-Vis-IR extinction spectroscopy analysis to perform a preliminary analysis of arsenate solution (Figure 6). Application of the LSPR chip method showed the linear range of arsenate in the bioleaching process from 1.0 μM to 100.0 μM, and a detection limit of 0.898 μM was obtained.

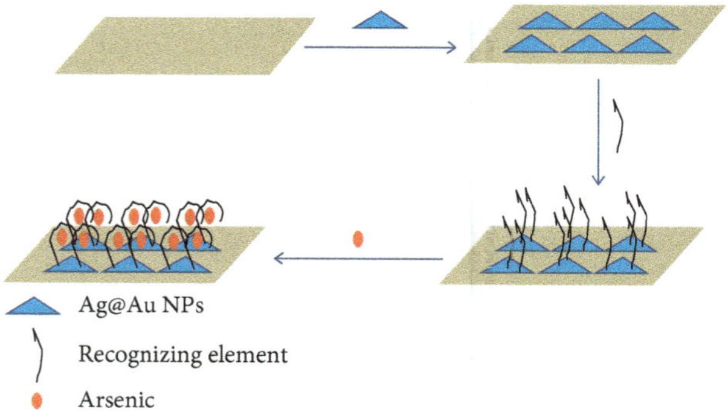

FIGURE 5: Mechanism flow chart of the preparation of the Ag@Au core-shell NP chip and its application in analysis.

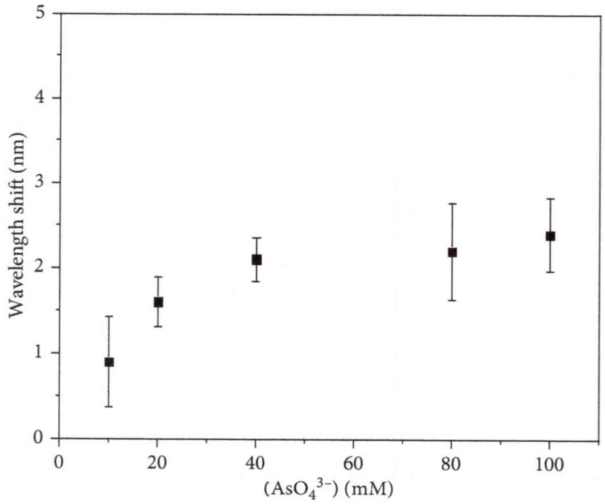

FIGURE 6: Preliminary analysis of arsenate by the LSPR chip.

4. Conclusion

Triangular Ag@Au core-shell NPs were successfully prepared, and Ag@Au triangular NP chips were prepared and characterized. The process of realgar bioleaching by *A. ferrooxidans* was studied. The prepared LSPR chip was applied to monitor the change in arsenic content in the bioleaching process. Additionally, it was found in the experimental process that the realgar bioleaching process by *A. ferrooxidans* could accelerate the oxidation of the ferrous ion. The resulting arsenic ion can be detected by a type of new chips, which provides an alternative method of monitoring by observing changes in arsenic content in the bioleaching process.

Conflicts of Interest

The authors declare that there are no conflicts of interest.

Authors' Contributions

Both authors approve the final version of the manuscript.

Acknowledgments

This work was supported by the Technology Program of Gansu Province (Grant Nos. 1604FKCA110 and 1606RJZA087), the Fundamental Research Funds for the Central Universities of China (Grant No. lzujbky-2017-197), the Project of Lanzhou City for Innovative and Entrepreneurial Talents (Grant No. 2017-RC-73), and the Science and Technology Project of Lanzhou City (Grant Nos. 2015-3-93, 2016-3-75, 2017-4-122, and 2018-4-59).

References

[1] J. H. Zhang, X. Zhang, Y. Q. Ni, X. J. Yang, and H. Y. Li, "Bioleaching of arsenic from medicinal realgar by pure and mixed cultures," *Process Biochemistry*, vol. 42, no. 9, pp. 1265–1271, 2007.

[2] P. Chen, L. Yan, F. F. Leng et al., "Bioleaching of realgar by *Acidithiobacillus ferrooxidans* using ferrous iron and elemental sulfur as the sole and mixed energy sources," *Bioresource Technology*, vol. 102, no. 3, pp. 3260–3267, 2011.

[3] S. Peng, C. Peng, W. Dong et al., "Realgar transforming solution displays anticancer potential against human hepatocellular carcinoma HepG2 cells by inducing ROS," *International Journal of Oncology*, vol. 50, no. 2, pp. 660–670, 2017.

[4] X. Wang, X. Zhang, Z. Xu, Z. Wang, X. Yue, and H. Li, "Reversal effect of arsenic sensitivity in human leukemia cell line K562 and K562/ADM using Realgar Transforming Solution," *Biological and Pharmaceutical Bulletin*, vol. 36, no. 4, pp. 641–648, 2013.

[5] P. Chen, L. Yan, X. Yue, and H. Li, "Optimal parameters for bioleaching of realgar using *Acidithiobacillus ferrooxidans* under different growth conditions and mathematical analysis," *Biocatal Biotransformation*, vol. 31, no. 1, pp. 33–41, 2013.

[6] K. M. Mayer and J. H. Hafner, "Localized surface plasmon resonance sensors," *Chemical Reviews*, vol. 111, no. 6, pp. 3828–3857, 2011.

[7] E. Petryayeva and U. J. Krull, "Localized surface plasmon resonance: nanostructures, bioassays and biosensing—a review," *Analytica Chimica Acta*, vol. 706, no. 1, pp. 8–24, 2011.

[8] S. Szunerits and R. Boukherroub, "Sensing using localised surface plasmon resonance sensors," *Chemical Communications*, vol. 48, no. 72, pp. 8999–9010, 2012.

[9] L. B. Sagle, L. K. Ruvuna, J. A. Ruemmele, and R. P. Van Duyne, "Advances in localized surface plasmon resonance spectroscopy biosensing," *Nanomedicine*, vol. 6, no. 8, pp. 1447–1462, 2011.

[10] K. M. Mayer and J. H. Hafner, "Localized surface plasmon resonance sensors," *Chemical Reviews*, vol. 111, no. 6, pp. 3828–3857, 2011.

[11] T. K. Sau, A. L. Rogach, F. Jäckel, T. A. Klar, and J. Feldmann, "Properties and applications of colloidal nonspherical noble metal nanoparticles," *Advanced Materials*, vol. 22, no. 16, pp. 1805–1825, 2010.

[12] M. Rycenga, C. M. Cobley, J. Zeng et al., "Controlling the synthesis and assembly of silver nanostructures for plasmonic applications," *Chemical Reviews*, vol. 111, no. 6, pp. 3669–3712, 2011.

[13] K. A. Homan, M. Souza, R. Truby et al., "Silver nanoplate contrast agents for in vivo molecular photoacoustic imaging," *ACS Nano*, vol. 6, no. 1, pp. 641–650, 2012.

[14] M. Kim, Y. W. Lee, D. Kim et al., "Reshaping nanocrystals for tunable plasmonic substrates," *ACS Applied Materials and Interfaces*, vol. 4, no. 9, pp. 5038–5043, 2012.

[15] Q. Zhang, J. Ge, T. Pham et al., "Reconstruction of silver nanoplates by UV irradiation: tailored optical properties and enhanced stability," *Angewandte Chemie International Edition*, vol. 48, no. 19, pp. 3516–3519, 2009.

[16] Q. Zhang, N. Li, J. Goebl, Z. Lu, and Y. Yin, "A systematic study of the synthesis of silver nanoplates: is citrate a "magic" reagent?," *Journal of the American Chemical Society*, vol. 133, no. 46, pp. 18931–18939, 2011.

[17] W. P. Hall, S. N. Ngatia, and R. P. Van Duyne, "LSPR biosensor signal enhancement using nanoparticle-antibody conjugates," *Journal of Physical Chemistry C*, vol. 115, no. 5, pp. 1410–1414, 2011.

[18] Y. Cui, B. Ren, J. L. Yao, R. A. Gu, and Z. Q. Tian, "Synthesis of AgcoreAushell bimetallic nanoparticles for immunoassay based on surface-enhanced Raman spectroscopy," *Journal of Physical Chemistry B*, vol. 110, no. 9, pp. 4002–4006, 2006.

[19] H. He, X. Xu, H. Wu, and Y. Jin, "Enzymatic plasmonic engineering of Ag/Au bimetallic nanoshells and their use for sensitive optical glucose sensing," *Advanced Materials*, vol. 24, no. 13, pp. 1736–1740, 2012.

[20] Y. Ni, D. Wan, and K. He, "16S rDNA and 16S-23S internal transcribed spacer sequence analyses reveal inter- and intraspecific *Acidithiobacillus* phylogeny," *Microbiology*, vol. 154, no. 8, pp. 2397–2407, 2008.

[21] P. Chen, L. Yan, Q. Wang, Y. Li, and H. Li, "Surface alteration of realgar (As_4S_4) by *Acidithiobacillus ferrooxidans*," *International Microbiology*, vol. 15, no. 1, pp. 9–15, 2012.

[22] Q. Zhang, Y. Hu, S. Guo, J. Goebl, and Y. Yin, "Seeded growth of uniform Ag nanoplates with high aspect ratio and widely tunable surface plasmon bands," *Nano Letters*, vol. 10, no. 12, pp. 5037–5042, 2010.

Development of UV Spectrophotometric Procedures for Determination of Amlodipine and Celecoxib in Formulation: Use of Scaling Factor to Improve the Sensitivity

Mahesh Attimarad,[1] Katharigatta Narayanaswamy Venugopala,[1,2] Bandar E. Aldhubiab,[1] Anroop Balachandran Nair,[1] Nagaraja SreeHarsha,[1] Shinu Pottathil,[3] and Sabah H. Akrawi[1]

[1]*Department of Pharmaceutical Sciences, College of Clinical Pharmacy, King Faisal University, Al-Hofuf, Al-Ahsa 31982, Saudi Arabia*
[2]*Department of Biotechnology and Food Technology, Durban University of Technology, Durban 4001, South Africa*
[3]*Department of Biomedical Sciences, College of Clinical Pharmacy, King Faisal University, Al-Hofuf, Al-Ahsa 31982, Saudi Arabia*

Correspondence should be addressed to Mahesh Attimarad; mattimarad@kfu.edu.sa

Academic Editor: Sarfaraz Ahmed Mahesar

FDA has recently approved a new fixed-dose combination of amlodipine besylate (AMD) and celecoxib (COX) for the treatment of hypertension and osteoarthritis. No analytical method has been reported for analysis of these two analytes so far. Hence, to monitor the quality and quantity in the formulation of AMD and COX a simple, accurate, precise, economical, and eco-friendly spectroscopic analytical method has been established. The first method involves the determination of AMD and COX by the first derivative UV spectroscopic method with scaling factor 10. However, the second method was based on the direct measurement of UV absorbance of AMD at 364.3 nm and ratio first derivative UV spectroscopic method for COX. Both methods showed good linearity in the range of 5 to 40 μg/ml for COX, whereas AMD showed linearity in the range of 0.5 to 10 μg/ml in first derivative method with scaling factor 10 and 1 to 10 μg/ml in the second method with good correlation coefficient ($R^2 > 0.998$). Both the methods were validated for LOD, LOQ, accuracy, precision, recovery studies, and stability as per the ICH guidelines, and the validated results were well within the acceptable range. Both the methods were successfully utilized for the determination of AMD and COX in the presence of each other in the formulation, and statistically compared between the proposed methods. Therefore, the proposed procedures can be utilized for regular quality control studies.

1. Introduction

Hypertension and osteoarthritis (OA) are major health problems in the middle- and old-age population. Generally, these two diseases coexist, and 40% of OA patients were diagnosed for hypertension [1]. However, the treatment for these two diseases is challenging, because of the adverse effects of NSAIDs on the blood pressure [2]. Nevertheless, celecoxib (COX, Figure 1(a)) was found to be a better choice because it has low risk on blood pressure than ibuprofen, naproxen, or other NSAIDs, and COX has low gastrointestinal and kidney toxicity [3, 4]. Hence, FDA has recently approved a fixed-dose combination of calcium channel blocker, amlodipine besylate, and selective COX-2 inhibitor, celecoxib, for the treatment of hypertension and OA [5].

Amlodipine besylate (AMD, Figure 1(b)), a long-acting calcium channel blocker, is a dihydropyridine derivative extensively used for the management of hypertension and angina pectoris [6]. Several analytical methods were described in the literature for quantification of AMD alone and

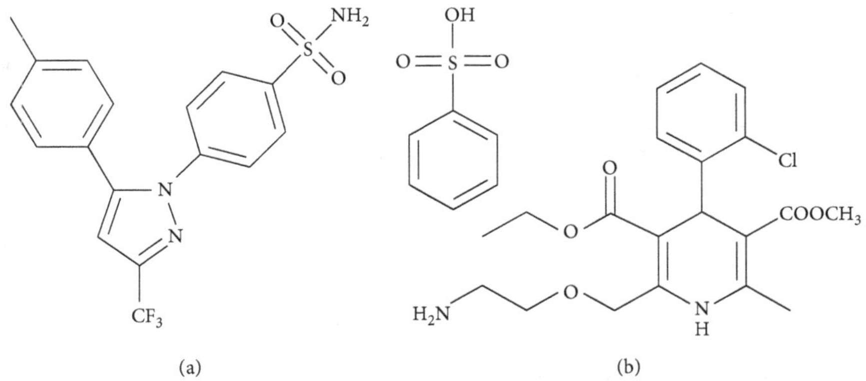

FIGURE 1: Chemical structure of celecoxib (a) and amlodipine besylate (b).

in combination with other drugs from formulation and biological fluids. These analytical methods include spectrophotometric methods [7–9], spectrofluorometric methods [10], HPTLC [11], RP-HPLC [12–14], LCMS/MS [15], capillary electrophoresis, and electrochemical methods [16–23]. Celecoxib, (COX) selective cox-2 inhibitor, is a potent analgesic, antipyretic, and anti-inflammatory drug. COX is far better than other traditional NSAIDs in terms of gastrointestinal safety profile, by sustaining the physiological amount of prostaglandins in the stomach and kidney. Hence, it is generally preferred in chronic inflammatory diseases such as OA [24, 25] and also been used in chemodefensive activity in different cancers [26]. The medicinal application of COX has been increasing continuously; hence, different analytical methods have also been reported, such as spectrophotometry [27, 28], spectrofluorometry [28, 29], RPHPLC [30–32], LCMS/MS, [33, 34], and capillary electrophoresis methods [35]. However, no analytical technique has been reported for simultaneous assessment of AMD and COX in recently approved pharmaceutical formulation.

UV spectroscopic methods are simple, economical, and accurate. However, for multicomponent formulations consisting of analytes showing a complete overlap of UV absorption spectra, it becomes difficult to estimate the analytes in the presence of each another using zero-order UV spectra without prior separation. However, derivatization of zero-order spectra will allow us to analyze these components in the presence of each other, by measuring the peak amplitude of derivative spectra at zero crossings for one of the analyte. Furthermore, derivatization of ratio spectra can also be used to avoid the interference by another component and tablet excipients [36–40]. Hence, an attempt has been made in this proposal to develop two economical, simple, accurate, and precise derivative UV spectrophotometric methods for simultaneous determination of AMD and COX in pharmaceutical formulations.

2. Experimental

2.1. Chemicals and Reagents. Pure samples of amlodipine besylate (99.1%) and celecoxib (98.9%) were procured from the Sigma Aldrich (Germany). Analytical-grade ethanol was purchased from Sigma Aldrich. The distilled water prepared in our laboratory by Milli Q (Millipore, USA) was used throughout the experiments. Amlodipine tablet (10 mg/tablet) and celecoxib capsule (200 mg/capsule) were purchased from the local market. Tablets consisting of AMD : COX in a concentration of 2.5 : 200, 5 : 200, and 10 : 200 mg/tablet, respectively, were prepared in laboratory using 8 mg of lactose, 5 mg of magnesium stearate, 10 mg microcrystalline cellulose, 12.5 mg talc, and 10 mg sodium starch glycolate per tablet.

2.2. Instruments. Shimadzu UV-Vis spectrophotometer (1700) has been used to record the UV spectra of AMD and COX using 10 mm quartz cuvettes. Scanning of the samples was performed at a speed of 50 nm/min by adjusting the slit width to 1 nm. At high concentration levels of analytes, spectra showed high noise; hence, spectra were smoothened with 8 nm. Calibrated volumetric flasks were used for preparations of stock and working standards and sample solutions.

2.3. Preparation of Stock and Working Solutions. Stock solutions of AMD and COX standards were prepared by transferring precisely balanced 100 mg of AMD and COX to two separate 100 ml measuring flasks consisting of 50 ml of ethanol. AMD was freely soluble in ethanol, whereas to dissolve COX, the measuring flask was mixed thoroughly for 5 min with the help of sonicator, and then the final volume was adjusted to the 100 ml with ethanol. The working standards were prepared by diluting these solutions with ethanol: water (50% V/V) as a solvent.

2.4. Preparation of AMD and COX Sample Solutions Using Formulation and Laboratory-Prepared Tablets. Twenty tablets of AMD (10 mg/tablet) were weighed, powdered, and the average weight was calculated. Twenty capsules of COX (200 mg/Capsule) were opened and contents were weighed separately, the average weight was calculated, and then mixed. Both AMD and COX powder equivalent to 10 mg of AMD and 200 mg of COX were transferred to 100 ml measuring flask. Around 75 ml of ethanol was transferred and sonicated for 15 min at room temperature to dissolve the analytes into the solvent. The solution was filtered and the filter paper was washed with fresh ethanol and the final volume was adjusted with ethanol. Furthermore, the solution was diluted with 50% ethanol to bring the concentration in the range of calibration curve. Similarly, laboratory-prepared tablets were powdered and sample solutions were prepared using the above-mentioned procedure and analyzed.

3. Procedure

3.1. First Derivative Spectrophotometric Method (First Method). Sufficient amount of AMD and COX standard stock solutions were transferred into two separate series of 10 ml volumetric flasks, to get concentration in the range of 0.5–10 μg/ml of AMD (0.5, 2, 4, 6, 8, and 10 μg/ml), and 5 to 40 μg/ml of COX (5, 10, 15, 20, 30 and 40 μg/ml). All these solutions were subjected to the absorption measurement in the range of 200 nm to 400 nm to record the UV absorption spectra using 50% v/v ethanol/water solution as blank. The first derivative spectra were computed for all these spectra using 4 nm as $\Delta\lambda$ with scaling factor 10. The peak amplitude of AMD spectra at zero crossings for COX and peak amplitude for COX spectra at zero crossings for AMD were recorded. The calibration curves were constructed between these amplitude values against the corresponding concentration of AMD and COX separately and respective regression equations were constructed.

3.2. Ratio First Derivative Spectroscopy (Second Method). Sufficient amount of AMD stock solutions was transferred to a sequence of 10 ml measuring flasks containing 20 μg/ml of COX to get the concentration in the level of 1 to 10 μg/ml AMD (1, 2, 4, 6, 8, 10 μg/ml). Similarly, COX standard stock solution was added to a sequence of 10 ml measuring flasks consisting of 2 μg/ml of AMD to get the concentration in the range of 5 to 40 μg/ml of COX (5, 10, 15, 20, 30, 40 μg/ml). UV absorbance spectra were recorded for combined AMD and COX solutions and stored. The absorbance of AMD was measured at 363.4 nm and a calibration curve was constructed for AMD by plotting a graph between absorbance versus concentration. Separately, 2 μg/ml solution of AMD was prepared and the UV spectrum was recorded. The second series of absorbance spectra of AMD and COX were converted to ratio spectra of COX by dividing the combined spectra with UV spectrum of AMD (2 μg/ml). Resulting ratio spectra were converted into the first derivative of ratio spectra of COX by derivatizing with 4 nm as $\Delta\lambda$. The peak amplitude at 286.7 nm was measured, and the graph was constructed by plotting a graph between amplitude value and corresponding concentration. Regression equations were also calculated from calibration curves.

3.3. Application of UV Spectroscopic Methods to the Formulation. Aliquot of sample solution was diluted with 50% ethanol solution to get the working sample solution containing both AMD and COX 1 : 20 μg/ml, 1 : 40 μg/ml, 0.5 : 40 μg/ml, respectively. UV absorption spectra were recorded in the range of 200 nm–400 nm. For the first method, the normal UV spectra of sample solutions were converted into the first derivative using 4 nm as $\Delta\lambda$ with a scaling factor of 10. The concentrations of AMD and COX were determined by recording the peak amplitude at zero-crossing wavelength and corresponding regression equations. In the second method, for determination of AMD, absorbance was measured at 364.3 nm and concentration was calculated using regression equitation. For determination of the concentration of COX, the normal UV spectrum of the formulation was divided by UV spectrum of AMD (2 μg/ml) and the resulted ratio spectrum was converted into a first derivative spectrum using 4 nm as $\Delta\lambda$. The peak amplitude was measured at 286.7 nm, and the concentration of COX was calculated using the corresponding regression equation. Furthermore, both the methods were compared using Student's "*t*" test and "*F*" test to know the difference between the analysis results of both methods.

4. Results and Discussion

The UV absorption spectra for AMD and COX in ethanol/water (50% v/v) solution are shown in Figure 2. The UV spectra of COX were completely overlapped by UV spectra of AMD, making it difficult to determine COX in presence of AMD without prior separation by direct UV spectrophotometry. Hence, two derivative UV spectrophotometric methods were developed for simultaneous determination of AMD and COX in formulations. The derivatization of UV spectra could allow us to determine the analytes in the presence of excipients and also remove the interference from the overlapping analyte.

Both the analytes were soluble in ethanol and in the concentration of calibration curve range working standard solutions were clear solution in the 50% ethanol solution. Hence, 50% ethanol solution was selected as a solvent to develop simple, economical, and eco-friendly UV-spectroscopic methods.

The first method was based on the conversion of normal spectra into the first derivative spectra by differentiating the absorbance spectra of a sample by wavelength by wavelength. Derivative spectra allow us to separate overlapped spectra and also eliminate the interference by excipients and other analytes. This increases the resolution of multicomponent spectra and enhances specificity and sensitivity. Furthermore, this feature allows for the determination of the concentration of one component in the presence of others by measuring the amplitude of derivative spectra at zero-crossing wavelengths, where one of the components will have zero absorbance [36, 37].

The fixed-dose formulations of AMD and COX are available in three different ratios: 2.5 : 200, 5 : 200, and 10 : 200 mg/tablet, respectively. Due to low concentration of AMD in the formulation compared to COX, at zero-crossing wavelength for COX, AMD showed negligible absorbance up to 2 μg/ml due to low amplitude of first derivative spectra. Hence, it was difficult to determine simultaneously both

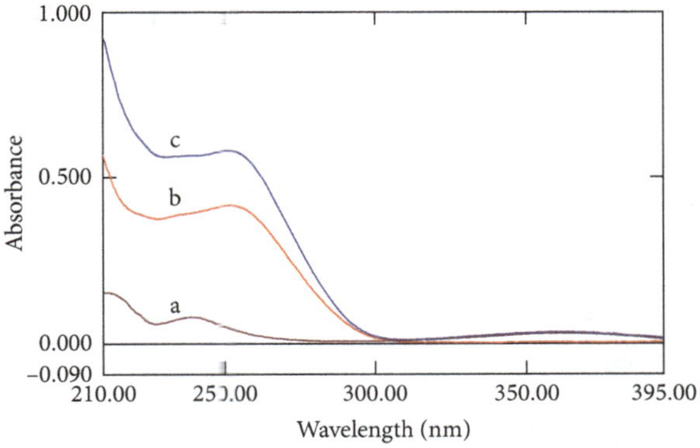

FIGURE 2: Normal UV absorption spectra of AMD (a), COX (b), and mixture (c).

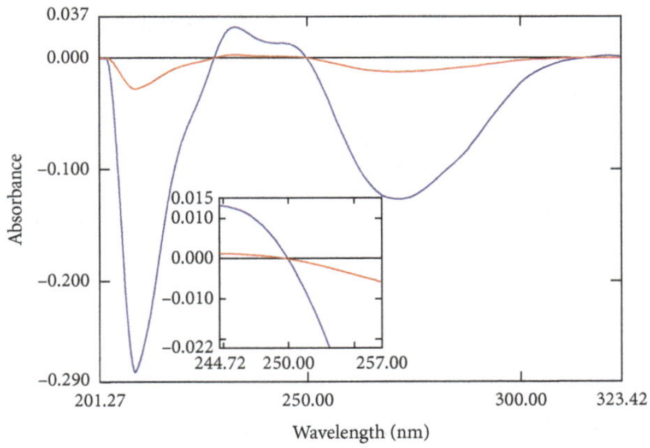

FIGURE 3: First derivative spectra of AMD (0.5 μg/ml) and COX (10 μg/ml) with scaling factor 1 (red) and scaling factor 10 (blue). Enlarged part shows the absorbance difference at 252 nm.

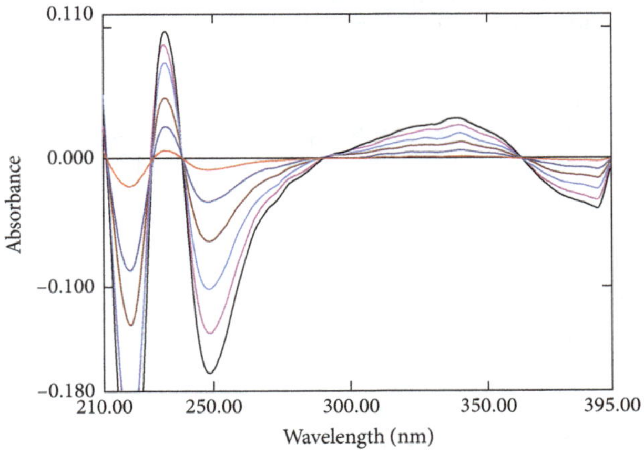

FIGURE 4: The first-order derivative spectra of AMD 0.5 to 10 μg/ml.

analytes in a formulation containing 2.5 : 200 mg and 5 : 200 mg of AMD and COX. Therefore, with an increase in the concentration, AMD proportionally increases the concentration of COX. Absorption of COX solution, with a concentration above 40 μg/ml, was more than 2.5 and showed noise. Hence, scaling factor of 10 was used during the conversation of spectra to first derivative spectra, which increases the amplitude of the first derivative spectra and allowed to determine the low concentration of AMD from 0.5 μg/ml (Figure 3).

Absorbance spectra were recorded for AMD and COX solutions and converted into first derivative spectra with a scaling factor 10 using 4 nm as $\Delta\lambda$. Different wavelengths 2, 4, 8, and 10 nm were applied as $\Delta\lambda$. However, 4 nm was found to be optimal with scaling factor 10; hence, 4 nm was used as $\Delta\lambda$ throughout the experiment. The first derivative spectra of AMD showed 2 maxima at 233.16 nm and 338.34 nm and 3 minima at −219.68 nm, −249.22 nm, and −393.55 nm (Figure 4); however, COX showed zero crossings at −252.0 nm at which AMD had some absorption amplitude (Figure 5). The first derivative spectra of COX showed 1 maximum at 245.59 nm and 2 minima at −209.8 and −272.08 nm (Figure 6). However, AMD showed zero crossings at −210.56 nm and −289.4 nm where COX had some absorbance (Figure 5). However, the amplitude was good at −210.5 nm but linearity was not good, whereas at −289.4 nm, amplitude is low but showed good linearity with excellent regression coefficient. Hence, wavelengths of−252.0 nm and −289.4 nm were selected for quantification of AMD and COX, respectively. Furthermore, analytes showed the same absorbance in pure and in presence of each other at zero-crossing wavelength (Figure 5).

The second method involves the direct measurement of AMD and ratio derivative method for COX measurement. Since UV spectra of COX showed no absorbance in the range of 320 nm to 400 nm, where AMD had some absorbance, a calibration curve was constructed for AMD by measuring the absorbance at 363.4 nm (Figure 7). However, the spectra of COX was completely overlapped by AMD. Hence, ratio derivative spectroscopic method [38–40] has been adopted for quantification of COX, which is based on the following principle. For two compounds (P and Q) with no chemical interaction between the compounds and which obey Beers' law for each compound, the following equation (1) can be constructed:

$$A_M = \varepsilon_P C_P + \varepsilon_Q C_Q, \quad (1)$$

where A_M is the absorbance of P and Q mixture, ε_P and ε_Q are molar absorptivity of compounds P and Q, respectively, and C_P and C_Q are the concentration of compounds P and Q, respectively. Dividing the above equation with absorbance equation of one of the pure compound solution spectra ($A_{Q^\circ} = \varepsilon_{Q^\circ} C_{Q^\circ}$) results into the following equation (2):

$$\frac{A_M}{A_{Q^\circ}} = \frac{A_P}{A_{Q^\circ}} + \frac{C_Q}{C_{Q^\circ}}. \quad (2)$$

The expression C_Q/C_{Q° is constant, which can be eliminated by converting the ratio spectra to derivative spectra. This derivatization generates several maxima and minima, and measurement of peak amplitude at these maxima and minima wavelengths allows us to determine the concentration of one component in the presence of

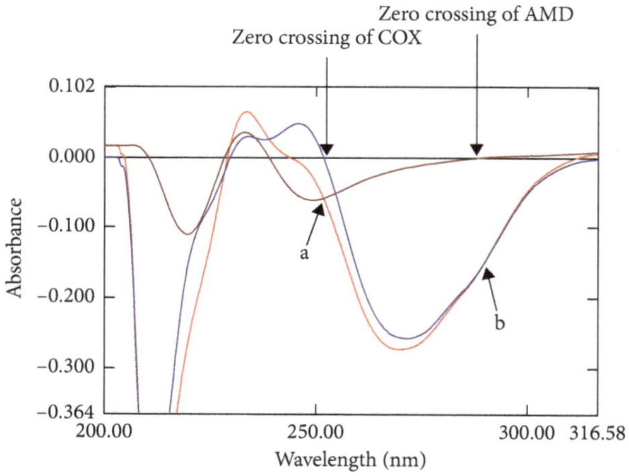

Figure 5: The first-order derivative spectra of AMD (brown), COX (blue), and a mixture of AMD and COX (red). (a) AMD (4 μg/ml) showing the same absorbance in pure and mixture at zero crossings for COX. (b) COX (20 μg/ml) showing the same absorbance in pure and mixture at zero crossings of AMD.

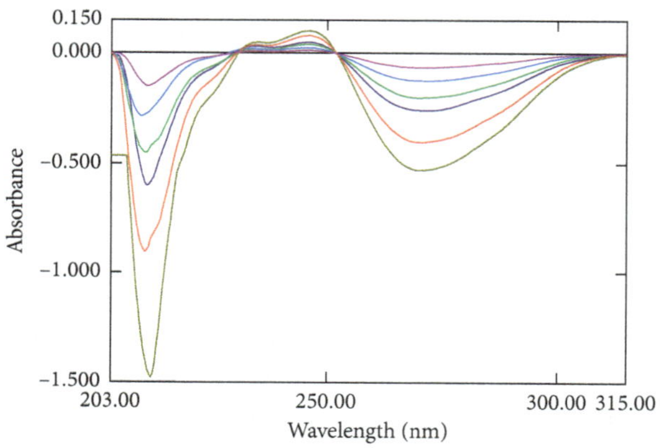

Figure 6: The first-order derivative spectra of COX 5 to 40 μg/ml.

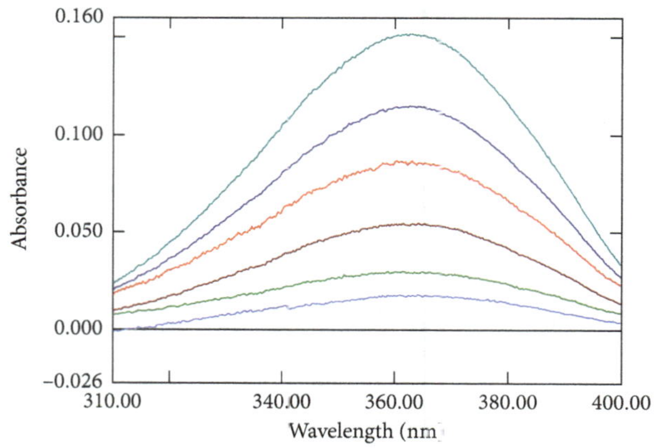

Figure 7: Normal spectra of AMD 1 to 10 μg/ml (second method).

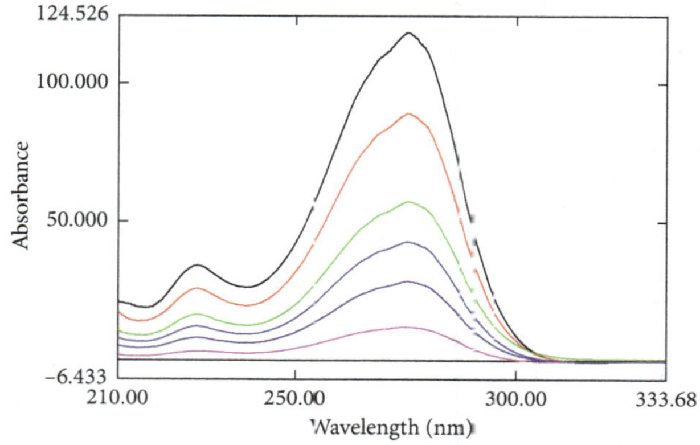

Figure 8: Ratio spectra of COX 5 to 40 μg/ml using AMD 2 μg/ml solution spectrum as a divisor.

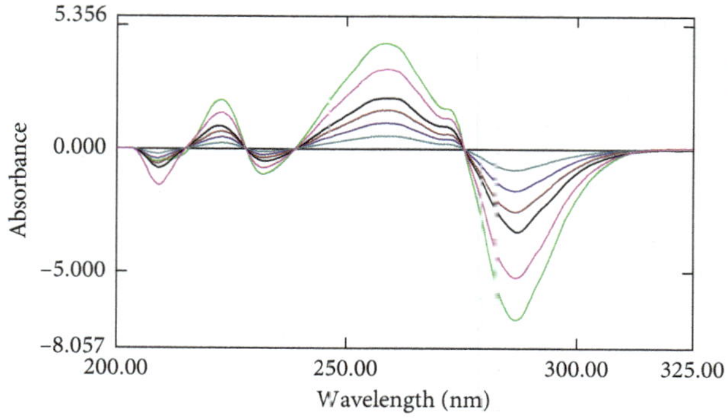

Figure 9: First-order ratio derivative spectra of COX 5 to 40 μg/ml using 4 nm as $\Delta\lambda$.

interfering another compound and excipients [38–40]. In this method, UV absorbance spectra of a mixture of AMD and COX were divided by the UV absorbance spectra of AMD solution to generate ratio spectra of COX in the concentration range of 5 to 40 μg/ml (Figure 8). Different concentration solutions of AMD were studied (0.5, 2, 4, 8, and 10 μg/ml) as a devisor; however, devisor concentration of 2 μg/ml was selected due to good sensitivity. Ratio spectra were converted into first derivative spectra using 4 nm as $\Delta\lambda$. First derivative spectra (Figure 9) showed two maxima at 222.92 nm and 258.24 nm and three minima at −209.10 nm, −232.35, and −286.7 nm. The peak amplitude measured at these wavelengths was directly proportional to the concentration of COX. However, the amplitude at 222.92 nm, −209.10 nm, and 232.35 nm was low, whereas at 286.7 nm showed good recovery of COX in the laboratory mixed samples with low RSD and good linearity with excellent regression coefficient ($r^2 > 0.999$) when compared to amplitude measured at 258.24 nm. Hence, 286.7 nm was selected for analysis of COX. The peak amplitude was measured at 286.7 nm at different concentrations, and a calibration curve was constructed by plotting a graph between amplitude versus concentration of COX. Alternatively, the linearity equation was constructed from the calibration curve.

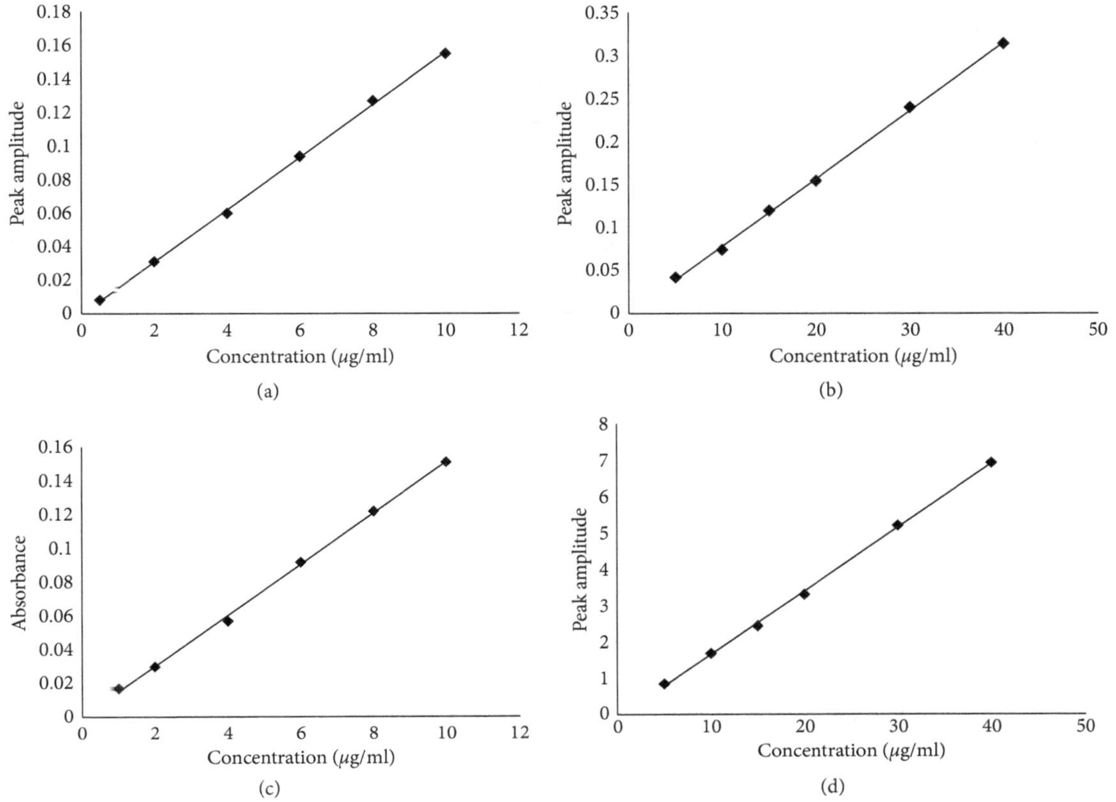

FIGURE 10: Calibration curves of AML 0.5 to 10 μg/ml, first derivative method (a), 1 to 10 μg/ml in second method (normal absorption at 363.4 nm) (c), COX 5 to 40 μg/ml, first derivative method (b), and second method (first derivative of ratio spectra) (d).

4.1. Method Validation. Proposed spectroscopic procedures were validated adhering to the ICH guiding principles for linearity, the limit of detection and quantification, accuracy, precision, recovery, selectivity, and stability.

4.2. Linearity. The linearity range was determined for both analytes by both methods by analyzing the different concentration solutions. AMD exhibited excellent linearity in the concentration of 0.5 to 10 μg/ml in the first method and 1 to 10 μg/ml in the second method with good correlation coefficient ($r^2 > 0.998$). COX was linear in the range of 5 to 40 μg/ml with an excellent correlation coefficient ($r^2 > 0.999$) in both methods. The calibration curves for both analytes by both methods are shown in Figure 10. The linearity range, regression equations, and correlation coefficients are tabulated in Table 1.

4.3. Limits of Detection and Limits of Quantification. The sensitivity of the spectroscopic methods was established by determining the limits of detection (LOD) and limits of quantification (LOQ). LOD and LOQ were determined according to the ICH guidelines. LOD was determined by 3.3 p/m, whereas LOQ was determined by 10 p/m, where p represents the standard deviation of calibration curve and m represents the slope of the curve. The low LOD and LOQ values showed the good sensitivity of the methods (Table 1).

TABLE 1: Regression equations and validation parameters results for AMD and COX.

Parameters	First derivative method		Second method	
Drugs	AMD	COX	AMD	COX
Wave length (nm)	−252.0	−289.4	363.4	−286.7
Linearity range (μg/ml)	0.5–10	5–40	1–10	5–40
LOD (μg/ml)	0.14	0.45	0.31	0.34
LOQ (μg/ml)	0.42	1.24	0.92	0.99
Slop (m)	0.0157	0.0079	0.0151	0.1753
Intercept (c)	0.0004	−0.0004	0.0006	−0.0947
Correlation coefficient (r^2)	0.9992	0.9990	0.9985	0.9991

4.4. Precision and Accuracy. The precision of the methods was assessed by interday and intraday precision (Supplementary material File (available here)). For intraday precision, the three different concentrations of both analytes were analyzed by an optimized method in triplicate on the same day. For interday precision, the same solutions were analyzed for three successive days. The precision was expressed as % RSD, whereas accuracy was expressed as % relative error and tabulated in Table 2. The low % RSD (<2.0) and low % RE indicate good precision and accuracy of both methods.

4.5. Recovery Studies. Recovery studies of the spectroscopic methods were evaluated by the standard addition method

Table 2: Precision and accuracy data.

	Amount of drug (μg/mL)	Interday Amount found mean ($n=3$) ± SD	% RSD	% RE	Intraday Amount found mean ($n=9$) ± SD	% RSD	% RE
First derivative method							
AMD	1	1.01 ± 0.01	0.51	0.99	0.99 ± 0.01	1.02	−1.01
	4	3.93 ± 0.05	0.41	−1.78	4.01 ± 0.04	1.00	0.25
	8	7.94 ± 0.18	1.88	−0.76	7.9 ± 0.15	1.90	−1.27
COX	5	4.96 ± 0.05	1.01	−0.81	5.08 ± 0.06	1.18	1.57
	20	20.16 ± 0.38	1.51	0.79	19.76 ± 0.31	1.58	−1.25
	40	39.4 ± 0.66	1.7	−1.52	40.08 ± 0.59	1.47	0.20
Second method							
AMD	1	0.99 ± 0.01	1.02	−1.01	0.98 ± 0.01	1.02	−2.00
	4	4.06 ± 0.07	1.14	1.48	3.94 ± 0.06	2.04	−2.04
	8	7.92 ± 0.14	1.97	−1.01	7.86 ± 0.17	2.16	−1.78
COX	5	4.91 ± 0.03	1.01	−1.83	5.04 ± 0.07	1.39	0.79
	20	19.67 ± 0.21	1.39	−1.68	19.81 ± 0.33	1.67	−0.96
	40	39.85 ± 0.71	1.33	−0.38	39.4 ± 0.69	1.75	−1.52

SD: standard deviation; % RSD: percent relative standard deviation; % RE: percent relative error.

Table 3: Recovery study results by the standard addition method.

	First derivative method Amount added (μg/ml)	% Recovery	Second method Amount added (μg/ml)	% Recovery
AMD	1	98.87	1	98.49
	2	100.94	2	101.12
	3	99.09	3	99.44
	Across mean	99.63		99.68
	% RSD	1.14		1.33
COX	5	101.23	5	99.17
	10	98.77	10	101.39
	15	99.06	15	99.28
	Across mean	99.69		99.95
	% RSD	1.34		1.25

(Supplementary material File (available here)). Three different concentration solutions of AMD (1, 2, and 3 μg/ml) and COX (5, 10, and 15 μg/ml) were added separately to the previously analyzed formulation solution consisting of 1 μg/ml of AMD and 20 μg/ml of COX. Then, the total amount of AMD and COX were determined using the corresponding regression equations, and the amount of added analyte concentrations was computed. The accuracy was expressed as percentage recovery and % RSD (Table 3). The percentage recovery was in the range of 98.70% to 101.83% with low % RSD (<2%), which indicates the best accuracy of the methods.

4.6. Stability Studies. Standard stock solutions and working standard solutions were stored in the refrigerator at 4°C and analyzed on a daily basis for 7 days. No variation in the concentration of both analytes was observed even on the 7th day, which indicates the stability of both analytes.

4.7. Application of Proposed Methods for Determination AMD and COX from the Formulation Mixture and Laboratory-Prepared Tablets. Both UV spectroscopic procedures were effectively utilized for quantification of AMD and COX from the formulation mixture and laboratory prepared tablets (Supplementary material File (available here)). Fixed-dose combination of AMD and COX was not available in the local market; hence, separate formulations of AMD and COX were mixed in the proposition to get the concentration of coformulation, and tablets were prepared in laboratory and analyzed. The analysis results (Table 4) were in agreement with the amount of AMD and COX in the formulation. Further recovery studies results (Table 3) were also in agreement with the amount of AMD and COX added to the previously analyzed mixture. The validity of the methods was assessed by determining the accuracy of the methods by the standard addition method. The analysis results (Tables 3 and 4) with low % RSD showed the excellent accuracy and precision of the methods. In addition, it also proved the absence of excipients' interference in the analysis of analytes from formulations. Furthermore, the determined "t" and "F" values were less than the critical "t" and "F" values (Table 5), indicating that there is no significant difference in the analysis results between the two methods.

Table 4: Determination of AMD and COX from formulation mixture and tablets.

	Formulation ratio (mg/Tab)	First derivative method % Recovery		Second method % Recovery	
		Formulation mixture	Tablets	Formulation mixture	Tablets
AMD	10	98.41	98.54	99.08	98.55
COX	200	99.29	101.87	98.90	99.03
AMD	5	99.21	101.08	100.36	100.89
COX	200	98.58	98.14	98.49	101.38
AMD	2.5	98.70	100.98		
COX	200	101.37	98.06		

Table 5: Statistical results of the comparison between the two proposed methods.

Parameters	AMD ($n = 6$)	COX ($n = 6$)
Average (% recovery)	99.58	100.09
% RSD	0.69	1.36
T (2.570)[a]	1.385	0.621
F (5.050)[b]	2.171	1.034

[a]The critical "t" value. [b]Critical "F" value at ($P = 0.05$).

5. Conclusions

Two simple UV spectroscopic procedures have been established for determination of AMD and COX in the presence of each other and excipients. These are the first analytical methods reported for determination of AMD and COX in the presence of each other. Both methods are simple, economical, eco-friendly, accurate, and precise. The first derivative spectroscopic method can determine both the analytes in a binary mixture in two steps: derivatization and measurement of amplitude at zero-crossing wavelength. On the other hand, the second method involves direct measurement of AMD and COX, which has been analyzed in three steps: division for ratio, derivatization, and measurement of amplitude; however, ratio derivatization gives many maxima and minima to determine the amount of analyte in the presence of another analyte. Finally, both the methods have been successfully applied for the determination of AMD and COX from formulation mixture and laboratory-prepared tablets. The statistical comparison results confirm that there is no significant difference between the two proposed methods.

Conflicts of Interest

The authors have no conflicts of interest.

Acknowledgments

The authors acknowledge the Deanship of Scientific Research at King Faisal University for the financial support under Nasher Track (Grant No. 186160).

Supplementary Materials

File 1: first derivative spectra of AMD and COX for precision and accuracy. File 2: (A) normal spectra of AMD and COX in a ratio 2 : 5, 4 : 20, and 8 : 40 μg/ml. Enlarged part for AMD (2, 4, 8 μg/ml) for measuring absorption at 363.4; (B) first derivative of ratio spectra of C5, 20, and 40 μg/ml for precision and accuracy. File 3: first derivative of Mixture of AMD and COX in a ratio 1 : 20 (formulation), 2 : 25, 3 : 30, and 4 : 35 μg/ml, respectively, for formulation and recovery studies. File 4: (A) ratio spectra of COX 20, 25, 30, and 35 μg/ml using AMD (2 μg/ml) solution spectrum as devisor; (B) first derivative of ratio spectra of COX 20, 25, 30, and 35 μg/ml using 4 nm as $\Delta\lambda$ for formulation and recovery studies. File 5: (A) normal spectra of AMD and COX in a ratio (a) 0.5 : 20, (b) 1 : 20, and (c) 0.5 : 40 μg/ml; (B) first derivative spectra of AMD and COX in a ratio (a) 0.5 : 20, (b) 1 : 20, and (c) 0.5 : 40 μg/ml; (C) (first method) ratio spectra of tablet solution of (a) COX 20 μg/ml and (b) COX 40 μg/ml using AMD (2 μg/ml) solution spectrum as devisor; (D) first derivative of ratio spectra of tablet solution of (a) COX 20 μg/ml and (b) COX 40 μg/ml using 4 nm as $\Delta\lambda$. (*Supplementary Materials*)

References

[1] P. Verdecchia, F. Angeli, G. Mazzotta et al., "Treatment strategies for osteoarthritis patients with pain and hypertension," *Therapeutic Advances in Musculoskeletal Disease*, vol. 2, no. 4, pp. 229–240, 2010.

[2] A. Hirayama, N. Tanahashi, H. Daida et al., "Assessing the cardiovascular risk between celecoxib and nonselective nonsteroidal antiinflammatory drugs in patients with rheumatoid arthritis and osteoarthritis," *Circulation Journal*, vol. 78, no. 1, pp. 194–205, 2014.

[3] F. K. Chan, A. Lanas, J. Scheiman, M. F. Berger, H. Nguyen, and J. L. Goldstein, "Celecoxib versus omeprazole and diclofenac in patients with osteoarthritis and rheumatoid arthritis (CONDOR): a randomised trial," *The Lancet*, vol. 376, no. 9736, pp. 173–179, 2010.

[4] C. Walker, "Are all oral COX-2 selective inhibitors the same? A consideration of celecoxib, etoricoxib, and diclofenac," *International Journal of Rheumatology*, vol. 2018, Article ID 1302835, 12 pages, 2018.

[5] S. M. Smith and R. M. Cooper-DeHoff, "Fixed-dose combination amlodipine/celecoxib (consensi) for hypertension and osteoarthritis," *The American Journal of Medicine*, vol. 132, no. 2, pp. 172–174, 2019.

[6] S. C. Sweetman, Martindale, The Complete Drug Reference, The Pharmaceutical Press, London, UK, 37th edition, 2011.

[7] N. T. Q. Trang, N. V. Hop, N. D. G. Chau, and T. B. Tran, "Simultaneous determination of amlodipine, hydrochlorothiazide, and valsartan in pharmaceutical products by a combination of full spectrum measurement and kalman filter algorithm," *Advances in Materials Science and Engineering*, vol. 2019, Article ID 5719651, 9 pages, 2019.

[8] N. Usharani, K. Divya, and V. V. S. Ashrtiha, "Development and validation of UV-derivative spectroscopic and RP-HPLC methods for the determination of amlodipine besylate and valsartan in tablet dosage form and comparison of the developed methods by student's *T*-test," *Indian Journal of Pharmaceutical Education and Research*, vol. 51, no. 4s, pp. s776–s782, 2017.

[9] M. Desavathu, P. Raghuveer, and P. Sarada, "Insertion of dual drugs of hypertension in gelatin pockets for chrono pharmacotherapy and its evaluation by in-vitro and ex- vivo studies," *Indian Journal of Pharmaceutical Education and Research*, vol. 52, no. 4s, pp. s184–s196, 2018.

[10] H. M. Abdel-Wadood, N. A. Mohamed, and A. M. Mahmoud, "Validated spectrofluorometric methods for determination of amlodipine besylate in tablets," *Spectrochimica Acta Part A: Molecular and Biomolecular Spectroscopy*, vol. 70, no. 3, pp. 564–570, 2008.

[11] J. J. Pandya, M. Sanyal, and P. S. Shrivastav, "Simultaneous densitometric analysis of amlodipine, hydrochlorothiazide, lisinopril, and valsartan by HPTLC in pharmaceutical formulations and human plasma," *Journal of Liquid Chromatography & Related Technologies*, vol. 40, no. 9, pp. 467–478, 2017.

[12] A. M. Brondi, J. S. Garcia, and M. G. Trevisan, "Development and validation of a chromatography method using tandem UV/charged aerosol detector for simultaneous determination of amlodipine besylate and olmesartan medoxomil: application to drug-excipient compatibility study," *Journal of Analytical Methods in Chemistry*, vol. 2017, Article ID 4878316, 8 pages, 2017.

[13] I. Kasagić-Vujanović, B. Jančić-Stojanović, and D. Ivanović, "Investigation of the retention mechanisms of amlodipine besylate, bisoprolol fumarate, and their impurities on three different ILIC columns," *Journal of Liquid Chromatography & Related Technologies*, vol. 41, no. 9, pp. 523–531, 2018.

[14] D. Sangeetha and M. K. Vadlamudi, "Stability-indicating method for the determination of assay and quantification of impurities in amlodipine-atorvastatin combination dosage form by RP-HPLC," *Journal of Liquid Chromatography & Related Technologies*, vol. 40, no. 11, pp. 576–598, 2017.

[15] L. Wang, W. Liu, Z. Zhang, and Y. Tian, "Validated LC-MS/MS method for the determination of amlodipine enantiomers in rat plasma and its application to a stereoselective pharmacokinetic study," *Journal of Pharmaceutical and Biomedical Analysis*, vol. 158, pp. 74–81, 2018.

[16] P. E. Irden, I. H. Taşdemir, C. Kaçar, and E. Kiliç, "Simultaneous determination of valsartan and Amlodipine besylate in human serum and pharmaceutical dosage forms by voltammetry," *International Journal of Electrochemical Science*, vol. 9, pp. 2208–2220, 2014.

[17] E. Arkan, Z. Karimi, M. Shamsipur, and R. Saber, "An electrochemical senor for determination of amlodipine besylate based on graphene–chitosan nano composite film modified glassy carbon electrode and application in biological and pharmaceutical samples," *Journal of Pharmaceutical Sciences*, vol. 3, pp. 99–107, 2014.

[18] G. R. Mansano, A. P. P. Eisele, L. H. Dall'Antonia, S. Afonso, and E. R. Sartori, "Electroanalytical application of a boron-doped diamond electrode: improving the simultaneous voltammetric determination of Amlodipine and valsartan in urine and combined dosage forms," *Journal of Electroanalytical Chemistry*, vol. 738, pp. 188–194, 2015.

[19] A. S. Rajpurohit, D. K. Bora, and A. K. Srivastava, ""Simultaneous determination of Amlodipine and losartan using an iron metal–organic framework/mesoporous carbon nanocomposite-modified glassy carbon electrode by differential pulse voltammetry," *Analytical Methods*, vol. 10, no. 45, pp. 5423–5438, 2018.

[20] E. Djaalab, M. Samar, S. Zougar, and R. Kherrat, "Electrochemical biosensor for the determination of amlodipine besylate based on gelatin-polyaniline iron oxide biocomposite film," *Catalysts*, vol. 8, no. 6, p. 233, 2018.

[21] E. Mircia, T. Balaci, G. Hancu, V. Ion, and A. Cârje, "Simultaneous determination of amlodipine and atorvastatin by capillary electrophoresis from fixed pharmaceutical formulations," *FARMACIA*, vol. 64, no. 3, pp. 398–402, 2016.

[22] M. Attimarad, N. Sreeharsha, B. E. Al-Dhubaib, A. B. Nair, and K. N. Venugopala, "Capillary electrophoresis: MEKC assay method for simultaneous determination of olmesartan medoxomil, Amlodipine besylate and hydrochlorothiazide in tablets," *Indian Journal of Pharmaceutical Education and Research*, vol. 50, no. 2, pp. 188–195, 2016.

[23] M. Adriana, H. Gabriel, R. Alexandru Vlad, S. Ştefana, R. Soare, and H. Kelemen, "Simultaneous determination of Amlodipine and telmisartan from pharmaceutical products by way of capillary electrophoresis," *Current Issues in Pharmacy and Medical Sciences*, vol. 29, no. 1, pp. 42–46, 2016.

[24] S. E. Nissen, N. D. Yeomans, D. H. Solomon et al., "Cardiovascular safety of celecoxib, naproxen, or ibuprofen for arthritis," *New England Journal of Medicine*, vol. 375, no. 26, pp. 2519–2529, 2016.

[25] F. Angeli, M. Trapasso, S. Signorotti, P. Verdecchia, and G. Reboldi, "Amlodipine and celecoxib for treatment of hypertension and osteoarthritis pain," *Expert Review of Clinical Pharmacology*, vol. 11, no. 11, pp. 1073–1084, 2018.

[26] R. E. Harris, G. A. Alshafie, H. Abou-Issa et al., "Chemoprevention of breast cancer in rats by celecoxib, a cyclooxygenase 2 inhibitor," *Cancer Research*, vol. 60, no. 8, pp. 2101–2103, 2000.

[27] R. N. Saha, C. Sajeev, P. R. Jadhav, S. P. Patil, and N. Srinivasan, "Determination of celecoxib in pharmaceutical formulations using UV spectrophotometry and liquid chromatography," *Journal of Pharmaceutical and Biomedical Analysis*, vol. 28, no. 3-4, pp. 741–751, 2002.

[28] A. A. Gouda, M. I. Kotb El-Sayed, A. S. Amin, and R. El Sheikh, "Spectrophotometric and spectrofluorometric methods for the determination of non-steroidal anti-inflammatory drugs: a review," *Arabian Journal of Chemistry*, vol. 6, no. 2, pp. 145–163, 2013.

[29] R. N. Saha, S. Chandran, P. R. Jadhav, and P. B. Kharwade, "Rapid and sensitive spectrofluorimetric method for the estimation of celecoxib and flurbiprofen," *Indian Journal of Pharmaceutical Sciences*, vol. 68, no. 1, pp. 20–25, 2006.

[30] J. Emami, R. Fallah, and A. Ajami, "A rapid and sensitive HPLC method for the analysis of celecoxib in human plasma: application to the pharmacokinetic studies," *DARU Journal of Pharmaceutical Sciences*, vol. 16, pp. 211–217, 2008.

[31] H. R. Bapatu, R. Maram, and S. Murthy, "Stability-indicating HPLC method for quantification of celecoxib and diacerein

along with its impurities in capsule dosage form," *Journal of Chromatographic Science*, vol. 53, no. 1, pp. 144–153, 2015.

[32] O. S. S. Chandana and R. Ravichandrababu, "Stability indicating HPLC method for celecoxib related substances in solid dosage forms," *International Journal of Research in Pharmaceutical Sciences*, vol. 7, no. 1, pp. 10–18, 2017.

[33] M.-S. Park, W.-S. Shim, S.-V. Yim, and K.-T. Lee, "Development of simple and rapid LC-MS/MS method for determination of celecoxib in human plasma and its application to bioequivalence study," *Journal of Chromatography B*, vol. 902, pp. 137–141, 2012.

[34] N. P. Patel, M. Sanyal, P. S. Shrivastav, and B. N. Patel, "Estimation of celecoxib in human plasma by rapid and selective LC-MS/MS method for a bioequivalence study," *International Journal of Pharmacy and Pharmaceutical Sciences*, vol. 10, no. 10, pp. 16–22, 2018.

[35] M. K. Srinivasu, D. Sreenivas Rao, and G. O. Reddy, "Determination of celecoxib, a COX-2 inhibitor, in pharmaceutical dosage forms by MEKC," *Journal of Pharmaceutical and Biomedical Analysis*, vol. 28, no. 3-4, pp. 493–500, 2002.

[36] A. H. Kamal, S. F. El-Malla, and S. F. Hammad, "A Review on UV spectrophotometric methods for simultaneous multicomponent analysis," *European Journal of Pharmaceutical and Medical Research*, vol. 3, pp. 348–360, 2016.

[37] M. Attimarad, M. Shahzad Chohan, and A. Ahmed Balgoname, "Simultaneous determination of moxifloxacin and flavoxate by RP-HPLC and ecofriendly derivative spectrophotometry methods in formulations," *International Journal of Environmental Research and Public Health*, vol. 16, no. 7, p. 1196, 2019.

[38] F. Salinas, J. J. B. Nevado, and A. E. Mansilla, "A new spectrophotometric method for quantitative multicomponent analysis resolution of mixtures of salicylic and salicyluric acids," *Talanta*, vol. 37, no. 3, pp. 347–351, 1990.

[39] A. Afkhami and M. Bahram, "Successive ratio-derivative spectra as a new spectrophotometric method for the analysis of ternary mixtures," *Spectrochimica Acta Part A: Molecular and Biomolecular Spectroscopy*, vol. 61, no. 5, pp. 869–877, 2005.

[40] N. M. Bhatt, V. D. Chavada, M. Sanyal, and P. S. Shrivastav, "Manipulating ratio spectra for the spectrophotometric analysis of diclofenac sodium and pantoprazole sodium in laboratory mixtures and tablet formulation," *The Scientific World Journal*, vol. 2014, Article ID 495739, 10 pages, 2014.

18

Identification of *Tilletia foetida*, *Ustilago tritici*, and *Urocystis tritici* Based on Near-Infrared Spectroscopy

Yaqiong Zhao,[1] Feng Qin,[1] Fei Xu,[2] Jinxing Ma,[1] Zhenyu Sun,[3] Yuli Song,[2] Longlian Zhao,[4] Junhui Li,[4] and Haiguang Wang[1]

[1]*College of Plant Protection, China Agricultural University, Beijing 100193, China*
[2]*Institute of Plant Protection, Henan Academy of Agricultural Sciences, Zhengzhou 450002, China*
[3]*Institute of Plant Protection, Gansu Academy of Agricultural Sciences, Lanzhou 730070, China*
[4]*College of Information and Electrical Engineering, China Agricultural University, Beijing 100083, China*

Correspondence should be addressed to Haiguang Wang; wanghaiguang@cau.edu.cn

Academic Editor: Alessandra Durazzo

Identifying plant pathogens for disease diagnosis and disease control strategy making is of great significance. In this study, based on near-infrared spectroscopy, a method for identifying three kinds of pathogens causing wheat smuts, including *Tilletia foetida*, *Ustilago tritici*, and *Urocystis tritici*, was investigated. Based on the acquired near-infrared spectral data of the teliospore samples of the three pathogens, pathogen identification models were built in different spectral regions using distinguished partial least squares (DPLS), backpropagation neural network (BPNN), and support vector machine (SVM). Satisfactory identification results were achieved using the DPLS, BPNN, and SVM models built in each of the 22 spectral regions. By contrast, the modeling effects of DPLS and SVM were better than those of BPNN. The modeling ratio of the training set to the testing set affected the identification results of the BPNN models more than those obtained using the DPLS and SVM models. In this study, a rapid, accurate, and nondestructive method was provided for plant pathogen identification, and some basis was provided for disease diagnosis, pathogen monitoring, and disease control. Moreover, some methodological references and supports were provided for identification of quarantine wheat smut fungi in plant quarantine.

1. Introduction

The occurrence and epidemics of plant diseases can cause severe yield losses and quality decline of agricultural products and sometimes lead to serious social problems. It is crucial to prevent and control plant diseases in a timely manner for ensuring normal agricultural production and food security. Accurate and rapid identification of plant diseases and the causal agents is the most important prerequisite of disease monitoring and early warning, which are the bases for prevention and control of plant diseases. Nowadays, morphological identification methods and molecular marker-based identification techniques are commonly used for pathogen identification. For example, morphological identification methods [1–5] and molecular biological identification methods [2–4, 6–11] have been developed for identification of the causal agents of many plant smuts with important economic significance in agricultural production. However, morphological identification methods highly depend on personnel experience, and it is difficult to clearly observe very small pathogens using ordinary optical microscopes, and thus identification errors easily occur. Molecular biological identification methods are highly accurate but generally require professional instruments and reagents, professional personnel to operate, and a relatively long time to achieve identification results [12, 13]. In addition, nondestructive identification of pathogens cannot be implemented using molecular biological methods. Therefore, a new method to solve the problems in identification of plant pathogens is urgently required. To ensure the quality and safety of agricultural products and the safe production of wheat, a rapid, accurate, and nondestructive identification method for three kinds of wheat smut fungi based on near-infrared spectroscopy (NIRS) was explored in this present study.

NIRS, a rapid, low-cost, pollution-free, and nondestructive analytical technique, can be used to conduct both qualitative analysis and quantitative analysis of samples using material information contained in near-infrared spectra and has been widely applied in agriculture, food, petroleum, chemical, pharmacy, and other industries [14–24]. Since NIRS is an indirect analytical technique, when it is applied to perform qualitative or quantitative analysis of an unknown sample, it is necessary to obtain near-infrared spectral data of a large number of known samples and then to establish a qualitative or quantitative identification model using chemometric methods.

In recent years, NIRS has been applied to detection and identification of insect pests [25, 26] and plant diseases [27–29]. Wu et al. [27] built an early detection model for gray mold (caused by *Botrytis cinerea*) on eggplant leaves based on acquired visible and near-infrared spectra using backpropagation neural network (BPNN), and an identification accuracy of 88% was achieved. For the early and rapid detection of soybean pod anthracnose (caused by *Colletotrichum truncatum*), integrating the successive projections algorithm and least square support vector machine (LS-SVM), a disease detection model with an accuracy of 95.45% was built based on visible/near-infrared spectra by Feng et al. [28]. To realize nondestructive detection of citrus huanglongbing (caused by *Candidatus liberibacter*), Liu et al. [29] acquired the near-infrared spectra (4000–9000 cm^{-1}) of three categories of citrus leaves including healthy leaves, nutrient-deficient leaves, and huanglongbing leaves, subsequently preprocessed the original spectral data using the methods including first derivative transformation, smoothing, and multiple scattered correction, and then built a LS-SVM model with an accuracy of 100% for identifying the three categories of leaves. At present, there are still few reports on the use of NIRS for identification and early monitoring of plant diseases and identification of plant pathogens.

In our previous studies, NIRS was used to implement early diagnosis and detection of wheat stripe rust caused by *Puccinia striiformis* f. sp. *tritici* (*Pst*) [30–32], qualitative discrimination of plant disease species [30, 31], disease severity assessment [33], and qualitative identification and quantitative analysis of *Pst* and *P. recondita* f. sp. *tritici* (*Prt*) based on NIRS [32, 34]. In combination with a molecular biology technique, real-time polymerase chain reaction, quantitative determination of the relative contents of *Pst* DNA in wheat leaves in incubation period, was implemented based on NIRS [32]. In addition, the research on quantitative determination of germinability of *Pst* urediospores based on NIRS was conducted [35], and the effect of ultraviolet B radiation on NIRS-based identification of *Pst* was investigated [36]. In particular, in the study of qualitative identification and quantitative analysis of *Pst* and *Prt*, based on the acquired near-infrared spectra of pure and mixed samples of the *Pst* and *Prt* urediospores, a qualitative identification model was built using distinguished partial least squares (DPLS) and a quantitative determination model was built using quantitative partial least squares [34]. The two models demonstrated good performance, and the qualitative identification and quantitative analysis of *Pst* and *Prt* were realized based on NIRS.

Based on the studies mentioned above, the rapid and accurate identification of three kinds of wheat smut fungi including *Tilletia foetida*, *Ustilago tritici*, and *Urocystis tritici* causing common bunt, loose smut, and flag smut on wheat, respectively, was investigated using NIRS technology in this study. Based on the acquired near-infrared spectra of the teliospores of *T. foetida*, *Ustilago tritici*, and *Urocystis tritici*, pathogen identification models were built with different modeling ratios of training sets and testing sets in different modeling spectral regions using DPLS, BPNN, and support vector machine (SVM). The aim of this study was to provide a rapid, accurate, and nondestructive method for the identification of multiple kinds of plant pathogens and to provide some basis for disease diagnosis, pathogen monitoring, and disease control measure making. In addition, some methodological references were provided for identifying quarantine wheat smut fungi in plant quarantine.

2. Materials and Methods

2.1. Materials. *T. foetida*, *Ustilago tritici*, and *Urocystis tritici* were used as experimental materials. The specimens of *T. foetida* and *Urocystis tritici* were collected from wheat growing areas in Henan Province, China in 2014 and 2015. The specimens of *Ustilago tritici* were collected from wheat growing areas in Gansu Province, China, in 2015. For *T. foetida* and *Urocystis tritici*, diseased wheat panicles were collected. For *Ustilago tritici*, diseased wheat stems, leaves, and leaf sheaths with typical symptoms of wheat flag smut were collected.

For *T. foetida*, each of the bunt ball from the diseased wheat panicles was cracked with an inoculation needle, and the released black powdery teliospores of approximately 7–12 mg were treated as a sample. There were 1071 bunt balls obtained in this study. For *Ustilago tritici*, each diseased wheat panicle was clamped with sterile forceps, then the black powdery teliospores were shacked off and landed on a piece of a smooth-surface weighing paper. The teliospores of 6-7 mg were treated as a sample, and a total of 389 teliospore samples of *Ustilago tritici* were obtained. For *Urocystis tritici*, 10 mg of teliospores from the diseased wheat stems, leaves, and leaf sheaths were taken as a sample. A total of 30 teliospore samples of *Urocystis tritici* were obtained.

2.2. Acquisition of Near-Infrared Spectral Data. Using a FT-NIR MPA spectrometer (Bruker, Germany), the near-infrared spectra of teliospore samples of *T. foetida*, *Ustilago tritici*, and *Urocystis tritici* were acquired. Each teliospore sample was placed into a sample cup (4 mm in diameter) for near-infrared spectral measurement using the integrating sphere diffuse reflectance method. Especially, the tightness of teliospores in the sample cup was kept as consistent as possible to reduce the errors that may be induced by different tightness. The spectral resolution and the number of scan processes in the measurement system were set as 8 cm^{-1} and 32, respectively. The near-infrared spectra in the range of 4000 to 12000 cm^{-1} were collected. Each teliospore sample was used for acquisition of only one near-infrared spectrum.

Totally, 1490 teliospore samples were measured and 1490 near-infrared spectra were acquired including 1071, 389, and 30 spectra of *T. foetida*, *Ustilago tritici*, and *Urocystis tritici*, respectively. After averaging the spectral data of *T. foetida*, *Ustilago tritici*, and *Urocystis tritici* according to their categories, the spectral curve of each category is shown in Figure 1. There were considerable differences between the average spectra in the range of 4000–12000 cm^{-1} except the range of 6000–7000 cm^{-1}. In particular, obvious differences could be found among the three average spectra in the range of 7200–12000 cm^{-1}.

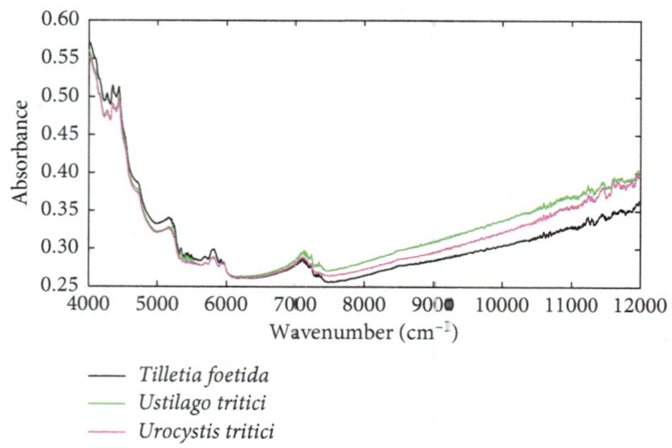

Figure 1: Spectral curves for the three kinds of fungi including *Tilletia foetida*, *Ustilago tritici*, and *Urocystis tritici*.

2.3. Establishment of Pathogen Identification Models Based on NIRS. The acquired near-infrared spectra had relatively strong random noises in high-frequency regions (Figure 1). To eliminate the effects of random noises in high-frequency regions and to select the suitable spectral region for modeling, the spectral region of 4000–12000 cm^{-1} and 21 spectral regions divided from the region of 4000–10000 cm^{-1}, including 4000–5000, 4000–6000, 4000–7000, 4000–8000, 4000–9000, 4000–10000, 5000–6000, 5000–7000, 5000–8000, 5000–9000, 5000–10000, 6000–7000, 6000–8000, 6000–9000, 6000–10000, 7000–8000, 7000–9000, 7000–10000, 8000–9000, 8000–10000, and 9000–10000 cm^{-1}, were taken as modeling spectral regions. In this study, the acquired near-infrared spectral data were processed using the two following methods: Data treatment method 1 and Data treatment method 2, and then pathogen identification models were built with three modeling ratios of training sets to testing sets (i.e., 3 : 1, 4 : 1 and 5 : 1) in the different spectral regions using three modeling methods including DPLS, BPNN, and SVM. The main steps for establishment of pathogen identification models are shown in Figure 2.

Data treatment method 1 was based on all the acquired near-infrared spectral data. The near-infrared spectra of each kind of pathogen were randomly selected according to modeling ratio (3 : 1, 4 : 1 or 5 : 1) and then were integrated together to form training set and testing set, respectively. The number of near-infrared spectra of each kind of pathogen in training sets and testing sets under different modeling ratios is shown in Table 1. With the modeling ratios of 3 : 1, 4 : 1, and 5 : 1, DPLS, BPNN, and SVM were used to build the pathogen identification models in the different modeling spectral regions, respectively. The identification accuracies of training and testing sets were used to evaluate the models. While building a DPLS model, the number of principal components was set as 1, 2, 3, . . ., or 10. The number of neurons in the hidden layer while building a BPNN model was set as 10, 50, or 100. Each BPNN model was built using the neural network toolbox of the software MATLAB 8.2 (MathWorks, Natick, MA, USA) with the default settings except for changes in the number of neurons in the hidden layer and randomly taking 90% of data from the training set for training and the remaining 10% for validation. Using C-SVM in LIBSVM package developed by the Chih-Jen Lin group from Taiwan, China [37], SVM models for pathogen identification were built with radial basis function as the kernel function. The optimal values of penalty parameter C and kernel function parameter g for each SVM model were searched using grid search algorithm in the range of 2^{-10} to 2^{10} with 0.8 as the searching step. Identification accuracies were calculated by 3-fold cross validation on the training set at all points within the grid. When the identification accuracy of the training set was the highest, the values of C and g were selected as the optimal parameters of the SVM model.

For Data treatment method 2, to reduce the bias that may be induced by the great differences among the numbers of samples of the three kinds of pathogens, the near-infrared spectra of the samples of *T. foetida* were randomly divided into 10 groups that were labeled as TFG1, TFG2, TFG3, . . ., and TFG10, and the near-infrared spectra of the samples of *Ustilago tritici* were also randomly divided into 10 groups that were labeled as UTG1, UTG2, UTG3, . . ., and UTG10. Then, TFG1, UTG1, and all the near-infrared spectra of *Urocystis tritici* formed a data subset that was labeled as data subset 1. TFG2, UTG2, and all the near-infrared spectra of *Urocystis tritici* formed data subset 2. In such a way, a total of 10 data subsets (including data subset 1, data subset 2, data subset 3, . . ., and data subset 10) were formed. The near-infrared spectra of each kind of pathogen in each data subset were randomly selected according to modeling ratio (3 : 1, 4 : 1 or 5 : 1) and then were integrated together to form training set and testing set, respectively. The number of near-infrared spectra of each kind of pathogen in the training and testing sets for each data subset under different modeling ratios is shown in Table 2. Based on each data subset, using DPLS, BPNN, and SVM modeling methods as described above, the pathogen identification models were built in the different modeling spectral regions when the modeling ratios of the training sets to the testing sets were 3 : 1, 4 : 1 and 5 : 1. The average identification accuracies and the standard deviation (SD) of the identification accuracies of training sets and testing sets on the 10 data subsets were used to evaluate the

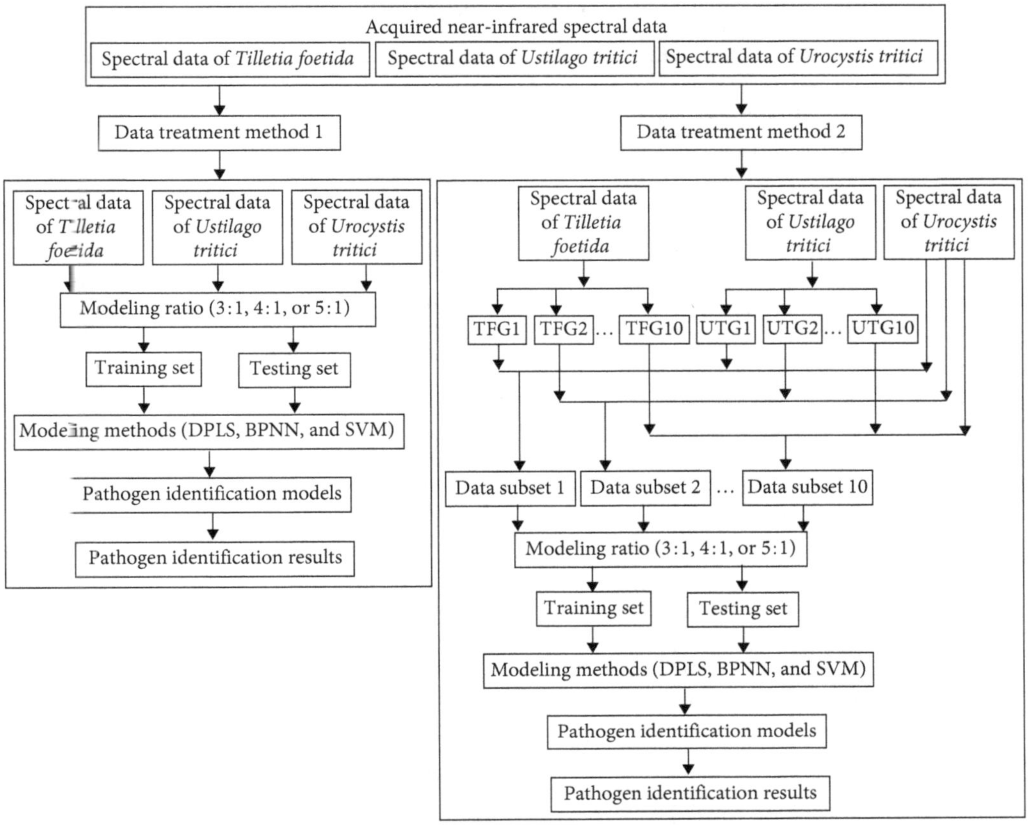

FIGURE 2: Work flow diagram of establishment of pathogen identification models based on near-infrared spectroscopy.

TABLE 1: Number of near-infrared spectra of each kind of pathogen in the training and testing sets under different modeling ratios when Data treatment method 1 was used.

Modeling ratio of training set to testing set	Training set				Testing set			
	Tilletia foetida	*Ustilago tritici*	*Urocystis tritici*	Total number of samples	*Tilletia foetida*	*Ustilago tritici*	*Urocystis tritici*	Total number of samples
3 : 1	803	291	22	1116	268	98	8	374
4 : 1	856	311	24	1191	215	78	6	299
5 : 1	892	324	25	1241	179	65	5	249

Note. There were 1490 near-infrared spectra including 1071, 389, and 30 spectra of *Tilletia foetida*, *Ustilago tritici*, and *Urocystis tritici*, respectively.

modeling effect. The value of SD can be calculated using the following formula

$$SD = \sqrt{\frac{\sum_{i=1}^{n} (A_i - \overline{A})^2}{n}}, \quad (1)$$

where SD is the standard deviation of the identification accuracies of training sets or testing sets on n data subsets, A_i is the identification accuracy of the training set or the testing set on the ith data subset, $\overline{A}$ is the average of the identification accuracies of training sets or testing sets on n data subsets, and n is the number of the data subsets. In this study, n was equal to 10.

For each modeling method (DPLS, BPNN, or SVM), when Data treatment method 1 or Data treatment method 2 was used, the optimal modeling results obtained with each modeling ratio of the training set to the testing set in each near-infrared spectral region were selected. When the DPLS models were built for pathogen identification, the optimal modeling results were selected from the identification results obtained with different numbers of principal components in a modeling spectral region, according to the identification accuracies of training sets and testing sets (Data treatment method 1) or the average identification accuracies and the SD values for training sets and test sets on the 10 data subsets (Data treatment method 2). If the same identification accuracies for training sets and testing sets were obtained when different numbers of principal components were used, the modeling effect with the minimum number of principal components was optimal. When the BPNN models were built, the optimal modeling results were selected from the results obtained with different numbers of neurons in the

Table 2: Number of near-infrared spectra of each kind of pathogen in the training and testing sets for each data subset under different modeling ratios when Data treatment method 2 was used.

Modeling ratio of training set to testing set	Data subset	Training set				Testing set			
		Tilletia foetida	*Ustilago tritici*	*Urocystis tritici*	Total number of samples	*Tilletia foetida*	*Ustilago tritici*	*Urocystis tritici*	Total number of samples
3 : 1	1	80	29	22	131	27	10	8	45
3 : 1	2	80	29	22	131	27	10	8	45
3 : 1	3	80	29	22	131	27	10	8	45
3 : 1	4	80	29	22	131	27	10	8	45
3 : 1	5	80	29	22	131	27	10	8	45
3 : 1	6	80	29	22	131	27	10	8	45
3 : 1	7	80	29	22	131	27	10	8	45
3 : 1	8	80	29	22	131	27	10	8	45
3 : 1	9	80	29	22	131	27	10	8	45
3 : 1	10	81	28	22	131	27	10	8	45
4 : 1	1	85	31	24	140	22	8	6	36
4 : 1	2	85	31	24	140	22	8	6	36
4 : 1	3	85	31	24	140	22	8	6	36
4 : 1	4	85	31	24	140	22	8	6	36
4 : 1	5	85	31	24	140	22	8	6	36
4 : 1	6	85	31	24	140	22	8	6	36
4 : 1	7	85	31	24	140	22	8	6	36
4 : 1	8	85	31	24	140	22	8	6	36
4 : 1	9	85	31	24	140	22	8	6	36
4 : 1	10	86	30	24	140	22	8	6	36
5 : 1	1	89	32	25	146	18	7	5	30
5 : 1	2	89	32	25	146	18	7	5	30
5 : 1	3	89	32	25	146	18	7	5	30
5 : 1	4	89	32	25	146	18	7	5	30
5 : 1	5	89	32	25	146	18	7	5	30
5 : 1	6	89	32	25	146	18	7	5	30
5 : 1	7	89	32	25	146	18	7	5	30
5 : 1	8	89	32	25	146	18	7	5	30
5 : 1	9	89	32	25	146	18	7	5	30
5 : 1	10	90	31	25	146	18	7	5	30

Note. There were 1490 near-infrared spectra including 1071, 389, and 30 spectra of *Tilletia foetida*, *Ustilago tritici*, and *Urocystis tritici*, respectively. There were 10 data subsets, and each data subset was composed of 176 spectra.

hidden layer in a modeling spectral region, according to the identification accuracies of training sets and testing sets (Data treatment method 1) or the average identification accuracies and the SD values for training sets and testing sets on the 10 data subset (Data treatment method 2). If the same identification accuracies for training and testing sets were obtained when different numbers of neurons in the hidden layer were used, the modeling effect with the minimum number of neurons was optimal.

Moreover, when the pathogen identification models were built using DPLS, BPNN, or SVM, a comprehensive comparison of the optimal modeling results obtained in different modeling spectral regions was performed. A comparison of the modeling results obtained with Data treatment method 1 and Data treatment method 2 was also conducted.

3. Results

3.1. Pathogen Identification Results of the DPLS, BPNN, and SVM Models Using Data Treatment Method 1. When the acquired near-infrared spectral data were processed with Data treatment method 1, very good performance with the identification accuracies of the training and testing sets of more than 97% was achieved using the DPLS models built with different modeling ratios in each of the 22 modeling spectral regions (Table 3). The identification accuracies of the training and testing sets reached 100% for the DPLS models built with different modeling ratios in the modeling spectral regions, except for the spectral regions of 4000–5000, 6000–7000, 8000–9000, 8000–10000, and 9000–10000 cm^{-1}. When the DPLS models were built in the narrow spectral regions, the identification accuracies of the training and testing sets could reach 100% only for the models with different modeling ratios in the two spectral regions of 5000–6000 and 7000–8000 cm^{-1}. The identification accuracies of the training and testing sets could reach 100% for the DPLS models built with different modeling ratios in the relatively wide spectral regions, except for the spectral region of 8000–10000 cm^{-1}. The results indicated that the width of modeling spectral region had certain effects on the identification results of the DPLS models. In addition, the results demonstrated that relatively low-frequency regions and relatively high-frequency regions in the acquired near-infrared spectra could decrease the identification

Table 3: The optimal results except those with both the identification accuracy of the training set and the identification accuracy of the testing set equal to 100% when the DPLS models for Data treatment method 1 were built with different modeling ratios in each modeling spectral region.

Spectral region (cm^{-1})	Modeling ratio of training set to testing set	The number of principal components	Identification accuracy of training set (%)	Identification accuracy of testing set (%)
4000–5000	3 : 1	9	99.91	99.73
4000–5000	4 : 1	9	100	99.67
4000–5000	5 : 1	10	99.92	100
6000–7000	3 : 1	10	99.46	98.66
6000–7000	4 : 1	9	99.33	99.33
6000–7000	5 : 1	9	99.44	98.80
8000–9000	3 : 1	8	98.21	98.13
8000–9000	4 : 1	9	98.40	98.33
8000–9000	5 : 1	9	98.39	97.99
8000–10000	3 : 1	10	99.91	99.47
8000–10000	4 : 1	8	99.92	99.33
8000–10000	5 : 1	9	99.84	99.60
9000–10000	3 : 1	9	98.03	97.33
9000–10000	4 : 1	7	97.98	97.66
9000–10000	5 : 1	9	97.99	97.59

accuracies of the training and testing sets. In particular, high-frequency regions with relatively wide ranges had some effects on the identification results of the models with different modeling ratios.

When the acquired near-infrared spectral data were processed with Data treatment method 1, the BPNN models built with different modeling ratios in each modeling spectral region performed well (Table 4). The optimal identification accuracies of the training and testing sets were 87.66% and 88.29%, respectively, when the BPNN model was built with the modeling ratio equal to 4 : 1 and the number of neurons in the hidden layer equal to 10 in the spectral region of 9000–10000 cm^{-1}. The identification accuracies of the training and testing sets reached more than 90% in other cases. When the BPNN models were built in the spectral region of 4000–12000 cm^{-1}, the identification accuracies of the training and testing sets for the models built with the modeling ratios of 3 : 1 and 5 : 1 reached 100%, and those for the model built with the modeling ratio equal to 4 : 1 and the number of neurons in the hidden layer equal to 50 were 99.83% and 100% respectively. When the BPNN models were built in the 21 spectral regions from the range of 4000 to 10000 cm^{-1}, the optimal identification accuracies of the training sets and the testing sets for the models with different modeling ratios reached 100%, except for the spectral regions of 4000–6000, 5000–6000, 5000–9000, 6000–7000, 6000–8000, 6000–9000, 6000–10000, 7000–8000, 8000–9000, 8000–10000, and 9000–10000 cm^{-1}. For the BPNN models built in each of the 21 spectral regions from the range of 4000–10000 cm^{-1} except the spectral regions of 6000–7000, 8000–9000, 8000–10000, and 9000–10000 cm^{-1}, at least when one modeling ratio was used, both the identification accuracy of the training set and that of the testing set reached 100%. For the BPNN models built in the narrow spectral regions, 4000–5000, 5000–6000, and 7000–8000 cm^{-1}, both the identification accuracies of the training and testing sets could reach 100%. However, when the BPNN models were built in the spectral regions of 5000–6000 and 7000–8000 cm^{-1}, the modeling ratio affected the modeling results. For the BPNN models built in the spectral region of 5000–6000 cm^{-1}, when the modeling ratios were 3 : 1, 4 : 1, and 5 : 1, the identification accuracies of the training sets were 100%, 97.98%, and 99.92%, respectively, and those of the testing sets were 100%, 97.99%, and 100%, respectively. For the BPNN models built with the modeling ratios of 3 : 1, 4 : 1, and 5 : 1 in the spectral region of 7000–8000 cm^{-1}, the identification accuracies of the training sets were 100%, 97.98%, and 100%, respectively, and those of the testing sets were 100%, 97.99%, and 100%, respectively.

When Data treatment method 1 was used, the SVM models that were built with different modeling ratios in each modeling spectral region presented very good performances, with the optimal identification accuracies of the training and testing sets of more than 99% (Table 5). For the SVM models that were built in the modeling spectral regions, except the spectral region of 9000–10000 cm^{-1}, the identification accuracies of 100% of the training and testing sets were achieved at least when one modeling ratio was used. When the SVM models were built in the narrow spectral regions of 4000–5000, 6000–7000, 7000–8000, and 8000–9000 cm^{-1}, both the identification accuracies of the training sets and those of the testing sets reached 100% when each of the three modeling ratios was used. For the SVM models built in the spectral region of 5000–6000 cm^{-1}, when the modeling ratios were 3 : 1 and 4 : 1, the optimal identification accuracies of the training sets were 99.91% and 100%, respectively, and those of the testing sets were 100% and 99.67%, respectively, and when the modeling ratio was 5 : 1, both the optimal identification accuracy of the training set and that of the testing set were 100%. For the SVM models built in the spectral region of 9000–10000 cm^{-1}, the optimal identification accuracy of the training set was 100% and that of the testing set was 99.73% when the modeling ratio was 3 : 1; when the modeling ratio was 4 : 1, the optimal identification accuracies of the training set and the testing set were 99.92% and 100%, respectively; and when the modeling ratio was 5 : 1,

TABLE 4: The optimal results except those with both the identification accuracy of the training set and the identification accuracy of the testing set equal to 100% when the BPNN models for Data treatment method 1 were built with different modeling ratios in each modeling spectral region.

Spectral region (cm^{-1})	Modeling ratio of training set to testing set	The number of neurons in the hidden layer	Identification accuracy of training set (%)	Identification accuracy of testing set (%)
4000–12000	4 : 1	50	99.83	100
4000–6000	4 : 1	10	97.98	97.99
4000–6000	5 : 1	10	97.99	97.99
5000–6000	4 : 1	10	97.98	97.99
5000–6000	5 : 1	100	99.92	100
5000–9000	3 : 1	10	98.03	97.86
6000–7000	3 : 1	100	98.03	97.59
6000–7000	4 : 1	50	97.98	97.99
6000–7000	5 : 1	10	97.99	97.59
6000–8000	3 : 1	50	98.03	97.86
6000–9000	4 : 1	10	97.98	97.99
6000–10000	3 : 1	10	98.03	97.86
6000–10000	4 : 1	10	97.98	97.99
7000–8000	4 : 1	10	97.98	97.99
8000–9000	3 : 1	50	98.03	97.86
8000–9000	4 : 1	50	97.98	97.99
8000–9000	5 : 1	50	97.99	97.99
8000–10000	3 : 1	100	98.03	97.86
8000–10000	4 : 1	50	97.98	97.99
8000–10000	5 : 1	10	97.99	97.99
9000–10000	3 : 1	50	92.11	91.98
9000–10000	4 : 1	10	87.66	88.29
9000–10000	5 : 1	50	97.58	97.19

TABLE 5: The optimal results except those with both the identification accuracy of the training set and the identification accuracy of the testing set equal to 100% when the SVM models for Data treatment method 1 were built with different modeling ratios in each modeling spectral region.

Spectral region (cm^{-1})	Modeling ratio of training set to testing set	Optimal parameters		Identification accuracy of training set (%)	Identification accuracy of testing set (%)
		C	g		
4000–12000	3 : 1	0.2500	6.9644	100	99.73
4000–6000	4 : 1	1.3195	64.0000	100	99.67
4000–10000	3 : 1	0.4353	6.9644	100	99.73
5000–6000	3 : 1	4.0000	36.7583	99.91	100
5000–6000	4 : 1	1.3195	337.7940	100	99.67
5000–10000	3 : 1	0.4353	12.1257	100	99.73
5000–10000	5 : 1	0.7579	111.4305	100	99.60
6000–8000	3 : 1	6.9644	64.0000	100	99.73
6000–8000	5 : 1	2.2974	588.1336	100	99.20
6000–9000	3 : 1	1.3195	21.1121	100	99.47
6000–9000	5 : 1	1.3195	337.7940	100	99.20
6000–10000	3 : 1	1.3195	12.1257	99.91	99.73
9000–10000	3 : 1	64.0000	111.4305	100	99.73
9000–10000	4 : 1	588.1336	4.0000	99.92	100
9000–10000	5 : 1	12.1257	337.7940	100	99.60

those of the training set and the testing set were 100% and 99.60%, respectively.

As described above, satisfactory results could be obtained using the pathogen identification models built using DPLS, BPNN, and SVM in different spectral regions. Compared with the DPLS and BPNN modeling methods, the identification accuracies of the training and testing sets could reach 100% in more spectral regions when the pathogen identification models were built using SVM. By contrast, the effects of the modeling ratio on the identification results of the models built using BPNN and SVM were greater than those on the identification results of the models built using DPLS.

3.2. Pathogen Identification Results of the DPLS, SVM, and BPNN Models Using Data Treatment Method 2. When Data treatment method 2 was used, very satisfactory modeling results were achieved using the DPLS models built on each data subset with different modeling ratios in each modeling spectral region. The average identification accuracies of the training sets and the testing sets on the 10 data subsets reached more than 98% (Table 6). Except for the models built with the modeling ratio of 3 : 1 in the spectral region of 6000–7000 cm^{-1} and the models built on each data subset with the three modeling ratios in the spectral region of 8000–9000 or 9000–10000 cm^{-1}, the average identification accuracies of the training sets and the testing sets on the 10 data subsets reached 100% for the DPLS models built on each data subset with different modeling ratios in each of the other modeling spectral regions. For the DPLS models built in the narrow spectral region of 4000–5000, 5000–6000, 6000–7000, or 7000–8000 cm^{-1}, both the average identification accuracies of the training and testing sets on the 10 data subsets reached 100%, when one or more than one modeling ratios were used. By contrast, there was some reduction in the average identification accuracies of the training sets and the testing sets on the 10 data subsets when the DPLS models were built in the narrow high-frequency region of 8000–9000 or 9000–10000 cm^{-1}, demonstrating that high-frequency regions had an impact on the modeling results. Although both the average identification accuracies of the training sets and those of the testing sets on the 10 data subsets did not reach 100% for the models built with the modeling ratio of 3 : 1 in the spectral region of 6000–7000 cm^{-1} and for those built with each modeling ratio in the spectral region of 8000–9000 or 9000–10000 cm^{-1}, the average identification accuracies were still very high, and the SD values of the identification accuracies for the 10 data subsets were very small. The results indicated that satisfactory identification performance could be achieved using the DPLS models based on NIRS and that the models had high stability.

When Data treatment method 2 was used, satisfactory identification results could be obtained using the BPNN models built on each data subset in each modeling spectral region (Table 7), but the modeling ratio had a great influence on the identification results of the models. For the BPNN models built in the spectral region of 4000–8000 cm^{-1}, both the average identification accuracy of the training sets and that of the testing sets on the 10 data subsets reached 100% when the modeling ratio was 4 : 1 or 5 : 1. When the BPNN models were built with the modeling ratio of 5 : 1 in the spectral region of 5000–6000 cm^{-1} on the 10 data subsets, both the average identification accuracy of the training sets and that of the testing sets reached 100%. When the BPNN models were built with the modeling ratio of 3 : 1 in the spectral region of 5000–10000 cm^{-1} on the 10 data subsets, both the average identification accuracy of the training sets and that of the testing sets reached 100%. For the models built with the modeling ratio of 3 : 1 or 5 : 1 in the spectral region of 4000–5000 cm^{-1}, the ones built with the modeling ratio of 4 : 1 in the spectral region of 5000–6000 cm^{-1}, the ones built with the modeling ratio of 3 : 1 or 5 : 1 in the spectral region of 6000–7000 cm^{-1}, and the ones built with the modeling ratio of 3 : 1 in the spectral region of 8000–10000 cm^{-1}, the SD values of the identification accuracies of the training and testing sets were both relatively high, indicating poor modeling effects.

When Data treatment method 2 was used, the SVM models built on the 10 data subsets, with each modeling ratio in each modeling spectral region, presented very good performance with the averages of the identification accuracies of the training sets and the testing sets of more than 98% (Table 8). Very small SD values were obtained for the identification accuracies of the training and testing sets. For the SVM models built on the 10 data subsets with the modeling ratio of 5 : 1 in the spectral region of 4000–7000 cm^{-1}, the average identification accuracies of the training and testing sets were 99.93% and 100%, respectively, and the SD values for the training and testing sets were 0.002040 and 0, respectively. For the SVM models built on the 10 data subsets with the modeling ratio of 5 : 1 in the spectral region of 4000–12000 cm^{-1}, the average identification accuracies of the training and testing sets were 99.86% and 100%, respectively, and the SD values for the training and testing sets were 0.002720 and 0, respectively. For the SVM models built on the 10 data subsets with the modeling ratio of 4 : 1 in the spectral region of 5000–6000 cm^{-1}, the average identification accuracies of the training and testing sets were 99.86% and 100%, respectively, and the SD values for the training and testing sets were 0.004290 and 0, respectively. The identification results of the models in the three cases above were the best. By contrast, the worst identification results were obtained when the models were built on the 10 data subsets with the modeling ratio of 5 : 1 or 3 : 1 in the spectral region of 8000–9000 cm^{-1} and on the 10 data subsets with the modeling ratio of 4 : 1 in the spectral region of 9000–10000 cm^{-1}. For the models built in these three cases, the average identification accuracies of the training sets were 99.73%, 99.77%, and 99.57%, respectively, and the SD values for the training sets were 0.006261, 0.004893, and 0.005715, respectively; the average identification accuracies of the testing sets were 98.33%, 98.00%, and 98.06%, respectively, and the SD values for the testing sets were 0.02236, 0.01554, and 0.03298, respectively.

The results showed that the pathogen identification models built using DPLS performed better than those built using BPNN or SVM when Data treatment method 2 was used. The modeling effect of DPLS on each data subset was more stable than that of BPNN. When the pathogen identification models were built using BPNN in the spectral region of 4000–5000, 5000–6000, 6000–7000, or 8000–10000 cm^{-1}, greater SD values of the identification accuracies of the training and testing sets were obtained. Although the modeling effect of SVM was not as good as that of DPLS, very satisfactory identification accuracies of the training and testing sets were achieved using the SVM models built on each data subset.

3.3. Comparison of Modeling Results Obtained with Data Treatment Methods 1 and 2. The modeling effect using DPLS with Data treatment method 2 was better than that achieved with Data treatment method 1. When Data treatment method 2 was used, the average identification accuracies of the training and testing sets did not simultaneously reach 100% for the DPLS models built on the 10 data subsets with

Table 6: The optimal results except the average identification accuracies and the standard deviations of the identification accuracies for the 10 data subsets for the training sets and the testing sets were 100% and 0 when the DPLS models for Data treatment method 2 were built with different modeling ratios in each modeling spectral region.

Spectral region (cm^{-1})	Modeling ratio of training set to testing set	The number of principal components	Training sets		Testing sets	
			Average identification accuracy for the 10 data subsets (%)	Standard deviation of the identification accuracies for the 10 data subsets	Average identification accuracy for the 10 data subsets (%)	Standard deviation of the identification accuracies for the 10 data subsets
6000–7000	3 : 1	7	100	0	99.78	0.006660
8000–9000	3 : 1	6	100	0	99.78	0.006660
8000–9000	4 : 1	6	100	0	99.44	0.01112
8000–9000	5 : 1	6	100	0	99.67	0.009990
9000–10000	3 : 1	5	99.70	0.006994	98.45	0.01734
9000–10000	4 : 1	6	100	0	98.61	0.02241
9000–10000	5 : 1	5	99.66	0.01026	99.33	0.01332

each modeling ratio only in two spectral regions (8000–9000 and 9000–10000 cm^{-1}). When Data treatment method 1 was used, both the identification accuracy of the training set and that of the testing set were less than 100% for the DPLS models built with each modeling ratio in five spectral regions (4000–5000, 6000–7000, 8000–9000, 8000–10000, and 9000–10000 cm^{-1}).

When Data treatment method 1 was used, the optimal accuracies of the training and testing sets could reach more than 97% for the BPNN models built in different spectral regions. Among the identification results obtained using the BPNN models built with different modeling ratio in each spectral region, both the optimal accuracy of the training set and that of the testing set were less than 97%, only for the models built with the modeling ratio of 3 : 1 or 4 : 1 in the spectral region of 9000–10000 cm^{-1}. However, when Data treatment method 2 was used, relatively great differences were observed between the average identification accuracies of the training sets and between the average identification accuracies of the testing sets for the BPNN models built on the 10 data subsets in different modeling spectral regions. Among the selected optimal identification results, the range of the average identification accuracies of the training sets was 88.77–100% and the range of the average identification accuracies of the testing sets was 87.67–100%. In addition, when Data treatment method 2 was used, the modeling ratio greatly influenced the identification results of the BPNN models. Relatively large identification accuracy SD values of the training and testing sets were obtained when the BPNN models were built on the 10 data subsets with one or two modeling ratios in some spectral regions (4000–5000, 5000–6000, 6000–7000, and 8000–10000 cm^{-1}). This indicated that the poor modeling effects were achieved on some data subsets using BPNN and that there were relatively great differences both between the identification accuracies of the training sets and between the identification accuracies of the testing sets for the 10 BPNN models built with certain modeling ratio in certain modeling spectral region. For instance, when Data treatment method 2 was used and the BPNN models were built in the spectral region of 4000–5000 cm^{-1} on the 10 data subsets with a ratio of the training set to the testing set of 3 : 1, the optimal identification results (Figure 3) were achieved with the number of neurons in the hidden layer equal to 50. The SD values of the identification accuracies of the training and testing sets on the 10 data subsets were 0.1541 and 0.1508, respectively. Relatively low identification accuracies of the training and testing sets were obtained for the BPNN models built on data subset 1 and data subset 2. For the BPNN model built on data subset 1, the identification accuracies of the training and testing sets were 70.23% and 66.67%, respectively. For the BPNN model built on data subset 2, the identification accuracies of the training and testing sets were 54.20% and 57.78%, respectively. The results indicated that these two models presented poor performance in pathogen identification.

When Data treatment method 1 or Data treatment method 2 was used, the built SVM models presented very good performance with very high identification accuracies of both the training sets and the testing sets. In particular, when Data treatment method 2 was used, for the SVM models built on the 10 data subsets, very small SD values of the identification accuracies of the training and testing sets were obtained. It was demonstrated that very satisfactory identification results could be achieved using the SVM models built on each data subset and that this performance had very high stability.

4. Conclusions and Discussion

The application of NIRS to the identification of three kinds of wheat smut pathogens, including *T. foetida*, *Ustilago tritici*, and *Urocystis tritici*, was investigated in this study. The acquired near-infrared spectral data were processed using Data treatment method 1 and Data treatment method 2 to form the training and testing sets. Then, the DPLS, BPNN, and SVM models for pathogen identification were built in 22 spectral regions. Using Data treatment method 1 or Data treatment method 2, satisfactory identification results were achieved when each of the three modeling methods was used for model-building in different spectral regions. By contrast, the modeling effect of DPLS was the most stable, followed by that of SVM and that of BPNN was the worst. In this study, a rapid, accurate, and nondestructive method for the identification of the three kinds of wheat

TABLE 7: Pathogen identification results of the BPNN models built with different modeling ratios in each modeling spectral region when Data treatment method 2 was used.

Spectral region (cm^{-1})	Modeling ratio of training set to testing set	The number of neurons in the hidden layer	Training sets: Average identification accuracy for the 10 data subsets (%)	Training sets: Standard deviation of the identification accuracies for the 10 data subsets	Testing sets: Average identification accuracy for the 10 data subsets (%)	Testing sets: Standard deviation of the identification accuracies for the 10 data subsets
4000–12000	3:1	100	99.70	0.006994	99.78	0.006660
4000–12000	4:1	10	98.00	0.04237	97.22	0.04648
4000–12000	5:1	50	98.02	0.04058	97.67	0.05176
4000–5000	3:1	50	92.14	0.1541	92.00	0.1508
4000–5000	4:1	100	96.57	0.04367	96.39	0.04488
4000–5000	5:1	100	92.88	0.1175	93.33	0.1183
4000–6000	3:1	50	99.54	0.009770	99.56	0.01332
4000–6000	4:1	50	99.71	0.008580	99.44	0.01668
4000–6000	5:1	10	99.59	0.01024	99.33	0.01332
4000–7000	3:1	10	98.02	0.04279	97.78	0.04662
4000–7000	4:1	50	95.29	0.08590	95.00	0.08766
4000–7000	5:1	50	99.59	0.01024	99.67	0.009990
4000–8000	3:1	10	99.31	0.01617	99.33	0.01017
4000–8000	4:1	50	100	0	100	0
4000–8000	5:1	50	100	0	100	0
4000–9000	3:1	10	99.92	0.002280	99.78	0.006660
4000–9000	4:1	10	99.57	0.009156	99.17	0.01780
4000–9000	5:1	50	99.93	0.002040	99.67	0.009990
4000–10000	3:1	10	99.77	0.004893	99.78	0.006660
4000–10000	4:1	10	99.64	0.007314	99.44	0.01668
4000–10000	5:1	10	99.73	0.006261	100	0
5000–6000	3:1	50	99.54	0.01143	99.11	0.02038
5000–6000	4:1	50	95.14	0.1272	94.44	0.1400
5000–6000	5:1	50	100	0	100	0
5000–7000	3:1	10	99.62	0.009181	99.33	0.01421
5000–7000	4:1	10	99.29	0.01153	98.89	0.01844
5000–7000	5:1	10	97.67	0.05708	97.33	0.05925
5000–8000	3:1	10	99.85	0.004590	99.56	0.008880
5000–8000	4:1	50	99.86	0.004290	99.44	0.01112
5000–8000	5:1	10	99.80	0.006150	100	0
5000–9000	3:1	100	99.92	0.002280	99.78	0.006660
5000–9000	4:1	50	99.79	0.004573	99.44	0.01668
5000–9000	5:1	100	99.80	0.004381	99.67	0.009990
5000–10000	3:1	50	100	0	100	0
5000–10000	4:1	50	100	0	99.72	0.008340
5000–10000	5:1	50	99.66	0.01026	99.67	0.009990
6000–7000	3:1	50	94.28	0.1015	94.22	0.1010
6000–7000	4:1	50	95.72	0.08484	94.72	0.08094
6000–7000	5:1	50	88.77	0.1922	87.67	0.1700
6000–8000	3:1	50	100	0	99.78	0.006660
6000–8000	4:1	50	99.86	0.004290	99.44	0.01112
6000–8000	5:1	50	97.88	0.04379	97.00	0.06046
6000–9000	3:1	100	98.24	0.04069	98.00	0.04604
6000–9000	4:1	50	99.86	0.002840	99.44	0.01112
6000–9000	5:1	50	99.93	0.002040	99.67	0.009990
6000–10000	3:1	100	99.47	0.01186	99.33	0.01421
6000–10000	4:1	50	97.93	0.04233	96.67	0.04779
6000–10000	5:1	10	99.04	0.02435	99.33	0.01332
7000–8000	3:1	50	99.85	0.003040	99.33	0.01421
7000–8000	4:1	50	99.86	0.004290	99.44	0.01668
7000–8000	5:1	50	99.93	0.002040	99.00	0.02135
7000–9000	3:1	50	98.24	0.04040	97.33	0.04534
7000–9000	4:1	50	98.43	0.03805	97.22	0.04648
7000–9000	5:1	50	97.95	0.04029	97.33	0.05122
7000–10000	3:1	50	95.65	0.08542	95.56	0.08374

Table 7: Continued.

Spectral region (cm^{-1})	Modeling ratio of training set to testing set	The number of neurons in the hidden layer	Training sets: Average identification accuracy for the 10 data subsets (%)	Training sets: Standard deviation of the identification accuracies for the 10 data subsets	Testing sets: Average identification accuracy for the 10 data subsets (%)	Testing sets: Standard deviation of the identification accuracies for the 10 data subsets
7000–10000	4 : 1	50	99.29	0.01356	98.33	0.02833
7000–10000	5 : 1	50	99.66	0.006296	99.33	0.01332
8000–9000	3 : 1	50	97.79	0.04002	97.11	0.03854
8000–9000	4 : 1	50	98.00	0.03949	96.39	0.05423
8000–9000	5 : 1	50	97.74	0.04030	96.67	0.04945
8000–10000	3 : 1	10	94.12	0.1173	93.56	0.1207
8000–10000	4 : 1	10	96.14	0.06422	95.28	0.05700
8000–10000	5 : 1	10	97.13	0.04351	96.67	0.05579
9000–10000	3 : 1	50	93.89	0.08533	93.56	0.08401
9000–10000	4 : 1	50	97.36	0.05689	96.11	0.07049
9000–10000	5 : 1	50	97.74	0.04076	97.00	0.05261

smut fungi was provided, and methodological references were also provided for the identification of quarantine wheat smut fungi in the processes of plant quarantine.

A variety of wheat diseases are caused by smut fungi, including wheat common bunt, wheat loose smut, wheat flag smut, wheat dwarf bunt, and wheat Karnal bunt. The diseases caused by wheat smut fungi can cause serious yield losses in production and even losses of up to 100% under appropriate environmental conditions [2, 11, 38]. In addition, the flour quality can seriously decline due to contamination by the causal agents. Among wheat smut fungi, *Tilletia controversa* (the causal agent of wheat dwarf bunt) and *T. indica* (the causal agent of wheat Karnal bunt) are two kinds of important entry-exit quarantine pests in many countries. In the processes of import and export wheat quarantine, it is critical to detect and identify the species of wheat smut fungi and the categories of wheat bunt galls carried in wheat. At present, the identification of these pathogens is conducted mainly based on morphological observation [1–5] and molecular biological detection [2–4, 9, 11]. Morphological observation is usually performed by using light microscopy and scanning electron microscopy. Molecular biological methods based on molecular markers, such as intergenic spacer [9], internal transcribed spacer [11], and extension factor [11], have been used to identify wheat smut fungi. NIRS is a rapid and nondestructive analytical technique [14, 15, 18, 22]. In this study, a method based on NIRS was provided for identification of the three kinds of wheat smut fungi including *T. foetida*, *Ustilago tritici*, and *Urocystis tritici*. In future studies, NIRS can be used to identify other pathogens and an NIRS-based identification system can be developed for the identification of multiple kinds of plant pathogens. Moreover, direct detection of bunt gall and healthy wheat seeds using NIRS can be explored to provide technical support for rapid online detection of imported and exported wheat.

The numbers of acquired near-infrared spectra of the three kinds of pathogens obtained in this study were not equal. There was a considerable difference among the numbers, and the spectra of *Urocystis tritici* were fewer than those of the other two pathogens. Although all the acquired near-infrared spectra were divided into different data subsets using Data treatment method 2 and relatively perfect results were achieved using the built DPLS, BPNN, and SVM models, the bias in pathogen identification may still be caused by the unbalanced numbers of samples. In further studies, the collection of *Urocystis tritici* samples and the acquisition of near-infrared spectra of the corresponding samples can be strengthened to obtain more near-infrared spectra of the pathogen and the pathogen identification models can be further optimized to ensure the accurate identification stability of the established models.

Near-infrared spectra contain material information that is useful to identify samples, and some spectral bands associated with material signals play important roles in qualitative analysis [14, 15, 18, 22]. To the best of our knowledge, systematic studies on material composition of wheat smut fungi have not been reported. In the future, the studies on this aspect can be conducted to explore mechanisms to interpret the near-infrared spectra of different wheat smut fungi, which is conducive to selecting the suitable modeling spectral region for pathogen identification.

To reduce noise interference and improve the signal-to-noise ratio, the number of scan processes in the measurement system was set as 8 cm^{-1} when near-infrared spectra of pathogen samples were acquired in this study. In addition, when teliospores were placed into the sample cup, the tightness of the teliospores was kept as consistent as possible to reduce the differences that may be induced by different tightness. Based on original near-infrared spectral data, satisfactory modeling results were obtained using the models built in the simply divided modeling spectral regions in this study. Therefore, no mathematical methods were applied to preprocess the acquired original near-infrared spectral data and no statistical methods were used to perform the modeling spectral region selection. At present, there are a variety of spectral preprocessing methods and modeling spectral region selection methods [15, 22]. In future studies, original near-infrared spectral data can be preprocessed using the selected suitable method to eliminate background interference that may exist, a suitable method can be chosen

TABLE 8: Pathogen identification results of the SVM models built with different modeling ratios in each modeling spectral region when Data treatment method 2 was used.

Spectral region (cm^{-1})	Modeling ratio of training set to testing set	Training sets: Average identification accuracy for the 10 data subsets (%)	Training sets: Standard deviation of the identification accuracies for the 10 data subsets	Testing sets: Average identification accuracy for the 10 data subsets (%)	Testing sets: Standard deviation of the identification accuracies for the 10 data subsets
4000–12000	3:1	99.92	0.002280	99.78	0.006660
4000–12000	4:1	99.79	0.003254	99.17	0.02499
4000–12000	5:1	99.86	0.002720	100	0
4000–5000	3:1	100	0	99.78	0.006660
4000–5000	4:1	100	0	99.72	0.008340
4000–5000	5:1	100	0	99.67	0.009990
4000–6000	3:1	100	0	99.56	0.008880
4000–6000	4:1	99.86	0.004290	99.72	0.008340
4000–6000	5:1	100	0	99.67	0.009990
4000–7000	3:1	100	0	99.11	0.01776
4000–7000	4:1	100	0	99.72	0.008340
4000–7000	5:1	99.93	0.002040	100	0
4000–8000	3:1	100	0	99.11	0.01473
4000–8000	4:1	100	0	99.72	0.008340
4000–8000	5:1	100	0	99.67	0.009990
4000–9000	3:1	100	0	99.56	0.008880
4000–9000	4:1	100	0	98.89	0.02545
4000–9000	5:1	99.66	0.01026	98.67	0.03055
4000–10000	3:1	100	0	99.56	0.008880
4000–10000	4:1	99.86	0.002840	99.17	0.02499
4000–10000	5:1	99.73	0.006261	98.67	0.03055
5000–6000	3:1	99.70	0.009150	99.33	0.01421
5000–6000	4:1	99.86	0.004290	100	0
5000–6000	5:1	100	0	99.67	0.009990
5000–7000	3:1	100	0	99.33	0.01421
5000–7000	4:1	100	0	99.72	0.008340
5000–7000	5:1	100	0	99.67	0.009990
5000–8000	3:1	100	0	99.11	0.01088
5000–8000	4:1	100	0	98.89	0.01362
5000–8000	5:1	99.86	0.002720	98.67	0.02211
5000–9000	3:1	100	0	99.56	0.008880
5000–9000	4:1	100	0	99.72	0.008340
5000–9000	5:1	99.59	0.01233	98.67	0.03055
5000–10000	3:1	100	0	99.56	0.008880
5000–10000	4:1	99.86	0.002840	98.89	0.03333
5000–10000	5:1	100	0	99.67	0.009990
6000–7000	3:1	99.85	0.003040	99.34	0.01017
6000–7000	4:1	99.86	0.002840	98.33	0.03333
6000–7000	5:1	99.80	0.003116	99.00	0.01526
6000–8000	3:1	99.92	0.002280	98.89	0.01110
6000–8000	4:1	100	0	99.17	0.01274
6000–8000	5:1	99.93	0.002040	98.67	0.01631
6000–9000	3:1	99.92	0.002280	99.11	0.01473
6000–9000	4:1	99.86	0.002840	98.61	0.03345
6000–9000	5:1	99.86	0.002720	99.00	0.01526
6000–10000	3:1	99.92	0.002280	99.56	0.008880
6000–10000	4:1	99.86	0.002840	98.61	0.03345
6000–10000	5:1	99.93	0.002040	99.67	0.009990
7000–8000	3:1	100	0	99.11	0.01088
7000–8000	4:1	99.72	0.004734	98.89	0.02545
7000–8000	5:1	100	0	99.33	0.01332
7000–9000	3:1	99.77	0.006870	99.11	0.01088
7000–9000	4:1	99.86	0.002840	98.61	0.02241
7000–9000	5:1	100	0	99.33	0.01333
7000–10000	3:1	99.85	0.004590	99.33	0.01017

Table 8: Continued.

Spectral region (cm^{-1})	Modeling ratio of training set to testing set	Training sets: Average identification accuracy for the 10 data subsets (%)	Training sets: Standard deviation of the identification accuracies for the 10 data subsets	Testing sets: Average identification accuracy for the 10 data subsets (%)	Testing sets: Standard deviation of the identification accuracies for the 10 data subsets
7000–10000	4 : 1	99.86	0.002840	98.89	0.01844
7000–10000	5 : 1	99.86	0.004110	99.33	0.01332
8000–9000	3 : 1	99.77	0.004893	98.00	0.01554
8000–9000	4 : 1	99.86	0.002840	98.33	0.02224
8000–9000	5 : 1	99.73	0.006261	98.33	0.02236
8000–10000	3 : 1	99.92	0.002280	99.56	0.01332
8000–10000	4 : 1	99.79	0.004573	98.61	0.03345
8000–10000	5 : 1	99.80	0.004381	99.00	0.02135
9000–10000	3 : 1	99.70	0.005065	98.67	0.02266
9000–10000	4 : 1	99.57	0.005715	98.06	0.03298
9000–10000	5 : 1	99.52	0.005344	98.67	0.01631

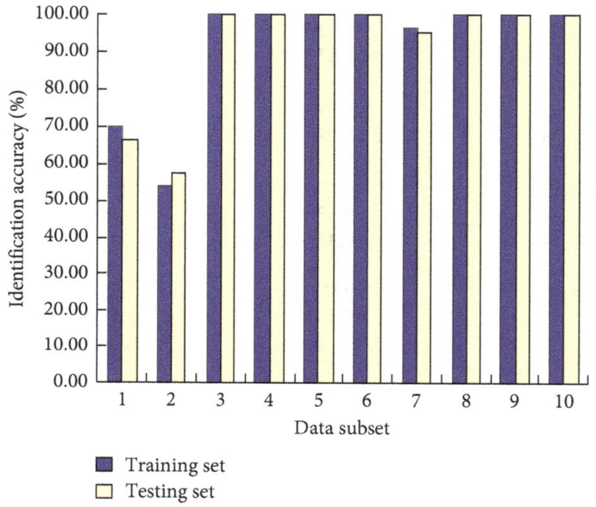

Figure 3: Pathogen identification results of the BPNN models built in the modeling spectral region of 4000–5000 cm^{-1} with a ratio of the training set to the testing set of 3 : 1 and the number of neurons in the hidden layer equal to 50 on each data subset when Data treatment method 2 was used.

to select the optimal modeling spectral region, and then the modeling results can be compared to the results obtained in this study to select the optimal modeling method.

The results obtained in this study demonstrated that the modeling effect may be affected by the spectral data used in modeling. For example, when Data treatment method 2 was used and the BPNN models were built with the modeling ratio of 3 : 1 or 5 : 1 in the spectral region of 4000–5000 cm^{-1}, with the modeling ratio of 4 : 1 in the spectral region of 5000–6000 cm^{-1}, with the modeling ratio of 3 : 1 or 5 : 1 in the spectral region of 6000–7000 cm^{-1}, or with the modeling ratio of 3 : 1 in the spectral region of 8000–10000 cm^{-1}, relatively small averages and large SD values of identification accuracies of the training and testing sets on the 10 data subsets were obtained. This was mainly due to the low identification accuracies of the training and testing sets for the model built on some data subset. In comparison, it was more likely to achieve satisfactory identification results when modeling was conducted in a relatively wide spectral region.

In this study, good identification results were obtained using the three modeling methods including DPLS, BPNN, and SVM. When Data treatment method 1 or Data treatment method 2 was used, a DPLS, BPNN, or SVM model with satisfactory identification accuracies of the training and testing sets could be obtained. In practical applications, one of the modeling methods can be used or more than one modeling methods can be used to synthetically verify identification results, according to the situation. However, in comparison, DPLS is the most stable, followed by SVM then BPNN. In this study, only three modeling methods (DPLS, BPNN, and SVM) were used to build the pathogen identification models. However, many modeling methods for qualitative identification analyses have been reported [15, 22, 24]. In further studies, other modeling methods can be used in an attempt for classification, discrimination, and identification of wheat smut fungi.

In this study, the identification of the three kinds of wheat smut fungi was investigated based on NIRS, aiming to provide a method for the rapid and nondestructive detection and identification of wheat smut fungi. According to the method proposed in this study, the teliospores of related pathogens can be obtained from wheat plants with the symptoms of wheat smuts in the field, then the near-infrared spectra can be acquired using a near-infrared spectrometer in the laboratory, and finally the samples can be identified using the built models. If possible, a portable near-infrared spectrometer can be developed for direct detection of wheat smut fungi in the field. In production practice, wheat smuts are mainly controlled by using seed treatment measures besides utilization of resistant cultivars. Pathogen identification can avoid reserving wheat seeds from the diseased fields and the further transfer of wheat seeds after reservation and can provide a basis for countermeasures. The results of this study provide a methodological reference for performing this task. Once wheat plants are infected by smut fungi, the hyphae expand inside the leaves, and during the later stage of disease progress, typical symptoms appear and teliospores are produced. Therefore, to perform early detection of the related diseases, studies on the detection of infected wheat plants based on NIRS should be conducted as well as studies on early detection of wheat stripe rust and wheat leaf rust mentioned above [30–32].

Conflicts of Interest

The authors declare that there are no conflicts of interest regarding the publication of this paper.

Authors' Contributions

Yaqiong Zhao and Feng Qin contributed equally to this paper.

Acknowledgments

This study was supported by the National Key Research and Development Program of China (2018YFD0200402), the National Key Technologies Research and Development Program of China (2012BAD19BA04), and the International Research Exchange Scheme of the Marie Curie Program of the 7th Framework Program (ref. PIRSES-GA-2013-612659).

Supplementary Materials

Table S1: pathogen identification results of the DPLS models built with different modeling ratios and numbers of principal components in each modeling spectral region when Data treatment method 1 was used. Table S2: pathogen identification results of the BPNN models built with different modeling ratios and numbers of neurons in the hidden layer in each modeling spectral region when Data treatment method 1 was used. Table S3: pathogen identification results of the DPLS models built on each data subset with different modeling ratios and numbers of principal components in each modeling spectral region when Data treatment method 2 was used. Table S4: pathogen identification results of the BPNN models built on each data subset with different modeling ratios and numbers of neurons in the hidden layer in each modeling spectral region when Data treatment method 2 was used. Table S5: pathogen identification results of the SVM models built on each data subset with different modeling ratios in each modeling spectral region when Data treatment method 2 was used. (*Supplementary Materials*)

References

[1] V. O. Stockwell and E. J. Trione, "Distinguishing teliospores of *Tilletia controversa* from those of *T. Cariesby* fluorescence microscopy," *Plant Disease*, vol. 70, no. 10, pp. 924–926, 1986.

[2] H. G. Wang, H. Y. Zhu, Z. H. Ma, L. Liu, and G. X. Zhang, "The progress and prospect in research on wheat dwarf bunt," *Review of China Agricultural Science and Technology*, vol. 7, no. 4, pp. 21–27, 2005.

[3] W. Dong, W. Q. Chen, and T. G. Liu, "Identification and detection methods of *Tilletia controversa* Kühn," *Plant Protection*, vol. 33, no. 6, pp. 128–131, 2007.

[4] L. Guo, "Taxonomy of smut fungi in China: a brief review," *Mycosystema*, vol. 34, no. 5, pp. 817–820, 2015.

[5] H. X. Yu, L. Gao, X. H. Kang, T. G. Liu, B. Liu, and W. Q. Chen, "Discrimination of teliospores in TCK and TFL with laser scanning confocal microscope," *China Plant Protection*, vol. 36, no. 2, pp. 5–8, 2016.

[6] J. G. Mcdonald, E. Wong, G. T. Kristiansson, and G. P. White, "Direct amplification by PCR of DNA from ungerminated teliospores of *Tilletia* species," *Canadian Journal of Plant Pathology*, vol. 21, no. 1, pp. 78–80, 1999.

[7] G. Bakkeren, J. W. Kronstad, and C. A. Levesque, "Comparison of AFLP fingerprints and ITS sequences as phylogenetic markers in Ustilaginomycetes," *Mycologia*, vol. 92, no. 3, pp. 510–521, 2000.

[8] R. D. Frederick, K. E. Snyder, P. W. Tooley et al., "Identification and differentiation of *Tilletia indica* and *T. walkeri* using the polymerase chain reaction," *Phytopathology*, vol. 90, no. 9, pp. 951–960, 2000.

[9] H. Liang, Y. L. Peng, G. Z. Zhang, W. Q. Chen, and T. G. Liu, "Amplification and sequence analysis of the rDNA intergenic spacer (rDNA-IGS) from three *Tilletia* species," *Acta Phytopathologica Sinica*, vol. 36, no. 5, pp. 407–412, 2006.

[10] Y. Y. Cao, L. F. Wang, L. P. Duan et al., "Development of a real-time fluorescence loop-mediated isothermal amplification assay for rapid and quantitative detection of *Ustilago maydis*," *Scientific Reports*, vol. 7, article 13394, 2017.

[11] K. G. Savchenko, L. M. Carris, J. Demers, D. S. Manamgoda, and L. A. Castlebury, "What causes flag smut of wheat?," *Plant Pathology*, vol. 66, no. 7, pp. 1139–1148, 2017.

[12] N. W. Schaad and R. D. Frederick, "Real-time PCR and its application for rapid plant disease diagnostics," *Canadian Journal of Plant Pathology*, vol. 24, no. 3, pp. 250–258, 2002.

[13] S. Sankaran, A. Mishra, R. Ehsani, and C. Davis, "A review of advanced techniques for detecting plant diseases," *Computers and Electronics in Agriculture*, vol. 72, no. 1, pp. 1–13, 2010.

[14] G. T. Xu, H. F. Yuan, and W. Z. Lu, "Development of modern near-infrared spectroscopic techniques and its applications," *Spectroscopy and Spectral Analysis*, vol. 20, no. 2, pp. 134–142, 2000.

[15] Y. L. Yan, L. L. Zhao, D. H. Han, and S. M. Yang, *Basis and Application of Near-infrared Reflectance Spectroscopy*, China Light Industry Press, Beijing, China, 2005.

[16] Q. Chen, J. Zhao, M. Liu, J. Cai, and J. Liu, "Determination of total polyphenols content in green tea using FT-NIR spectroscopy and different PLS algorithms," *Journal of Pharmaceutical and Biomedical Analysis*, vol. 46, no. 3, pp. 568–573, 2008.

[17] M. K. Ahmed and J. Levenson, "Application of near-infrared spectroscopy to the quality assurance of ethanol and butanol blended gasoline," *Petroleum Science and Technology*, vol. 30, no. 2, pp. 115–121, 2012.

[18] X. L. Chu and W. Z. Lu, "Research and application progress of near-infrared spectroscopy analytical technology in China in the past five years," *Spectroscopy and Spectral Analysis*, vol. 34, no. 10, pp. 2595–2605, 2014.

[19] H. Chen, Z. Lin, H. Wu, L. Wang, T. Wu, and C. Tan, "Diagnosis of colorectal cancer by near-infrared optical fiber

spectroscopy and random forest," *Spectrochimica Acta Part A: Molecular and Biomolecular Spectroscopy*, vol. 135, pp. 185–191, 2015.

[20] N. Liu, H. A. Parra, A. Pustjens, K. Hettinga, P. Mongondry, and S. M. van Ruth, "Evaluation of portable near-infrared spectroscopy for organic milk authentication," *Talanta*, vol. 184, pp. 128–135, 2018.

[21] M. Lopo, C. A. Teixeira dos Santos, R. N. M. J. Páscoa, A. R. Graça, and J. A. Lopes, "Near-infrared spectroscopy as a tool for intensive mapping of vineyards soil," *Precision Agriculture*, vol. 19, no. 3, pp. 445–462, 2018.

[22] C. Pasquini, "Near-infrared spectroscopy: a mature analytical technique with new perspectives—a review," *Analytica Chimica Acta*, vol. 1026, pp. 8–36, 2018.

[23] P. I. Monteiro, J. S. Santos, V. R. Alvarenga Brizola et al., "Comparison between proton transfer reaction mass spectrometry and near-infrared spectroscopy for the authentication of Brazilian coffee: a preliminary chemometric study," *Food Control*, vol. 91, pp. 276–283, 2018.

[24] D. Granato, P. Putnik, D. B. Kovačević et al., "Trends in chemometrics: food authentication, microbiology, and effects of processing," *Comprehensive Reviews in Food Science and Food Safety*, vol. 17, no. 3, pp. 663–677, 2018.

[25] J. Perez-Mendoza, J. E. Throne, F. E. Dowell, and J. E. Baker, "Chronological age-grading of three species of stored-product beetles by using near-infrared spectroscopy," *Journal of Economic Entomology*, vol. 97, no. 3, pp. 1159–1167, 2004.

[26] J. Perez-Mendoza, J. E. Throne, E. B. Maghirang, F. E. Dowell, and J. E. Baker, "Insect fragments in flour: relationship to lesser grain borer (Coleoptera: bostrichidae) infestation level in wheat and rapid detection using near-infrared spectroscopy," *Journal of Economic Entomology*, vol. 98, no. 6, pp. 2282–2291, 2005.

[27] D. Wu, L. Feng, C. Q. Zhang, and Y. He, "Early detection of gray mold (Cinerea) on eggplant leaves based on vis/near-infrared spectra," *Journal of Infrared and Millimeter Waves*, vol. 26, no. 4, pp. 269–273, 2007.

[28] L. Feng, S. S. Chen, B. Feng, F. Liu, Y. He, and B. G. Lou, "Early detection of soybean pod anthracnose based on spectrum technology," *Transactions of the Chinese Society of Agricultural Engineering*, vol. 28, no. 1, pp. 139–144, 2012.

[29] Y. D. Liu, H. C. Xiao, Q. Deng, Z. C. Zhang, X. D. Sun, and Y. S. Xiao, "Nondestructive detection of citrus greening by near-infrared spectroscopy," *Transactions of the Chinese Society of Agricultural Engineering*, vol. 32, no. 14, pp. 202–208, 2016.

[30] X. L. Li, Z. H. Ma, L. L. Zhao, J. H. Li, and H. G. Wang, "Early diagnosis of wheat stripe rust and wheat leaf rust using near-infrared spectroscopy," *Spectroscopy and Spectral Analysis*, vol. 33, no. 10, pp. 2661–2665, 2013.

[31] X. L. Li, F. Qin, L. L. Zhao, J. H. Li, Z. H. Ma, and H. G. Wang, "Detection of *Puccinia striiformis* f. sp. *tritici* latent infections in wheat leaves using near-infrared spectroscopy technology," *Spectroscopy and Spectral Analysis*, vol. 34, no. 7, pp. 1853–1858, 2014.

[32] Y. Q. Zhao, Y. L. Gu, F. Qin et al., "Application of near-infrared spectroscopy to quantitatively determine relative content of *Puccnia striiformis* f. sp. *tritici* DNA in wheat leaves in incubation period," *Journal of Spectroscopy*, vol. 2017, Article ID 9740295, 12 pages, 2017.

[33] X. L. Li, F. Qin, L. L. Zhao, J. H. Li, Z. H. Ma, and H. G. Wang, "Identification and classification of disease severity of wheat stripe rust using near-infrared spectroscopy technology," *Spectroscopy and Spectral Analysis*, vol. 35, no. 2, pp. 367–371, 2015.

[34] X. L. Li, Z. H. Ma, L. L. Zhao, J. H. Li, and H. G. Wang, "Application of near-infrared spectroscopy to qualitative identification and quantitative determination of *Puccinia striiformis* f. sp. *tritici* and *P. recondita* f. sp. *tritici*," *Spectroscopy and Spectral Analysis*, vol. 34, no. 3, pp. 643–647, 2014.

[35] Y. Q. Zhao, F. Qin, P. Cheng et al., "Quantitative determination of germinability of *Puccinia striiformis* f. sp. *tritici* urediospores using near-infrared spectroscopy technology," *Journal of Spectroscopy*, vol. 2015, Article ID 384214, 8 pages, 2015.

[36] P. Cheng, X. L. Li, F. Qin et al., "Effects of UV-B radiation on near-infrared spectroscopy and identification of *Puccinia striiformis* f. sp. *tritici*," *Journal of Spectroscopy*, vol. 2014, Article ID 751458, 9 pages, 2014.

[37] C. C. Chang and C. J. Lin, "LIBSVM," *ACM Transactions on Intelligent Systems and Technology*, vol. 2, no. 3, pp. 1–27, 2011.

[38] L. H. Purdy, "Flag smut of wheat," *The Botanical Review*, vol. 31, no. 4, pp. 565–606, 1965.

19

Prediction Model of the Key Components for Lodging Resistance in Rapeseed Stalk Using Near-Infrared Reflectance Spectroscopy (NIRS)

Jie Kuai,[1] Shengyong Xu,[2] Cheng Guo,[1] Kun Lu,[2] Yaoze Feng,[2] and Guangsheng Zhou[1]

[1]*College of Plant Science & Technology, Huazhong Agricultural University, Wuhan 430070, China*
[2]*College of Engineering, Huazhong Agricultural University, Wuhan 430070, China*

Correspondence should be addressed to Guangsheng Zhou; zhougs@mail.hzau.edu.cn

Academic Editor: Tino Hofmann

The chemical composition of rape stalk is the physiological basis for its lodging resistance. By taking the advantage of NIRS, we developed a rapid method to determine the content of six key composition without crushing the stalk. Rapeseed stalks in the mature stage of growth were collected from three cultivation modes over the course of 2 years. First, we used the near-infrared spectroscope to scan seven positions on the stalk samples and took their average to form the spectral data. The stalks were then crushed and sieved; then the ratio of carbon and nitrogen, ratio of acid-insoluble lignin and lignin, and the content of soluble sugar and cellulose were determined using the combustion method, weighing method, and colorimetric method, respectively. The partial least squares regression (PLSR) method was used to establish a prediction model between the spectral data and the chemical measurements, and all models were evaluated by an internal interaction verification and an external independent test set sample. To improve the accuracy of the model and reduce the computing time, some optimization methods have been applied. Some outliers were removed, and then the data were preprocessed to determine the best spectral information band and the optimal principal component number. The results showed that elimination of outliers effectively improved the precision of the prediction model and that no spectral pretreatment method exhibited the highest prediction accuracy. In summary, the NIRS-based prediction model could facilitate the rapid nondestructive detection in the key components of rapeseed stalk.

1. Introduction

Rapeseed is one of the most widely cultivated oilseeds crops worldwide, and it is the main source of vegetable oil and protein. In the field, stalk lodging is the key limiting factor for the further improvement of rapeseed yield. Similar to other crops, the content of lignin and cellulose in the rapeseed stalk is closely related to its lodging resistance [1]. In addition, rapeseed stalk is an important renewable resource; the rapid quantification of its major components effectively improves the utilization rate of the stalk [2]. Determination of the main components of stalk lodging resistance in regard to massive rapeseed breeding practice can accelerate the breeding of rapeseed with lodging resistance. However, the commonly used chemical testing methods are time- and labor-consuming with low efficiency. On this basis, a rapid method to evaluate the key components of lodging resistance is desired. Such an advance could provide technical support for the study of lodging resistance in rapeseed.

The near-infrared reflectance spectroscopy (NIRS) technology is rapid and highly efficient and has been applied in the agriculture and many other crops, such as in rice [3], citrus [4], and rapeseed [5]. With the NIRS and multiple linear regression methods, Ding et al. established a quantitative analytical model of glucose and fructose to evaluate their content in honey [6]. Bagherpour et al. researched an optical method based on near-infrared spectroscopy (900–1600 nm) to determine the content of soluble solid and sucrose in sugar beet [7]. Rungpichayapichet et al. developed calibration models for firmness, total soluble solids (TSS), titratable acidity (TA), and ripening index (RPI) using PLSR analysis. The results indicated that NIRS can be used as a reliable nondestructive technique for mango quality assessment [8]. Mathison et al. adopted NIRS for the prediction of the nutritive value of straw with 195 samples of barley straw. They showed that NIRS was a useful method for predicting chemical composition of straw and estimating its ruminal degradability characteristics [9]. Kaur et al. developed a calibration equation for oil content using NIRS in *Brassica juncea* and *Brassica napus*. The reference values of oil content were generated by nuclear magnetic resonance (NMR). The predicted model was validated in the case of *B. juncea* with an r^2 value 0.85. The results indicated that NIRS could be used satisfactorily for rapid determination of oil content in *B. juncea* [10]. Teye et al. assessed the feasibility of measuring total fat content in cocoa beans by using Fourier transform near-infrared (FT-NIR) spectroscopy based on a systematic study of spectral variable selection via multivariate regression. Experimental results showed that the model based on the novel Si-SVMR algorithm was superior to the others [11]. Shetty and Gislum presented a method of NIRS combined with chemometrics which was used to quantify fructan concentration in samples from seven grass species. The PLSR approach was used on the full spectra to model NIR spectroscopy data [12]. NIRS predictive equations were used to provide accurate high-throughput phenotyping of seed content, opening new perspectives in gene identification following QTL mapping and genome-wide association studies [13].

Tremendous efforts have been made in the study on the straw composition using multispectral technology. Niu et al. used hyperspectral imaging technology to rapidly determine the content of five elements: nitrogen, carbon, hydrogen, sulfur, and oxygen, in the straw of maize, rice, wheat, and rapeseed. Their analysis was achieved through applying the optimal method combined with a competitive adaptive reweighted sampling-partial least squares algorithm (CARS-PLS). Their results indicated that the nitrogen and oxygen contents in the straw could be effectively assessed via their methodology [14]. Li et al. used NIRS combined with PLSR to establish the optimal quantitative model for the biocomponents (nitrogen, carbon, hydrogen, sulfur, and oxygen) of straw [15]. XUE investigated an online analysis method of proximate component (moisture, ash, volatile matter, and fixed carbon) and lignocellulose components (cellulose, semicellulose, and lignin) of coarse crushed corn stover using NIRS. After optimized pretreatment, all the NIRS models were effectively developed using the PLSR method [16]. Sheng et al. provided models of three types of agroforestry biomass (moisture, ash, volatile, and fixed carbon) and caloric value of pine, China fir, and cotton stalk via spectral techniques, indicates that the traditional industrial analysis methods could be completely replaced by visible near-infrared spectroscopy in order to achieve the rapid determination of components and calorific value of the agroforestry biomass [17]. Fu et al. collected and prepared samples of various varieties of rice straw from different locations and created a quantitative analysis model based on the stepwise multiple linear regression (SMLR) of near-infrared spectroscopy, PLSR, and principal component regression (PCR). Data suggested that their model was capable of rapidly testing the content of soluble sugar in rice straw [18]. Lohr developed a partial least squares calibration models for glucose, fructose, sucrose, and starch in leaves of chrysanthemum and pelargonium cuttings by using a stepwise enzymatic-photometric method [19].

Overall, many studies have been carried out to evaluate the content of stem and leaf components in crops using multispectral analysis, and they can effectively overcome the primary drawbacks of traditional chemical testing methods (time consumption and the tedium of the procedures). However, in most of the cases, samples were smashed before the evaluation in these studies. It is difficult to control the homogeneity of material when the sample is smashed, which may increase the error of spectral scanning. In addition, the overall process is usually complicated, and the smashed material is not able to be used after the study is complete. Currently, a method to determine the key components of lodging resistance in rapeseed by spectral analysis without smashing and sieving the rapeseed samples remains unavailable. In the present study, we propose a fast, accurate, and nondestructive NIRS-based method to predict six components including carbon, nitrogen, lignin, acid quality lignin, soluble sugar, and cellulose in noncrushed rapeseed stalk. Our method provides important and fundamental data for the selection of rapeseed with strong lodging resistance and allows the stalk to be used after analysis for other purposes.

2. Materials and Methods

2.1. Experimental Materials. The plant materials were obtained from the experimental station of Huazhong Agricultural University in both 2017 and 2018. Rapeseeds were bunch planted in September and thinned at the third to fifth leaf stages. The field was flat with medium fertility, and the previous crop was rice harvested during early September. The plots were 10 m × 2 m. *Brassica napus* hybrids HYZ62 and FY520, common Swede-type rapeseeds HH901 and HS5, and the DH population consisting of 150 lines (TN) were used as plant materials. Samples were collected during the maturity stage. Plants representing the average yield and lodging status were selected. After removing the roots of samples, the plants were dried and bagged separately. The sample will be dried twice at 105°C and at 80°C until achieving constant weight separately. The cultivation methods in the present study were as follows:

(1) Three-factor split plot experiment with three replicates: the four rapeseed varieties served as the main plot, fertilizer type served as the primary split plot, and the level of N fertilizer (0 kg/hm^2, 180 kg/hm^2, and 360 kg/hm^2 of pure nitrogen), P (0 kg/hm^2, 120 kg/hm^2, and 240 kg/hm^2 of P_2O_5), and K (0 kg/hm^2, 150 kg/hm^2, and 300 kg/hm^2 of K_2O) served as the secondary split plot, so a total of 108 treatments with different NPK ratios were carried out.

(2) Three-factor split plot experiment with three replicates: the four rapeseed varieties served as the main plot, the levels of N fertilizer served as the primary split plot (120 kg/hm^2, 240 kg/hm^2, and 360 kg/hm^2 and urea with 46.7% nitrogen content was used as N source), and three planting densities (15 × 104 plants/hm^2, 30 × 104 plants/hm^2, and 45 × 104 plants/hm^2) served as the secondary split plot. Phosphorus (calcium superphosphate with 12% P content), potassium (KCl with 60% K), and borate fertilizer were applied once as the base fertilizer with 150 kg/hm^2 for P and K and 7.5 kg/hm^2 of borax.

(3) Complete random block design with three replicates: each variety was planted one row with a length of 250 cm, a row spacing of 30 cm, and a plant spacing of 21 cm. Compound fertilizer (15 : 15 : 15) of 750 kg/hm^2 with 7.5 kg/hm^2 of borax was used as base fertilizer, and 75 kg/hm^2 of urea was additionally applied during the seedling stage. The filed management was carried out as regular practice.

A Fourier near-infrared spectroscope BRUKERFT-NIR (VECTOR33N, Bruker Inc., Germany) was equipped with PbS detector, quartz rotating sample cup, gold-plated integrating sphere, and OPUS analysis software with the wave number of 12000 cm^{-1}–4000 cm^{-1} and the resolution of 8 cm^{-1}. The near-infrared spectroscope was preheated for 20 min before scanning at room temperature. The sample was put into a sample cell at the same position each time and then covered by the gold-plated integrating sphere to avoid light leak. To ensure the accuracy and integrity of the scanning procedure seven positions of each sample would be scanned for seven spectra, and then the average of the seven spectra data was taken to build the model. Using a blade, the sample was evenly cut into three parts, and then the middle piece of each part was cut off. As shown in Figure 1(a), samples demarcated as 1 and 2 were the end faces of the stalk. 3 and 4 were the fresh cut surface to avoid variation errors due to long time exposures in the air, and 5–7 were cross sections of the stalk. Figure 1(b) shows the spectra of the samples. All spectral preprocessing was performed in the OPUS software (OPUS 7.0, BRUKER OPTICS, Germany) The SPASS software (IBM SPSS Statistics 24.0, IBM, USA) was adopted for data analysis and mapping.

The stalks were dried after scanning and then ground using a universal high-speed smashing machine (FW100, Tianjin TAISITE Instrument Co., LTD, China) to measure their chemical components. Carbon and nitrogen were determined using an elemental analyzer (Elementar vario max cn, Germany). Oxygen was added to the samples of 250 mg for 90 s, and various forms of nitrogen and carbon in the sample were converted to stable nitrogen and carbon dioxide by the combustion method (the combustion tube of 900°C and the reduction tube of 830°C), and then the contents were tested by an infrared detection module. The ratio of acid-insoluble lignin and lignin was determined using the sulfuric acid method [20]. The cellulose ratio was evaluated by colorimetry [21]. The ratio of soluble sugar was detected by the anthrone colorimetry method [22]. The results of the chemical analysis are shown in Table 1.

2.2. Constructive Process of the Model. The accuracy of model creation is an important step to ensure the predicted results are close to the true chemical values in the plant matter [23]. The mean spectra were obtained by averaging the seven spectra scanned using OPUS software, and the model was further established using the PLSR method and the obtained chemical data. We conducted the following procedures.

2.2.1. Partitioning of Correction Set and Validation Set. The experimental data were arranged randomly. The division ratio of this paper is approximately 3 : 1, which is very close to the traditional division ratio of 7 : 3. Figure 2 shows the partition of correction set and validation set of the stalk components in rapeseed. Of these, CA and VA represented the calibration set and validation set, respectively. The data in Table 2 indicated that the samples set exhibited the desired uniformity.

2.2.2. Elimination of Outliers. Changes in environmental factors are also important, such as temperature, humidity, and air flow rate, which affect the accuracy of the spectrum [24]. Therefore, eliminating outliers from the sample set is crucial. In the present study, the outliers were removed based on the principle of predicting concentration residuals [25]. Leave-one-out cross validation was performed each time a sample was deleted, and the samples that increased the correlation between the true value and predicated value (R) while decreasing the root-mean-square error of cross validation (RMSECV) were considered outliers. About 1/6 to 1/5 of the samples in the prediction set were classified as outliers and eliminated from the sample set.

2.2.3. Optimization of Spectral Preprocessing. A total of 11 methods were employed to preprocess the spectra after eliminating the outliers, including elimination constant offset, straight line subtraction, vector normalization, min-max normalization, multiplication scatter correction, the first-order derivative correction, the second-order derivative correction, the first-order derivative and straight line subtraction, first-order derivative and vector normalization, first-order derivative and multiplication scattering correction (MSC), and no preprocessing [26]. Using the sample after removing the outlier, the optimal processing method was selected by comparing R and RMSECV.

2.2.4. Spectral Band Optimization. Choosing the right band can improve the accuracy of the model and reduce the running time. After determining the optimal spectral preprocessing, the spectrum was divided into 45 bands, and the correlation coefficient and the mean square deviation of internal verification at different bands were observed. The

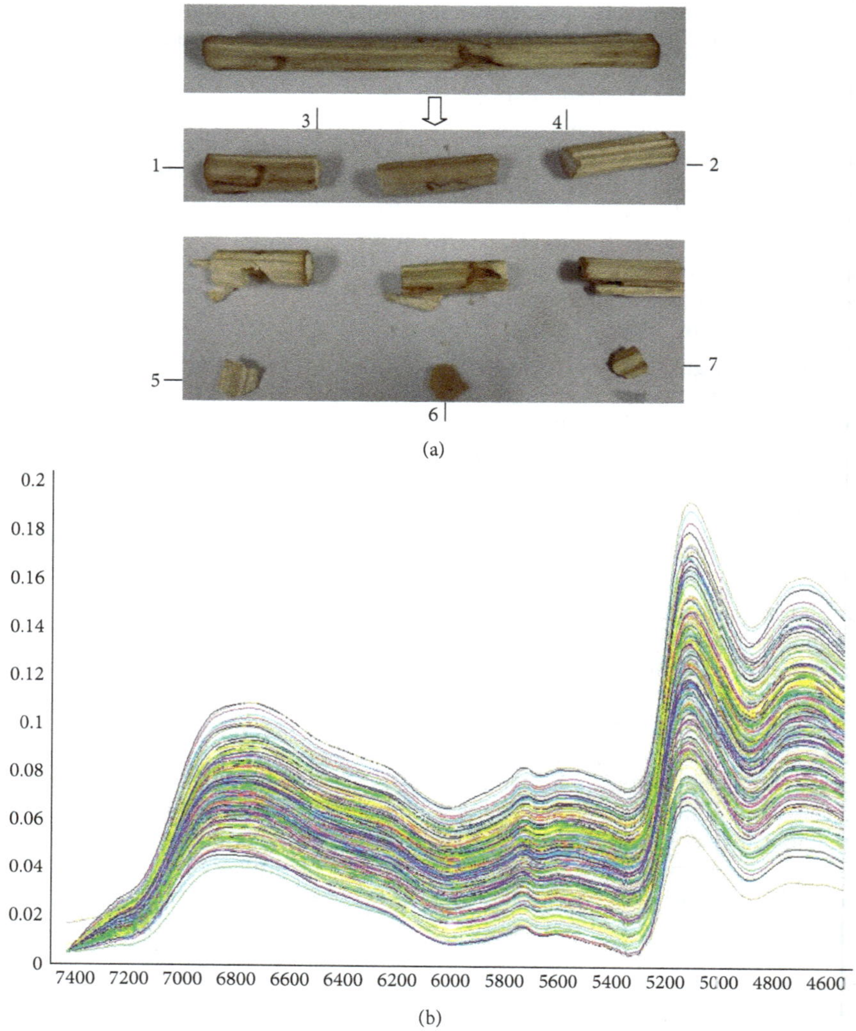

FIGURE 1: (a) Schematic of the scanning position on rapeseed stalks. (b) Spectra of the samples.

TABLE 1: Statistical results of six rapeseed stem components.

Content	Number	Maximum (%)	Minimum (%)	Mean (%)	Standard deviation	Coefficient of variation
Carbon	340	46.37	41.13	44.18	1.05	0.024
Nitrogen	340	0.68	0.12	0.34	0.09	0.260
Lignin	240	34.96	18.64	25.27	3.41	0.135
Acid quality lignin	240	38.51	13.69	23.92	4.58	0.192
Soluble sugar	200	5.68	1.52	3.18	0.94	0.295
Cellulose	200	38.27	9.69	23.47	6.23	0.265

optimal wave number was chosen as the one with the smallest RMSECV.

2.2.5. Determination of the Optimal Principal Component Number. Determining the optimal principal component can reduce the dimension. It not only lowers the number of input data needed to create a model but also speeds up the operation of the model with improved prediction accuracy. During the model creation, 1–10 were tested, respectively, as the principal component number, and the value showing the maximum R and minimum RMSECV of the PLSR model was determined as the optimal principal component number.

2.2.6. Validation of the Calibration Model. To verify whether the accuracy of the model meets the requirement of component prediction, the spectra of the samples from the validation set were introduced into the established model to obtain

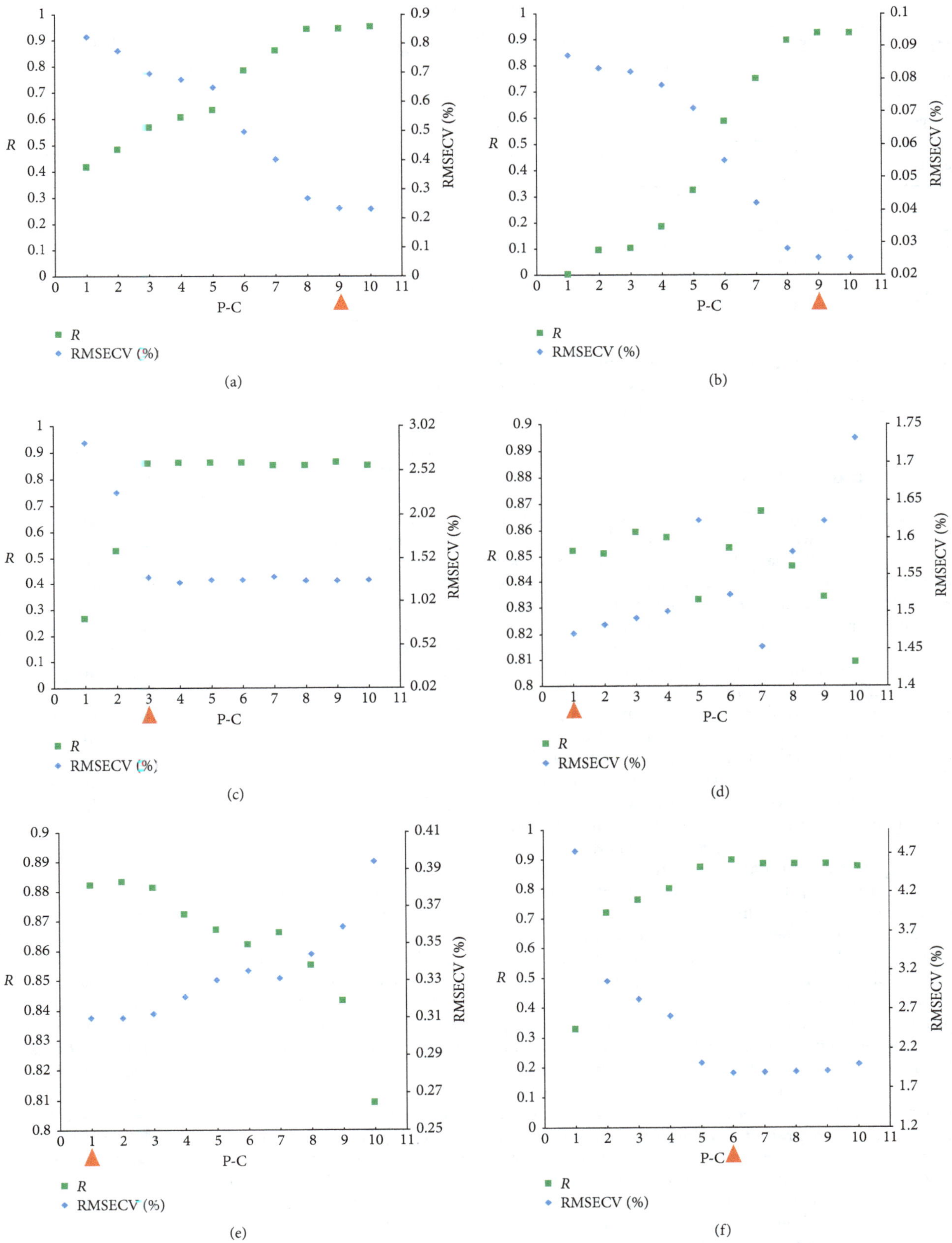

Figure 2: Relationship between principal component number and R/RMSECV of the prediction model.

TABLE 2: Division of the calibration set and validation set.

Index	Set type	Quantity of samples	Maximum value (%)	Minimum value (%)	Mean value (%)	Standard deviation
Carbon	CA	250	46.37	41.13	44.17	1.03
	VA	90	46.37	41.23	44.21	1.12
Nitrogen	CA	250	0.68	0.12	0.34	0.08
	VA	90	0.68	0.13	0.33	0.10
Lignin	CA	180	34.96	18.64	25.27	3.16
	VA	60	34.95	19.64	25.25	4.08
Acid quality lignin	CA	180	34.80	14.83	23.65	4.12
	VA	60	38.51	13.70	24.72	5.71
Soluble sugar	CA	150	5.68	1.52	3.17	0.88
	VA	50	5.55	1.53	3.21	1.12
Cellulose	CA	150	38.27	9.69	23.42	5.80
	VA	50	38.27	11.36	23.63	7.43

predication values (predicted by the model) and true values (from the chemical test). These values were used to calculate the R and RMSECV between prediction values and standard values.

3. Results and Analysis

3.1. Effect of Eliminating Outliers. In the modeling based on NIRS, there were two types of outliers in the calibration set. One type was the sample with significant difference between the chemical determination value and the predicted value, which may be caused by the large error of chemical measurement, the spectral measurement error, or the error during data entry. These abnormal samples must be eliminated before modeling. The other type was high-leverage samples. Comparing with other samples, this type sample contains extreme components and far away from the average of the overall sample. These abnormal samples were obviously not useful for global modeling because of destroying the uniformity of the sample distribution. But those were useful for enriching the calibration set and improving the accuracy of subsequent samples. The presented method in this paper adopted the eliminating outlier method as shown in the literature [24]. In the literature, the proportion of eliminating outliers reached 36.7%. Considering the balance performance between global modeling and local modeling, the eliminating rate was set at approximately 20%. The eliminating rate of carbon, nitrogen, soluble sugar, and cellulose was 20%. The eliminating rate of lignin and acid quality Lignin was 16.7%. The remaining samples were then used to create the model using the PLSR method. Compared to the results obtained without eliminating outliers, using the principal component number of nine for the carbon calibration set, R between standard value and predicted value was increased from 0.960 to 0.973, and RMSECV was decreased from 0.286% to 0.240% with a reduction of predictive residual to 0.600%. The principal component number of nitrogen calibration set remained at nine, R increased from 0.929 to 0.96, RMSECV was reduced from 0.030% to 0.024%, and the predictive residual lowered to 0.6%. The principal component number of lignin calibration set remained as three, R elevated from 0.860 to 0.923, RMSECV reduced from 1.610% to 1.250%, and predictive residual lowered to 0.6%. The principal component number of acid-insoluble lignin calibration set remained as one, R was increased from 0.893 to 0.928, and RMSECV was reduced from 1.850% to 1.480%, and in this satuation the predictive residual was 4.000%. The principal component number of soluble sugar calibration set changed from one to nine, exhibiting an increase of R from 0.910 to 0.939, the reduction of RMSECV from 0.900% to 0.310%, and the decrease of predictive residual to 3.000%. The principal component number of celluloses in the calibration set was unchanged to 6 with the increase of R from 0.917 to 0.945, the decrease of RMSECV from 2.310% to 1.880%, and the predictive residual lowered to 5.000%. In conclusion, the accuracy of the prediction model was improved slightly after eliminating outliers in calibration set. The increase rate of R is between 1.3% and 6.3%, and the decrease rate of RMSECV is between 0.026% and 0.59%. The prediction model may also have the capability to modeling accuracy without eliminating outliers.

3.2. Optimization of the Spectral Preprocessing. When using the full band spectrum to build the model, the spectral information content is large, resulting in enormous computation and processing time. Occasionally, there are other factors such as collinearity in spectral information, which leads to the incapability to extract relevant information from the models with full optical spectrum. Such concerns can be avoided after optimizing the spectral band. The comparison of 11 spectral pretreatments for the prediction models of carbon content is shown in Table 3. The R and RMSECV indicated that the best approach was without spectral preprocessing, and their values were 0.973 and 0.240%, respectively. The suboptimal choice was a first-order derivative, and R and RMSECV were 0.942 and 0.356%, respectively. R and RMSECV of the first-order derivative were decreased by 0.031 and 0.116%, respectively. The least desirable processing was vector normalization, in which R was the minimum and RMSECV was the maximum. Overall, the best way to build the model for carbon content was with

TABLE 3: The result of carbon spectrum using eleven spectral preprocessing methods.

Number	Pretreatment method	Wave number (cm^{-1})	Principal component number	R	RMSECV (%)
1	No spectral pretreatment	7501.7–4246.5	9	0.973	0.240
2	First-order derivative	7501.7–4246.5	10	0.942	0.356
3	Straight line subtraction	7501.7–4597.5	10	0.928	0.394
4	Constant offset elimination	7501.7–4597.5	10	0.923	0.407
5	First-order derivative + straight line subtraction	7501.7–4246.5	9	0.922	0.411
6	First-order derivative + MSC	7501.7–4246.5	8	0.916	0.424
7	First-order derivative + vector normalization	7501.7–4246.5	8	0.915	0.426
8	Second-order derivative	7501.7–4246.5	9	0.911	0.436
9	Multiplication scattering correction	7501.7–6097.8 5453.7–4246.5	6	0.908	0.443
10	Min-max normalization	6799.8–6097.8 5453.7–4246.5	8	0.897	0.469
11	Vector normalization	7501.7–4246.5	9	0.896	0.469

no spectral preprocessing and/or first-order derivative processing. The same approach was applied to another five elements, and no preprocessing was optimal for all in each case.

Generally, samples must be ground and placed in a sample cell for the determination of components in straw. For example, when evaluating the nutrients in corn stalk, the best spectral preprocessing of soluble sugar and acid detergent fiber is the first-order derivative and MSC [2]. However, in the present study, no spectral pretreatment was the optimal preprocessing for all spectra, and only the optimal number of principal component number and spectral range differed, indicating that the original spectrogram obtained by scanning was the best way to build a model. This may be due to (1) in the present study, the scanning object is an intact rapeseed stalk without treatments, such as crushing and sieving, which reduced operation processes and possible errors, and (2) the gold-plated integrating ball on top of the sample cell fills the gap between rapeseed stalk and the cell, allowing stalks with various diameters to hold good positioning in the machine, which improved the homogeneity of samples and prevents the interference of natural light.

3.3. Optimization of Spectral Band. Using the approach of no spectral preprocessing and PLSR to build the model, the full spectrum was divided into 45 bands. The relationship of wave number and RMSECV in the prediction model of N from our measurements is shown in Table 4. In these models for nitrogen, the wave number of 7501.7 cm^{-1}–5449.8 cm^{-1} and 4601.3 cm^{-1}-4246.5 cm^{-1} exhibited the minimal RMSECV of 0.024%. The worst/highest RMSECV value of 0.081% was obtained at 11998.9 cm^{-1}–7497.9 cm^{-1}. Our data suggested that selecting the appropriate waveband can effectively improve the accuracy of the model. The same method was applied for the other 5 elements, and the optimal wave number was 7501.7 cm^{-1}–4246.5 cm^{-1}, 7501.7 cm^{-1}–6097.8 cm^{-1}, and 4601.3 cm^{-1}–4423.9 cm^{-1}; 11998.9 cm^{-1}–7497.9 cm^{-1} and 5453.7 cm^{-1}–4246.5 cm^{-1}; 11998.9 cm^{-1}–5449.8 cm^{-1}; 7501.7 cm^{-1}–5449.8 cm^{-1}; and 4601.3 cm^{-1}–4246.5 cm^{-1} for carbon, lignin, acid-insoluble lignin, soluble sugar, and cellulose, respectively.

3.4. Determination of the Principal Component Number. Without spectral preprocessing, at the best wave number, the principal compoment numer will be determined with the maximum R and minimum RMSECV. The prediction model of carbon content at the waveband of 7501.7 cm^{-1}–4246.5 cm^{-1} with no preprocessing is shown in Figure 2(a), where the x-axis P-C represents principal component number. Within the number of principal components of 1–8, R was elevated as the increase in the number, and RMSECV was gradually decreased, indicating that the extracted spectral information was gradually completed during the establishment of the model. When the number of principal component number reached 10, R was lower and RMSECV was higher than those with the number of 9, suggesting the existence of over fitting when excessive information was extracted. The optimal number of principal component number was 9 with the highest R between true value and predicted value of 0.973 and the RMSECV of 0.240%. Therefore, the best number of principal components was determined as 9. The same approach was applied to the other five contents as shown in Figures 2(b)~2(f), and the optimal numbers of principal component number were 9, 9, 3, 1, 1, and 6, respectively. In generally, the R/RMSECV is decreased or increased gradually and continuously. The relationship between the number and R/RMSECV changes regularly, except for the model of acid quality lignin. However, Figure 2(d) shows that when the principal component number was 5 or 7, R/RMSECV is fluctuating drastically and deviated from the regular law. Therefore, 7 was regarded as an outlier and then 1 was considered as the best principal component number.

3.5. Validation of the Corrected Model. The predication results of all six components are shown in Table 5. The models were evaluated based on the previous method [27] in which the correlation coefficient of the calibration set R served as the

Table 4: The r elationship b etween s pectral w ave n umber and RMSECV for nitrogen.

Number	Wave number (cm^{-1})	RMSECV (%)
1	7501.7–5449.8 4601.3–4246.5	0.024
2	7501.7–4246.5	0.025
3	7501.7–5449.8 4601.3–4423.9	0.030
4	6101.7–4246.5	0.030
5	6475.8–5449.8 4601.3–4246.5	0.031
6	7501.7–6097.8 4601.3–4246.5	0.033
7	6475.8–5449.8 4601.3–4423.9	0.033
8	7501.7–6475.8 4601.3–4246.5	0.036
9	7501.7–6097.8 5453.7–4246.5	0.036
10	7501.7–5449.8 4423.9–4246.5	0.037
11	7501.7–6475.8 4423.9–4246.5	0.038
12	6101.7–4597.5	0.040
13	6101.7–5449.8 4601.3–4246.5	0.041
14	7501.7–4597.5	0.041
15	5453.7–4246.5	0.042
16	7501.7–5449.8	0.042
17	7501.7–5449.8	0.042
18	6475.8–5449.8 4423.9–4246.5	0.044
19	7501.7–6097.8 5453.7–4597.5	0.050
20	6475.8–5449.8	0.050
21	7501.7–6475.8 4601.3–4423.9	0.051
22	4601.3–4246.5	0.052
23	4601.3–4246.5	0.052
24	7501.7–6097.8	0.056
25	6101.7–5449.8	0.058
26	7501.7–6475.8	0.059
27	11998.9–7497.9 11998.9–4246.5	0.060
28	6101.7–4246.5	0.061
29	5453.7–4597.5	0.061
30	4601.3–4423.9	0.063
31	11998.9–5449.8 11998.9–6097.8	0.062
32	4601.3–4246.5 5453.7–4246.5	0.063
33	4423.9–4246.5	0.064
34	11998.9–6097.8 4601.3–4246.5	0.065
35	11998.9–7497.9 6101.7–5449.8 4601.3–4246.5	0.065
36	11998.9–4597.5	0.066
37	11998.9–7497.9 5453.7–4246.5	0.066
38	11998.9–7497.9 6101.7–4597.5	0.0657
39	11998.9–7497.9 4601.3–4246.5	0.066
40	11998.9–6097.8 5453.7–4597.5	0.068
41	11998.9–7497.9 5453.7–4597.5	0.069
42	11998.9–5449.8	0.071
43	11998.9–7497.9 6101.7–5449.8	0.072
44	11998.9–6097.8	0.072
45	11998.9–7497.9	0.081

criteria. If $0.900 < R < 0.950$, the correction model was considered successful, and if $R > 0.950$, the correction model was considered extremely successful. In the present study, the correction model of carbon and nitrogen content was extremely successful, while the models for lignin, acid-insoluble lignin, soluble sugar, and cellulose were successful. RMSECV and root-mean-square error of prediction (RMSEP) are also factors used to evaluate the model, both of which are generally less than one when creating a model. Combining all three values, the models of carbon, nitrogen, and soluble sugar were extremely successful, while others were successful, and all models met the requirement of prediction.

4. Conclusion

In the present study, we used 340 samples to build a model for carbon and nitrogen, 240 samples for lignin and acid-insoluble lignin, and 200 samples for cellulose and soluble sugar. As compared to the previous similar studies, the sample size in the present study was larger, the treatments were more various, and the applicability was wider. The content of carbon, nitrogen, acid-insoluble lignin, soluble lignin, and cellulose obtained using chemical methods was comparable to those from [28, 29], indicating these chemical methods were reliable, so the data could serve as a good base line value.

Overall, the tested materials in this experiment were highly representative, and the above models met the needs of component determination. Compared to sample crushing, this approach ensures the integrity of rapeseed samples by avoiding damage to the rapeseed stalk during the grinding process, so the tested samples can be further used for other purposes. If the content of samples from various locations or different years is desired, other components can be added to modify the model to meet wider applicability requirements.

Table 5: The test result of the model.

Index	Pretreatment method	Wave number (cm^{-1})	Principal component number	Calibration set		Validation set	
				R	RMSECV (%)	R	RMSEP (%)
Carbon	No spectral pretreatment	7501.7–4246.5	9	0.973	0.240	0.887	0.348
Nitrogen	No spectral pretreatment	7501.7–5449.8 4601.3–4246.5	9	0.959	0.024	0.875	0.033
Lignin	No spectral pretreatment	7501.7–6097.8 5453.7–4246.5	3	0.923	1.250	0.760	1.691
Acid-insoluble lignin	No spectral pretreatment	11998.9–7497.9 5453.7–4246.5	1	0.928	1.480	0.766	2.448
Soluble sugar	No spectral pretreatment	11998.9–6097.8	1	0.939	0.310	0.833	0.428
Cellulose	No spectral pretreatment	7501.7–5449.8	6	0.945	1.880	0.863	2.600

In addition, the problems and solutions encountered in creating the model herein can provide reference for other similar studies.

Disclosure

Kuai Jie and Xu Shengyong are co-first authors.

Conflicts of Interest

The authors declare that they have no conflicts of interest.

Acknowledgments

This work was supported by the National Key Research and Development Program of China (2018YFD1000904).

References

[1] Y. Zhang, B. Tian, G. Shi, L. Huang, and L. Yang, "Effects of the key enzymes activity on the stem lodging resistance in *Brassicanapus L.*," *Chinese Agricultural Science Bulletin*, vol. 27, no. 18, pp. 135–138, 2011.

[2] J. Xue, L. Han, Z. Yang, G. Shen, and Y. Liu, "On-line measurement of nutrient content of corn stover using NIRS," *Transactions of the Chinese Society for Agricultural Machinery*, vol. 47, no. 7, pp. 216–223, 2016.

[3] S. Oh, M. C. Lee, M. C. Yu, S. Lee, and M. Rauf, "Protein contents in collected rice landraces germplasm using near-infrared reflectance spectroscopy (NIRS)," *Korean Journal of Plant Resources*, vol. 30, no. 4, pp. 450–465, 2017.

[4] I. Torres, D. Pérez-Marín, M.-J. De la Haba, and M.-T. Sánchez, "Developing universal models for the prediction of physical quality in citrus fruits analysed on-tree using portable NIRS sensors," *Biosystems Engineering*, vol. 153, pp. 140–148, 2017.

[5] R. Font, M. d. Río-Celestino, E. Cartea, and A. de Haro-Bailón, "Quantification of glucosinolates in leaves of leaf rape (*Brassica napus* ssp. *pabularia*) by near-infrared spectroscopy," *Phytochemistry*, vol. 66, no. 2, pp. 175–185, 2005.

[6] J. Ding, Q. Zhang, S. Li, and H. Liu, "Fast detection of glucose and fructose content in honey using NIR spectroscopy," *Spectroscopy and Spectral Analysis*, vol. 36, no. 10, pp. 197-198, 2016.

[7] H. Bagherpour, S. Minaei, N. M. Abdollahian, and F. M. E. Khorasani, "Non-destructive determination of sugar content in root beet by near infrared spectroscopy (NIRS)," *Iranian Journal of Food Science and Technology*, vol. 12, no. 46, pp. 219–228, 2015.

[8] P. Rungpichayapichet, B. Mahayothee, M. Nagle, P. Khuwijitjaru, and J. Muller, "Robust NIRS models for non-destructive prediction of postharvest fruit ripeness and quality in mango," *Postharvest Biology & Technology*, vol. 111, pp. 31–40, 2016.

[9] G. W. Mathison, H. Hsu, R. Soofi-Siawash et al., "Prediction of composition and ruminal degradability characteristics of barley straw by near infrared reflectance spectroscopy," *Canadian Journal of Animal Science*, vol. 79, no. 4, pp. 519–523, 1999.

[10] B. Kaur, M. K. Sangha, and G. Kaur, "Development of near-infrared reflectance spectroscopy (NIRS) calibration model for estimation of oil content in *Brassica juncea* and *Brassica napus*," *Food Analytical Methods*, vol. 10, no. 1, pp. 1–7, 2016.

[11] E. Teye and X. Huang, "Novel prediction of total fat content in cocoa beans by FT-NIR spectroscopy based on effective spectral selection multivariate regression," *Food Analytical Methods*, vol. 8, no. 4, pp. 945–953, 2015.

[12] N. Shetty and R. Gislum, "Quantification of fructan concentration in grasses using NIR spectroscopy and PLSR," *Field Crops Research*, vol. 120, no. 1, pp. 31–37, 2011.

[13] S. Jasinski, A. Lecureuil, M. Durandet, P. Bernardmoulin, and P. Guerche, "ArabidopsisSeed content QTL mapping using high-throughput phenotyping: the assets of near infrared spectroscopy," *Frontiers in Plant Science*, vol. 7, no. 3, pp. 1682–1693, 2016.

[14] Z. Niu, X. Li, and H. Gao, "Rapid detection of element content in straw biomass using hyperspectral imaging technology," *Transactions of the Chinese Society of Agricultural Engineering*, vol. 30, no. 22, pp. 181–187, 2014.

[15] X. Liu, K. Zhu, Z. Niu, and X. Cheng, "NIRS rapid detection o f elemental analysis for straw biomass based on PLS

algorithm," *Journal of Huazhong Agricultural University*, vol. 34, no. 2, pp. 131–135, 2015.

[16] J. Xue, Z. Yang, L. Han, Y. Liu, Y. Liu, and C. Zhou, "On-line measurement of proximates and lignocellulose components of corn stover using NIRS," *Applied Energy*, vol. 137, pp. 18–25, 2015.

[17] K. Sheng, Y. Shen, H. Yang, W. Wang, and W. Luo, "Rapid determination of componential contents and calorific value of selected agricultural biomass feedstocks using spectroscopic technology," *Spectroscopy and Spectral Analysis*, vol. 32, no. 10, pp. 2805–2809, 2012.

[18] M. Fu, M. Liu, Z. Niu, Y. Fang, and Y. Xiang, "Rapidly detecting the content of soluble sugar in rice straw with near infrared reflectance spectroscopy," *Journal of Huazhong Agricultural University*, vol. 35, no. 2, pp. 115–121, 2016.

[19] D. Lohr, P. Tillmann, U. Druege, S. Zerche, T. Rath, and E. Meinken, "Non-destructive determination of carbohydrate reserves in leaves of ornamental cuttings by near-infrared spectroscopy (NIRS) as a key indicator for quality assessments," *Biosystems Engineering*, vol. 158, pp. 51–63, 2017.

[20] G. Fan, L. Zhou, Q. Lu et al., "Determination of lignin content in four common crop straws by sulfuric acid method," *Jiangsu Agricultural Sciences*, vol. 5, no. 5, pp. 395–398, 2010.

[21] Y. Pei, Y. Li, Y. Zhang et al., "G-lignin and hemicellulosic monosaccharides distinctively affect biomass digestibility in rapeseed," *Bioresource Technology*, vol. 203, pp. 325–333, 2016.

[22] B. Xie, Z. Rao, and J. Xiang, "Evaluation of uncertainty of the determination of soluble sugar in tobacco by anthrone colorimetry," *China Measurement & Test*, vol. 41, no. 1, pp. 10–13, 2015.

[23] C. Jia, S. Zhu, and G. Zhao, "Rapid determination of total protein and wet gluten in commercial wheat flour using siSVR-NIR," *Food Chemistry*, vol. 221, pp. 1939–1946, 2017.

[24] J. Moros, I. Llorca, M. L. Cervera, A. Pastor, S. Garrigues, and M. de la Guardia, "Chemometric determination of arsenic and lead in untreated powdered red paprika by diffuse reflectance near-infrared spectroscopy," *Analytica Chimica Acta*, vol. 613, no. 2, pp. 196–206, 2008.

[25] S. de la Guardia, Y. Wang, and J. Wu, "Outlier sample eliminating criterions and methods for building calibration model of near infrared spectroscopy analysis," *Transactions of The Chinese Society of Agricultural Machinery*, vol. 35, no. 4, pp. 115–119, 2004.

[26] Å. Rinnan, F. van den Berg, and S. B. Engelsen, "Review of the most common pre-processing techniques for near-infrared spectra," *TrAC Trends in Analytical Chemistry*, vol. 28, no. 10, pp. 1201–1222, 2009.

[27] D. F. Malley, C. McClure, P. D. Martin, K. Buckley, and W. P. McCaughey, "Compositional analysis of cattle manure during composting using a field-portable near-infrared spectrometer," *Communications in Soil Science & Plant Analysis*, vol. 36, no. 4-6, pp. 455–475, 2005.

[28] T. Liu and C. Guan, "Grey relational analysis between lodging iidex and biochemistry components of stem agronomic characteristics in rapeseed (*Brassica napus* L)," *Chinese Journal of Oil Crop Sciences*, vol. 30, no. 2, pp. 152–156, 2008.

[29] J. Kuai, X.-Z. Du, M. Hu et al., "Effect of symbiotic periods and plant densities on growth and yield of rape-seed intercropping cotton," *Acta Agronomica Sinica*, vol. 42, no. 4, pp. 591–599, 2016.

20

Chronological Classification of Ancient Mortars Employing Spectroscopy and Spectrometry Techniques

M. Ramacciotti,[1,2] S. Rubio,[2] G. Gallello,[1,2,3] M. Lezzerini,[4,5] S. Columbu,[6] E. Hernandez,[7] A. Morales-Rubio,[2] A. Pastor,[2] and M. de la Guardia[2]

[1]*Department of Prehistory, Archaeology and Ancient History, Faculty of Geography and History, University of Valencia, Av. Blasco Ibáñez 28, 46010 Valencia, Spain*
[2]*Department of Analytical Chemistry, Faculty of Chemistry, University of Valencia, C/ Dr. Moliner 50, 46100 Burjassot, Valencia, Spain*
[3]*Department of Archaeology, University of York, King's Manor, York YO1 7EP, UK*
[4]*Department of Earth Sciences, University of Pisa, Via S. Maria 53, 56126 Pisa, Italy*
[5]*Applied and Laser Spectroscopy Laboratory, ICCOM-CNR, Via G. Moruzzi 1, 56124 Pisa, Italy*
[6]*Department of Chemical and Geological Sciences, University of Cagliari, Via Trentino 51, 09127 Cagliari, Italy*
[7]*Sagunto Archaeological Museum, C/ del Castillo 23, 46500 Sagunto, Valencia, Spain*

Correspondence should be addressed to G. Gallello; gianni.gallello@york.ac.uk

Academic Editor: Javier Garcia-Guinea

Forty-two mortar samples, from two archaeological excavations located in Sagunto (Valencian Community, Spain), were analysed by both portable energy dispersive X-ray fluorescence spectroscopy (pED-XRF) and inductively coupled plasma mass spectrometry (ICP-MS) to determine major and minor elements and traces including rare earth elements (REEs). Collected data were crossed with those previously obtained from Sagunto Castle mortars, and principal component analysis (PCA) was applied to discriminate the construction phases of the unearthed buildings. REE permitted to ascribe most of the masonries to the Roman Imperial period. Moreover, a statistical model was built by employing partial least squares discriminant analysis (PLS-DA) in order to classify the mortars from Roman Imperial period and from Islamic period due to the problematic overlapping between these two phases. Results confirmed the effectiveness of the developed indirect chronology method, based on REE data, to discriminate among historic mortars from different construction periods on a wide scale including different Sagunto archaeological sites.

1. Introduction

Mortar is a building material composed essentially of binder and aggregate fractions and, in some cases, of additives of different types [1–3]. In particular, lime mortar had a key function in Roman architecture: Roman people made their walls and structures with this material adding reactive materials, like pozzolan materials (i.e., *pulvis puteolanus* and *cocciopesto*), to give a hydraulic character to the mortars [4–8]. The use of mortar in architecture is documented during the Middle Ages and in the following historical periods [6–13].

The chemical analysis by statistical approach, together with mineralogical and petrographic characterization of ancient mortars and polished stones, has shown to be a useful tool in the interpretation of the construction phases of several archaeological sites and historical complexes [14–20]. Moreover, ancient mortars are subjected to decay phenomena which also must be detected and evaluated for conservation issues [21–24].

TABLE 1: Sample description including area and building of sampling sites and mortar type.

Sample	Archaeological excavation	Building	Mortar type
C05	Los Huertos Street	Circus MSU1005	Lime mortar
C06	Los Huertos Street	Circus MSU1006	Lime mortar
C07	Los Huertos Street	Circus MSU1007	Lime mortar
C08	Los Huertos Street	Circus MSU1008	Lime mortar
C09	Los Huertos Street	Circus MSU1009	Lime mortar
C10a	Los Huertos Street	Circus MSU1010a	Lime mortar
C10b	Los Huertos Street	Circus MSU1010b	Lime mortar
C20	Los Huertos Street	Circus MSU1020	Lime mortar
C26	Los Huertos Street	Circus MSU1026	Lime mortar
CLC	Los Huertos Street	Cloaca	Lime mortar
EW	Los Huertos Street	Eastern Wall	Lime mortar
JMN	Los Huertos Street	Northern Jamb	Lime mortar
NR	Los Huertos Street	Noria	Lime mortar
NW	Los Huertos Street	Northern Wall	Lime mortar
P15	Los Huertos Street	Pilaster MSU3015	Lime mortar
P16	Los Huertos Street	Pilaster MSU3016	Lime mortar
P17	Los Huertos Street	Pilaster MSU3017	Lime mortar
P18	Los Huertos Street	Pilaster MSU3018	Lime mortar
R3a	Los Huertos Street	Room 3 MSU1032	Lime mortar
R3b	Los Huertos Street	Room 3 MSU1050	Lime mortar
UN1	Los Huertos Street	Unknown MSU3038	Lime mortar
UN2	Los Huertos Street	Unknown MSU6006	Lime mortar
WLL	Los Huertos Street	Well	Lime mortar
RS10A	Railroad Station	MSU1010a	Lime mortar
RS10B	Railroad Station	MSU1010b	Lime mortar
RS10C	Railroad Station	MSU1010c	Lime mortar
RS11A	Railroad Station	MSU1011a	Lime mortar
RS11B	Railroad Station	MSU1011b	Lime mortar
RS11C	Railroad Station	MSU1011c	Lime mortar
RS24A	Railroad Station	MSU1024a	Lime mortar
RS24B	Railroad Station	MSU1024b	Lime mortar
RS24C	Railroad Station	MSU1024c	Lime mortar
RS25A	Railroad Station	MSU1025a	Lime mortar
RS25B	Railroad Station	MSU1025b	Lime mortar
RS25C	Railroad Station	MSU1025c	Lime mortar
RSLW	Railroad Station	Long Wall	Lime mortar
R1a	Los Huertos Street	Room 1 MSU3029	Earth mortar
R1b	Los Huertos Street	Room 1 MSU3030	Earth mortar
R1c	Los Huertos Street	Room 1 MSU3039	Earth mortar
R2a	Los Huertos Street	Room 2 MSU3062	Earth mortar
R2b	Los Huertos Street	Room 2 MSU3063	Earth mortar
R2d	Los Huertos Street	Room 2 MSU3064	Earth mortar

Note. MSU: masonry stratigraphic unit.

This paper shows the results of the analyses carried out on ancient mortars collected from buildings discovered during two recent archaeological excavations at Sagunto, a town located in the Eastern Spain, ca. 30 km north of Valencia, close to the Costa del Azahar on the Mediterranean Sea. Sagunto is well known in the world for its complex history, and its area has been occupied since the Iberian Age. During the Roman period, Sagunto was interested in constructing important buildings such as the Circus and the Theatre. Thereafter, Sagunto was occupied by Islamic people, and during the Modern Ages, the city was involved in the Napoleonic Wars [25].

The collected mortar samples were analysed by both portable energy dispersive X-ray fluorescence spectroscopy (pED-XRF) and by an inductively coupled plasma mass spectrometer (ICP-MS) to determine their major elements, minor elements, and trace elements. The mineral element concentration and, in particular, the measured contents of rare earth elements (REE) of these samples were compared with those previously obtained from Sagunto Castle [26] by applying principal component analysis (PCA) and partial least squares discriminant analysis (PLS-DA) to determine the construction periods.

2. Materials and Methods

2.1. Sampling. The studied mortars (Table 1) were sampled during two archaeological excavations located in the city of

FIGURE 1: Map of the city of Sagunto with the area of sampling (the satellite picture was obtained from Google Earth while the map of Sagunto from the Institut Cartogràfic Valencià, http://www.icv.gva.es).

Sagunto (Figure 1). Twenty-nine samples were collected during the archaeological excavation of Los Huertos St., an area characterized by the presence of buildings and materials from the Roman Imperial phase and from the Islamic occupation in the Middle Ages. Twenty-three samples of lime mortars were collected from these buildings: nine samples come from the Circus (C), a building dating back to the Roman Imperial period, one from the cloaca (CLC), one from a jamb (JMN), two from two different walls (Eastern Wall, EW, and Northern Wall, NW), one from a noria (NR), and four from a pilaster (P). Six samples of earth mortar come from two rooms that the archaeologists interpreted as may belong to the Islamic occupation (Room 1, samples R1 and Room 2, samples R2), and in this study, they were analysed as control samples. Two samples of lime mortar were collected from Room 3 (R3), two from an unidentified building (UN), and one from a well (WLL). Thirteen samples of lime mortars were collected from different masonries during the emergency archaeological excavation of an area close to Sagunto's Railroad Station (RS named samples). Data of twenty-three lime mortar samples from different buildings of the Castle of Sagunto, studied by Gallello et al. [26], were taken into account as the calibration set to perform data analysis and to date the structures of the above-quoted excavations. Part of the calibration samples come from masonries dating back to the Roman Imperial Period: five samples from the Theatre (TR) and one from the Curia (CUR); the other samples come from masonries dated back to the Islamic occupation phase: four samples from the first part of the Islamic Wall (MI), one of the wall which was considered modern before the study (MM), one from a Hermitage (ERM), two from two Islamic reworks of the Torre Central Estudiantes (TCE), and two from the *tabernae* of the Imperial Forum. Moreover, two samples from the Curia (CUR) and five samples from the Basilica of the Imperial Forum (FBI), whose dating was uncertain between the above-quoted two phases, were added to the dataset.

2.2. Major and Minor Elements Determination. All the analyses were carried out on each entire mortar sample previously pulverized ($D_{max} < 63\,\mu m$) and homogenized through Agatha mortar and pestle. Major and minor element concentrations were obtained by using a S1 Titan energy dispersive portable X-ray fluorescence spectrometer (pED-XRF) from Bruker (Kennewick, Washington DC, USA) equipped with an Rh X-ray tube and X-Flash® SDD. Geochem-trace calibration was used to perform the quantitative analyses, and

Table 2: Accuracy of pED-XRF analysis evaluated through the use of CRM (soil NIM GBW07408 and limestone NCS DC60108a) reference samples. As regards NCS DC60108a, not detected Ti concentration and uncertainty of Al_2O_3 determination in the limestone sample are related to the sensitivity of the spectrometer.

Element	NIM GBW07408		NCS DC60108a	
	Certified	Measured	Certified	Measured
Al_2O_3	11.92 ± 0.15	12.06 ± 0.81	0.33 ± 0.03	0.61 ± 0.19
SiO_2	58.61 ± 0.13	52.87 ± 3.71	2.09 ± 0.06	2.76 ± 0.05
CaO	8.27 ± 0.12	8.60 ± 0.28	51.61 ± 0.15	50.31 ± 3.15
Ti	0.38 ± 0.01	0.37 ± 0.01	0.009 ± 0.001	N/D
Fe	3.13 ± 0.03	3.27 ± 0.09	0.12 ± 0.02	0.15 ± 0.01

Note. Certified and measured values of the analyzed elements. The values are expressed as weight percentages (wt.%). N/D: not detected.

Table 3: ICP-MS parameters employed for the mineral analysis.

Instrumental conditions	
Vacuum pressure (Pa)	5.8×10^{-6}
Flow of plasma gas Ar (L/min)	1.0
RF power (W)	1550
RF coupling (V)	1.80
Sampling depth (mm)	8.0
Nebulizer pump (rps)	0.3
Lens	
Extraction lens 1 (V)	0
Extraction lens 2 (V)	−200
Omega lens (V)	9.2
Omega lens deviation (V)	−120
Sample parameters	
Carrier flow (L/min)	0.35
Work mode	HMI-8
Integration parameters	
Acquisition mode	Spectrum
Per peak points	1
Replicates	3
Per replicas' readings	100
Dwell time UMA (s)	0.5–1

S1 Sync software from Bruker was employed to measure Al_2O_3, SiO_2, CaO, Ti, and Fe. The accuracy of the measurements was verified by using the following certified reference materials: soil NIM GBW07408 and limestone NCS DC60108a (Table 2). All the reading's standard errors range between 1 and 5 wt.%, except for Al_2O_3 measured on NCS DC60108a, whose standard deviation increases up to 22 wt.% for concentration less than 0.5 wt.% probably due to the low sensitivity of the instrument for the determination of this element.

2.3. Trace Elements and REE Determination. Previously pulverized and homogenized samples of each entire mortar were prepared for the inductively coupled plasma mass spectrometry (ICP-MS) analysis. The mortars digestion method was developed by Gallello et al. [26] in order to provide reproducible and comparable results compatible with the sensitivity of the analytical method employed. The wet digestion consisted in the addition of 1.35 mL of HCl (37%) and 0.45 mL of HNO_3 (69%) to ca. 0.15 g of each

Table 4: Analytical features of mineral elements determination in mortar samples by ICP-MS.

Element	Mass	LOD	LOQ	R^2
La	139	0.0009	0.003	0.9994
Ce	140	0.0009	0.003	0.9993
Pr	141	0.0003	0.0010	0.9991
Nd	142	0.002	0.007	0.9994
Sm	152	0.0002	0.0007	0.9992
Eu	151	0.0005	0.0017	0.9998
Gd	158	0.004	0.015	0.9993
Tb	159	0.00011	0.0004	0.9989
Dy	162	0.0007	0.002	0.9991
Ho	165	7.4E-05	0.0002	0.9991
Er	166	0.0003	0.0011	0.9989
Tm	169	0.00017	0.0006	0.9986
Yb	172	0.0006	0.002	0.9989
Lu	175	0.00015	0.0005	0.9990
Sc	45	0.02	0.08	0.9999
Y	89	0.0013	0.004	0.9999
Ba	138	0.002	0.008	0.9992
Bi	209	0.006	0.018	0.9991
Cd	111	0.003	0.011	0.9998
Cr	52	0.4	1.3	0.9999
Co	59	0.008	0.03	0.9998
Cu	63	0.009	0.03	0.9998
Pb	207	0.12	0.4	0.9993
Li	7	0.0016	0.005	0.9996
Mn	55	0.17	0.58	0.9998
Mo	95	0.012	0.04	0.9998
Ni	60	0.009	0.03	0.9990
Sr	88	0.011	0.04	0.9996
Tl	205	0.0018	0.006	0.9988
V	51	1.2	4	0.9998
Zn	64	0.2	0.8	0.9994
Rh*	103			

Note. Mass, detection limit (LOD), quantification limit (LOQ), and R^2 of 31 elements detected in the studied samples being LOD and LOQ expressed as μg/g for all elements. *Internal standard.

sample in glass tubes, which were heated in a boiling water bath for about 40 min. Then, the solutions were poured into plastic tubes and brought up to 25 mL with purified water. The concentration of the following elements: Ba, Bi, Cd, Cr, Co, Cu, Pb, Li, Mn, Mo, Ni, Sr, Tl, V, and Zn, REE (La, Ce, Pr, Nd, Sm, Eu, Gd, Tb, Dy, Ho, Er, Tm, Yb, and Lu), Sc, and Y were determined. Two multielement stock solutions for the ICP analysis in 5% HNO_3, containing the abovequoted elements at a concentration of 1000 mg/L, were used as stock standards for calibration. 5 mL volumetric flasks were used in adding the corresponding volume of standard solutions 0.15 mL of HNO_3, 0.45 mL of HCl, and the water necessary to reach the final volume. The concentration of trace elements ranges from 1 μg/L to 600 μg/L except for REE, Y, and Sc that ranges from 1 μg/L to 100 μg/L. The measurement accuracy was verified by using the certified reference materials soil NIM GBW07408 and limestone NCS DC60108a. As the internal standard, 50 μL of a 1000 mg/L Rh solution was added to each sample and to each calibration standard. The analyses were performed through an Agilent 7900 inductively coupled plasma mass spectrometer. The measurement conditions that are shown in Table 3 shows the

ICP-MS parameters employed for the analyses, and Table 4 shows the main analytical features obtained for the measured mass of each considered isotope, including the instrumental detection and quantification limits (LOD and LOQ), and the coefficient of determination (R^2) of the corresponding calibration lines.

2.4. Statistical Data Processing. The PCA models were built by using data obtained from a total of forty-two analysed samples, and the set of data obtained from samples dated to the Roman Imperial period and the Islamic occupation phase obtained previously by Gallello et al. [26]. Major and minor elements and trace elements including REE were employed as variables. This technique was used to explore large geochemical datasets by reducing the number of variables and providing a deep insight into the structure of the variance of the dataset. Data were processed through mean center and autoscale prior to modelling, and the obtained model was cross validated through leave-one-out method.

The PLS-DA model was built employing twenty-two samples as a calibration set: nine samples from the Circus (Los Huertos Street excavation) and samples studied by Gallello et al. [26] including five from the Roman Theatre, TR, built during the Roman Imperial Period and eight samples from Islamic masonries (two samples from the Imperial Forum tabernae, TFI, four samples from the first part of the Islamic Wall, MI, and two from the Torre Central Estudiantes, TCE). Five of these samples (C07, C08, C10b, and two TR samples) were randomly selected as an internal validation set to test the model (data not shown). Sixteen variables were employed (REE, Y, and Sc). Finally, thirty-seven mortar samples dated back to Roman Imperial period, Islamic occupation phase, and also some uncertain samples were included as the test set to predict their class. Data were preprocessed employing mean centering and autoscale, and "leave one out" cross validation method was employed for the evaluation of results. Data analysis was carried out using the PLS Toolbox 6.5 for Eigenvector Research Inc., (Wenatchee, WA, USA) running in Matlab R2014b from MathWorks Inc., (Natick, MA, USA).

3. Results and Discussion

3.1. Geochemical Results. The analytical results of major and minor elements as well as trace elements and REE data are reported in the Supplementary Materials (Annexes 1 and 2).

Lime mortar samples from the Railroad Station and from the Circus area buildings have comparable amounts of Al_2O_3 and SiO_2 which range from 0.44 to 1.05 wt.% and from 11.30 to 26.97 wt.%, respectively, except for samples RS25A (Al_2O_3: 3.02 wt.%; SiO_2: 34.36 wt.%) and RS11C (Al_2O_3: 5.17 wt.%; SiO_2: 36.72 wt.%), which are both from the Railroad Station and show particularly high values of these two chemical components. On the contrary, Railroad Station's samples have lower concentration of CaO (30 ± 4 wt.%) and higher one of Fe (1.3 ± 0.1 wt.%) and Ti (0.12 ± 0.01 wt.%) than Los Huertos Street mortars (CaO: 40 ± 3 wt.%; Fe: 0.99 ± 0.14 wt.%; Ti: 0.08 ± 0.02 wt.%). The samples of earth mortars from Room 1 (R1a, R1b, and R1c) and Room 2 (R2a, R2b, and R2d) show the lowest values of CaO (15.3 ± 1.1 wt.%) and the highest ones of Al_2O_3 (10.9 ± 1.1 wt.%), SiO_2 (37.8 ± 1.7 wt.%), Fe (2.4 ± 0.3 wt.%), and Ti (0.25 ± 0.02 wt.%).

The lime mortar samples from Los Huertos Street and from the Railroad Station have comparable concentrations in almost all the measured trace elements. In particular, it is worth noticing that the two groups of samples show similar REE total amounts ($\sum$REE) (Los Huertos Street: 20 ± 4 μg/g; Railroad Station: 20 ± 2 μg/g) and Sc concentrations (Los Huertos Street: 1.1 ± 0.3 μg/g; Railroad Station: 1.1 ± 0.1 μg/g), while samples from Los Huertos Street have a slightly higher concentration of Y than the ones from the Railroad Station excavation (4.1 ± 0.8 μg/g and 3.5 ± 0.4 μg/g resp.). Concerning the earth mortars from Room 1 and Room 2, they have comparable contents of REE, Y, and Sc. REE total amounts go from 37 μg/g to 53 μg/g, and Y and Sc range from 5 μg/g to 10 μg/g and from 2 μg/g to 3 μg/g, respectively.

To give a deep insight into the mechanisms that influence the amount of REE in the mortar samples, it is interesting to look at the correlation among the REE and analysed elements for each group of mortars. Table 5 shows the Pearson correlation coefficient (ρ) considering as variables REE total amounts and the indicated elements for each group of mortars considered separately (see Supplementary Materials Annex 2). Only statistically significant data were taken into account given $p = 0.05$. REE concentration of Los Huertos Street lime mortar samples show positive correlations, as well as for Sc ($\rho = 0.75$), which has a chemical behaviour similar to lanthanides, with Al_2O_3 ($\rho = 0.57$), Tl ($\rho = 0.50$),

Table 5: Pearson correlation coefficient (ρ) between $\sum$REE and each measured element for the three groups of mortar.

Element	Railroad Station	Los Huertos Street	
	Lime mortars	Lime mortars	Earth mortars
Al_2O_3	−0.21	**0.57**	0.61
SiO_2	−0.45	0.21	−0.23
CaO	−0.08	**−0.43**	−0.7
Fe	0.18	0.3	**0.76**
Ti	−0.1	0.28	**0.81**
^{209}Bi	−0.21	0.23	**0.95**
^{207}Pb	0.41	0.03	−0.09
^{205}Tl	−0.42	**0.5**	**0.97**
^{138}Ba	0.15	0.06	**0.85**
^{111}Cd	0.01	0.02	0.58
^{95}Mo	−0.04	−0.29	**0.95**
^{89}Y	**0.59**	0.25	0.6
^{88}Sr	0	**−0.41**	**0.76**
^{64}Zn	0.37	−0.24	**0.93**
^{63}Cu	−0.19	−0.22	**0.9**
^{60}Ni	0.32	0.15	**0.89**
^{59}Co	**0.51**	**0.51**	**0.96**
^{55}Mn	0.23	−0.06	**0.94**
^{52}Cr	0.48	**0.55**	**0.94**
^{51}V	**0.56**	0.24	**0.88**
^{45}Sc	0.49	**0.75**	**0.71**
^{7}Li	0.19	0.35	**0.82**

Note. Statistically significant values of ρ are given in bold ($p = 0.05$).

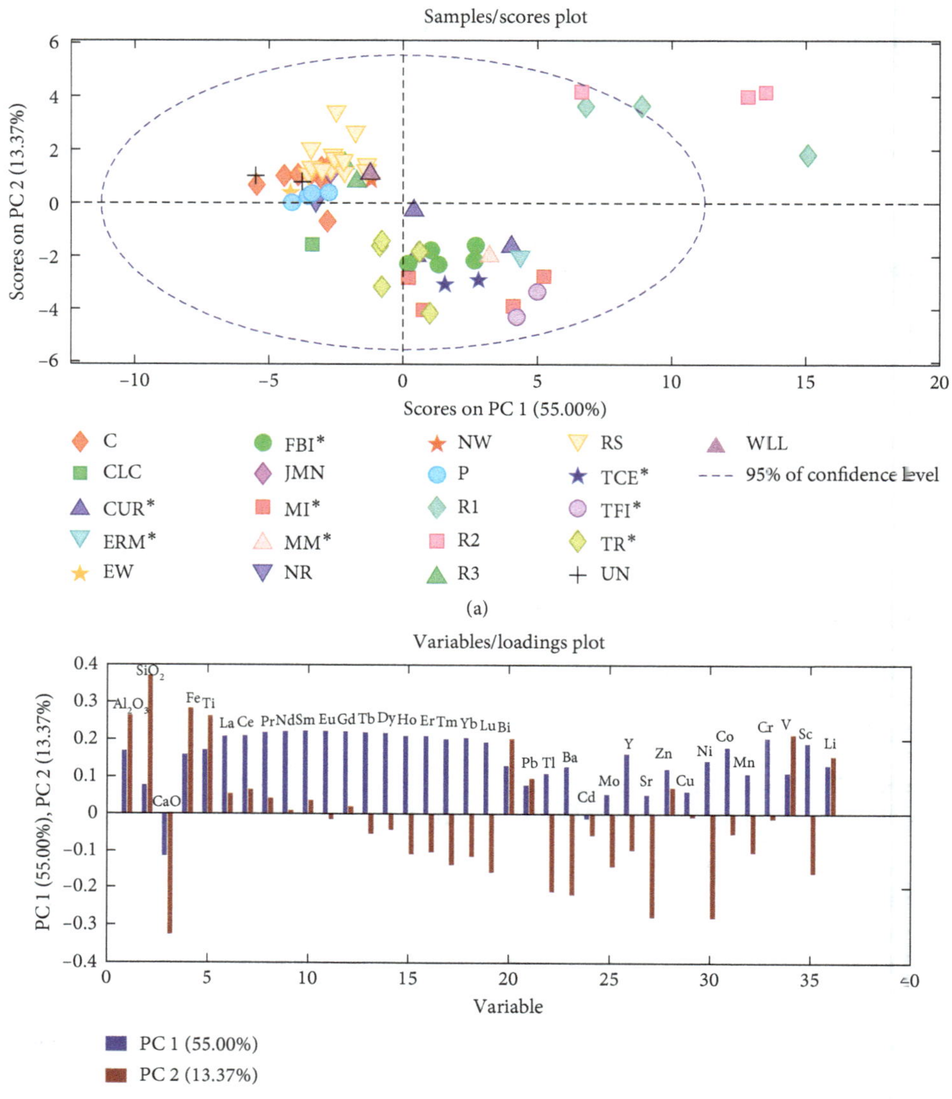

FIGURE 2: PCA study employing all the measured element concentrations. Scores (a) and loadings (b) plot of PC1 and PC2. The samples marked with a star (*) are related to Sagunto Castle buildings [17]. ERM = Hermitage; FBI = Imperial Basilica; FRW = Wester Republican Forum; MI = Islamic Wall (1st section); MII = Islamic Wall (2nd section); MM = Modern Wall; SMR = Republican Wall; TCE = Torre Central Estudiantes; TFI = Imperial Forum tabernae; TMP = Republican Diana's Temple; TR = Roman Theatre.

Co ($\rho = 0.51$), and Cr ($\rho = 0.55$) and negative correlations with Ca ($\rho = -0.43$) and Sr ($\rho = -0.41$), which suggests that the major contribution to the amount of REE comes from limestone clay impurities or from aluminosilicates and silicate rock clasts present in the aggregate fraction [27]. Concerning the lime mortar samples from the Railroad Station excavation, significant correlations were not detected among major elements and REE. However, significant positive correlations were found with Y ($\rho = 0.59$), Co ($\rho = 0.51$) and V ($\rho = 0.56$). As regards earth mortars (R1 and R2) from Los Huertos Street, in this case, the REE total amount has high positive correlations with most of the trace elements (for Bi, Tl, Ba, Mo, Sr, Zn, Cu, Ni, Co, Mn, Cr, V, Sc, and Li, ρ goes from 0.71 to 0.97) and, among the major elements, with Fe ($\rho = 0.76$) and Ti ($\rho = 0.81$), suggesting that the major contribution in lanthanides comes probably from clay fraction, other silicate minerals, and silicate rock clasts.

3.2. Chemometrics for Construction Phase Discrimination. The identification of the construction phases of the buildings was conducted by comparing the mortars from the Railroad Station and Los Huertos Street to the ones from Sagunto Castle published by Gallello et al. [25], and whose construction periods were retraced in the same work. In particular, this previous study was focused on REE concentrations due to their proved effectiveness in

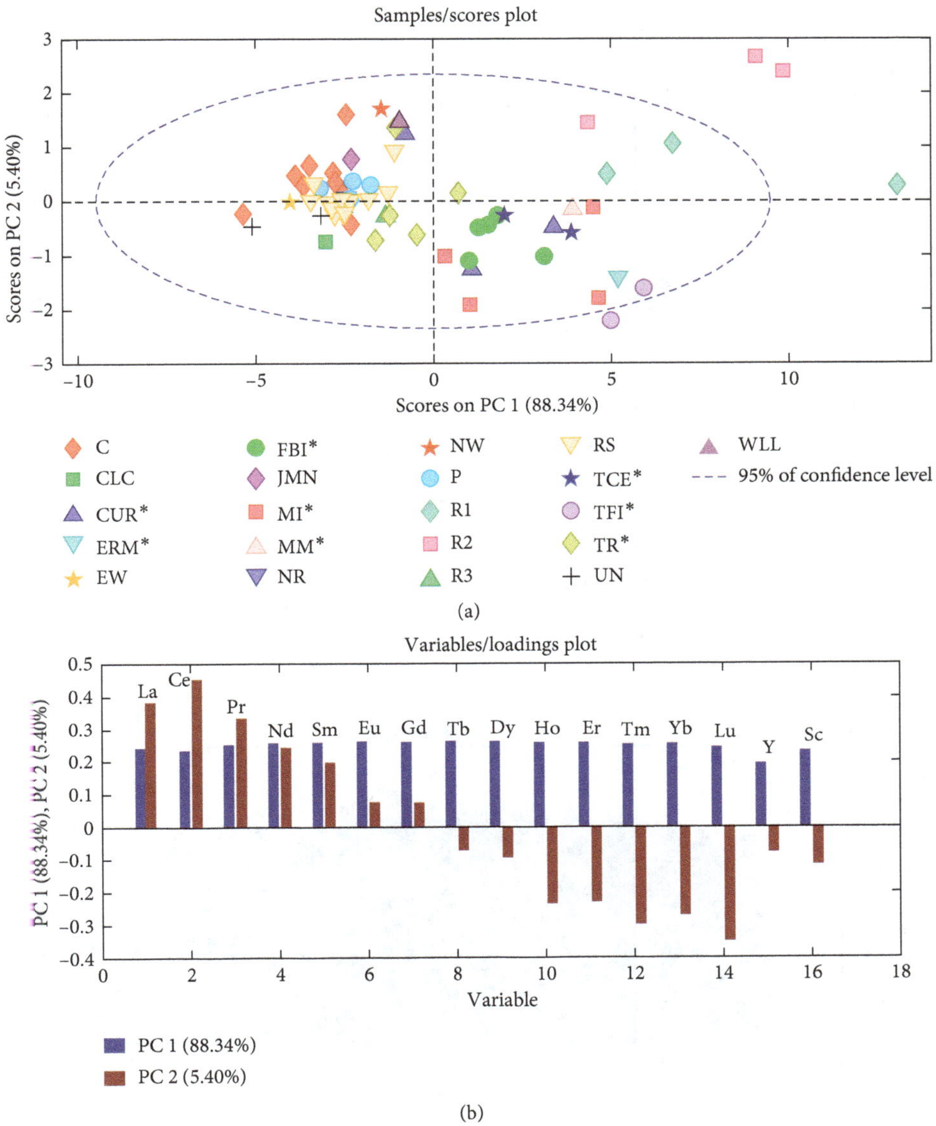

FIGURE 3: PCA study employing only REE, Y, and Sc concentrations. Scores (a) and loadings (b) plot of PC1 and PC2. The legend for Sagunto Castle samples' classes (*) is indicated in Figure 2.

archaeometric studies about provenance and raw materials of lithic and lithoic archaeological artifacts [28–30].

To compare the samples from the excavations to the Imperial and Islamic mortars from Sagunto Castle [26], the principal component analysis (PCA) was carried using all variables (i.e., elements) (Figure 2). Figure 2(b) shows the contribution of the variables in PC1 (55.00% of samples' variance) and PC2 (13.37% of samples' variance). Figure 2(a) shows the position of each sample in a diagram where the x-axis corresponds to PC1 coordinates and y-axis to PC2. Most of the samples of Los Huertos Street and all Railroad Station are grouped together in the left side of the plot, while the samples dated back both to Imperial Roman period and Islamic period from de Castle are grouped in the right part of the graph. The samples of earth mortars from Room 1 (R1a, R1b, and R1c) and Room 2 (R2a, R2b, and R2d) of Los Huertos Street are plotted in the upper right side of the diagram, at the limit of the confidence interval or out of it.

In Figure 3(b), the contribution of the REE employed as variables is shown. PC1 explains 88.34% of the data variance and PC2 explains 5.40%. Lime mortar samples from Los Huertos Street and all Railroad Station excavations are grouped with the mortars from the Roman Theatre (TR) (Figure 3(a)), suggesting that the dating to the Roman Imperial period was also confirmed by the archaeological data while the earth mortar samples are plotted in PC1 and PC2 positive directions. Finally, a separation between mortars from Imperial Period (PC1 negative direction), and mortars from the Islamic Period, located in the left lower area of PC can be appreciated.

Since the discrimination of lime mortars from Roman Imperial and Islamic buildings seems to be the most

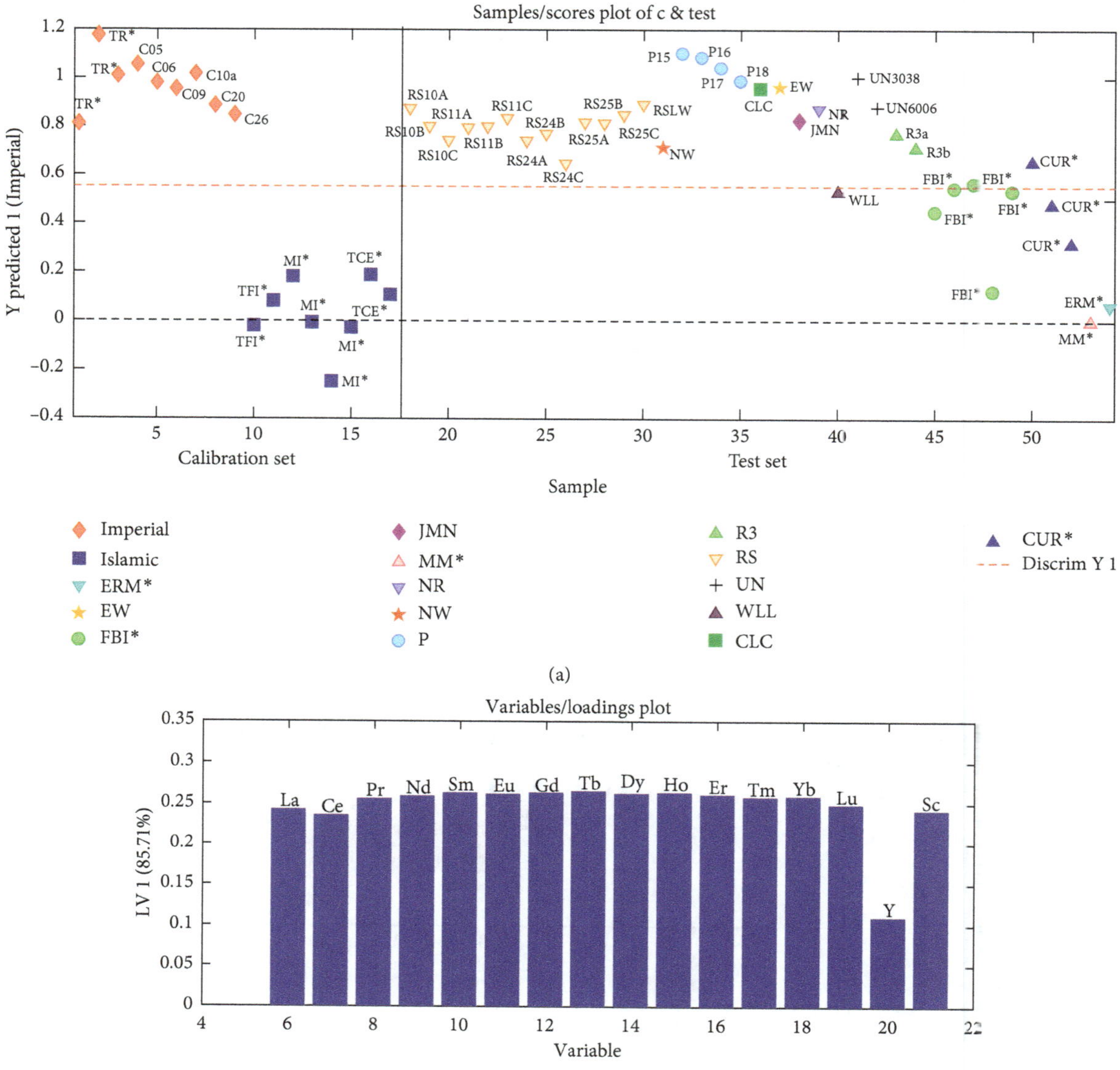

Figure 4: PLS-DA results. (a) Calibration set and predicted class membership of the other samples. (b) VIP scores of the PLS-DA model. The legend for Sagunto Castle classes (*) is indicated in Figure 2.

problematic issue, the principal least squares discriminant analysis (PLS-DA) was employed to build a statistical model to classify uncertain samples. The calibration set and validation set were established by using samples from buildings and masonries that were previously classified by Gallello et al. [26] and by archaeological data as Roman Imperial (C and TR) or as Islamic (MII, TFI, and TCE) (Figure 4(a)). Figure 4(b) shows the contribution of each variable to the model. As it can be seen, the model clearly separates between mortars of the two different construction phases. Subsequently, the model was applied to a test set composed of uncertain Roman Imperial period or Islamic mortars (Figure 4(a)). The PLS-DA results confirmed that all the mortars from Los Huertos Street and Railroad Station are classified as Roman Imperial period materials except for sample WLL that has been classified as Islamic, which is coherent with the archaeological data that confirm the occupation of the area during the Islamic phase. As observed, on the contrary of Los Huertos Street, in Railroad Station samples, major elements are not correlated with REE (Table 5), and this could suggest that the mortars were made in a different way or this is an indicator of recycling building materials, and, according to the archaeological wall stratigraphy interpretation, the Railroad Station structures could be dated after the Imperial period. Thus, maybe they were built during the Late Roman Age. On the contrary, one mortar sample from the Basilica of the Imperial Forum (FBI SC) and one mortar from the Curia testify the presence of both Roman Imperial masonries and Islamic reworks in Roman buildings due to the heavy reemployment of the Imperial Forum area during the Islamic occupation [26].

The use of REE, Y, and Sc as variables has shown their high effectiveness in the discrimination of different kinds of mortars to retrace the construction history of archaeological sites and monuments. The comparison among mortars coming from different archaeological excavations and monuments of the same geographical area seems to be a very viable method to solve chronological issues related to the construction phases of both structures unearthed during archaeological excavations and historical complexes.

4. Conclusion

The study of the historic mortars from two archaeological excavations (i.e., Los Huertos Street and Railroad Station) located in the city of Sagunto permitted us to solve issues that had remained unanswered in the previous studies on these building materials.

The use of multivariate statistics employing REE as variables allowed us to classify the lime mortars belonging to the Imperial Roman period, in line with the archaeological data; however, the case of the Railroad Station structure needs deeper studies to confirm the chronology as Imperial Roman or Late Roman Ages. Also, it was possible to establish the chronological phase of some masonries of Sagunto Castle whose period of construction was uncertain and that can be attributed to the phase of the Islamic occupation or to the Roman Imperial period. The PLS-DA model was helpful to discriminate among mortars of uncertain attribution and that can be useful for possible future studies in Sagunto area.

The correlation among REE and major/minor and trace elements shows that the mechanisms that allowed REE to be discriminating parameters between Sagunto mortars belonging to different periods are related to chemical weathering processes involving limestone clay impurities, aluminosilicates, and silicate rock clasts present in the aggregate fraction. Therefore, further works need to be developed focusing the understanding of the chemical processes that are influencing the REE distributions.

Summarizing, the present study has confirmed the effectiveness of REE data in building materials as markers of different construction phases and has proved the usefulness of the application of the ancient mortar analysis on a geographical scale wider than a single archaeological excavation or monuments.

Conflicts of Interest

The authors declare that there are no conflicts of interest with any institution or funding body.

Acknowledgments

The authors acknowledge also the analytical support of the Central Service for Experimental Research (SCSIE) of the University of Valencia, and of Dr. Laredo Ortíz and Dr. Soriano Vega. The authors would like to thank all the archaeologists working in Railroad Station and Los Huertos Street excavations, who have contributed the realization of this study and especially the Archaeologist at the Sagunto's Council, Ms. Carla Flors. Authors acknowledge the financial support of Generalitat Valenciana (PROMETEO project II/2014/077) and Ministerio de Economia y Competitividad-Feder (Project CTQ 2014-52841-P and Project CTQ 2012-38635). Gianni Gallello acknowledges the financial support of the European Commission (Project H2020-MSCA-IF-2015-704709-MATRIX).

References

[1] A. Moropoulou, A. Bakolas, and K. Bisbikou, "Investigation of the technology of historic mortars," *Journal of Cultural Heritage*, vol. 1, no. 1, pp. 45–58, 2000.

[2] M. Theodoridou, I. Ioannou, and M. Philokyprou, "New evidence of early use of artificial pozzolanic material in mortars," *Journal of Archaeological Science*, vol. 40, no. 8, pp. 3263–3269, 2013.

[3] M. Franzini, L. Leoni, M. Lezzerini, and F. Sartori, "The mortar of the "Leaning Tower" of Pisa: the product of a medieval technique for preparing high-strength mortars," *European Journal of Mineralogy*, vol. 12, no. 6, pp. 1151–1163, 2000.

[4] F. Marra, A. Danti, and M. Gaeta, "The volcanic aggregate of ancient Roman mortars from the Capitoline Hill: petrographic criteria for identification of Rome's "pozzolans" and historical implications," *Journal of Volcanology and Geothermal Research*, vol. 308, pp. 113–126, 2015.

[5] M. Lezzerini, M. Ramacciotti, F. Cantini et al., "Archaeometric study of natural hydraulic mortars: the case of the Late Roman Villa dell'Oratorio (Florence, Italy)," *Archaeological and Anthropological Science*, vol. 9, no. 4, pp. 603–615, 2017.

[6] M. Lezzerini, S. Pagnotta, S. Raneri et al., "Examining the reactivity of volcanic ash in ancient mortars by using a micro-chemical approach," *Mediterranean Archaeology and Archaeometry*, In press.

[7] S. Columbu and A. M. Garau, "Mineralogical, petrographic and chemical analysis of geomaterials used in the mortars of Roman Nora theatre (south Sardinia, Italy)," *Italian Journal of Geosciences*, vol. 136, no. 2, pp. 238–262, 2017.

[8] S. Columbu, "Petrographic and geochemical investigations on the volcanic rocks used in the Punic-Roman archaeological site of Nora (Sardinia, Italy)," *Earth Environmental Sciences*, In press.

[9] J. Elsen, A. Brutsaert, M. Deckers, and R. Brulet, "Microscopical study of ancient mortars from Tournai (Belgium)," *Materials Characterization*, vol. 53, no. 2–4, pp. 289–294, 2004.

[10] C. Borges, A. Santos Silva, and R. Veiga, "Durability of ancient lime mortars in humid environment," *Construction and Building Materials*, vol. 66, pp. 606–620, 2014.

[11] M. Lezzerini, S. Legnaioli, G. Lorenzetti, V. Palleschi, and M. Tamponi, "Characterization of historical mortars from the bell tower of St. Nicholas church (Pisa, Italy)," *Construction and Building Materials*, vol. 69, pp. 203–214, 2014.

[12] D. Miriello, F. Antonelli, C. Apollaro et al., "New data about the ancient mortars from the archaeological site of Kyme (Turkey): compositional characterization," *Minerals Engineering*, vol. 84, pp. 497–517, 2015.

[13] S. Columbu, F. Sitzia, and G. Ennas, "The ancient pozzolanic mortars and concretes of Heliocaminus baths in Hadrian's Villa (Tivoli, Italy)," *Archaeological and Anthropological Sciences*, vol. 9, no. 4, pp. 523–553, 2017.

[14] F. Carò, M. P. Riccardi, and M. T. Mazzilli Savini, "Characterization of plasters and mortars as a tool in archaeological studies: the case of Lardirago Castle in Pavia, Northern Italy," *Archaeometry*, vol. 50, no. 1, pp. 85–100, 2008.

[15] E. Arizio, R. Piazza, W. R. L. Cairns, L. Apollonia, and A. Botteon, "Statistical analysis on ancient mortars: a case study of the Balivi Tower in Aosta (Italy)," *Construction and Building Materials*, vol. 47, pp. 1309–1316, 2013.

[16] C. Corti, L. Ramazzi, R. Bugini et al., "Thermal analysis and archaeological chronology: the ancient mortars of the site of Baradello (Como, Italy)," *Thermochimica Acta*, vol. 572, pp. 71–84, 2013.

[17] F. Marra, E. D'Ambrosio, M. Gaeta, and M. Mattei, "Petrochemical identification and insights on chronological employment of the volcanic aggregates used in ancient Roman mortars," *Archaeometry*, vol. 58, no. 2, pp. 177–200, 2016.

[18] A. M. Conte, L. Corda, D. Esposito, and E. Giorgi, "Characterization of mortars from the medieval Abbey of Cerrate (southern Italy)," *Journal of Archaeological Science: Reports*, vol. 12, pp. 463–479, 2017.

[19] S. Columbu, "Provenance and alteration of pyroclastic rocks from the Romanesque Churches of Logudoro (north Sardinia, Italy) using a petrographic and geochemical statistical approach," *Applied Physics A*, vol. 123, p. 165, 2017.

[20] S. Columbu, F. Sitzia, and G. Verdiani, "Contribution of petrophysical analysis and 3D digital survey in the archaeometric investigations of the Emperor Hadrian's Baths (Tivoli, Italy)," *Rendiconti Lincei*, vol. 26, no. 4, pp. 455–474, 2015.

[21] I. Papayianni and M. Stefanidou, "Durability aspects of ancient mortars of the archaeological site of Olynthos," *Journal of Cultural Heritage*, vol. 8, no. 2, pp. 193–196, 2007.

[22] I. Papayianni, V. Pachta, and M. Stefanidou, "Analysis of ancient mortars and design of compatible repair mortars: the case study of Odeion of the archaeological site of Dion," *Construction and Building Materials*, vol. 40, pp. 84–92, 2013.

[23] M. Lezzerini, F. Antonelli, S. Columbu et al., "The documentation and conservation of the cultural heritage: 3D laser scanning and GIS techniques for thematic mapping of the stonework of the façade of St. Nicholas Church (Pisa, Italy)," *International Journal of Architectural Heritage: Conservation, Analysis, and Restoration*, vol. 10, no. 1, pp. 9–19, 2016.

[24] S. Columbu, M. Palomba, F. Sitzia, and M. R. Murgia, "Geochemical and mineral-petrographic studies of stones and mortars from the Romanesque Saccargia Basilica (Sardinia, Italy) to define their origin and alteration," *Italian Journal of Geosciences*, vol. 137, no. 3, pp. 1–27, 2018.

[25] P. P. Ripollés, *Opulentissima Saguntum*, Bancaja, Sagunto, Spain, 2004.

[26] G. Gallello, M. Ramacciotti, M. Lezzerini et al., "Indirect chronology method employing rare earth elements to identify Sagunto Castle mortar construction periods," *Microchemical Journal*, vol. 132, pp. 251–261, 2017.

[27] H. R. Rollins, *Using Geochemical Data: Evaluation, Presentation, Interpretation*, Pearson Education Limited, London, UK, 1993.

[28] G. Gallello, T. Orozco, A. Pastor, M. de la Guardia, and J. Bernabeu, "Regional provenance of dolerite prehistoric objects through mineral analysis," *Microchemical Journal*, vol. 124, pp. 167–174, 2016.

[29] A. Hein, P. M. Day, P. S. Quinn, and V. Kilikoglou, "The geochemical diversity of Neogene clay deposits in Crete and its implications for provenance studies of Minoan pottery," *Archaeometry*, vol. 46, no. 3, pp. 357–384, 2004.

[30] S. Cagno, M. Mendera, T. Jeffries, and K. Janssens, "Raw materials for post-medieval Tuscan glassmaking: new insight from LA-ICP-MS analyses," *Journal of Archaeological Science*, vol. 37, no. 12, pp. 3030–3036, 2010.

Application of Infrared Spectroscopy for Functional Compounds Evaluation in Olive Oil

Sarfaraz Ahmed Mahesar,[1] **Massimo Lucarini**,[2] **Alessandra Durazzo**,[2] **Antonello Santini**,[3] **Anja I. Lampe**,[4] **and Johannes Kiefer**[4]

[1]*National Centre of Excellence in Analytical Chemistry, University of Sindh, Jamshoro-76080, Pakistan*
[2]*CREA-Research Centre for Food and Nutrition, Via Ardeatina 546, Rome 00178, Italy*
[3]*Department of Pharmacy, University of Napoli Federico II, Via D. Montesano 49, 80131 Napoli, Italy*
[4]*Technische Thermodynamik, Universität Bremen, Badgasteiner Str. 1, 28359 Bremen, Germany*

Correspondence should be addressed to Sarfaraz Ahmed Mahesar; sarfaraz.mahesar@usindh.edu.pk

Academic Editor: Jose S. Camara

Olive oil is a liquid fat obtained from the fruit of *Olea europaea*, a plant belonging to the *Oleaceae* family, which is widely cultivated and diffused in the Mediterranean area. It is largely produced and used since antiquity. It is mainly used and consumed as food but also as key ingredient in a wide variety of cosmetic products, e.g., to moisturize and nourish dry skin. In the last few decades, olive oil has received much attention as compared to the other seed-obtained oils as well as to the animal fats due to many functional compounds with positive effects on health. To maintain the genuine picture of olive oil, it is essential to assure its authenticity and quality. The presence of bioactive compounds, which characterize the olive oil owing to their antioxidant properties, can be assessed by spectroscopic and chromatographic methods. Currently, spectroscopic techniques combined with chemometric data analysis represent one of the most promising detection methods in the food sector. They offer rapid, versatile, and inexpensive data collection and analyses. The main advantages include the limited and simple sample preparation and the possibility to get spectra directly from the production line. Infrared spectroscopy (mid- and near-infrared) coupled to chemometrics is considered as powerful, fast, accurate, and nondestructive analytical tool for rapid and precise determination of the bioactive compounds content, as well of their bioactivities, i.e., antioxidant properties. These techniques represent a valid alternative to the existing conventional methods of analysis, e.g., based on chromatography and mass spectrometry. Indeed, the present review focuses on the application of infrared spectroscopy for functional compounds evaluation in olive oil.

1. Introduction

Olive oil is a high-quality natural product, typical from the Mediterranean area, where it has been produced for millennia, and it is obtained from the fruits of *Olea europaea* L. [1–4]. The quality of the starting matrix, i.e., the olives, is a crucial parameter to obtain a high-quality product [5]. The quality can be affected, for example, by fungal occurrence and mycotoxin contamination. In particular, the contamination with microfungi capable of producing secondary metabolites (mycotoxins) that potentially affect health is a growing source of concern [6–12]. Consequently, considerable research interest towards more environmentally compatible, field-deployable, and sustainable fungi control systems was prompted over the last couple of years [13–15].

The process of extracting the oil from the seeds is a crucial step as well. There are various methods for this purpose and they can give different yield and quality grades [16]. As an example, extra-virgin olive oil (EVOO) is obtained using mechanical pressure devices. The lower-grade products are typically obtained from a variety of extraction methods including high-pressure extraction, solvent extraction, heat treatment, esterification, and refining. The lower-grade olive oils (e.g., lampante olive oil, pomace oil, refined pomace oil, and refined olive oil) are often characterized by a comparatively high acidity. For instance, according to the current regulations, the acidity of extra-virgin olive oil must not exceed 0.8%, while olive oil and lampante olive oil contain 2% and >2%, respectively. Refining processes lower the acidity to about 0.3%. In blends of virgin and refined olive oils the free acidity is typically about 1%. Olive pomace oil obtained by extraction with a solvent also has acidity around 1%, while the refined olive pomace oil can have an acidity of only 0.3%.

EEC Regulation 29/2012 determines the compulsory and optional information to include on olive oil labels, supplementing the general rules relating to foodstuffs and their marketing in EEC Regulation 2000/2013. The chemical composition of olive oil is characterized mainly by triacylglycerols, monoacylglycerols, diacylglycerols, and free fatty acids [17, 18], as well as fatty acid derivatives (i.e., phospholipids and waxes) [19, 20]. Moreover, there is a diverse group of compounds such as sterols, aliphatic alcohols, chlorophylls, carotenoids, and hydrocarbons [21–23]. The bioactive compounds in olive oil play an imperative role in the authentication and genuineness, quality and purity, traceability, and more importantly to the health point of view [24–33].

Olive oil is a highly interesting product from the nutraceutical perspective [34–42]. It exhibits sensory properties due to the presence of substances with functional activity, such as carotenoids, chlorophylls, tocopherols, and phenolic compounds. These substances are often missing in other common seed-based oils [43]. Moreover, olive oil is characterized by high levels of natural antioxidants, which are associated to potential beneficial properties to humans [44–49]. Phenolic compounds are the main bioactive compounds in olive oil and can be divided into lipophilic and hydrophilic phenols. As reported by El Riachy et al. [50], lipophilic phenols such as tocopherols [51] can be commonly found in other vegetable oils. On the contrary, most hydrophilic phenols found in olive oil are exclusive and/or characteristic of the *Olea europaea* species giving rise to a chemotaxonomic peculiarity [52].

A recent work by Garcia and co-workers [53] divided the phenolic compounds with biological activity into four groups: simple phenols, secoiridoid derivatives, lignans, and flavones. Oleuropein, hydroxytyrosol, tyrosol, and ligstroside are examples of specific phenols, which are usually present in olive oil. In addition, caffeic acid, vanillic acid, and syringic acid are characteristic components of olive oil [54, 55]. The phenolic composition and its total concentration in olive oil have an impact on the characteristic taste, which is bitter and pungent. It depends on different factors, mainly on the degree of ripening of the fruit, the cultivar, the geographic origin [56–61], and the processing [62, 63]. Phenolic compounds are important antioxidants and act as radical scavengers in biological systems, consequently diminishing the lipid oxidation rate [64]. Phenolic compounds are also responsible for the olive oil stability during storage because of their antioxidant activity. With reference to the tocopherols in olive oil, α-tocopherol is the major constituent while β- and γ-tocopherols are present in minor quantities [65, 66]. Figure 1 shows the chemical structure of α-tocopherol.

The greenish color of virgin olive oil is due to the presence of chlorophyll pigments. Amongst the chlorophyll compounds, pheophytin A is found in high amounts in olive oils. On the contrary, lutein and β-carotene are the major "yellow" coloring pigments responsible for the yellowish color of virgin olive oil. Due to the antioxidant nature and pro-oxidant activity of carotenoids and chlorophylls, respectively, they both play a crucial role in the oxidative stability during storage [67]. Moreover, phenols and tocopherols, together with carotenoids, provide excellent oxidative stability to olive oils and have a synergic anticarcinogenic and antioxidant activity.

The main lipid components of olive oil are unsaturated fatty acids such as mono-(oleic) and di-glycerides (linoleic acid). Both are bioactive compounds and play a major role in human health [68]. Besides this, triterpene squalene is also considered as a functional compound in the unsaponifiable fraction of olive oil. The content of such functional compounds is contributing to or even determining the quality and positive sensory properties especially in extra-virgin olive oil.

Many instrumental analytical methods have been developed for the determination of functional compounds in olive oil, mainly using mass spectrometry and chromatography. However, most of these methods are expensive, time-consuming, nonenvironmentally friendly and also may require highly qualified staff for the instrumental analyses. On the contrary, spectroscopic methods such as ultraviolet, visible, near-infrared, and mid-infrared spectroscopy have been successfully employed as rapid, nondestructive, real-time techniques capable of on/offline monitoring of various quality parameters of olive oil. Their benefits include minimum sample preparation and good measurement accuracy with less or no waste generation. This review article presents an updated overview of the infrared spectroscopic methods in combination with chemometric data analysis to determine the functional compounds in olive oil.

2. Basics of Infrared Spectroscopy (Theory and Instrumentation)

2.1. Theoretical Principles. Infrared (IR) and near-infrared (NIR) spectroscopy are based on the absorption of photons by molecules. Absorption can take place when the energy of the photon to be absorbed matches the energy difference between two molecular rotovibrational energy levels. The mid-infrared spectral range is normally specified as 2.5–50 μm (200–4000 cm^{-1}), where the fundamental vibrational modes are located. On the contrary, the near-infrared covers the range 0.75–2.5 μm (4000 to 13333 cm^{-1}) containing overtone and combination bands. A general requirement for allowing the absorption of a photon is that an oscillating dipole moment be produced during the vibrational motion of the molecule. This prerequisite can be expressed mathematically by the following equation:

$$\left(\frac{\partial\mu}{\partial q}\right) \neq 0, \qquad (1)$$

where μ is the dipole moment and q is the normal coordinate. When this equation is fulfilled, the molecule or a specific vibrational mode is called IR-active.

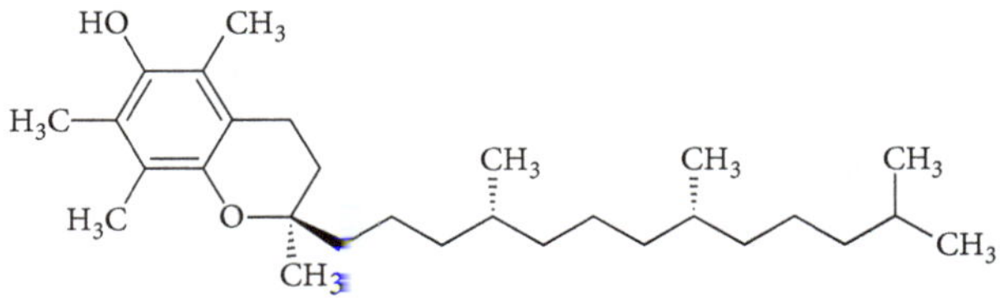

FIGURE 1: The structure of α-tocopherol.

The observed absorbance A, i.e., negative logarithmic transmission (intensity I of the light transmitted divided by the incident intensity I_0) of a sample of thickness d is given by the Beer–Lambert law:

$$A = -\log\left(\frac{I}{I_0}\right) = \varepsilon_\lambda c d, \quad (2)$$

with the extinction coefficient ε, which depends on the wavelength λ, and the concentration c of absorbing molecules. The IR spectrum eventually comprises the individual contributions from all the different vibrations originating from all the molecules in the sample. Consequently, it represents a true molecular fingerprint [69–71].

2.2. Instrumentation. As mentioned above, IR and NIR spectroscopy are absorption-based methods and hence the radiation needs to match the energy difference of the molecular transitions of interest. In other words, sources and detectors in the mid- and near-infrared are required. In the near-IR, Xenon and tungsten lamps are typically used as light sources and the detectors are commonly based on indium gallium arsenide (InGaAs) semiconductor materials. This allows us to perform spectroscopic analysis at wavelengths up to about 2.5 μm. The common silicon-based detectors that can be found in digital cameras are not suitable at wavelengths longer than 1.1 μm. Typical sources of mid-infrared radiation are glowing black-body radiators. Thermal detectors as well as semiconductors based on MCT and $LiTaO_3$ are suitable for detecting the radiation.

Spectral dispersion of the signal is normally achieved by an interferometric analysis employing a Michelson interferometer. The obtained interferogram is then Fourier transformed to yield the spectrum in the frequency domain giving rise to the common name Fourier-transform infrared (FTIR) spectroscopy. Most commercial instruments are FTIR spectrometers.

There is a variety of approaches to achieve an interaction between the radiation and the sample. The most common method is the transmission measurement. The sample is placed in-between two transmissive plates, and the transmitted intensity in the infrared region is recorded as a function of wavelength. Choosing the appropriate sample thickness, however, is crucial in this approach. A very thick sample will result in strong lines being saturated since almost all the photons around the peak center wavelength are absorbed. Such peaks manifest as broadened bands without the typical Gaussian or Lorentzian shape. This may affect the data evaluation. For liquids, typical sample thicknesses lie in the range 25–500 μm. Therefore, it is needless to say that the preparation of the sample is crucial as even dust particles may have significant influence. Common window materials include KBr, CaF_2, ZnSe, and NaCl, which are transmissive for the IR radiation. Conventional borosilicate or fused silica quartz glasses cannot be employed.

The method referred to as attenuated total reflection (ATR) FTIR spectroscopy is easier to use. In such an experiment, the IR radiation is propagating in a transmissive high-refractive-index material (often ZnSe or a diamond crystal), called the internal reflection element (IRE). The sample has a lower refractive index and is in contact with the surface of the crystal. The radiation undergoes total internal reflection at the angle α at this surface, so that the evanescent field can interact with the sample. As a consequence, the reflected beam is attenuated and carries the spectroscopic information.

Both approaches are illustrated schematically in Figure 2. An advantage of the transmission arrangement lays in the well-defined sample thickness, which is given simply by the distance between the plates where the sample is placed. On the opposite, in ATR experiments, the effective path length depends on the penetration depth, d_p, of the evanescent wave. This is a complex function of the wavelength and must be taken into account when ATR FTIR spectra are interpreted quantitatively [72]. The penetration depth is approximately 1/5 of the wavelength and thus consists on order of a few micrometers in the mid-infrared wavelength spectrum. The resulting effective path length, d_{eff}, is usually the actual parameter of interest. It represents the corresponding sample thickness that would lead to the same absorption in a transmission experiment [73]. It is not a physical quantity and therefore it cannot be determined in a straightforward manner [72–74].

In the analysis of food and food products like olive oil, the ATR approach is highly beneficial since it does not require sample preparation and hence the measurements can be performed very quickly. The measurement time per sample is typically significantly below one minute. The data analysis depends on the information desired. Chemometric approaches are the methods of choice when the aim is to quantify constituents such as unsaturated fatty acids or other bioactive compounds. Cross-correlation techniques can be applied to compare the recorded spectrum with a library in order to find out whether or not the sample is genuine. This is often sufficient in order to identify adulterated and counterfeit products. Example applications will be discussed in more detail in the following. In order to give an impression, the FTIR spectra of 17 different olive oil samples from a local supermarket in Bremen, Germany, are displayed in Figure 3. The individual spectra are very similar to each other. This highlights the potential of FTIR spectroscopy for the authentication of olive oils. The most prominent signature at 1745 cm^{-1} is the characteristic C=O stretching vibration, which can be assigned to the fatty acids and ester groups.

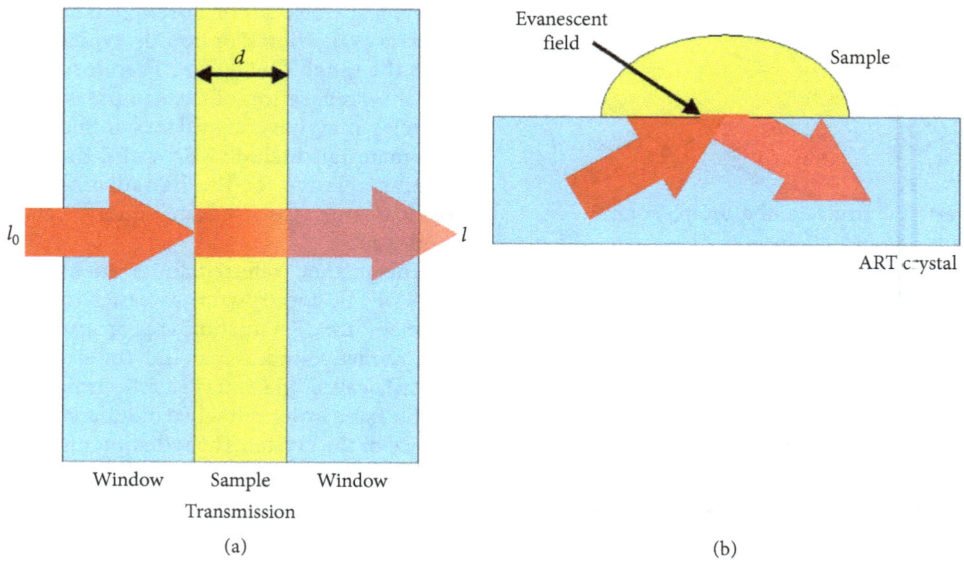

FIGURE 2: Schematic concepts of transmission and ATR spectroscopy. (a) Transmission. (b) ATR.

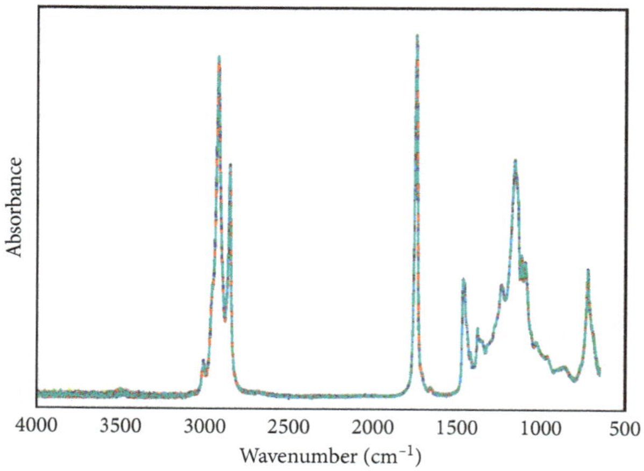

FIGURE 3: FTIR spectra of 17 different olive oil samples.

3. Application of Infrared Spectroscopy to Olive Oil: Focus on Functional Compounds Evaluation

For many years, analytical methods such as TLC, HPLC, GC-MS, and LC-MS/MS were the gold standard to identify and quantify the bioactive compounds (phenols, squalene, tocopherol, and fatty acids) in olive oils [75, 76]. Nowadays, as pointed out previously, the evaluation of bioactive compounds should be integrated in a multidisciplinary study approach for food research. This calls for the combination of advanced and innovative analytical techniques and sophisticated statistical methods for the data analysis in food science [77–79]. NIR and MIR techniques are rapidly emerging in this field. The bioactive compounds and antioxidant properties can be predicted by the spectroscopic methods. In particular, infrared spectroscopy coupled to chemometrics is considered as a powerful, fast, accurate, and nondestructive analytical tool for rapid and precise determination of the content of bioactive compounds as well as their bioactivities, i.e., antioxidant properties. It represents a versatile alternative to the conventional methods. Further advantages in the context of food analysis include the rapid and inexpensive acquisition of data with nil or minimal sample preparation and the possibility to record spectra at or even in the production line. Besides a lot of studies on adulteration and discrimination of olive oil as well as monitoring contaminations, process or recovery by infrared spectroscopy [80–90], the focus here is on the evaluation of functional components.

3.1. Fatty Acids. Being the main constituents of olive oils, fatty acids and their triglycerides dominate the spectra of olive oils. Mailer [91] collected 216 olive oil samples throughout the Australian olive-growing areas to develop a NIR spectroscopy calibration for a range of quality parameters including the major fatty acids in olive oil. Multiple correlation coefficients squared (R^2) for minor fatty acids such as stearic acid (0.86) and linolenic acid (0.85) were reported relatively low because of the very narrow concentration range, while the major fatty acids oleic (0.99) and linoleic (1.00) provided high levels of accuracy compared with the reproducibility of the reference method.

It has been reported that authentication of virgin olive oil is time-consuming and requires expertise in the use of sophisticated instruments. To overcome this issue, Galtier et al. [92] evaluated the chemometric treatment of NIR spectra for the assessment of fatty acids and triacylglycerols in French virgin olive oil samples ($n = 125$) for their classification (PLS1-DA) into geographically closed registered designations of origin (RDOs). The chemometric treatments of NIR spectra revealed that regression vectors of each RDO were correlated to one or more specific components with similar results obtained by conventional techniques according to their cultivar compositions. For the very first time, Maggio et al. [93] reported a rapid method to determine the fatty acid profile of olive oils using FTIR spectroscopy with the ATR sampling technique as described in Section 2. Calibration models were built using PLS regression. The fatty acid calibration models (oleic acid, linoleic acid, SFA, MUFA, and PUFA) were constructed in the spectral range from 3033 to 700 cm^{-1} using PLS. The values obtained for the correlation coefficient for all models were in the range of 0.96–0.99. The LODs of the FTIR-chemometry methods were reported for oleic acid 3.0%, linoleic acid 0.5%, saturated fatty acids 1.3%, monounsaturated fatty acids 0.3%, and polyunsaturated fatty acids 0.3%, with recovery rates in the range 98–103%. The proposed method showed results comparable to the one obtainable by using the official method, but with the advantage of being more rapid and less expensive. Another study by Gurdeniz et al. described the classification and quantification of Turkish extra-virgin olive oil on the basis of fatty acids by using mid-infrared spectroscopy in combination with chemometric methods [94]. The PCA successfully distinguished the southern olive oil samples from the northern samples, while PLS detected a correlation between the spectral data and the fatty acid profile. The values obtained for the correlation coefficient were 0.83, 0.97, 0.97, 0.83, and 0.69 for stearic, oleic, linoleic, arachidic, and linolenic acids, respectively. Dupuy et al. [95] investigated the potential use of NIR and MIR spectroscopy for the quantitative analysis of triacylglycerols and fatty acids for the identification of the RDO of extra-virgin olive oils. PLS and PLS-DA regressions were separately applied to the two spectral ranges. MIR spectroscopy was found to perform better. It was further reported that RDO identification by DA using a multiblock method showed proficiency.

The quantification of oxidized fatty acids in virgin olive oil using mid-infrared spectroscopy coupled with multiple linear regression has been described by Lerma-García et al. [96]. For the prediction of oxidized fatty acid, the entire mid-infrared region was divided into 25 subregions. The subregions affected by oxidation were identified using multiple linear regression with a cube root data transformation that successfully predicted oxidized fatty acids. The value obtained for correlation coefficient was 0.944 with a prediction error of 17%. Uncu and Ozen [97] studied some important quality parameters (oxidative stability, color pigments, fatty acid profile, and phenolic composition) of olive oils by FTIR spectroscopy as one of the vibrational spectroscopic methods: it is possible to predict successfully the oxidative stability, the major fatty acids (palmitic (0.87), oleic (0.94), and linoleic acid (0.97), saturated (0.91), monounsaturated (0.94), and polyunsaturarted fatty acids (0.97)), and the chlorophyll content and some phenolic compounds of the samples.

Casale et al. [98] authenticated the Italian protected designation of origin (PDO) extra-virgin olive oil Chianti Classico using UV-vis, NIR and MIR spectroscopy coupled with a chemometric data analysis. Unequal class models (UNEQ) and soft independent modelling of class analogy (SIMCA) along with PLS were employed individually and jointly on the fatty acid composition data. PLS successfully predicted the oleic and linoleic acid content in the extra-virgin olive oils. Similarly, Inarejos-García et al. [99] investigated the degree of unsaturation of virgin olive oil by NIR spectroscopy. They reported an excellent performance of the multivariate algorithms in particular for the fatty acid profile (r-multiple coefficient of determination = 0.84–0.96). In a different approach, Valli et al. [100] pointed out new parameters for the detection of low quality of extra-virgin olive oil. These parameters were based on the total amount of fatty acids methyl and ethyl esters Σ (FAMEs + FAEEs) and their ratio (FAEEs/FAMEs) applying PLS to MIR spectra. The spectroscopic data showed a good agreement between predicted and actual reference values of calibration data sets. The limit of detection and quantification was reported to be 8.8 and 29.3 $mg{\cdot}kg^{-1}$ for Σ (FAMEs + FAEEs) with R^2 (0.98). On the contrary, the ratio of FAEEs/FAMEs showed R^2 (0.83).

Recently, the European Union regulation for olive oil set a value of fatty acids ethyl ester contents (35 $mg{\cdot}kg^{-1}$) in extra-virgin olive oils. To evaluate the fatty acid alkyl esters in extra-virgin olive oil, Cayuela [101] recently developed predictive models for the determination of total fatty acid alkyl esters based on NIR spectroscopy. The external validation exercise of FAAEs, FAEEs, and FAMEs using the NIR spectrometer showed SEP values of 33.6 and 54.5, 25.6, and 67.2, 18.7 with $r = 0.91$ and 0.93, 0.89, and 0.88, 0.92, respectively, for two validation sets.

3.2. Phenolic Compounds, Tocopherols, and Squalenes. Several studies applied chemometrics to spectroscopic data in order to develop models and predict the content of minor components/functional compounds such as phenolic compounds and tocopherols in olive oils [1, 102, 103]. The recent chapter of Lagouri et al. [102] summarizes the optical UV-Vis-NIR-MIR spectroscopic tools and chemometrics used for the monitoring of olive oil functional compounds. Fatty acids, phenols, tocopherols, carotenoids, and squalene were considered with a focus on the bioactivity, chemistry, and processing. Inarejos-García et al. [99] investigated minor components, sensory characteristics, and quality of virgin olive oil by NIR spectroscopy. Their multivariate algorithms allowed reasonably accurate measurements, in particular of hydroxytyrosol derivatives (r-multiplecoefficient of determination = 0.86–0.88) and C6 alcohols (r-multiple coefficient of determination = 0.69–0.80). FTIR spectroscopy, on the contrary, was used by Gouvinhas et al. [104]. They carried out a chemometric analysis of FTIR spectra from extra-virgin olive oils of three cultivars from Portugal (cultivars "Cobranco," "Galega," and "Picual") at different maturation stages. To create a model for the discrimination of the olive oil samples, principal component analysis and discriminant analysis were utilised. The need for such sophisticated data analysis can also be seen in Figure 3, which shows that the spectra of olive oils are very similar. For the quantification of individual compounds, Gouvinhas et al. [104] developed regression-based calibration and validation models. This allowed the determination of the total phenolic content, *ortho*-diphenols, flavonoids, and antioxidant activity during the ripening process. For *ortho*-diphenols and flavonoids, the slope R^2 and the RMSE were higher than 0.98 and lower than 0.03, respectively, whereas the phenolic content had the lowest multiple coefficient of determination in the calibration model ($R^2 = 0.94$). The same authors reported a slope R^2 and the RMSE of 0.93 and 0.04, respectively, for the determination of the antioxidant activity. Their quality measures became worse when data were cross validated ($R^2 = 0.86$, RMSECV = 0.05).

Mora-Ruiz et al. [105] applied NIR and MIR spectroscopy to study polar phenolic compounds of virgin olive oil and their impact on the oil quality. They obtained satisfactory multivariate test set validation algorithms for total polar phenolic (TPP) compounds (*r*-coefficient of determination = 0.91), hydroxytyrosol and tyrosol secoiridoid derivatives (HtyrSec, TyrSec; $r = 0.91$ and 0.92, respectively) by NIR spectroscopy. Moreover, the authors pointed out that, in contrast to the NIR data, the chemometric analysis of the MIR spectra gave no satisfactory validation models ($r = 0.43$, 0.54, and 0.66 for HtyrSec, TyrSec, and TPP). This was unexpected as the calibration algorithms for MIR actually gave an even better correlation than NIR ($r > 0.96$ for all the polar phenolics studied). On the contrary, the optimization of phenolic compounds extraction from EVOO was achieved applying a response surface methodology as proposed by Fratoddi et al. [106].

Also, very good MIR results for the quantification of virgin olive oil phenolic compounds were reported by Hirri et al. [107]. They obtained values for the correlation coefficient and the root mean square errors of prediction of 0.99 and 0.11, respectively. This study underlined also that the spectral region in the range 3050–600 cm^{-1} was useful for predicting the total polyphenol content. Bellincontro et al. [108] also focused on the total phenols. They developed a portable NIR-AOTF tool for the on-field and non-destructive measurement of specific and total phenols in olives for oil production. Models were developed for the main phenolic compounds (e.g., oleuropein, verbascoside, and 3,4-DHPEA-EDA) and total phenols by PLS. The results in terms of R^2 for the calibration, prediction, and cross validation ranged between 0.930 and 0.998, 0.874 and 0.942, and 0.837 and 0.992, respectively. A recent preliminary study of Trapani et al. [109] applied cost-saving NIR based on a discrete filter system for the rapid measurement of total phenolic content and oleuroperin of olive fruits, in addition to the moisture, oil, and sugar. Although PLS models built for the latter ones were satisfactory, the instrument did not prove itself suitable for obtaining predictive models for phenolic compound contents. This is probably because the necessary wavelengths in the section of absorbance from 1100 to 1400 nm were not covered. Further studies in this direction are necessary and requested for enabling the possible application and use.

Concerning tocopherols, Cayuela and García [110] recently reported the classification of olive oils based on α-tocopherol and total tocopherol content using NIR spectroscopy. Other studies utilised pigments such as carotenoids and chlorophyll derivatives in extra-virgin olive oil as parameters/index of authenticity and quality. Pigments in oil matrices can be identified and quantified by chromatographic techniques after a pretreatment of the samples such as extraction and/or saponification, or by spectroscopic techniques without pretreatment of the samples [111]. Another work by the same authors quantified the main pigments, i.e., β-carotene, lutein, pheophytin A, and pheophytin B, in several extra-virgin olive oils produced from a blend of three cultivars (Moraiolo, Frantoio, and Leccino) typical of Tuscany (Italy) in three different years. For this purpose, they employed a method based on the mathematical analysis of the near-ultraviolet-visible absorption spectra of the oils [112]. Besides a good prediction of fatty acid ethyl ester and wax, Uncu et al. [113] quantified lutein, pheophytin A, and their derivatives and total xanthophylls by FTIR + UV/vis with a range of R^2_{cv} of 0.71–0.85, R^2_{pred} of 0.70–0.84, and RPD = 1.5–2.5 values but the prediction of the rest of the pigments were poor ($R^2_{cv} = 0.60$–0.76, R^2_{pred}: 0.42–0.62, and RPD = 1.2–1.5). Along the same line, Borello and Domenici [114] compared two near-UV/vis spectroscopic techniques for the determination of pigments, i.e., total carotenoids, and total chlorophyll derivatives in virgin and extra-virgin olive oils. Although UV/vis spectroscopy is very tempting for analysing the pigments, NIR and MIR spectroscopy typically gives a more specific molecular fingerprint of oil. Cayuela and García [115] developed rapid NIR techniques using PLS analysis for measuring the squalene content of olive oil. The external validation exercise for estimating squalene was performed with the squared validation regression coefficient (r^2) 0.83 and residual predictive deviation of 2.31.

3.3. Antioxidants. The assessment of interactions between natural active compounds and other food matrix components represent the bioactivity of a food. The first indicator of potential health benefits is commonly given by the antioxidant properties. The comprehension of concerted and synergistic actions, of antagonist interactions, or of no effect of biologically active compounds should be considered in this context. The same is true for the factors influencing the overall effects such as the peculiar combination of antioxidants and the structure of the food matrix [77]. Lu and Rasco [116] reviewed and summarized the recent applications of infrared spectroscopy for the development of models for the prediction of antioxidant properties in food, in addition to qualitative and quantitative analysis of antioxidant compounds. It is also worthy of mentioning the review article of Cozzolino [117], who concluded that infrared spectroscopy is a versatile analytical tool for the quantitative determination of antioxidant properties in agricultural products, foods, and plants.

Cerretani et al. [118] reported the rapid FTIR determination of phenolics and antioxidant activity of olive oil, in addition to water. For total phenols, good results were obtained considering the spectrum from 3610 to 816 cm^{-1}, while the range from 3707 to 1105 cm^{-1} was most suitable for ABTS+. Satisfactory LOD values were achieved by the FTIR-chemometry approach: 12.5 (mg gallic acid/kg oil) for total phenols and 0.76 (mmol Trolox/kg oil) for ABTS+. The authors concluded that the results obtained with the chemometric data analysis are comparable to those obtained using the official reference method. This highlights the potential of the spectroscopic technique as an alternative to the standard procedure for routine analysis or control at-line of production processes.

4. Conclusion

NIR and MIR spectroscopy coupled with chemometric data analysis has shown considerable potential for the determination of functional compounds present in olive oil. Recent applications have demonstrated to perform the assay of antioxidant activity (tocopherols, phenols, and squalene), to allow the discrimination of geographical origins, to facilitate the assessment of sensory attributes, and to determine the fatty acid profile, which is characteristic of the oil matrix. IR spectroscopic methods represent an interesting alternative to the conventional analytical methods as they do not require a pretreatment of the samples. Moreover, they offer a high sample throughput and significant time saving. In addition, these techniques are considerably more environmentally friendly since no solvents or carrier gases are used. These new approaches based on mid- and near-IR absorption spectroscopy are therefore interesting alternatives, for example, for rapid screening purposes.

In the future, IR spectroscopic methodology in combination with chemometric data evaluation will possibly overtake the conventional methodologies. In some cases, this may require the change of the regulatory frameworks, which are currently naming only conventional analytical tools for certain official purposes. However, the convincing performance of the spectroscopic approaches with respect to authentication and comprehensive quantitative analysis will certainly make a strong case in the regulatory and political processes. Without a doubt, the methods reviewed in this paper will soon be extensively used in the oil processing industries as a routine method to evaluate the characteristic parameters of olive oil.

Conflicts of Interest

The authors declare that they have no conflicts of interest.

References

[1] M. Casale and R. Simonetti, "Review: near infrared spectroscopy for analysing olive oils," *Journal of Near Infrared Spectroscopy*, vol. 22, no. 2, pp. 59–80, 2014.

[2] D. Boskou, "Olive fruits, table olives, and olive oil bioactive constituents," in *Olive and Olive Oil Bioactive Constituents*, D. Boskou, Ed., pp. 1-30, AOCS Press, Urbana, IL, USA, 1st edition, 2015.

[3] F. P. Zito, B. Polese, L. Vozzella et al., "Good adherence to mediterranean diet can prevent gastrointestinal symptoms: a survey from Southern Italy," *World Journal of Gastrointestinal Pharmacology and Therapeutics*, vol. 7, no. 4, pp. 564–571, 2016.

[4] C. M. Lăcătuşu, E. D. Grigorescu, M. Floria, A. Onofriescu, and B. M. Mihai, "The mediterranean diet: from an environment-driven food culture to an emerging medical prescription," *International Journal of Environmental Research and Public Health*, vol. 16, no. 6, p. 942, 2019.

[5] L. Conte, A. Bendini, E. Valli et al., "Olive oil quality and authenticity: a review of current EU legislation, standards, relevant methods of analyses, their drawbacks and recommendations for the future," *Trends in Food Science & Technology*, 2019, In press.

[6] A. Srobarova, J. A. Teixeira da Silva, G. Kogan, A. Ritieni, and A. Santini, "Beauvericin decreases cell viability of wheat," *Chemistry and Biodiversity*, vol. 6, pp. 1208–1215, 2009.

[7] P. Mikušová, A. Šrobárová, M. Sulyok, and A. Santini, "Fusarium fungi and associated metabolites presence on grapes from Slovakia," *Mycotoxin Research*, vol. 29, no. 2, pp. 97–102, 2013.

[8] R. Bhat and K. R. N. Reddy, "Challenges and issues concerning mycotoxins contamination in oil seeds and their edible oils: updates from last decade," *Food Chemistry*, vol. 215, pp. 425–437, 2017.

[9] K. Otto, A. Santini, and S. Oancea, "Recent aflatoxin survey data in milk and milk products: a review," *International Journal of Dairy Technology*, vol. 70, no. 3, pp. 320–331, 2017.

[10] P. Mikusová, A. Ritieni, A. Santini, G. Juhasová, and A. Srobárová, "Contamination by moulds of grape berries in Slovakia," *Food Additives & Contaminants Part A*, vol. 27, no. 5, pp. 738–747, 2010.

[11] A. Santini, G. Meca, S. Uhlig, and A. Ritieni, "Fusaproliferin, beauvericin and enniatins: occurrence in food—a review," *World Mycotoxin Journal*, vol. 5, no. 1, pp. 71–81, 2012.

[12] P. Mikušová, A. Santini, A. Ritieni, J. Pavlokin, and A. Šrobárová, "Berries contamination by microfungi in Slovakia vineyard regions: impact of climate conditions on microfungi biodiversity," *Revista Iberoamericana de Micologia*, vol. 29, no. 3, pp. 126–131, 2012.

[13] A. Cimmino, A. Andolfi, M. C. Zonno et al., "Chenopodolin: a phytotoxic unrearranged ent-pimaradiene diterpene produced by phoma chenopodicola, a fungal pathogen for *Chenopodium album* biocontrol," *Journal of Natural Products*, vol. 76, no. 7, pp. 1291–1297, 2013.

[14] A. Medina, S. Mohale, N. I. P. Samsudin, A. Rodriguez-Sixtos, A. Rodriguez, and N. Magan, "Biocontrol of mycotoxins: dynamics and mechanisms of action," *Current Opinion in Food Science*, vol. 17, pp. 41–48, 2017.

[15] M. F. Abdallah, M. Ameye, S. De Saeger, K. Audenaert, and G. Haesaert, "Biological control of mycotoxigenic fungi and their toxins: an update for the pre-harvest approach," in *Fungi and Mycotoxins—Their Occurrence, Impact on Health and the Economy as Well as Pre- and Postharvest Management Strategies*, P. Berka Njobeh, Ed., Intech Open, London, UK, 2018.

[16] S. Vichi, A. Romero, J. Tous, and J. Caixach, "The activity of healthy olive microbiota during virgin olive oil extraction influences oil chemical composition," *Journal of Agricultural and Food Chemistry*, vol. 59, no. 9, pp. 4705–4714, 2011.

[17] S. Cicerale, X. A. Conlan, A. J. Sinclair, and R. S. J. Keast, "Chemistry and health of olive oil phenolics," *Critical Reviews in Food Science and Nutrition*, vol. 49, no. 3, pp. 218–236, 2009.

[18] A. Salvo, A. Rotondo, G. L. La Torre, N. Cicero, and G. Dugo, "Determination of 1,2/1,3-diglycerides in Sicilian extra-virgin olive oils by 1H-NMR over a one-year storage period," *Natural Product Research*, vol. 31, no. 7, pp. 822–828, 2017.

[19] L. Schwingshackl, A. M. Lampousi, M. P. Portillo, D. Romaguera, G. Hoffmann, and H. Boeing, "Olive oil in the

prevention and management of type 2 diabetes mellitus: a systematic review and meta-analysis of cohort studies and intervention trials," *Nutrition and Diabetes*, vol. 7, no. 4, p. e262, 2017.

[20] E. Alves, M. R. M. Domingues, and P. Domingues, "Polar lipids from olives and olive oil: a review on their identification, significance and potential biotechnological applications," *Foods*, vol. 7, no. 7, p. 109, 2018.

[21] E. M. Yubero-Serrano, J. Lopez-Moreno, F. Gomez-Delgado, and J. Lopez-Miranda, "Extra virgin olive oil: more than a healthy fat," *European Journal of Clinical Nutrition*, 2018, In press.

[22] G. Serreli and M. Deiana, "Biological relevance of extra virgin olive oil polyphenols metabolites," *Antioxidants*, vol. 7, no. 12, p. 170, 2018.

[23] V. Francisco, C. Ruiz-Fernández, V. Lahera et al., "Natural molecules for healthy lifestyles: oleocanthal from extra virgin olive oil," *Journal of Agricultural and Food Chemistry*, vol. 67, no. 14, pp. 3845–3853, 2019.

[24] D. L. García-González, R. Aparicio-Ruiz, and R. Aparicio, "Virgin olive oil—chemical implications on quality and health," *European Journal of Lipid Science and Technology*, vol. 110, no. 7, pp. 602–607, 2008.

[25] M. Aparicio-Soto, M. Sánchez-Hidalgo, M. Á. Rosillo, M. L. Castejón, and C. Alarcón-de-la-Lastra, "Extra virgin olive oil: a key functional food for prevention of immune-inflammatory diseases," *Food & Function*, vol. 7, no. 11, pp. 4492–4505, 2016.

[26] S. Rigacci and M. Stefani, "Nutraceutical properties of olive oil polyphenols. An itinerary from cultured cells through animal models to humans," *International Journal of Molecular Science*, vol. 17, no. 6, p. 843, 2016.

[27] M. Piroddi, A. Albini, R. Fabiani et al., "Nutrigenomics of extra-virgin olive oil: a review," *Biofactors*, vol. 43, no. 1, pp. 17–41, 2017.

[28] B. Saibandith, J. P. E. Spencer, I. R. Rowland, and D. M. Commane, "Olive polyphenols and the metabolic syndrome," *Molecules*, vol. 22, no. 7, p. 1082, 2017.

[29] A. Alkhatib, C. Tsang, and J. Tuomilehto, "Olive oil nutraceuticals in the prevention and management of diabetes: from molecules to lifestyle," *International Journal of Molecular Science*, vol. 19, no. 7, p. 2024, 2018.

[30] A. M. Borzì, A. Biondi, F. Basile, S. Luca, E. S. D. Vicari, and M. Vacante, "Olive oil effects on colorectal cancer," *Nutrients*, vol. 11, no. 1, p. 32, 2018.

[31] M. Crespo, J. Tomé-Carneiro, A. Dávalos, and F. Visioli, "Pharma-nutritional properties of olive oil phenols. Transfer of new findings to human nutrition," *Foods*, vol. 7, no. 6, p. 90, 2018.

[32] A. Foscolou, E. Critselis, and D. Panagiotakos, "Olive oil consumption and human health: a narrative review," *Maturitas*, vol. 118, pp. 60–66, 2018.

[33] O. Garcia-Martinez, C. Ruiz, A. Gutierrez-Ibanez, R. Illescas-Montes, and L. Melguizo-Rodriguez, "Benefits of olive oil phenolic compounds in disease prevention," *Endocrine, Metabolic and Immune Disorders Drug Targets*, vol. 18, no. 4, pp. 333–340, 2018.

[34] A. Santini, E. Novellino, V. Armini, and A. Ritieni, "State of the art of ready-to-use therapeutic food: a tool for nutraceuticals addition to foodstuff," *Food Chemistry*, vol. 140, no. 4, pp. 843–849, 2013.

[35] R. Andrew and A. A. Izzo, "Principles of pharmacological research of nutraceuticals," *British Journal of Pharmacology*, vol. 174, no. 11, pp. 1177–1194, 2017.

[36] A. Santini and E. Novellino, "Nutraceuticals—shedding light on the grey area between pharmaceuticals and food," *Expert Review of Clinical Pharmacology*, vol. 11, no. 6, pp. 545–547, 2018.

[37] A. Santini, S. M. Cammarata, G. Capone et al., "Nutraceuticals: opening the debate for a regulatory framework," *British Journal of Clinical Pharmacology*, vol. 84, no. 4, pp. 659–672, 2018.

[38] A. Santini, G. C. Tenore, and E. Novellino, "Nutraceuticals: a paradigm of proactive medicine," *European Journal of Pharmaceutical Sciences*, vol. 96, pp. 53–61, 2017.

[39] P. Daliu, A. Santini, and E. Novellino, "From pharmaceuticals to nutraceuticals: bridging disease prevention and management," *Expert Review of Clinical Pharmacology*, vol. 12, no. 1, pp. 1–7, 2018.

[40] A. Durazzo, "Extractable and non-extractable polyphenols: an overview," in *Non-Extractable Polyphenols and Carotenoids: Importance in Human Nutrition and Health. Food Chemistry, Function and Analysis, No. 5*, F. Saura-Calixto and J. Pérez-Jiménez, Eds., p. 37, Royal Society of Chemistry, London, UK, 2018.

[41] A. Durazzo, L. D'Addezio, E. Camilli et al., "From plant compounds to botanicals and back: a current snapshot," *Molecules*, vol. 23, no. 8, p. 1844, 2018.

[42] A. Durazzo and M. Lucarini, "A Current shot and rethinking of antioxidant research strategy," *Brazilian Journal of Analytical Chemistry*, vol. 5, no. 20, pp. 9–11, 2018.

[43] F. Aouidi, N. Dupuy, J. Artaud et al., "Rapid quantitative determination of oleuropein in olive leaves (*Olea europaea*) using mid-infrared spectroscopy combined with chemometric analyses," *Industrial Crops and Products*, vol. 37, no. 1, pp. 292–297, 2012.

[44] M. Jansen and J. Birch, "Composition and stability of olive oil following partial crystallization," *Food Research International*, vol. 42, no. 7, pp. 826–831, 2009.

[45] E. N. Frankel, "Chemistry of extra virgin olive oil: adulteration, oxidative stability, and antioxidants," *Journal of Agricultural and Food Chemistry*, vol. 58, no. 10, pp. 5991–6006, 2010.

[46] E. N. Frankel, "Nutritional and biological properties of extra virgin olive oil," *Journal of Agricultural and Food Chemistry*, vol. 59, no. 3, pp. 785–792, 2009.

[47] M. Celano, V. Maggisano, S. M. Lepore, D. Russo, and S. Bulotta, "Secoiridoids of olive and derivatives as potential coadjuvant drugs in cancer: a critical analysis of experimental studies," *Pharmacological Research*, vol. 142, pp. 77–86, 2019.

[48] J. M. Moran, O. Leal-Hernandez, R. Roncero-Martin, and J. D. Pedrera-Zamorano, "Antitumor perspectives of oleuropein," *Journal of Food Science*, vol. 84, no. 3, p. 384, 2019.

[49] P. Przychodzen, R. Wyszkowska, M. Gorzynik-Debicka, T. Kostrzewa, A. Kuban-Jankowska, and M. Gorska-Ponikowska, "Anticancer potential of oleuropein, the polyphenol of olive oil, with 2-methoxyestradiol, separately or in combination, in human osteosarcoma cells," *Anticancer Research*, vol. 39, no. 3, pp. 1243–1251, 2019.

[50] M. El Riachy, F. Priego-Capote, L. León, L. Rallo, and M. D. Luque de Castro, "Hydrophilic antioxidants of virgin olive oil. Part 1: hydrophilic phenols: a key factor for virgin olive oil quality," *European Journal of Lipid Science and Technology*, vol. 113, no. 6, pp. 678–691, 2013.

[51] C. L. Huang and B. E. Sumpio, "Olive oil, the mediterranean diet, and cardiovascular health," *Journal of the American College of Surgeons*, vol. 207, no. 3, pp. 407–416, 2008.

[52] J. Chandran, N. Nayana, and P. Nisha, "Phenolics in vegetable oils," in *Compounds in Food: Characterization and Analysis Food Analysis & Properties, Chapter 21*, L. M. L. Nollet and J. A. Gutierrez-Uribe, Eds., p. 407, Taylor & Francis Group, Abingdon, UK, 2018.
[53] A. García, E. Rodríguez-Juan, G. Rodríguez-Gutiérrez, J. J. Rios, and J. Fernández-Bolaños, "Extraction of phenolic compounds from virgin olive oil by deep eutectic solvents (DESs)," *Food Chemistry*, vol. 197, pp. 554–561, 2016.
[54] B. Bayram, B. Ozcelik, G. Schultheiss, J. Frank, and G. Rimbach, "A validated method for the determination of selected phenolics in olive oil using high-performance liquid chromatography with coulometric electrochemical detection and a fused-core column," *Food Chemistry*, vol. 138, no. 2-3, pp. 1663–1669, 2013.
[55] G. Cioffi, M. S. Pesca, P. De Caprariis, A. Braca, L. Severino, and N. De Tommasi, "Phenolic compounds in olive oil and olive pomace from Cilento (Campania, Italy) and their antioxidant activity," *Food Chemistry*, vol. 121, no. 1, pp. 105–111, 2010.
[56] C. Conde, S. Delrot, and H. Gerós, "Physiological, biochemical and molecular changes occurring during olive development and ripening," *Journal of Plant Physiology*, vol. 165, no. 15, pp. 1545–1562, 2008.
[57] A. Gómez-Rico, M. D. Salvador, and G. Fregapane, "Virgin olive oil and olive fruit minor constituents as affected by irrigation management based on SWP and TDF as compared to ET_c in medium-density young olive orchards (*Olea europaea* L. cv. Cornicabra and Morisca)," *Food Research International*, vol. 42, no. 8, pp. 1067–1076, 2009.
[58] R. Amarowicz, R. Carle, G. Dongowski et al., "Influence of postharvest processing and storage on the content of phenolic acids and flavonoids in foods," *Molecular Nutrition and Food Research*, vol. 53, no. 2, pp. S151–S183, 2009.
[59] O. Köseoğlu, D. Sevim, and P. Kadiroğlu, "Quality characteristics and antioxidant properties of Turkish monovarietal olive oils regarding stages of olive ripening," *Food Chemistry*, vol. 212, pp. 628–634, 2016.
[60] V. Di Stefano and M. G. Melilli, "Effect of storage on quality parameters and phenolic content of Italian extra-virgin olive oils," *Natural Product Research*, vol. 21, pp. 1–9, 2019.
[61] J. Giacometti, Č. Milin, F. Giacometti, and Z. Ciganj, "Characterisation of monovarietal olive oils obtained from croatian cvs. drobnica and buza during the ripening period," *Foods*, vol. 7, no. 11, p. 188, 2018.
[62] R. Romano, A. Giordano, L. Le Grottaglie et al., "Volatile compounds in intermittent frying by gas chromatography and nuclear magnetic resonance," *European Journal of Lipid Science and Technology*, vol. 115, no. 7, pp. 764–773, 2013.
[63] D. Naviglio, R. Romano, F. Pizzolongo et al., "Rapid determination of esterified glycerol and glycerides in triglyceride fats and oils by means of periodate method after transesterification," *Food Chemistry*, vol. 102, no. 1, pp. 399–405, 2007.
[64] D. Boskou, M. Tsimidou, and G. Blekas, "Olive oil composition," in *Olive Oil: Chemistry and Technology*, D. Boskou, Ed., AOCS Press, Champaign, IL, USA, 2006.
[65] V. N. Kamvissis, E. G. Barbounis, N. C. Megoulas, and M. A. Koupparis, "A novel photometric method for evaluation of the oxidative stability of virgin olive oils," *Journal of AOAC International*, vol. 91, no. 4, pp. 794–801, 2008.
[66] R. Ambra, F. Natella, S. Lucchetti, V. Forte, and G. Pastore, "α-Tocopherol, β-carotene, lutein, squalene and secoiridoids in seven monocultivar Italian extra-virgin olive oils," *International Journal of Food Sciences and Nutrition*, vol. 68, no. 5, pp. 538–545, 2017.
[67] E. Psomiadou and M. Tsimidou, "Pigments in Greek virgin olive oils: occurrence and levels," *Journal of the Science of Food and Agriculture*, vol. 81, no. 7, pp. 640–647, 2001.
[68] P. Viola and M. Viola, "Virgin olive oil as a fundamental nutritional component and skin protector," *Clinics in Dermatology*, vol. 27, no. 2, pp. 159–165, 2009.
[69] B. Schrader, *Infrared and Raman Spectroscopy*, VCH Verlagsgesellschaft, Weinheim, Germany, 1995.
[70] P. R. Griffiths and J. A. De Haseth, *Fourier Transform Infrared Spectrometry*, Wiley, Weinheim, Germany, 2nd edition, 2007.
[71] H. Haken and H. C. Wolf, *Molecular Physics and Elements of Quantum Chemistry*, Springer, Berlin, Germany, 1995.
[72] J. Kiefer, N. H. Rasul, P. K. Ghosh, and E. Von Lieres, "Surface and bulk porosity mapping of polymer membranes using infrared spectroscopy," *Journal of Membrane Science*, vol. 452, pp. 152–156, 2014.
[73] L. A. Averett, P. R. Griffiths, and K. Nishikida, "Effective path length in attenuated total reflection spectroscopy," *Analytical Chemistry*, vol. 80, no. 8, pp. 3045–3049, 2008.
[74] N. J. Harrick and F. K. du Pré, "Effective thickness of bulk materials and of thin films for internal reflection spectroscopy," *Applied Optics*, vol. 5, no. 11, pp. 1739–1743, 1966.
[75] C. I. G. Tuberoso, I. Jerković, M. Maldini, and G. Serreli, "Phenolic compounds, antioxidant activity, and other characteristics of extra virgin olive oils from Italian autochthonous varieties Tonda di Villacidro, Tonda di Cagliari, Semidana, and Bosana," *Journal of Chemistry*, vol. 2016, Article ID 8462741, 7 pages, 2016.
[76] D. Trombetta, A. Smeriglio, D. Marcoccia et al., "Analytical evaluation and antioxidant properties of some secondary metabolites in northern Italian mono- and multi-varietal extra virgin olive oils (EVOOs) from early and late harvested olives," *International Journal of Molecular Sciences*, vol. 18, no. 4, p. 797, 2017.
[77] A. Durazzo, "Study approach of antioxidant properties in foods: update and considerations," *Foods*, vol. 6, p. E17, 2017.
[78] A. Durazzo, J. Kiefer, M. Lucarini et al., "An Innovative and Integrated Food Research Approach: spectroscopy applications to milk and a case study of a milk-based dish," *Brazilian Journal of Analytical Chemistry*, vol. 5, no. 18, pp. 12–27, 2018.
[79] A. Durazzo, J. Kiefer, M. Lucarini et al., "Qualitative analysis of traditional Italian dishes: FTIR approach," *Sustainability*, vol. 10, no. 11, p. 4112, 2018.
[80] M. de Luca, W. Terouzi, G. Ioele et al., "Derivative FTIR spectroscopy for cluster analysis and classification of Morocco olive oils," *Food Chemistry*, vol. 124, no. 3, pp. 1113–1118, 2011.
[81] A. P. la Mata, A. Dominguez-Vidal, J. M. Bosque-Sendra, A. Ruiz-Medina, L. Cuadros-Rodríguez, and M. J. Ayora-Cañada, "Olive oil assessment in edible oil blends by means of ATR-FTIR and chemometrics," *Food Control*, vol. 23, no. 2, pp. 449–455, 2012.
[82] A. Rohman and Y. B. Che Man, "Quantification and classification of corn and sunflower oils as adulterants in olive oil using chemometrics and FTIR Spectra," *Scientific World Journal*, vol. 2012, Article ID 250795, 6 pages, 2012.
[83] O. Jović, T. Smolić, I. Primožič, and T. Hrenar, "Spectroscopic and chemometric analysis of binary and ternary edible oil mixtures: qualitative and quantitative study," *Analytical Chemistry*, vol. 88, no. 8, pp. 4516–4524, 2016.

[84] M. De Luca, D. Restuccia, M. L. Clodoveo, F. Puoci, and G. Ragno, "Chemometric analysis for discrimination of extra virgin olive oils from whole and stoned olive pastes," *Food Chemistry*, vol. 202, pp. 432–437, 2016.
[85] A. M. Jiménez-Carvelo, M. T. Osorio, A. Koidis, A. González-Casado, and L. Cuadros-Rodríguez, "Chemometric classification and quantification of olive oil in blends with any edible vegetable oils using FTIR-ATR and Raman spectroscopy," *LWT*, vol. 86, pp. 174–184, 2017.
[86] K. Georgouli, J. Martinez Del Rincon, and A. Koidis, "Continuous statistical modelling for rapid detection of adulteration of extra virgin olive oil using mid infrared and Raman spectroscopic data," *Food Chemistry*, vol. 217, pp. 735–742, 2017.
[87] M. H. abadi Sherahi, F. Shahidi, F. Tabatabai Yazdi, and S. M. Bagher Hashemi, "Effect of *Lactobacillus plantarum* on olive and olive oil quality during fermentation process," *LWT—Food Science and Technology*, vol. 82, pp. 572–580, 2018.
[88] I. Durán Merás, J. Domínguez Manzano, D. Airado Rodríguez, and A. Muñoz de la Peña, "Detection and quantification of extra virgin olive oil adulteration by means of autofluorescence excitation-emission profiles combined with multi-way classification," *Talanta*, vol. 178, pp. 751–762, 2018.
[89] L. Hadhoum, G. Burnens, K. Loubar, M. Balistrou, and M. Tazerout, "Bio-oil recovery from olive mill wastewater in sub-/supercritical alcohol-water system," *Fuel*, vol. 252, pp. 360–370, 2019.
[90] O. Uncu and B. Ozen, "A comparative study of mid-infrared, UV-Visible and fluorescence spectroscopy in combination with chemometrics for the detection of adulteration of fresh olive oils with old olive oils," *Food Control*, vol. 105, pp. 209–218, 2019.
[91] R. J. Mailer, "Rapid evaluation of olive oil quality by NIR reflectance spectroscopy," *Journal of the American Oil Chemists' Society*, vol. 81, no. 9, pp. 823–827, 2004.
[92] O. Galtier, N. Dupuy, Y. Le Dréau et al., "Geographic origins and compositions of virgin olive oils determinated by chemometric analysis of NIR spectra," *Analytica Chimica Acta*, vol. 595, no. 1-2, pp. 136–144, 2007.
[93] R. M. Maggio, T. S. Kaufman, M. D. Carlo et al., "Monitoring of fatty acid composition in virgin olive oil by Fourier transformed infrared spectroscopy coupled with partial least squares," *Food Chemistry*, vol. 114, no. 4, pp. 1549–1554, 2009.
[94] G. Gurdeniz, B. Ozen, and F. Tokatli, "Comparison of fatty acid profiles and mid-infrared spectral data for classification of olive oils," *European Journal of Lipid Science and Technology*, vol. 112, no. 2, pp. 218–226, 2010.
[95] N. Dupuy, O. Galtier, D. Ollivier, P. Vanloot, and J. Artaud, "Comparison between NIR, MIR, concatenated NIR and MIR analysis and hierarchical PLS model. Application to virgin olive oil analysis," *Analytica Chimica Acta*, vol. 666, no. 1-2, pp. 23–31, 2010.
[96] M. J. Lerma-García, E. F. Simó-Alfonso, A. Bendini, and L. Cerretani, "Rapid evaluation of oxidised fatty acid concentration in virgin olive oil using Fourier-transform infrared spectroscopy and multiple linear regression," *Food Chemistry*, vol. 124, no. 2, pp. 679–684, 2011.
[97] O. Uncu and B. Ozen, "Prediction of various chemical parameters of olive oils with Fourier transform infrared spectroscopy," *LWT—Food Science and Technology*, vol. 63, no. 2, pp. 978–984, 2015.
[98] M. Casale, P. Oliveri, C. Casolino et al., "Characterisation of PDO olive oil Chianti Classico by non-selective (UV-visible, NIR and MIR spectroscopy) and selective (fatty acid composition) analytical techniques," *Analytica Chimica Acta*, vol. 712, pp. 56–63, 2012.
[99] A. M. Inarejos-García, S. Gómez-Alonso, G. Fregapane, and M. D. Salvador, "Evaluation of minor components, sensory characteristics and quality of virgin olive oil by near infrared (NIR) spectroscopy," *Food Research International*, vol. 50, no. 1, pp. 250–258, 2013.
[100] E. Valli, A. Bendini, R. M. Maggio et al., "Detection of low-quality extra virgin olive oils by fatty acid alkyl esters evaluation: a preliminary and fast mid-infrared spectroscopy discrimination by a chemometric approach," *International Journal of Food Science & Technology*, vol. 48, no. 3, pp. 548–555, 2013.
[101] J. A. Cayuela, "Rapid NIR determination of alkyl esters in virgin olive oil," *Grasas Y Aceites*, vol. 68, no. 2, p. 195, 2017.
[102] V. Lagouri, V. Manti, and T. Gimisis, "Optical non-destructive UV-Vis-NIR-MIR spectroscopic tools and chemometrics in the monitoring of olive oil functional compounds: bioactivity, chemistry and processing," in *Book: Olives and Olive Oil as Functional Foods: Bioactivity, Chemistry and Processing. Chapter 11*, A. Kiritsakis and S. Fereidoon, Eds., , Shahidi Wiley and Sons Ltd., Hoboken, NY, USA, 1st edition, 2017.
[103] N. Nenadis and M. Z. Tsimidou, "Perspective of vibrational spectroscopy analytical methods in on-field/official control of olives and virgin olive oil," *European Journal of Lipid Science and Technology*, vol. 119, no. 1, article 1600148, 2017.
[104] I. Gouvinhas, J. M. M. M. de Almeida, T. Carvalho, N. Machado, and A. I. R. N. A. Barros, "Discrimination and characterisation of extra virgin olive oils from three cultivars in different maturation stages using Fourier transform infrared spectroscopy in tandem with chemometrics," *Food Chemistry*, vol. 174, pp. 226–232, 2015.
[105] M. E. Mora-Ruiz, P. Reboredo-Rodríguez, M. Desamparados Salvador et al., "Assessment of polar phenolic compounds of virgin olive oil by NIR and mid-IR spectroscopy and their impact on quality," *European Journal of Lipid Science and Technology*, vol. 119, no. 1, article 160009, 2017.
[106] I. Fratoddi, M. Rapa, G. Testa, I. Venditti, F. A. Scaramuzzo, and G. Vinci, "Response surface methodology for the optimization of phenolic compounds extraction from extra virgin olive oil with functionalized gold nanoparticles," *Microchemical Journal*, vol. 138, pp. 430–437, 2018.
[107] A. Hirri, M. Bassbasi, S. Souhassou, F. Kzaiber, and A. Oussama, "Prediction of polyphenol fraction in virgin olive oil using mid-infrared attenuated total reflectance attenuated total reflectance accessory-mid-infrared coupled with partial least squares regression," *International Journal of Food Properties*, vol. 19, no. 7, pp. 1504–1512, 2016.
[108] A. Bellincontro, A. Taticchi, M. Servili, S. Esposto, D. Farinelli, and F. Mencarelli, "Feasible application of a portable NIR-AOTF tool for on-field prediction of phenolic compounds during the ripening of olives for oil production," *Journal of Agricultural and Food Chemistry*, vol. 60, no. 10, pp. 2665–2673, 2012.
[109] S. Trapani, M. Migliorini, L. Cecchi et al., "Feasibility of filter-based NIR spectroscopy for the routine measurement of olive oil fruit ripening indices," *European Journal of Lipid Science and Technology*, vol. 119, no. 6, article 1600239, 2017.
[110] J. A. Cayuela and J. F. García, "Sorting olive oil based on alpha-tocopherol and total tocopherol content using

near-infra-red spectroscopy (NIRS) analysis," *Journal of Food Engineering*, vol. 202, pp. 79–88, 2017.
[111] C. Lazzerini, M. Cifelli, and V. Domenici, "Pigments in extra-virgin olive oil: authenticity and quality," in *Products from Olive Tree, Chapter 6*, D. Boskou and M. L. Clodoveo, Eds., pp. 99–114, InTech, Rijeka, Croatia, 1st edition, 2016.
[112] C. Lazzerini and V. Domenici, "Pigments in extra-virgin olive oils produced in Tuscany (Italy) in different years," *Foods*, vol. 6, no. 4, p. 25, 2017.
[113] O. Uncu, B. Ozen, and F. Tokatli, "Use of FTIR and UV-visible spectroscopy in determination of chemical characteristics of olive oils," *Talanta*, vol. 201, pp. 65–73, 2019.
[114] E. Borello and V. Domenici, "Determination of pigments in virgin and extra-virgin olive oils: a comparison between two near UV-vis spectroscopic techniques," *Foods*, vol. 8, no. 1, p. 18, 2019.
[115] J. A. Cayuela and J. F. García, "Nondestructive measurement of squalene in olive oil by near infrared spectroscopy," *LWT*, vol. 88, pp. 103–108, 2018.
[116] X. Lu and B. A. Rasco, "Determination of antioxidant content and antioxidant activity in foods using infrared spectroscopy and chemometrics: a review," *Critical Review of Food Science and Nutrition*, vol. 52, no. 10, pp. 853–875, 2011.
[117] D. Cozzolino, "Infrared spectroscopy as a versatile analytical tool for the quantitative determination of antioxidants in agricultural products, foods and plants," *Antioxidants*, vol. 4, no. 3, pp. 482–497, 2015.
[118] L. Cerretani, A. Giuliani, R. M. Maggio, A. Bendini, T. Gallina Toschi, and A. Cichelli, "Rapid FTIR determination of water, phenolics and antioxidant activity of olive oil," *European Journal of Lipid Science and Technology*, vol. 112, no. 10, pp. 1150–1157, 2010.

Apple Variety Identification Using Near-Infrared Spectroscopy

Caihong Li,[1] Lingling Li,[1] Yuan Wu,[1] Min Lu,[1] Yi Yang,[1,2] and Lian Li[1]

[1]*School of Information Science and Engineering, Lanzhou University, Lanzhou 730000, China*
[2]*Silk Road Economic Belt Research Center, Lanzhou University, Lanzhou 730000, China*

Correspondence should be addressed to Yuan Wu; hjf6318567@gmail.com

Academic Editor: Wee Chew

Near-infrared (NIR) spectra of apple samples were submitted in this paper to principal component analysis (PCA) and successive projections algorithm (SPA) to conduct variable selection. Three pattern recognition methods, backpropagation neural network (BPNN), support vector machine (SVM), and extreme learning machine (ELM), were applied to establish models for distinguishing apples of different varieties and geographical origins. Experimental results show that ELM models performed better on identifying apple variety and geographical origin than others. Especially, the SPA-ELM model could reach 98.33% identification accuracy on the calibration set and 96.67% on the prediction set. This study suggests that it is feasible to identify apple variety and cultivation region by using NIR spectroscopy.

1. Introduction

China is one of the main fruit-producing and fruit-consuming countries in the world. In recent years, both the area for fruit cultivation and the yield of fruits were increased continuously. Since apple has advantages of high nutrient value, high storage, and short supply cycle, it has become one of the four major fruits in the world. In 2014, the apple cultivation area and production in China reached 22724 km^2 and 40.93 million tons, respectively, according to the Food and Agriculture Organization (FAO) report. The overall quality of apple could be determined by external attributes (such as size, colour, and texture) and internal attributes (such as soluble solid content (SSC), total acid content (TAC), and vitamins). These attributes are greatly affected by the variety and cultivation region of apple. Different apple has specific firmness, crispness, juiciness, and taste. Apple fruits that are firm, crispy, juicy, and tasteful are more popular among consumers [1]. Moreover, apples of different varieties or geographical origins could be easily mixed during harvesting and marketing. Therefore, effective and reliable technologies that could identify apple variety and geographical origin are demanded urgently by sellers and consumers.

To determine the variety of fruits, many methodologies have been explored, such as deoxyribonucleic acid (DNA) analysis [2, 3], gas chromatography (GC) analysis [4], and amino acid composition [5]. However, these methods are always containing a considerable amount of time, manual work, and sample preparation [6].

NIR spectroscopy, as a rapid nondestructive detection method, has been proven effective in determining the internal quality attributes of various agricultural products and food [7–11]. Compared with conventional chemical and physical analytical technologies, NIR spectroscopy has the advantages of easy operation, fast detecting speed, and nondestructive measurement. NIR spectral data have information that are concerning the relative proportions of C-H, N-H, and O-H bands which are the main structural constituents of organic molecules [12, 13]. Many studies explored the application of NIR spectroscopy on SSC measurement of fruits, such as apple [14, 15], kiwifruit [16], and grape [17].

The main object of this study is to explore the feasibility of applying NIR spectroscopy for distinguishing apple variety and geographical origin. NIR diffuse reflectance spectra of 6 different apples (Fuji apple from Shandong, Fuji apple from Shaanxi, Red Star apple from Shandong, Red Star apple from Shaanxi, Gala apple from Shandong, and Gala apple from Shaanxi) were collected. Two variable selection algorithms were used, and three modelling methods were applied to establish models for distinguishing apples of different species and geographical origins. The capabilities of different models to identify apple variety and cultivation region were also investigated.

2. Materials and Methods

2.1. Sample Preparation. 100 Fuji apples (50 from Shandong and 50 from Shaanxi), 100 Red Star apples (50 from Shandong and 50 from Shaanxi), and 100 Gala apples (50 from Shandong and 50 from Shaanxi) were purchased from two main local markets (Table 1). The retailers guaranteed the apples' varieties and cultivation regions, and the skins of these samples were smooth and perfect. Before measurements, all samples were placed in airtight polyethylene bags and stored in a refrigerator to keep at cold temperature (4 ± 1°C) for 2 days. After storage, the apples were taken away from refrigerator, washed with clean water, wiped dry, and kept at room temperature (24 ± 2°C) for about 3 hours, and then they were used for measurements.

In this study, 40 Fuji apples (20 from Shandong and 20 from Shaanxi), 40 Red Star apples (20 from Shandong and 20 from Shaanxi), and 40 Gala apples (20 from Shandong and 20 from Shaanxi) were selected randomly as the prediction set, and the remaining 180 apples were used as the calibration set. The calibration set was used to establish models, and the prediction set was used to evaluate the performances of the established models.

Table 1: Source of apple samples.

Variety	Origin	Total fruit samples	Calibration set	Prediction set
Fuji	Shandong	50	30	20
Fuji	Shaanxi	50	30	20
Red Star	Shandong	50	30	20
Red Star	Shaanxi	50	30	20
Gala	Shandong	50	30	20
Gala	Shaanxi	50	30	20
Total		300	180	120

2.2. Spectra Acquisition. The NIR spectra were collected in diffuse reflectance mode using Ocean Optics USB2000-VIS-NIR-ES spectrometer (Ocean optics, USA), equipped with HL-2000 tungsten halogen light sources (Ocean optics, USA), and optical fibre reflection Probes (QR600-7-VIS-NIR, Ocean optics, USA). The NIR diffuse reflectance spectra were collected between 400 and 1021 nm with an internal of about 0.33 nm, which resulted in 1888 variables for each spectrum. Software Ocean View (Ocean optics, USA) was used to collect and transform the spectra. All measurements were undertaken at room temperature (24 ± 2°C).

Before measuring the spectra, the spectrometer was turned on for at least 1 hour for warming up. When measuring the spectra, the NIR optical fibre probe was kept close to the surface of apple in order to avoid surface reflectance and air interference. For each intact apple sample, the diffuse reflectance spectra was obtained at 15 different points that were randomly chosen along the equator, and each location was scanned 10 times for spectra measurements. Thus, a total of 150 scans were averaged to represent the spectral data for one apple and then used in further data analysis.

2.3. Principal Component Analysis. PCA is a very effective data mining technique and has been widely used for dealing with spectra data [18]. The principle of PCA is to reduce data dimension and orthogonalize the original multidimensional data to obtain a set of values of linearly uncorrelated variables to minimize the chance of overfitting and improve the speed of training procedure. This set of values is called principal components (PCs). In the case of small size of data set, PCs are computed simultaneously from a single matrix decomposition. The first PC accounts for as much of the variability in the original data as possible, and each subsequent component has lower variance possible than the preceding one. PCA could transform high-dimensional data into lower dimensional data without losing original information [19]. In most cases, PCs have much smaller variable numbers, yet still provide information that could describe most of variance of the original data.

2.4. Successive Projections Algorithm. SPA is a forward variable selection method, and its purpose is to select wavelengths whose information is minimally redundant [20]. In the process of SPA, the first variable of spectral data is used as the initial variable x_0 and then incorporates a new one at each iteration, until the preset number N is reached. At the next iteration, the second variable of spectral data is selected as the initial variable x_1, until all variables are used as the initial one. SPA has been successfully applied to select variables in spectra [21, 22].

For a classification problem, the number of extracted wavelengths could be determined by G value (validation cost) [23]:

$$G = \frac{1}{L}\sum_{i=1}^{L} g_i, \tag{1}$$

where L is the number of the validation samples, and g_i is defined as follows:

$$g_i = \frac{r^2(x_i, \mu_{Ii})}{\min_{Ij \neq Ii} r^2(x_i, \mu_{Ii})}, \tag{2}$$

where the numerator $r^2(x_i, \mu_{Ii})$ is the squared Mahalanobis distance between x_i (a vector of the ith sample at the selected wavelengths by SPA) and μ_{Ii} (the mean of the Ii-th variety apples at each wavelength). This numerator represents the squared Mahalanobis distance between x_i and the closest wrong variety. g_i should be as small as possible, which means x_i is close to its true variety and distant from other varieties.

2.5. Backpropagation Neural Network. The BPNN model is considered as the most widely used type of artificial neural network (ANN) model [24]. BPNN has the properties of forward propagation of the signals and backpropagation of the errors. The general structure of BPNN has three layers: an input layer, a hidden layer, and an output layer (Figure 1). The weights and thresholds in the architecture could be adjusted automatically according to the backpropagation of errors.

2.6. Support Vector Machine. The SVM, which is proposed by Vapnik and his colleagues, is a statistical method based on Vapnik-Chervonenkis (VC) dimension theory and the structural risk minimization (SRM) principle [25–27]. The target of SVM is to obtain the best compromise between complexity and learning ability of the model based on limited sample information. SVM has been applied in a considerable amount of studies in NIR spectroscopy as a discrimination method [28–30]. In this experiment, the radical basis function (RBF) was introduced as the kernel function for SVM models because RBF could reduce the computational complexity of training process. Cost parameter, which determines the trade-off between minimizing the training error and minimizing model complexity, and gamma parameter, which implicitly defines the nonlinear mapping from input space to certain higher dimensional space, are two main parameters in RBF kernel function. In the establishing processing of SVM models, the most important step is optimizing these two parameters [28].

2.7. Extreme Learning Machine. ELM, proposed by Huang [31], is a single-hidden layer feed forward neural network (SLFNs). ELM has advantages of learning fast with higher generalization performance and overcoming difficulties that are common in traditional gradient-based learning algorithms. The general structure of ELM is composed of three layers: an input layer, a hidden layer, and an output layer. It is important to select the number of neurons in the hidden layer for establishing a reliable ELM-based model because the number of neurons in the hidden layer influences the robustness and performance of the ELM model greatly. ELM randomly chooses and tunes the weights between the input layer and the hidden layer using continuous probability density function and sets the weights between the hidden layer and the output layer analytically.

2.8. Data Processing. In this study, all calculations were performed in Matlab R2014b (The Mathworks Inc., USA) under Windows 7 with 3.6 GHz CPU and 4 GB memory. Before the multivariate analysis, the raw spectral data (R) were converted to absorbance values ($\log(1/R)$). To reduce the noise, Savitzky–Golay smoothing method was used, and the segment size was set to be 5. Then multivariate scatter correction (MSC), which modifies the additive and multiplicative effects [32], was applied on the denoised data. PCA and SPA were implemented to extract essential information from the whole spectral regions respectively, and then fewer variables were applied as the input for constructing models. Three different algorithms were used to establish models to identify apple varieties and geographical origins. These three methods were BPNN, SVM, and ELM. Figure 2 shows the flowchart of this procedure.

In this experiment, the performance of the models was evaluated through identification accuracy. The identification accuracy was defined as

$$\text{accuracy} = \frac{\text{number of (TPs + TNs)}}{\text{number of (TPs + TNs + FPs + FNs)}} \times 100\%, \tag{3}$$

where TP indicates true positive, which represents a positive sample was classified as a positive example; TN indicates true negative, which represents a negative sample was classified as a negative example; FP indicates false positive, which represents a negative sample was classified as a positive example; and FN indicates false negative, which represents a positive sample was classified as a negative example. The higher the classification accuracy reached, the better performance the model obtained.

3. Results and Discussion

3.1. NIR Spectra of Apple Samples. Figure 3 shows the raw diffuse reflectance spectra of the experimented apple samples between 400 and 1021 nm. There are many crossovers and overlappings among these samples. The overall trends of the 300 apple spectra were similar. The absorbance spectra of all apple samples were similar, and there were no significant spectral differences among spectra of different varieties and geographical origins. Therefore, it was difficult to discriminate the variety and geographical origin directly based on the raw diffuse reflectance spectra. It is necessary to establish some models for identifying the apple variety and geographical origin.

3.2. Characteristic Wavelength Selection. A large amount of factors in the input vector may involve some irrelevant information and deteriorate the performance of models. In this study, we need to apply variable selection algorithms to the spectral data that had been preprocessed by Savitzky–Golay smoothing method and MSC. For PCA, Table 2 lists the contribution rates and accumulative contribution rates of the first 8 PCs. It showed that the first PC provided the most contribution (87.47%) and the second PC contributed to 7.14%. The first two PCs could represent more than 94% of the original spectral data. In this experiment, since more variables involving in the process of constructing models may slow down the calculation speed, the first 6 PCs, whose accumulative contribution rate reached 99.27%, were selected as the input to develop models based on the criterion for an increment of explained variance lower than 0.25% [33].

For SPA, Figure 4 presents the changing G value with the number of wavelengths. The overall trend of G value decreased with the increased number of wavelengths. The best

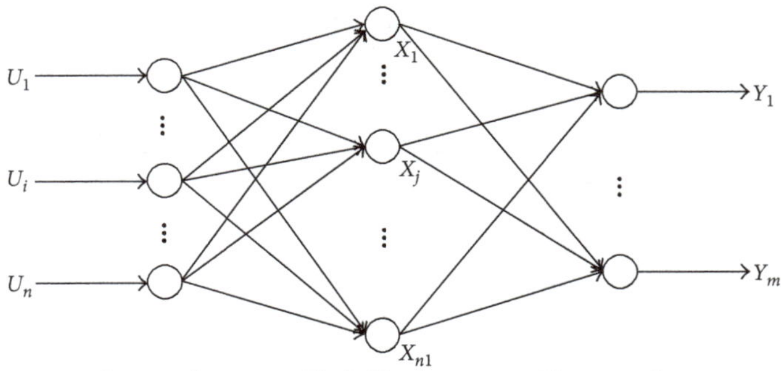

FIGURE 1: The general structure of BPNN model.

number of wavelengths could be achieved when the decrease rate of G value was lower than 5%, while the decrease was less than 0.01 [34]. In this experiment, the best number of wavelengths was 10. The selected wavelengths were 530.36, 547.81, 564.45, 587.67, 606.53, 624.58, 652.45, 665.77, 677.99, and 996.84 nm, respectively.

3.3. Classification Models. For the BPNN models, the optimal number of neurons in the hidden layer was calculated by using leave-one-out cross validation in the range from 10 to 50 (large number of neurons in the hidden layer had no performance improvement in the experiment). Table 3 lists the optimal number of the neurons in the hidden layer of BPNN models. The results for apple variety and geographical origin identification by BPNN models are given in Table 4. Obviously, the average performance was poor. Only identification accuracy of SPA-BPNN model on the calibration set was more than 50%.

For the SVM models, the optimal values of cost and gamma parameters were selected through the leave-one-out cross validation method. Table 5 lists the optimal values of cost and gamma parameters of SVM models. Table 6 gives the identification accuracy for apple variety and geographical origin by using SVM models. It can be seen that PCA-SVM model outperformed SPA-SVM model with average accuracy 98.89% on the calibration set and 85.83% on the prediction set.

When the variables extracted by PCA and SPA were used as the inputs for ELM models, the best numbers of the neurons in the hidden layer are listed in Table 7. The identification results of ELM models are shown in Table 8. It showed that all ELM models could efficiently distinguish apple variety and geographical origin for both the calibration set and the prediction set with accuracy more than 90%. In particular, the identification accuracy of SPA-ELM model could reach more than 95% for both the calibration set and the prediction set.

When comparing all the established models, it could be found that ELM models had better performance than BPNN models and SVM models. It suggested that suitable intelligent algorithm could improve prediction performance. PCA-SVM model could identify Red Star apple from

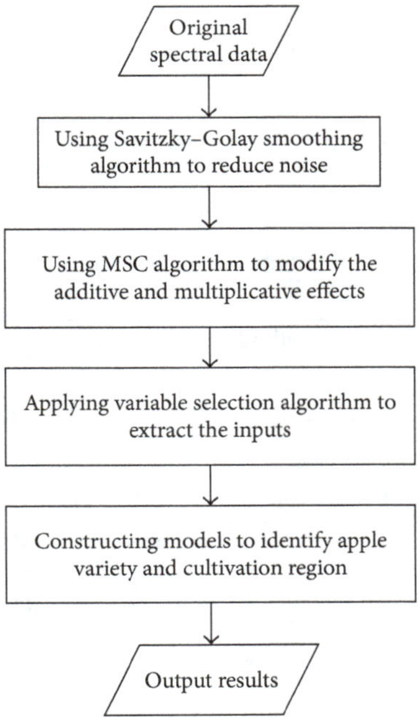

FIGURE 2: The flowchart of data processing.

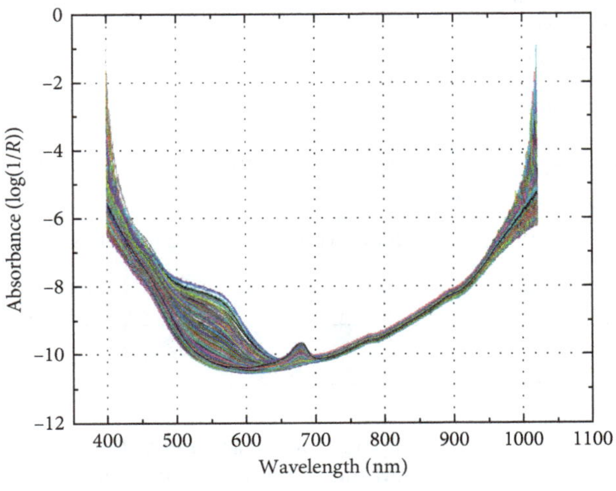

FIGURE 3: NIR spectra of all measured apples between 400 and 1021 nm.

TABLE 2: The contribution and accumulative contribution rates of the first 8 PCs of the spectra.

PCs	PC1	PC2	PC3	PC4	PC5	PC6	PC7	PC8
Contribution rate (%)	87.47	7.14	2.97	0.87	0.52	0.30	0.14	0.08
Accumulative contribution rate (%)	87.47	94.61	97.58	98.45	98.97	99.27	99.41	99.49

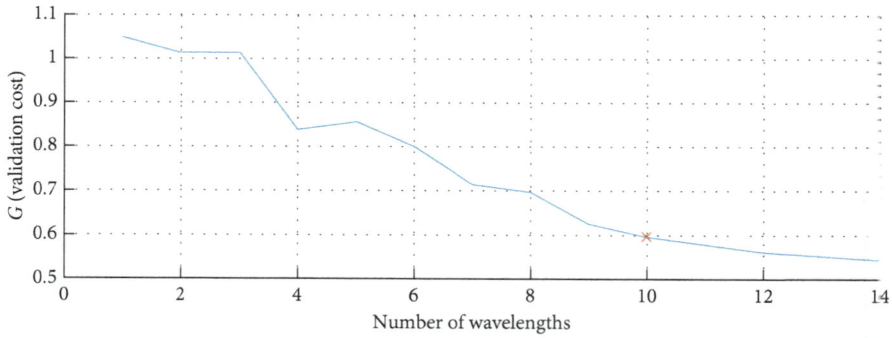

FIGURE 4: The G values under different numbers of wavelengths. The red × means the selected number of wavelengths.

TABLE 3: The optimal number of neurons in the hidden layer of BPNN models.

Dimension reduction methods	The number of neurons
PCA	17
SPA	21

TABLE 4: Classification accuracies of BPNN models.

Set	Calibration set		Prediction set	
	PCA (%)	SPA (%)	PCA (%)	SPA (%)
Fuji apple from Shandong	80	76.67	45	40
Fuji apple from Shaanxi	0	0	0	0
Red Star apple from Shandong	43.33	53.33	55	20
Red Star apple from Shaanxi	86.67	70	30	30
Gala apple from Shandong	10	43.33	5	65
Gala apple from Shaanxi	60	66.67	45	75
Average	46.67	51.67	30	38.33

TABLE 5: The optimal values of cost and gamma parameters of SVM models.

Dimension reduction methods	Cost value	Gamma value
PCA	5.66	2
SPA	36.76	0.76

Shandong and Gala apple from Shandong with 100% accuracy, while SPA-ELM model has the ability of identifying Fuji apple from Shandong, Fuji apple from Shaanxi, and Red Star apple from Shandong with 100% accuracy.

TABLE 6: Classification accuracies of SVM models.

Set	Calibration set		Prediction set	
	PCA (%)	SPA (%)	PCA (%)	SPA (%)
Fuji apple from Shandong	100	93.33	75	45
Fuji apple from Shaanxi	96.67	93.33	55	80
Red Star apple from Shandong	100	90	100	65
Red Star apple from Shaanxi	96.67	93.33	95	90
Gala apple from Shandong	100	93.33	100	90
Gala apple from Shaanxi	100	73.33	90	65
Average	98.89	89.44	85.83	72.50

TABLE 7: The best number of the neurons in the hidden layer of ELM models.

Dimension reduction methods	Number
PCA	70
SPA	20

4. Conclusion

In this study, the experiment results indicated that it was possible to develop a nondestructive technique to discriminate apple variety and cultivated region using NIR spectroscopy. PCA and SPA were applied to conduct variable selection of the spectra. BPNN, SVM, and ELM were used to establish models for distinguishing apple variety and geographical origin by using NIR diffuse

TABLE 8: Classification accuracies of ELM models.

Set	Calibration set		Prediction set	
	PCA (%)	SPA (%)	PCA (%)	SPA (%)
Fuji apple from Shandong	100	100	95	100
Fuji apple from Shaanxi	96.67	100	95	100
Red Star apple from Shandong	100	100	80	100
Red Star apple from Shaanxi	100	100	95	90
Gala apple from Shandong	100	96.67	95	100
Gala apple from Shaanxi	100	93.33	95	90
Average	99.44	98.33	92.50	96.67

reflectance spectroscopy. The results obtained from the experiments showed that it was feasible to identify apple variety and geographic origin using NIR spectroscopy in the range between 400 and 1021 nm, and the techniques were nondestructive. ELM models could give better identification accuracies than BPNN models and SVM models. For ELM models, PCA and SPA could both provide good performances. In particular, SPA-ELM could offer 98.33% accuracy for the calibration set and 96.67% accuracy for the prediction set. Therefore, this pattern would be suitable and effective for variety and geographical origin identification of apple fruits.

Conflicts of Interest

The authors declare that there are no conflicts of interest regarding the publication of this paper.

Acknowledgments

This study was supported by CERNET Innovation Project (Grant no. NGII20150603), Science and Technology Innovation Project of Foshan City, China (Grant no. 2015IT100095), the Fundamental Research Funds for the Central Universities (Grant no. lzujbky-2016-br03), and Science and Technology Planning Project of Guangdong Province, China (Grant no. 2016B010108002).

References

[1] F. R. Harker, E. M. Kupferman, A. B. Marin, F. Anne Gunson, and C. M. Triggs, "Eating quality standards for apples based on consumer preferences," *Postharvest Biology and Technology*, vol. 50, no. 1, pp. 70–78, 2008.

[2] E. Sheikha, A. Farag, A. Condur, D. D. Le Nguyen, G. Loiseau, and D. Montet, "Determination of fruit origin by using 26S rDNA fingerprinting of yeast communities by PCR-DGGE: preliminary application to Physalis fruits from Egypt," *Yeast*, vol. 26, no. 10, pp. 567–573, 2009.

[3] A. F. El Sheikha, I. Métayer, and D. Montet, "A biological bar code for determining the geographical origin of fruit by using 28S rDNA fingerprinting of fungal communities by PCR-DGGE: an application to Physalis fruits from Egypt," *Food Biotechnology*, vol. 25, no. 2, pp. 115–129, 2011.

[4] S. Nojima, C. Linn, and W. Roelofs, "Identification of host fruit volatiles from flowering dogwood (*Cornus florida*) attractive to dogwood-origin *Rhagoletis pomonella* flies," *Journal of Chemical Ecology*, vol. 29, no. 10, pp. 2347–2357, 2003.

[5] F. Licciardello, G. Muratore, C. Avola et al., "Geographical origin assessment of orange juices by comparison of free aminoacids distribution," in *Proceedings of II International Symposium on Citrus Biotechnology*, vol. 892, pp. 389–394, Catania, Italy, December 2009.

[6] B. M. Nicolai, K. Beullens, E. Bobelyn et al., "Nondestructive measurement of fruit and vegetable quality by means of NIR spectroscopy: a review," *Postharvest Biology and Technology*, vol. 46, no. 2, pp. 99–118, 2007.

[7] K. G. Adamopoulos and A. M. Goula, "Application of near-infrared reflectance spectroscopy in the determination of major components in taramosalata," *Journal of Food Engineering*, vol. 63, no. 2, pp. 199–207, 2004.

[8] L. Čurda and O. Kukačková, "NIR spectroscopy: a useful tool for rapid monitoring of processed cheeses manufacture," *Journal of Food Engineering*, vol. 61, no. 4, pp. 557–560, 2004.

[9] H. Cen, Y. Bao, Y. He, and D. W. Sun, "Visible and near infrared spectroscopy for rapid detection of citric and tartaric acids in orange juice," *Journal of Food Engineering*, vol. 82, no. 2, pp. 253–260, 2007.

[10] L. J. Janik, D. Cozzolino, R. Dambergs, W. Cynkar, and M. Gishen, "The prediction of total anthocyanin concentration in red-grape homogenates using visible-near-infrared spectroscopy and artificial neural networks," *Analytica Chimica Acta*, vol. 594, no. 1, pp. 107–118, 2007.

[11] I. Kavdir, R. Lu, D. Ariana, and M. Ngouajio, "Visible and near-infrared spectroscopy for nondestructive quality assessment of pickling cucumbers," *Postharvest Biology and Technology*, vol. 44, no. 2, pp. 165–174, 2007.

[12] D. Cozzolino, M. J. Kwiatkowski, M. Parker et al., "Prediction of phenolic compounds in red wine fermentations by visible and near infrared spectroscopy," *Analytica Chimica Acta*, vol. 513, no. 1, pp. 73–80, 2004.

[13] L. Xie, X. Ye, D. Liu, and Y. Ying, "Quantification of glucose, fructose and sucrose in bayberry juice by NIR and PLS," *Food Chemistry*, vol. 114, no. 3, pp. 1135–1140, 2009.

[14] Y. Wu, L. Li, L. Liu, and Y. Liu, "Nondestructive measurement of internal quality attributes of apple fruit by using NIR spectroscopy," *Multimedia Tools and Applications*, pp. 1–17, 2017.

[15] E. Bertone, A. Venturello, R. Leardi, and F. Geobaldo, "Prediction of the optimum harvest time of "Scarlet"apples using DR-UV–Vis and NIR spectroscopy," *Postharvest Biology and Technology*, vol. 69, pp. 15–23, 2012.

[16] S. Lee, S. C. Kim, K. C. Seong, C. H. Kim, Y. C. Um, and S. K. Lee, "Quality prediction of kiwifruit based on near infrared spectroscopy," *Korean Journal of Horticultural Science and Technology*, vol. 30, no. 6, pp. 709–717, 2012.

[17] V. González-Caballero, D. Pérez-Marín, M. I. López, and M. T. Sánchez, "Optimization of NIR spectral data management for quality control of grape bunches during on-vine ripening," *Sensors*, vol. 11, no. 6, pp. 6109–6124, 2011.

[18] G. James, D. Witten, T. Hastie et al., *An Introduction to Statistical Learning*, Springer, New York, USA, 2013.

[19] P. Oliveri, M. C. Casolino, M. Casale, L. Medini, F. Mare, and S. Lanteri, "A spectral transfer procedure for application of

a single class-model to spectra recorded by different near-infrared spectrometers for authentication of olives in brine," *Analytica Chimica Acta*, vol. 761, pp. 46–52, 2013.

[20] M. C. U. Araújo, T. C. B. Saldanha, R. K. H. Galvao et al., "The successive projections algorithm for variable selection in spectroscopic multicomponent analysis," *Chemometrics and Intelligent Laboratory Systems*, vol. 57, no. 2, pp. 65–73, 2001.

[21] D. Liu and W. Guo, "Identification of kiwifruits treated with exogenous plant growth regulator using near-infrared hyperspectral reflectance imaging," *Food Analytical Methods*, vol. 8, no. 1, pp. 164–172, 2015.

[22] B. Zhang, S. Fan, J. Li et al., "Detection of early rottenness on apples by using hyperspectral imaging combined with spectral analysis and image processing," *Food Analytical Methods*, vol. 8, no. 8, pp. 2075–2086, 2015.

[23] M. J. C. Pontes, R. K. H. Galvao, M. C. U. Araújo et al., "The successive projections algorithm for spectral variable selection in classification problems," *Chemometrics and Intelligent Laboratory Systems*, vol. 78, no. 1-2, pp. 11–18, 2005.

[24] A. M. Mouazen, B. Kuang, J. De Baerdemaeker, and H. Ramon, "Comparison among principal component, partial least squares and back propagation neural network analyses for accuracy of measurement of selected soil properties with visible and near infrared spectroscopy," *Geoderma*, vol. 158, no. 1-2, pp. 23–31, 2010.

[25] I. Belousov, S. A. Verzakov, and J. von Frese, "Applicational aspects of support vector machines," *Journal of Chemometrics*, vol. 16, no. 8–10, pp. 482–489, 2002.

[26] U. Thissen, M. Pepers, B. Üstün, W. J. Melssen, and L. M. C. Buydens, "Comparing support vector machines to PLS for spectral regression applications," *Chemometrics and Intelligent Laboratory Systems*, vol. 73, no. 2, pp. 169–179, 2004.

[27] V. Vapnik, *The Nature of Statistical Learning Theory*, Springer Science & Business Media, Berlin, Germany, 2013.

[28] F. Chauchard, R. Cogdill, S. Roussel, J. M. Roger, and V. Bellon-Maurel, "Application of LS-SVM to non-linear phenomena in NIR spectroscopy: development of a robust and portable sensor for acidity prediction in grapes," *Chemometrics and Intelligent Laboratory Systems*, vol. 71, no. 2, pp. 141–150, 2004.

[29] T. Sun, H. Lin, H. Xu, and Y. Ying, "Effect of fruit moving speed on predicting soluble solids content of "Cuiguan"pears (*Pomaceae pyrifolia* Nakai cv. Cuiguan) using PLS and LS-SVM regression," *Postharvest Biology and Technology*, vol. 51, no. 1, pp. 86–90, 2009.

[30] F. Liu, Y. He, L. Wang, and G. Sun, "Detection of organic acids and pH of fruit vinegars using near-infrared spectroscopy and multivariate calibration," *Food and Bioprocess Technology*, vol. 4, no. 8, pp. 1331–1340, 2011.

[31] G. B. Huang, Q. Y. Zhu, and C. K. Siew, "Extreme learning machine: theory and applications," *Neurocomputing*, vol. 70, no. 1–3, pp. 489–501, 2006.

[32] P. Geladi, D. MacDougall, and H. Martens, "Linearization and scatter-correction for near-infrared reflectance spectra of meat," *Applied Spectroscopy*, vol. 39, no. 3, pp. 491–500, 1985.

[33] M. Urbano-Cuadrado, M. L. De Castro, P. M. Pérez-Juan, J. García-Olmo, and M. A. Gómez-Nieto, "Near infrared reflectance spectroscopy and multivariate analysis in enology: determination or screening of fifteen parameters in different types of wines," *Analytica Chimica Acta*, vol. 527, no. 1, pp. 81–88, 2004.

[34] W. Guo, J. Gu, D. Liu, and L. Shang, "Peach variety identification using near-infrared diffuse reflectance spectroscopy," *Computers and Electronics in Agriculture*, vol. 123, pp. 297–303, 2016.

23

Determination of Processed Salmon Components Sticking to Polyethylene Terephthalate Coatings of Containers by FT-IR and Raman Vibrational Spectroscopy

E. Zumelzu,[1] **A. Sanz**,[2] **J. Medina**,[2] **and F. Rull**[2]

[1]*Institute of Materials, Austral University of Chile, P.O. Box 567, 5111187 Valdivia, Chile*
[2]*Department of Condensed Matter Physics, University of Valladolid, 47002 Valladolid, Spain*

Correspondence should be addressed to E. Zumelzu; ezumelzu@uach.cl

Academic Editor: Renata Diniz

Earlier studies determined that portions of salmon were strongly sticking to the polymer coating of the container walls after emptying the cans. In this sense, this work performed high- and low-frequency spectral characterizations of fresh salmon muscle, fat and skin by Fourier-transform infrared (FT-IR) and Raman spectroscopy analyses to elucidate which components were effectively sticking to the underlying coating. The spectral analyses evidenced that the bands of skin and muscle were clearly distinctive. However, less perceptible contrasts were observed between fat and muscle until band 1700 cm^{-1}, but above this limit, the minor spectral changes detected were sufficient to characterize both salmon components. The new spectral bands for skin occurred at 1030, 1202, and 1336 cm^{-1}. Differences in intensity were also observed for band ν(C=O) at 1744 cm^{-1}, even though it appeared in all components of the salmon. The bands for the ν(C–H) and ν(O–H) vibrations in the high-frequency region were the same, but the intensities and profiles were different. The similarities between the spectra of fresh salmon muscle and residues sticking to the polymer layers were substantial, corroborating that this is in fact the main component sticking to the polymer surface coating of industrial food cans.

1. Introduction

Vibrational spectroscopy is a useful and well-recognized technique to investigate the changes in chemical composition and molecular structure of proteins, carbohydrates, nucleic acids, cell membranes, and tissues. Molecular vibrations are defined as infrared (IR) active when the dipolar momentum of the molecule changes as the molecule vibrates, whereas vibrations are considered Raman active when the polarizability of the molecule changes as the molecule vibrates [1–3]. On the other hand, one of the main reasons for using Raman is that the spectrum delivers a structural fingerprint of the molecules analyzed. This has been evident with the development of surface-sensitive techniques for molecular analysis such as the surface-enhanced Raman spectroscopy (SERS), a highly sensitive method to detect nucleic acids, analytes, cells, and tissues [4].

The advance of other modern instrumental techniques based on vibrational, visible, near-infrared (NIR), mid-infrared (MIR), and Raman spectroscopy has led to a variety of applications in the food and beverage industry, especially in the alcoholic beverage industry [5]. Lately, analytical tools such as Raman spectroscopy and FT-IR have also been employed as useful noninvasive techniques given the global interest in food quality, production, and safety and have been shown suitable for practical use in food research [6]. In fact, there have been developments in the analysis of fish oils prone to oxidation and in the study of structural changes of fish proteins during comminution and gelation [7, 8].

Thus, vibrational spectroscopy techniques and especially Fourier-transform infrared (FT-IR) and Raman are nondestructive and noninvasive valuable characterization methods to elucidate material compositions, molecular structural changes, and environmental effects on biological and engineering materials in a variety of applications of the aquaculture industry [9–16].

Earlier studies in this research line have evaluated a composite consisting of a multilayered electrolytic chromium-coated steel (ECCS) laminate protected by a polyethylene terephthalate (PET) coating employed in salmon canning. The sticking of postmortem salmon to the container wall surface protected by PET and the effects on the polymer coating were studied, with a special focus on its degradation and hence on its use as a recyclable input in other applications [17]. Other works have established the presence of compounds in the diet of salmons and the existence of peptides that adhere to the PET polymer, leading to changes in the color of the container walls thus affecting the consumer's visual perception of quality [18].

The aforementioned studies have made evident the strong sticking and adherence of salmon muscles to the polymer and the influence of its degree of freshness and oxidation on the PET coating of the container, among other aspects. In addition, another work evaluated the effects of thermal sterilization of food containers and storage on muscle degradation inside the food can and considered all the muscle deposits on the polymer layer. Morphological characterizations by scanning electron microscopy (SEM) were also performed, but a thorough discrimination by vibrational spectroscopy to determine whether the sticking product was either muscle, fat, or skin residues or a combination of them was not fully done.

Thus, the present work is an original research aimed at determining by vibrational techniques that the material sticking to the polymer surface consists, indeed, of salmon muscle. Therefore, we will compare the results from earlier works for canned salmon muscle adhered to the polymer surface of container walls with fresh muscle, fat, and skin by FT-IR and Raman analyses.

This comparison is important both for the efficient discrimination of salmon muscle as the fraction sticking to the container and for highlighting the characterization techniques as useful tools to analyze the interactions between biological materials and industrial polymers employed in food containers, since the quality of products involves not only the production processes but also the correct identification of fish products.

2. Materials and Methods

The abovementioned studies cited in References were performed in metal-polymer containers consisting of a PET coating protecting an ECCS plate with an average thickness of 0.20 mm; the chemical composition (%wt) of steel in these samples was 0.074% C, 0.260% Mn, 0.021% P, 0.016% S, 0.012% Si, 0.032% Al, 45 ppm N, and Fe (remaining percentage). The steel plate was coated by a layer generated by electrolytic deposition and consisted of a 0.01 μm thick chromium (Cr^0) and chromium oxide (Cr_2O_3) with an average thickness of 0.01 μm and a total chromium content of 101.33 mg/m^2. This ECCS plate had a 30 μm thick protective layer of PET formed by polymerized units of dimethyl terephthalate and ethylene glycol.

The main components of canned salmon are muscle, fat, and a minimum portion of skin. The muscle has a fibrillar morphology, the fat is homogeneous, and the skin is grey and scaly, being easy to characterize by microscopy observations; we are especially interested in contrasting these observations with vibrational spectroscopy analyses to validate the accuracy of the characterization techniques [17].

Several samples from recently captured salmon specimens and of the same type than those employed for canning purposes by an aquaculture plant from Puerto Montt, Chile, were selected. The muscle, fat, and skin were separately analyzed, and their spectra were compared to earlier results from salmon residues sticking to container walls.

The samples were characterized using the analytical and structural techniques FT-IR and micro-Raman spectroscopy. The IR spectra were measured on different spots in the XY position with a Perkin Elmer Spectrum 100 FT-IR spectrometer supplied with a Universal ATR Sampling Accessory. These points were very close to those considered for Raman spectroscopy analyses in order to establish differences or similarities between the results of both techniques.

The micro-Raman analyses employed two back-illuminated fiber-coupled spectrometers featuring 633 nm and 785 nm wavelengths. The experiment considered mainly the data acquired at 785 nm. This device consisted of a multimode fiber laser BWTEK BRM-OEM-785 (785 nm), a Raman head BWTEK BAC100-785E, and an objective lens Zeiss Epiplan 50x/0.50 infinite/0 44 28 50 with a focal length of 6.5 mm. The maximum output of the laser through this lens was approximately 165 mW; the laser spot diameter measured on the samples was 48 microns.

The Raman spectrometer was a BWTEK Prime T BTC661E-785CUST, covering a spectrum in the range 0–3000 cm^{-1}, with a spectral resolution of 5 cm^{-1}, and a CCD Hamamatsu S10141-1107S operating at −10°C.

In addition, the analyses employed a WITec's alpha300 R confocal Raman imaging microscope, featuring a 785 nm wavelength laser, operated with an ultra-high-throughput spectrometer (UHTS), and accommodating VIS-NIR wavelengths with a sensitivity of 530–1100 nm

3. Results and Discussion

The FT-IR spectra depicted in Figure 1 correspond to fresh salmon muscle samples exhibiting various degrees of flesh colors with the purpose of establishing their relevant spectral features and were compared with deposits adhered to the PET polymer coating of food cans.

No important spectral variations were observed in the intensity and width of peaks for bands up to 3500 cm^{-1} regarding this component of salmon. The characterization was well-defined, and thus it was selected as a control for the fresh salmon muscle.

Figure 2 considers the most representative spectra of the series of tests performed for the characterization of fat from fresh salmon samples.

The FT-IR analyses showed certain wavelength similarities between salmon fat and muscle until band 1700 cm^{-1};

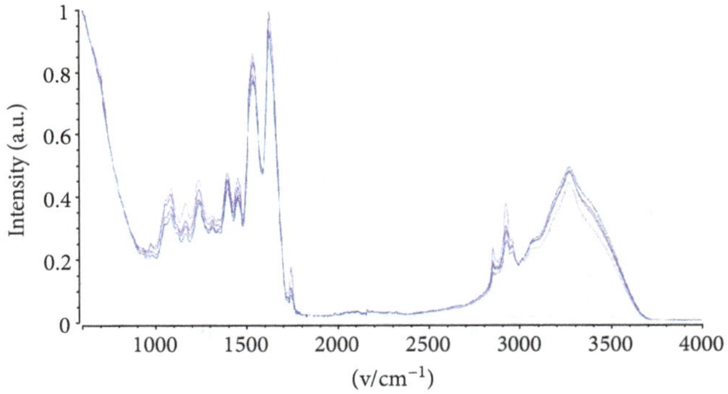

FIGURE 1: FT-IR spectra of fresh salmon muscle sampled from areas exhibiting two distinctive flesh colors: brown and pink.

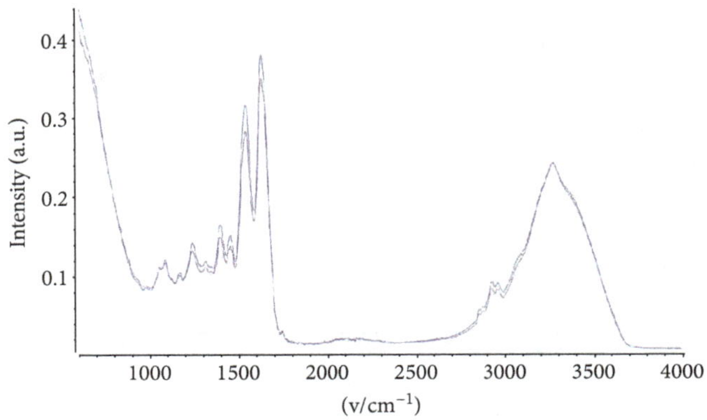

FIGURE 2: FT-IR spectra of fat from fresh salmon samples displaying a broad range of bandwidths.

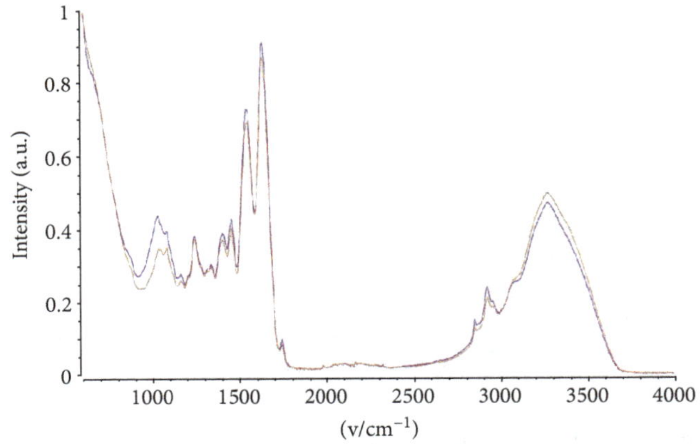

FIGURE 3: FT-IR spectra of the outer skin surface from fresh salmon sampled on two different spots.

however, above this limit, there were minor spectral changes but enough to differentiate both components.

In turn, Figures 3 and 4 characterize the skin of fresh salmon; no significant spectral changes can be observed when comparing the spectra to those in Figures 1 and 2.

The analyses of spectra from salmon skin samples showed no evidence of variations in characterization on both the outer and inner surfaces, and their composition was homogeneous with no presence of contaminants or environmental effects on them before canning.

The whole set of former analyses was made to compare the spectra of all components of fresh salmon and superimpose the plots in such a way as to visualize any differences based on the wavelength bands studied and represented in Figure 5.

From these results, it is possible to say that the spectra are slightly different, depending on the salmon component examined, but every component under study showed very homogeneous spectra. This is highly positive when trying to identify which component of salmon effectively adheres

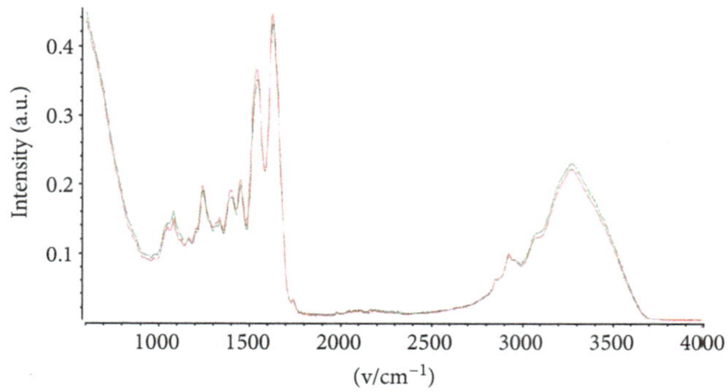

Figure 4: FT-IR spectra of the inner skin surface from fresh salmon samples.

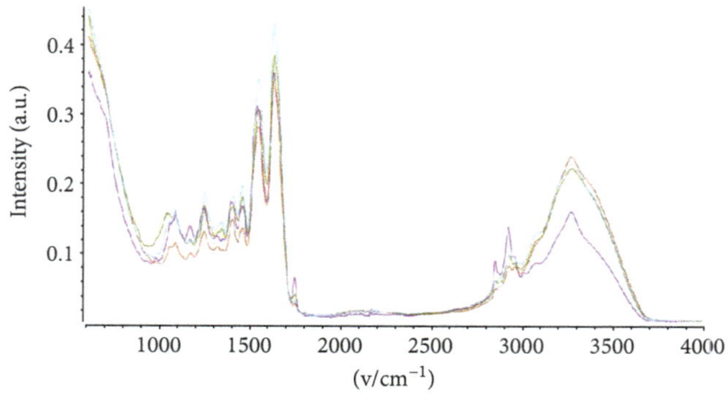

Figure 5: FT-IR spectral comparison of the different fresh salmon components. Brown spectrum belongs to salmon muscle, red spectrum to fat, and dark and light green to the outer and inner skin surfaces, respectively.

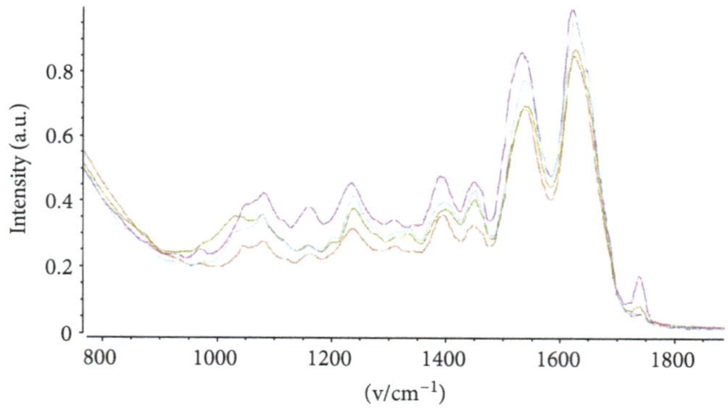

Figure 6: Comparison of low-frequency FT-IR spectra of different fresh salmon components.

to the polymer coating and thereby validates studies already performed and cited in References of this work [17, 18]. The following figures detail the regions of interest for comparison purposes. Thus, it can be seen from the low-frequency spectra in Figure 6 that the dissimilar bands are produced mainly by the salmon skin and muscle, whereas there are hardly any differences between those of fat and muscle. The skin shows new bands at 1030, 1202, and 1336 cm^{-1}. Differences in intensity are also observed for band ν(C=O) at 1744 cm^{-1}, although it appears in all components of the salmon.

Figure 7 shows that the bands for the ν(C–H) and ν(O–H) vibrations located in the high-frequency region are the same, but the intensities and profiles are different.

The second step of this study consisted of a comparison between the spectra from regions that in our former studies had shown a reduced or very low contamination by salmon adhered to the PET layer [17] and the salmon spectra individually characterized in this work.

As it can be noticed in Figure 8, the spectra from areas of salmon adherence to the PET polymer match well with those from fresh salmon muscle. It can also be seen that

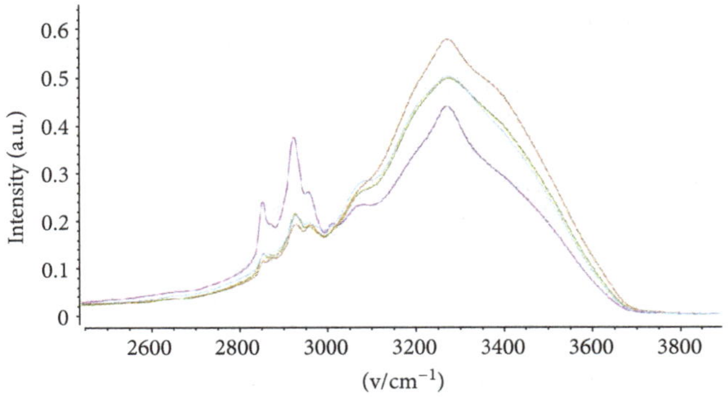

FIGURE 7: Comparison of high-frequency FT-IR spectra of different fresh salmon components.

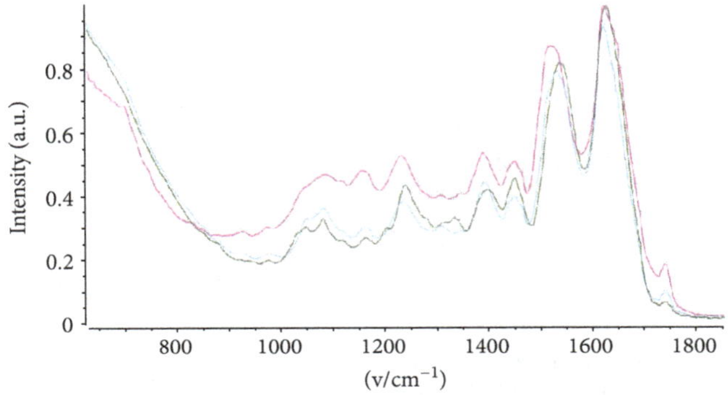

FIGURE 8: Comparison between the low-frequency FT-IR spectra of surface polymer coatings from areas of salmon adherence (pink spectrum) and those of salmon muscle (light blue spectrum) and skin (green spectrum).

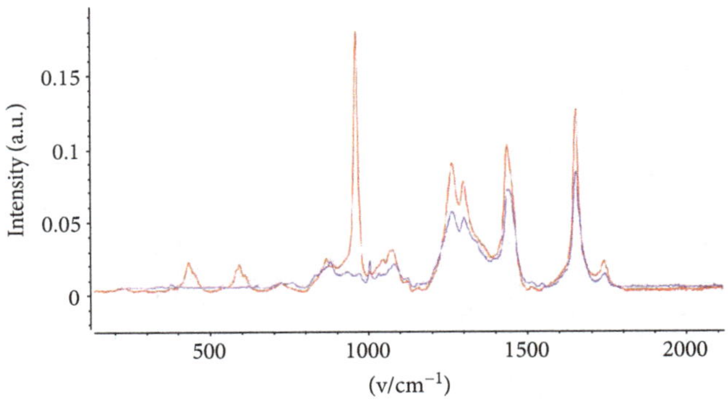

FIGURE 9: Raman spectra of salmon muscle deposits adhered to the food container walls (blue) and fresh salmon muscle (red).

the adherence spectra show wider bands than those of fresh salmon, and the amide II band located at 1544 cm^{-1} in fresh salmon clearly shifts to 1525 cm^{-1} in adhered salmon. This indicates a degradation of salmon in contact with the polymer coating due to the thermal treatment undergone during the canning process.

Regarding the Raman spectroscopy analyses and characterizations made during this validation study, Figure 9 gives details of the comparison between salmon residues adhered to the PET polymer surface [17] and fresh salmon muscle.

The Raman spectra for both types of samples are similar, confirming the earlier results obtained by FT-IR analyses.

4. Conclusions

A comprehensive original study of fresh salmon components, that is, muscle, fat, and skin, was performed by vibrational spectroscopy evaluations in order to compare and validate earlier studies on residual salmon deposits adhered to PET polymer coatings of ECCS plates, leading to changes in the appearance of the protective layers and limiting their recycling potential as ecomaterials.

The results allowed us to elucidate that the main residue that adhered to the container walls of food cans consisted in fact of salmon muscle, given the correspondence between the low-frequency FT-IR and Raman spectral analyses from fresh and canned samples.

The techniques employed are powerful and straightforward tools to validate these analyses when compared to other methods such as electron microscopy studies that require further preparation of biological samples, and secondarily, they have a valuable contribution to quality control processes in food safety inspections.

Conflicts of Interest

The authors declare that they have no conflicts of interest.

Acknowledgments

The authors gratefully acknowledge the FONDECYT program of CONICYT Chile for the financial support through Grant no. 1130634, the FONDEQUIP program of CONICYT Chile for the financial support in the acquisition of equipment code no. EQM 160050, and the special contributions of the Austral University of Chile and University of Valladolid in Spain.

References

[1] R. J. Swain and M. M. Stevens, "Raman microspectroscopy for non-invasive biochemical analysis of single cells," *Biochemical Society Transactions*, vol. 35, no. 3, pp. 544–549, 2007.

[2] A. S. Sivam, D. Sun-Waterhouse, C. O. Perera, and G. I. N. Waterhouse, "Application of FT-IR and Raman spectroscopy for the study of biopolymers in breads fortified with fibre and polyphenols," *Food Research International*, vol. 50, no. 2, pp. 574–585, 2013.

[3] J. Felten, H. Hall, J. Jaumot, R. Tauler, A. de Juan, and A. Gorzsás, "Vibrational spectroscopic image analysis of biological material using multivariate curve resolution – alternating least squares (MCR-ALS)," *Nature Protocols*, vol. 10, no. 2, pp. 217–240, 2015.

[4] A. F. Chrimes, K. Khoshmanesh, P. R. Stoddart, A. Mitchell, and K. Kalantar-zadeh, "Microfluidics and Raman microscopy: current applications and future challenges," *Chemical Society Reviews*, vol. 42, no. 13, pp. 5880–5906, 2013.

[5] R. Schalk, D. Geoerg, J. Staubach, M. Raedle, F. J. Methner, and T. Beuermann, "Evaluation of a newly developed mid-infrared sensor for real-time monitoring of yeast fermentations," *Journal of Bioscience and Bioengineering*, vol. 123, no. 5, pp. 651–657, 2017.

[6] A. Nawrocka and J. Lamorska, "Determination of food quality by using spectroscopic methods," in *Chapter 14, Agricultural and Biological Sciences, Advances in Agrophysical Research*, S. Grundas and A. Stepniewski, Eds., Intech, Croatia, 2013.

[7] D. P. Killeen, S. N. Marshall, E. J. Burgess, K. C. Gordon, and N. B. Perry, "Raman spectroscopy of fish oil capsules: polyunsaturated fatty acid quantitation plus detection of ethyl esters and oxidation," *Journal of Agricultural and Food Chemistry*, vol. 65, no. 17, pp. 3551–3558, 2017.

[8] Y. Kobayashi, S. G. Mayer, and J. W. Park, "FT-IR and Raman spectroscopies determine structural changes of tilapia fish protein isolate and surimi under different comminution conditions," *Food Chemistry*, vol. 226, pp. 156–164, 2017.

[9] H. Henmi, M. Hata, and M. Takeuchi, "Studies on the carotenoids in the muscle of salmon—V. Combination of astaxanthin and canthaxanthin with bovine serum albumin and egg albumin," *Comparative Biochemistry and Physiology Part B: Comparative Biochemistry*, vol. 99, no. 3, pp. 609–612, 1991.

[10] R. Schweitzer-Stenner, T. Measey, A. Hagarman, F. Eker, and K. Griebenow, "Salmon calcitonin and amyloid β: two peptides with amyloidogenic capacity adopt different conformational manifolds in their unfolded states," *Biochemistry*, vol. 45, no. 9, pp. 2810–2819, 2006.

[11] J. H. Cheng, Q. Dai, D. W. Sun, X. A. Zeng, D. Liu, and H. B. Pu, "Applications of non-destructive spectroscopic techniques for fish quality and safety evaluation and inspection," *Trends in Food Science & Technology*, vol. 34, no. 1, pp. 18–31, 2013.

[12] N. K. Afseth, M. Bloomfield, J. P. Wold, and P. Matousek, "A novel approach for subsurface through-skin analysis of salmon using spatially offset Raman spectroscopy (SORS)," *Applied Spectroscopy*, vol. 68, no. 2, pp. 255–262, 2014.

[13] J. L. Xu, C. Riccioli, and D. W. Sun, "An overview on nondestructive spectroscopic techniques for lipid and lipid oxidation analysis in fish and fish products," *Comprehensive Reviews in Food Science and Food Safety*, vol. 14, no. 4, pp. 466–477, 2015.

[14] T. Jiao, N. Krstulović, B. Wu, X. Chen, and Q. Zhang, "Spectroscopy in materials chemistry," *Journal of Spectroscopy*, vol. 2015, Article ID 943894, 2 pages, 2015.

[15] B. Raskovic, R. Heinke, P. Rösch, and J. Popp, "The potential of Raman spectroscopy for the classification of fish fillets," *Food Analytical Methods*, vol. 9, no. 5, pp. 1301–1306, 2016.

[16] J. Hikima, M. Ando, H. Hamaguchi et al., "On-site direct detection of astaxanthin from salmon fillet using Raman spectroscopy," *Marine Biotechnology*, vol. 19, no. 2, pp. 157–163, 2017.

[17] E. Zumelzu, M. J. Wehrhahn, F. Rull, H. Pesenti, O. Muñoz, and R. Ugarte, "Evaluation of salmon adhesion on PET-metal interface by ATR, FT-IR, and Raman spectroscopy," *Journal of Spectroscopy*, vol. 2015, Article ID 835798, 7 pages, 2015.

[18] E. Zumelzu, M. J. Wehrhahn, F. Rull, H. Pesenti, O. Muñoz, and R. Ugarte, "Effect of the degree of muscle oxidation on sticking and degradation of polyethylene terephthalate (PET) layers in electrolytic chromium coated steels," *Journal of Coatings Technology and Research*, vol. 14, no. 3, pp. 673–682, 2017.

Permissions

All chapters in this book were first published in JS, by Hindawi Publishing Corporation; hereby published with permission under the Creative Commons Attribution License or equivalent. Every chapter published in this book has been scrutinized by our experts. Their significance has been extensively debated. The topics covered herein carry significant findings which will fuel the growth of the discipline. They may even be implemented as practical applications or may be referred to as a beginning point for another development.

The contributors of this book come from diverse backgrounds, making this book a truly international effort. This book will bring forth new frontiers with its revolutionizing research information and detailed analysis of the nascent developments around the world.

We would like to thank all the contributing authors for lending their expertise to make the book truly unique. They have played a crucial role in the development of this book. Without their invaluable contributions this book wouldn't have been possible. They have made vital efforts to compile up to date information on the varied aspects of this subject to make this book a valuable addition to the collection of many professionals and students.

This book was conceptualized with the vision of imparting up-to-date information and advanced data in this field. To ensure the same, a matchless editorial board was set up. Every individual on the board went through rigorous rounds of assessment to prove their worth. After which they invested a large part of their time researching and compiling the most relevant data for our readers.

The editorial board has been involved in producing this book since its inception. They have spent rigorous hours researching and exploring the diverse topics which have resulted in the successful publishing of this book. They have passed on their knowledge of decades through this book. To expedite this challenging task, the publisher supported the team at every step. A small team of assistant editors was also appointed to further simplify the editing procedure and attain best results for the readers.

Apart from the editorial board, the designing team has also invested a significant amount of their time in understanding the subject and creating the most relevant covers. They scrutinized every image to scout for the most suitable representation of the subject and create an appropriate cover for the book.

The publishing team has been an ardent support to the editorial, designing and production team. Their endless efforts to recruit the best for this project, has resulted in the accomplishment of this book. They are a veteran in the field of academics and their pool of knowledge is as vast as their experience in printing. Their expertise and guidance has proved useful at every step. Their uncompromising quality standards have made this book an exceptional effort. Their encouragement from time to time has been an inspiration for everyone.

The publisher and the editorial board hope that this book will prove to be a valuable piece of knowledge for researchers, students, practitioners and scholars across the globe.

List of Contributors

Tatiana Ramirez-Mora
Restorative Department, Faculty of Dentistry, University of Costa Rica, San Pedro Montes de Oca 11801, Costa Rica

Claudia Dávila-Pérez and Fernando Torres-Méndez
Endodontics Postgraduate Program, Faculty of Dentistry, University of San Luis Potosí, San Luis Potosí 78290, Mexico

Grettel Valle-Bourrouet
Inorganic Chemistry Department, School of Chemistry, University of Costa Rica, San Pedro Montes de Oca 11801, Costa Rica

Surendra P. Verma and Kailasa Pandarinath
Instituto de Energías Renovables, Universidad Nacional Autónoma de México, Temixco, Mor 62580, Mexico

Sanjeet K. Verma and Héctor López-Loera
División de Geociencias, Instituto Potosino de Investigación en Ciencia y Tecnología, Camino a la Presa San José # 2055, Col. Lomas 4a Sec., San Luis Potosí, SLP 78216, Mexico

M. Abdelaly Rivera-Gómez and John S. Armstrong-Altrin
Instituto de Ciencias del Mar y Limnología, Unidad de Procesos Oceánicos y Costeros, Universidad Nacional Autónoma de México, Circuito Exterior s/n, 04510 CDMX, Mexico

Darío Torres-Sánchez
Posgrado en Geociencias Aplicadas, Instituto Potosino de Investigación en Ciencia y Tecnología, Camino a la Presa San José # 2055, Col. Lomas 4a Sec., San Luis Potosí, SLP 78216, Mexico

Lorena Díaz-González
Centro de Investigación en Ciencias, Instituto de Investigación en Ciencias Básicas y Aplicadas, Universidad Autónoma del Estado de Morelos, Cuernavaca, Mor 62209, Mexico

Alejandra Amezcua-Valdez
Posgrado en Ingeniería, Instituto de Energías Renovables, Universidad Nacional Autónoma de México, Temixco, Mor 62580, Mexico

Beatriz Adriana Rivera-Escoto
División de Materiales Avanzados, Instituto Potosino de Investigación en Ciencia y Tecnología, Camino a la Presa San José # 2055, Col. Lomas 4a Sec., San Luis Potosí, SLP 78216, Mexico

Mauricio Rosales-Rivera
Doctorado en Ciencias, Instituto de Investigación en Ciencias Básicas y Aplicadas, Universidad Autónoma del Estado de Morelos, Cuernavaca, Mor 62209, Mexico

Fernando Velasco-Tapia
Universidad Autónoma de Nuevo León, Facultad de Ciencias de la Tierra, Ex–Hacienda de Guadalupe, Carretera Linares–Cerro Prieto km 8, Linares, N.L. 67700, Mexico

Saverio Latorrata, Renato Pelosato, Paola Gallo Stampino, Cinzia Cristiani and Giovanni Dotelli
Department of Chemistry, Materials and Chemical Engineering "Giulio Natta", Politecnico di Milano, Piazza Leonardo da Vinci 32, 20133 Milano, Italy

Guolin Shi
Research Center of TCM Information Engineering, Beijing University of Chinese Medicine, Beijing 100029, China

Bing Xu, Xinyuan Shi and Yanjiang Qiao
Research Center of TCM Information Engineering, Beijing University of Chinese Medicine, Beijing 100029, China
Beijing Key Laboratory for Production Process Control and Quality Evaluation of Traditional Chinese Medicine, Beijing 100029, China

Xin Wang
TianJin Children's Hospital, TianJin 300204, China

Zhong Xue
School of Pharmacy, Hebei Medical University, Shijiazhuang 050017, China

Robin Legner
Department of Chemistry, University Duisburg-Essen, Universitaetsstraße 2, 45141 Essen, Germany
Department of Chemistry and ILOC, Niederrhein University of Applied Sciences, Frankenring 20, 47798 Krefeld, Germany

Alexander Wirtz and Martin Jaeger
Department of Chemistry and ILOC, Niederrhein University of Applied Sciences, Frankenring 20, 47798 Krefeld, Germany

O. Z. Alekperov
Institute of Physics, Azerbaijan National Academy of Sciences, Huseyn Javid Avenue 33, AZ-1143 Baku, Azerbaijan

E. P. Nakhmedov
Institute of Physics, Azerbaijan National Academy of Sciences, Huseyn Javid Avenue 33, AZ-1143 Baku, Azerbaijan
Faculty of Physics, Moscow State University, Baku Branch, Str. Universitetskaya 1, AZ-1144 Baku, Azerbaijan

Charles L. Y. Amuah, Kwasi Nyandey, Jerry Opoku-Ansah and Peter Osei-Wusu Adueming
University of Cape Coast, College of Agriculture and Natural Sciences, School of Physical Sciences, Department of Physics, Laser and Fibre Optics Centre, Cape Coast, Ghana

Ernest Teye and Francis Padi Lamptey
University of Cape Coast, College of Agriculture and Natural Sciences, School of Agriculture, Department of Agricultural Engineering, Cape Coast, Ghana

Claudio Gardana and Paolo Simonetti
Department of Food, Environmental and Nutritional Sciences (DeFENS), Università Degli Studi di Milano, Via Celoria 2, 20133 Milano, Italy

Antonio Scialpi and Christian Fachechi
Specchiasol, R&D Section, Via Rizzi 1/3, 37012 Bussolengo, Italy

Nan Zhong
College of Engineering, South China Agricultural University, Guangzhou, China
Key Laboratory of Key Technology on Agricultural Machine and Equipment, South China Agricultural University, Ministry of Education, Guangzhou, China
Guangdong Provincial Key Laboratory of Food Quality and Safety, Guangzhou 510642, China

Ling Yang
School of Information Science and Technology, Zhongkai University of Agriculture and Engineering, Guangzhou 510225, China

Ting Wu
College of Engineering, South China Agricultural University, Guangzhou, China
Key Laboratory of Key Technology on Agricultural Machine and Equipment, South China Agricultural University, Ministry of Education, Guangzhou, China
Guangdong Provincial Key Laboratory of Food Quality and Safety, Guangzhou 510642, China
School of Information Science and Technology, Zhongkai University of Agriculture and Engineering, Guangzhou 510225, China

Ying Zhang, Ying Li, Wendong Li and Zigang Sun
Optics and Optoelectronics Laboratory, Ocean University of China, Qingdao, Shandong 266100, China

Yunfeng Bi
College of Electromechanical and Information Engineering, Shandong University, Weihai, Shandong 264209, China

Jonathan Andrade, Cristina Guimarães Pereira, Thamiris Ranquine, Cosme Antonio Azarias, Maria José Valenzuela Bell and Virgilio de Carvalho dos Anjos
Grupo de Engenharia e Espectroscopia de Materiais, Departamento de Física, Universidade Federal de Juiz de Fora, Rua José Lourenço Kelmer s/n São Pedro, 36036-900 Juiz de Fora, MG, Brazil

Meili Wang, Guanghai Shi and Qing Bai
State Key Laboratory of Geological Processes and Mineral Resources, China University of Geosciences, Beijing, China

Joe C. C. Yuan
Taidiam Technology (Zhengzhou) Co. Ltd., Zhengzhou, China

Wen Han
National Gems & Jewelry Technology Administrative Center, Beijing, China

Marc Gardegaront, Delphine Farlay, Olivier Peyruchaud and Hélène Follet
INSERM, Universitè de Lyon, UMR1033, F-69008 Lyon, France

En Yang, Shirong Ge and Shibo Wang
School of Mechatronic Engineering, China University of Mining and Technology, Xuzhou 221116, China

Katarzyna Włodarska and Ewa Sikorska
Faculty of Commodity Science, Poznań University of Economics and Business, al. Niepodległości 10, 61-875 Poznań, Poland

Igor Khmelinskii
Universidade do Algarve, FCT, DQB and CEOT, Campus de Gambelas, 8005-139 Faro, Portugal

Ruixiang Xu, Wenbin Zhao, Zhengrong Wu, Yuheng Liu, Wantong Ma and Peng Chen
School of Pharmacy, Lanzhou University, No. 199 Donggang West Road, Lanzhou 730020, China

Shenghui Xu
College of Earth and Environmental Sciences, Lanzhou University, Key Laboratory of Western China's Environmental Systems, Ministry of Education, Lanzhou 730000, China

Lei Yan
College of Life Science and Technology, Heilongjiang Bayi Agricultural University, Daqing 163319, China

Mahesh Attimarad, Bandar E. Aldhubiab, Anroop Balachandran Nair, Nagaraja SreeHarsha and Sabah H. Akrawi
Department of Pharmaceutical Sciences, College of Clinical Pharmacy, King Faisal University, Al-Hofuf, Al-Ahsa 31982, Saudi Arabia

Katharigatta Narayanaswamy Venugopala
Department of Pharmaceutical Sciences, College of Clinical Pharmacy, King Faisal University, Al-Hofuf, Al-Ahsa 31982, Saudi Arabia
Department of Biotechnology and Food Technology, Durban University of Technology, Durban 4001, South Africa

Shinu Pottathil
Department of Biomedical Sciences, College of Clinical Pharmacy, King Faisal University, Al-Hofuf, Al-Ahsa 31982, Saudi Arabia

Yaqiong Zhao, Feng Qin, Jinxing Ma and Haiguang Wang
College of Plant Protection, China Agricultural University, Beijing 100193, China

Fei Xu and Yuli Song
Institute of Plant Protection, Henan Academy of Agricultural Sciences, Zhengzhou 450002, China

Zhenyu Sun
Institute of Plant Protection, Gansu Academy of Agricultural Sciences, Lanzhou 730070, China

Longlian Zhao and Junhui Li
College of Information and Electrical Engineering, China Agricultural University, Beijing 100083, China

Jie Kuai, Cheng Guo and Guangsheng Zhou
College of Plant Science & Technology, Huazhong Agricultural University, Wuhan 430070, China

Shengyong Xu, Kun Lu and Yaoze Feng
College of Engineering, Huazhong Agricultural University, Wuhan 430070, China

G. Gallello
Department of Prehistory, Archaeology and Ancient History, Faculty of Geography and History, University of Valencia, Av. Blasco Ibáñez 28, 46010 Valencia, Spain
Department of Analytical Chemistry, Faculty of Chemistry, University of Valencia, C/ Dr. Moliner 50, 46100 Burjassot, Valencia, Spain
Department of Archaeology, University of York, King's Manor, York YO1 7EP, UK

M. Ramacciotti
Department of Prehistory, Archaeology and Ancient History, Faculty of Geography and History, University of Valencia, Av. Blasco Ibáñez 28, 46010 Valencia, Spain
Department of Analytical Chemistry, Faculty of Chemistry, University of Valencia, C/Dr. Moliner 50, 46100 Burjassot, Valencia, Spain

S. Rubio, A. Morales-Rubio, A. Pastor and M. de la Guardia
Department of Analytical Chemistry, Faculty of Chemistry, University of Valencia, C/Dr. Moliner 50, 46100 Burjassot, Valencia, Spain

M. Lezzerini
Department of Earth Sciences, University of Pisa, Via S. Maria 53, 56126 Pisa, Italy
Applied and Laser Spectroscopy Laboratory, ICCOM-CNR, Via G. Moruzzi 1, 56124 Pisa, Italy

S. Columbu
Department of Chemical and Geological Sciences, University of Cagliari, Via Trentino 51, 09127 Cagliari, Italy

E. Hernandez
Sagunto Archaeological Museum, C/del Castillo 23, 46500 Sagunto, Valencia, Spain

Sarfaraz Ahmed Mahesar
National Centre of Excellence in Analytical Chemistry, University of Sindh, Jamshoro-76080, Pakistan

Massimo Lucarini and Alessandra Durazzo
CREA-Research Centre for Food and Nutrition, Via Ardeatina 546, Rome 00178, Italy

Antonello Santini
Department of Pharmacy, University of Napoli Federico II, Via D. Montesano 49, 80131 Napoli, Italy

Anja I. Lampe and Johannes Kiefer
Technische Thermodynamik, Universität Bremen, Badgasteiner Str. 1, 28359 Bremen, Germany

Caihong Li, Lingling Li, Yuan Wu, Min Lu and Lian Li
School of Information Science and Engineering, Lanzhou University, Lanzhou 730000, China

Yi Yang
School of Information Science and Engineering, Lanzhou University, Lanzhou 730000, China
Silk Road Economic Belt Research Center, Lanzhou University, Lanzhou 730000, China

E. Zumelzu
Institute of Materials, Austral University of Chile, 5111187 Valdivia, Chile

A. Sanz, J. Medina and F. Rull
Department of Condensed Matter Physics, University of Valladolid, 47002 Valladolid, Spain

Index

www.ingramcontent.com/pod-product-compliance
Lightning Source LLC
LaVergne TN
LVHW080917210926
848485LV00004B/669
* 9 7 8 1 6 3 9 8 9 3 6 5 2 *